Handbook of Experimental Pharmacology

Volume 247

More information about this series at http://www.springer.com/series/164

Emily M. Jutkiewicz
Editor

Delta Opioid Receptor Pharmacology and Therapeutic Applications

Springer

Editor
Emily M. Jutkiewicz
Department of Pharmacology
University of Michigan Medical School
Ann Arbor, Michigan
USA

ISSN 0171-2004 ISSN 1865-0325 (electronic)
Handbook of Experimental Pharmacology
ISBN 978-3-030-06980-3 ISBN 978-3-319-95133-1 (eBook)
https://doi.org/10.1007/978-3-319-95133-1

This Springer imprint is published by the registered company Springer Nature Switzerland AG
The registered company address is: Gewerbestrasse 11, 6330 Cham, Switzerland

Preface

In the mid-to-late 1970s, opioids receptors were identified solely on their pharmacological, physiological, and behavioral properties. They were labeled as the mu-, kappa-, and delta-opioid receptors based on the compounds that likely activated the receptor subtypes or their tissue localization. Delta-opioid receptors were first identified in the mouse vas deferens and were also thought to be present in the brain based on the binding and activity of the endogenous opioid peptides met- and leu-enkephalin (Lord et al. 1977). Delta-opioid receptors were cloned later in 1992 from a cDNA library prepared from NG108-15 cells that expressed high levels of delta-opioid receptors (Evans et al. 1992). These advances permitted and initiated the study of the delta-opioid receptor system.

While all clinically used opioid analgesics activate the mu-opioid receptor subtype, there are currently no approved medications that bind selectively to the other opioid receptor subtypes. When delta-opioid receptors were first identified, it was hoped that this receptor would hold the key for the development of a non-addicting analgesic. However, activation of delta-opioid receptors failed to produce pain relief in animal models of acute and moderate-to-severe pain. Research on the delta-opioid receptor system has persisted over the decades in order to further understand its role in biology and physiology.

With new momentum in the field of opioid receptor pharmacology, the delta-opioid receptor has served as the basis for studying new avenues in receptor signaling and trafficking and for discovering the novel therapeutic potential of opioid systems. The chapters in this volume are authored by a small selection of investigators who have made critical advances in our understanding of the function and activity of the delta-opioid receptor system, including the receptor itself and its endogenous and exogenous ligands. The research described in this volume highlights advances in our understanding of delta-opioid receptor pharmacology, its relevance to disease states, and possibilities for drug discovery and development. There are many more researchers who could have contributed; alas, there are only so many pages. The contributions to this volume will describe:

- New discoveries in the design and synthesis of delta-opioid receptor ligands, including the generation of quinolinomorphinan structures that appear to lack the convulsant properties observed with the traditional piperazinyl benzamide

compounds. Also, the development of bifunctional opioid ligands and the pharmacology of delta-opioid receptor heteromers.
- Identification of distinct expression patterns of delta-opioid receptors in different neuronal subtypes, brain regions, and within cells, and how these patterns may be related to receptor function and/or specific disease states.
- Signaling pathways downstream of delta-opioid receptors, including G protein-dependent and -independent signaling pathways, which may contribute to the actions of different delta-opioid receptor agonists.
- Finally, the role of delta-opioid receptors and the potential therapeutic effects of delta-opioid receptor ligands in disease states, such as pain, Parkinson's disease, wound healing, neuro- and cardioprotective effects, depression, anxiety, and addiction.

I would like to thank Dr. James Barrett, Editor in Chief of the *Handbook of Experimental Pharmacology*, for contacting and encouraging me to organize this volume and for his patience in its preparation. Also, I would like to thank Susanne Dathe, Balamurugan Elumalai, and Anand Ventakachalam from Springer for overseeing the production of this volume. Finally, I would like to thank all the researchers who have studied various aspects of delta-opioid receptor pharmacology and delta-opioid receptor ligands over the decades and in particular the contributors to this volume.

Ann Arbor, MI, USA Emily M. Jutkiewicz

References

Evans CJ, Keith DE Jr, Morrison H, Magendzo K, Edwards RH (1992) Cloning of a delta opioid receptor by functional expression. Science 258:1952–1955

Lord JA, Waterfield AA, Hughes J, Kosterlitz HW (1977) Endogenous opioid peptides: multiple agonists and receptors. Nature 267:495–499

Contents

Part I

DOPR Ligands

Delta Opioid Receptor (DOR) Ligands and Pharmacology: Development of Indolo- and Quinolinomorphinan Derivatives Based on the Message-Address Concept

Akiyoshi Saitoh and Hiroshi Nagase

Contents

Abstract

The pharmacology of the delta opioid receptor (DOR) has lagged, mainly due to the lack of an agonist with high potency and selectivity in vivo. The DOR is now receiving increasing attention, and there has been progress in the synthesis of better novel ligands. The discovery of a selective receptor DOR antagonist, naltrindole (NTI), stimulated the design and synthesis of (±)TAN-67, which was designed based on the message-address concept and the accessory site theory. Intensive studies using (±)TAN-67 determined the DOR-mediated various pharmacological effects, such as antinociceptive effects for painful diabetic neuropathy and cardiovascular protective effects. We improved the agonist

A. Saitoh
Department of Neuropsychopharmacology, National Center of Neurology and Psychiatry, National Institute of Mental Health, Tokyo 187-8553, Japan

H. Nagase (✉)
International Institute for Integrative Sleep Medicine (WPI-IIIS), University of Tsukuba, 1-1-1, Tennodai, Tsukuba, Ibaraki 305-8575, Japan
e-mail: nagase.hiroshi.gt@u.tsukuba.ac.jp

E. M. Jutkiewicz (ed.), *Delta Opioid Receptor Pharmacology and Therapeutic Applications*, Handbook of Experimental Pharmacology 247, DOI 10.1007/164_2016_18

activity of TAN-67 to afford SN-28, which was modified to KNT-127, a novel compound that improved the blood–brain barrier permeability. In addition, KNT-127 showed higher selectivity for the DOR and had potent agonist activity following systemic administration. Interestingly, KNT-127 produced no convulsive effects, unlike prototype DOR agonists. The KNT-127 type derivatives with a quinolinomorphinan structure are expected to be promising candidates for the development of therapeutic DOR agonists.

Keywords
(−)TAN-67 • Analgesic • Antidepressant • Antitussive • Anxiolytic • TRK-850 • δ Opioid receptor

1 Introduction

The opioid system was generally classified into three types mu, delta, and kappa by pharmacological and molecular biological studies, and these three types are activated by a family of endogenous peptides: endorphins, endomorphins, enkephalins, and dynorphins, respectively. Presently, a mu opioid receptor (MOR) agonist (morphine) is best known as a remarkably strong analgesic for severe pain, such as cancer pain. However, its use is limited by severe adverse effects, such as constipation, respiratory depression, vomiting, dependence, and tolerance. To develop compounds without these adverse effects, intense efforts have been concentrated on investigating compounds selective for the delta opioid receptor (DOR) or kappa opioid receptor (KOR). Previously, it was determined that KOR activation results in dysphoria and this strongly limited the development of KOR agonists for clinical study. However, the only clinically useful KOR agonist, nalfurafine hydrochloride (Remitch®), was launched in Japan as an antipruritic drug for kidney dialysis patients in 2009 (Nagase and Fujii 2011). On the other hand, no compounds selective for the DOR have been clinically approved, despite many reports indicating that promising pharmacological effects are exerted via the DOR. The pharmacology of the DOR has lagged behind that of MOR and KORs, mainly due to the lack of an agonist with high potency and selectivity in vivo. The DOR is now receiving increasing attention, and there is progress in the synthesis of better novel ligands. In this review, we summarize the progress in the design and synthesis of DOR ligands with indolo- and quinolinomorphinan skeletons and in their pharmacological effects.

2 Design and Synthesis of Indolo- and Quinolinomorphinan Derivatives as Highly Selective Non-peptidic DOR Ligands

2.1 Rational Drug Design of (±)TAN-67

The discovery of the selective DOR antagonist, naltrindole (NTI), by Portoghese et al. (1988) was a breakthrough (Fig. 1a) in the investigation of non-peptidic ligands that preferentially bind the DOR. This was the first non-peptidic ligand with high affinity and selectivity for the DOR based on the indolomorphinan skeleton; in addition, the selective DOR antagonist NTI is still being used to investigate pharmacological activity of DOR agonists.

The intensive structure–activity relationship study of NTI to improve its blood brain barrier (BBB) permeability led to the synthesis of the DOR antagonists, TRK-850 and TRK-851, which were more potent in vivo (Fig. 2). The discovery of these compounds suggested endogenous opioid receptor networks existed at the cough reflex and they were modulated via DORs in the central nervous system; this finding led to Toray's clinical candidate for a therapeutic drug for chronic cough (Sakami et al. 2008). Kamei et al. discovered that endogenous DOR-mediated stimulation had an inhibitory role in the endogenous MOR- and KOR-mediated suppressive regulatory mechanisms of the cough reflex, using both capsaicin- and citric acid-induced cough models of rodents and/or guinea pigs (Kamei 1996). It is expected that DOR antagonists would be robust antitussive drugs, without MOR-mediated side effects (see the review by Nagase and Fujii (2011) for detailed discussion).

Message-address concept plays a role in the design of opioid receptor-type-selective ligands with a tyrosine structure that is essential for opioid activity from the viewpoint of endogenous opioid chemistry. Message part contributes to the intrinsic activity of opioid ligand. Address part participates in opioid receptor type selectivity (MOR address part, small side chain; DOR address part, larger side chain; and KOR address part, largest side chain). In this binding model, we designed a novel type of DOR ligands with tyrosine partial structures. Based on the binding model of NTI with the DOR, we proposed the three-centered binding sites at the morphinan moiety are the message part, which includes an ionic interaction (protonated 17-nitrogen), a π–π interaction (phenol ring), and a hydrogen bond (3-hydroxy group), while the additional pharmacophore for DOR selectivity, a π–π interaction of the benzene ring at the indole moiety of NTI (Fig. 1b), is the address part, which determines opioid receptor type selectivity. In addition to the binding model for NTI, we utilized the accessory site theory to design novel selective DOR agonists. Accessory site theory plays a role in explaining the structural difference between agonist and antagonist. The accessory part for an antagonist is a characteristic lipophilic moiety which disturbs the structural change of receptor induced by an agonist after binding with receptor. Interestingly, removal of the accessory part of an antagonist produces an agonist. We hypothesized that the 10-methylene bridge and 4,5-epoxy ring would prohibit the free rotation of the phenol ring and prevent the conformational change necessary for developing the

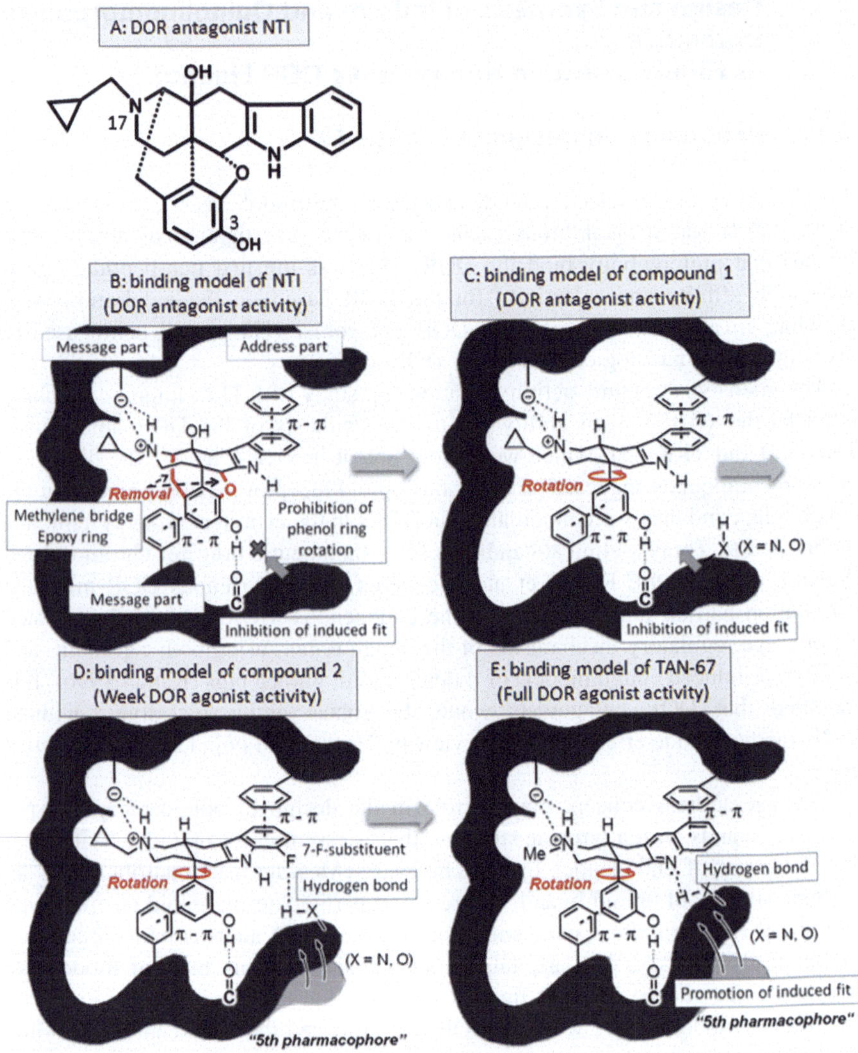

Fig. 1 Chemical structure formula and possible binding model of each ligand with the DOR

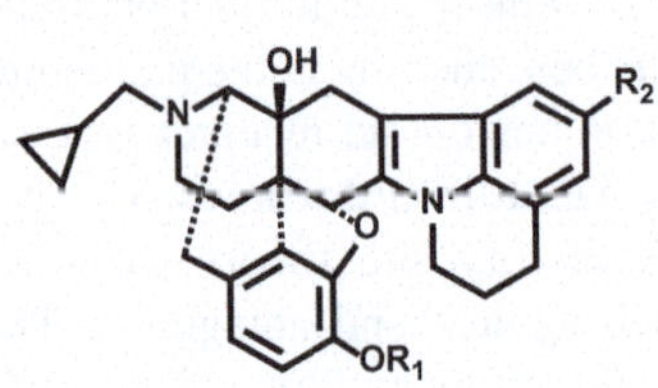

TRK-850: R_1 = Me, R_2 =H, TRK-851:R_1 = H, R_2 = F

Fig. 2 The structures of TRK-850 and TRK-851

agonist activity (Fig. 1b). On the basis of this hypothesis, we removed the 10-methylene bridge and 4,5-epoxy ring in NTI to afford compound **1** (Fig. 1c). Contrary to our expectation, compound **1** showed no agonist activity. However, after investigating the structure–activity relationship, we found that only compound **2** (Fig. 1d) with a 7-F substituent afforded week agonist activity. We postulated that the agonist activity of compound **2** might be derived from the hydrogen bond with the hydrogen donor site in the DOR. However, the position of the fluorine atom in compound **2** might be rather close to the outside, and the conformational change with the hydrogen bond would be insufficient, which would lead to week agonist activity (Fig. 1d). In the next step, we designed and synthesized a quinoline derivative with a decahydroisoquinoline skeleton to move the position of the lone electron pair for hydrogen bonding more inside the DOR binding site (Fig. 1e). The resulting quinoline derivative with the 17-cyclopropylmethyl substituent showed full agonist activity. After studying the elaborated structure–activity relationship, we obtained a highly selective and potent agonist for the DOR, (±)TAN-67, [(4aS*,12aR*)-4a-(3-hydroxyphenyl)-2-methyl-1,2,3,4,4a,5,12,12a-octahydropyrido[3,4-b]acridine]. The possible binding model of (±)TAN-67 with the DOR is shown in Fig. 1e.

In the radio-ligand competition assays, (±)TAN-67 showed high affinity for the DOR (K_i value = 1.12 nM) in the guinea pig cerebrum using ^{3}[H]DPDPE. In addition, it showed higher selectivity for the DOR with a 2,070-fold lower affinity for the MOR using ^{3}[H]DAMGO and a 1,600-fold lower affinity for the KOR using [^{3}H]ethylketocyclazocine (Nagase et al. 1998). Knapp et al. (1995) also reported that (±)TAN-67 showed a higher binding affinity (K_i = 0.647 nM) at the human DOR using [^{3}H]NTI and higher DOR binding selectivity (>1,000-fold) relative to the human MOR using [^{3}H]CTOP (Knapp et al. 1995). (±)TAN-67 produced a NTI-reversible inhibitory effect on the contraction of the mouse vas deferens with an IC_{50} value of 6.61 nM, suggesting that (±)TAN-67 showed agonist activity via the DOR (Nagase et al. 1998). (±)TAN-67 also showed a robust agonist activity (EC_{50} = 1.72 nM) for the inhibition of forskolin-stimulated cAMP accumulation at the human DOR expressed by intact Chinese hamster ovary (CHO) cells, but low potency (EC_{50} = 1,520 nM) at the human MOR expressed by the intact B82 mouse fibroblast cells (Knapp et al. 1995). These in vitro results from assays using human DOR and MOR showed that (±)TAN-67 has a similar binding affinity, selectivity, and potency as DPDPE, a representative prototype peptidic DOR ligand (Knapp et al. 1995).

2.2 Investigation of the Pharmacological Effects by TAN-67

(±)TAN-67 was the first non-peptidic DOR agonist designed on the basis of the accessory site hypothesis for NTI (Portoghese et al. 1990; Schwyzer 1977) and was included in a patent in 1991. After reporting the activities of (±)TAN-67, the many pharmacological effects induced via the DOR were investigated. Studies have identified the pharmacological effects induced by (±)TAN-67 and its derivatives, such as the antinociceptive effects (including treatment of painful diabetic neuropathy), cardiovascular protective effects, as well as effects on respiratory disorders (including antitussive effects), immunoregulatory functions, antidiuretic activity (including the treatment of urinary incontinence), psychiatric disorders (including antidepressant and anxiolytic effects), neurodegenerative diseases, and cancers [reviewed in Fujii et al. (2013)]. In this section, we review some promising pharmacological effects of (±)TAN-67.

2.2.1 Antinociceptive Effects

In early studies, the antinociceptive effects of (±)TAN-67 were evaluated in various mice pain models, such as the acetic acid abdominal constriction (writhing) test, hot-plate test, and tail-flick test. Subcutaneous (*s.c.*) administration of (±)TAN-67 at large doses of 30 and 100 mg/kg caused a significant decrease in the number of constrictions of mice in the acetic acid writhing test. The ED_{50} value was 31.4 mg/kg (95% confidence limits: 14.2–69.4 mg/kg) at 30 min after treatment (Kamei et al. 1995; Nagase et al. 1998). However, in the tail-flick test, *s.c.* administration of (±)TAN-67 produced no antinociceptive effects in mice (Kamei et al. 1995). Suzuki et al. reported that although the co-administration of (±)TAN-67 with morphine significantly shifted the morphine dose-response curve to the left, neither *s.c.* (40 mg/kg) nor intracerebroventricular (*i.c.v.*, 40 μg/mouse) administration of (±)TAN-67 alone produced inhibitory effects on the withdrawal latencies in the mouse hot-plate test. These results suggested that (±)TAN-67 had a low antinociceptive effect in the acetic acid writhing test and no effect in the tail-flick or hot-plate tests (Kamei et al. 1997; Kamei et al. 1995; Suzuki et al. 1995; Tseng et al. 1997). Thus, the higher potency and selectivity for the DOR of (±)TAN-67 that was observed in in vitro studies was not consistent with the findings from in vivo studies, which (±)TAN-67 produced no or weak antinociceptive effects.

To clarify the reason for this, we synthesized the optically active compounds (+) and (−)TAN-67 (Nagase et al. 2001). Interestingly, when (−)TAN-67 was given intrathecal (i.t., 8.9–89.4 nmol doses), there was dose- and time-dependent inhibition of the tail-flick response in mice 10 min after injection. The ED_{50} value was 17.1 nmol (95% confidence limits: 3.4–85.2 nmol). The increased tail-flick latencies of (−)TAN-67 in mice were completely antagonized by i.t. pretreatment with BNTX (a selective DOR antagonist), but not by CTOP (a selective MOR antagonist) or nor-BNI (a selective KOR antagonist), suggesting that (−)TAN-67 produced an antinociceptive effect mediated by the activation of the DOR, but not MOR or KOR (Tseng et al. 1997).

On the other hand, its enantiomer, (+)TAN-67, decreased the latencies of the tail-flick response in mice, even when smaller doses (1.8–8.9 nmol) were given i.t. When higher doses (17.9–89.4 nmol) were given i.t., (+)TAN-67 produced nociceptive-like behaviors, such as scratching and biting in mice (Tseng et al. 1997). This result suggested that the doses of (+)TAN-67 that produced a decrease of the tail-flick latencies were much smaller than the doses of (−)TAN-67 that produced antinociception. In addition, the severe nociception induced by

(+)TAN-67 was attenuated by i.t. pretreatment with baclofen (a selective $GABA_B$ receptor agonist), in the same manner as nociception was induced by the *N*-methyl-D-aspartate (NMDA) receptor antagonist MK801 (Yajima et al. 2000). And also, (+)TAN-67-induced nociception has been shown to be suppressed by a NK_1 receptor antagonist (Kamei et al. 1999). Taken together, these data suggested that the weak antinociceptive response of (±)TAN-67 was caused by its nociceptive effects, which physiologically antagonized the antinociceptive effects of (−)TAN-67. Based on these results, we proposed that (−)TAN-67 could produce DOR-selective antinociceptive effects.

2.2.2 Antinociceptive Effects for Painful Diabetic Neuropathy

Interestingly, we found that the systemic administration of (±)TAN-67 produced a significant inhibitory effect on the acetic acid abdominal constriction and tail-flick tests in diabetic mice, which are animal models for painful diabetic neuropathy (Kamei et al. 1995). The antinociceptive effects of (±)TAN-67 were greater than that in non-diabetic mice. These results supported the hypothesis by Kamei that mice with diabetes are hyperresponsive to DOR-mediated antinociception (Kamei et al. 1994). The *i.c.v.* administration of (−)TAN-67 (3–60 μg/mouse, *i.c.v.*) significantly increased the latencies in the tail-flick test in diabetic mice, which was different from the effects in non-diabetic mice (Kamei et al. 1997; Ohsawa et al. 1998). Interestingly, the *i.c.v.* pretreatment with both EGTA and ryanodine, which decreased the level of intracellular calcium, reduced (−)-TAN67-induced antinociception in diabetic mice. In contrast, the *i.c.v.* pretreatment with thapsigargin, a microsomal calcium-ATPase inhibitor, enhanced (−)-TAN67-induced antinociception in diabetic mice. These results suggested that the enhanced DOR agonist-induced antinociception in diabetic mice may be due to excessive intracellular calcium overload caused by dysfunctional calcium stores (Ohsawa et al. 1998). Actually, it was reported that the diabetic state affects intracellular calcium levels in neurons and various tissues (see review Fernyhough and Calcutt 2010; Levy et al. 1994). These findings suggested that DOR agonists, including (−)TAN-67 and its derivatives, should be considered as candidate therapeutic targets for treatment of painful diabetic neuropathy.

2.2.3 Cardiovascular Protective Effects

The cardiovascular protective effects of DOR agonists have been well established using (±)TAN-67, which showed potential to mimic the cardioprotective effect in ischemic preconditioning (IPC) in many animal species, including rats (Schultz et al. 1998), chicks (Huh et al. 2001), and dogs (Peart et al. 2003). Previously, in an in vivo rat model, (±)TAN-67 (10 mg/kg) significantly reduced infarct size (IS), expressed as a percent of the area at risk (IS/AAR), when the rats were intravenously infused for 15 min before occlusion and reperfusion periods (Schultz et al. 1998). In a dog model, Peart et al. (2003) also showed that (±)TAN-67, which was administered by intracoronary infusion for 30 min before left anterior descending coronary artery occlusion, produced significant reduction in IS/AAR, similar to that of IPC. Fryer et al. (1999) demonstrated that (±)TAN-67 (30 mg/kg) could induce the cardioprotective effect of IPC in rats 24–48 h following intraperitoneal (*i.p.*) administration. The IPC elicits both an acute and delayed phase of cardioprotection, where brief episodes of ischemia and reperfusion before a prolonged ischemic event limit myocardial cellular damage. These effects suggested that the DOR agonist (±)TAN-67 could induce both short-term and delayed cardioprotection.

The mechanisms for the cardioprotective effects against myocardial infarction induced by (±)TAN-67 are mediated by the activation of several molecular systems, including protein kinase C, ATP-sensitive potassium channels (K_{ATP}), tyrosine kinase, and extracellular signal-regulated kinase (Fryer et al. 2002). Taken together, it was suggested that activation of the DOR can mimic the cardioprotective effects of IPC in the heart; thus, DOR agonists, including (−)TAN-67 and its derivatives, may have therapeutic potential.

2.2.4 Anti-Alcohol Effects

Excessive ethanol consumption and alcohol addiction are serious threats to society, both socially as well as economically. The involvement of opioid receptors in ethanol consumption, as well as its rewards and dependence, has long been known (van Ree et al. 1999). Several studies indicated that the pharmacological activation of the DOR facilitates or inhibits ethanol consumption in rodents (van Rijn et al. 2010). Pharmacological blockades of the DOR by NTI have decreased ethanol consumption in rodents (Krishnan-Sarin et al. 1995a, b; Lê et al. 1993). Paradoxically, DOR knockout mice showed increased ethanol consumption (Roberts et al. 2001). Thus, the role of the DOR in alcohol intake is unclear.

It was shown that (±)TAN-67 (25 mg/kg, *s.c.*) decreased ethanol consumption in mice in a two-bottle choice self-administration test (van Rijn et al. 2010; van Rijn and Whistler 2009). Interestingly, (±)TAN-67 (25 mg/kg, *s.c.*) more effectively abolished the ethanol withdrawal-induced anxiety-like behaviors in mice that consumed ethanol than the typical anxiolytic, diazepam (van Rijn et al. 2010). These results suggested that (±)TAN-67 could decrease anxiety-like behaviors and be more effective than diazepam at reducing ethanol consumption (van Rijn et al. 2010). On the other hand, (±)TAN-67 increased the ethanol-induced place preference when the mice were injected prior to testing the conditioned place preference (CPP) to ethanol, while (±)TAN-67 produced no place preference by itself (van Rijn et al. 2010; van Rijn and Whistler 2009). Similar results were reported by Suzuki and colleague (Matsuzawa et al. 1999). Although (±)TAN-67 (20 mg/kg, *s.c.*) alone produced no effects on the CPP test in rats (Matsuzawa et al. 1999; Suzuki et al. 1996), (±)TAN-67 (20 mg/kg, *s.c.*) produced a significant ethanol-induced place preference when the rats were exposed to conditioned fear stress by an electrical foot shock in the place conditioning (Matsuzawa et al. 1999). These results may suggest that (±)TAN-67 produces the ethanol-induced place preference in the CPP test by decreasing the effects of aversive events (e.g., anxiolytic-like effects) during place conditioning. Conversely, another selective DOR agonist SNC80 reduced the rewarding effects of ethanol, which promote increased consumption (van Rijn et al. 2012).

Although further studies are necessary to elucidate the mechanisms of the dual efficacy of DOR agonists for ethanol consumption, (−)TAN-67 and its derivatives are expected to be interesting therapeutic targets for treatment-seeking alcoholics.

3 Synthesis of DOR Ligands That Improved the BBB Permeability

3.1 Rational Drug Design of KNT-127

Although the discovery of (±)TAN-67 greatly impacted the investigation of pharmacology via the DOR, the activity and permeability through the BBB was insufficient. Next, we tried to improve the potency and the ability to penetrate through the BBB. The key structural features of (−)TAN-67 are a freely rotatable phenol ring and a quinoline nitrogen. As shown in Fig. 3, (−)TAN-67 has neither the 4,5-epoxy ring nor the 10-methylene bridge. The phenol ring of (−)TAN-67 can rotate to a suitable position for induced fit, and the quinoline nitrogen can form a hydrogen bound with the DOR. We postulated that these binding interactions with the receptor would be sufficient to induce a structural change of the receptor, thereby inducing agonist activity. However, the resulting (−)TAN-67 afforded limited agonist activity. So, we tried to confirm if both the postulated accessory sites were necessary for full agonist activity. We found the 4,5-epoxy ring was not necessary, and this led us to design the compound SN-28, which only has the 10-methylene bridge (Fig. 3) (Nagase et al. 2009). The conformational analysis of (−)TAN-67 and SN-28 using the Conformational Analyzer with Molecular Dynamics And Sampling (CAMDAS) showed that the range of conformations available to SN-28 was almost the same as that of (−)TAN-67. This result suggested that SN-28 would have the ability to produce conformational change and therefore the presence of the 10-methylene bridge would not disturb the structural change of the receptor necessary for the agonist effect.

As we expected, SN-28 showed about 15 times more potent agonist activity than (−)TAN-67 and a sufficient selectivity for the DOR in vitro (Nagase et al. 2009). However, SN-28 when administrated *s.c.* at a dosage over 30 mg/kg showed no analgesic effects in the mouse acetic acid writhing test. On the other hand, the i.t. administration of SN-28 showed a strong analgesic effect in this test. This suggested that SN-28 would not penetrate through the BBB. To confirm this

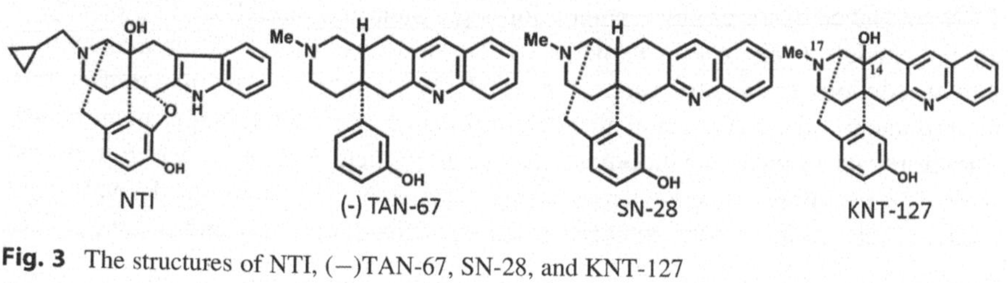

Fig. 3 The structures of NTI, (−)TAN-67, SN-28, and KNT-127

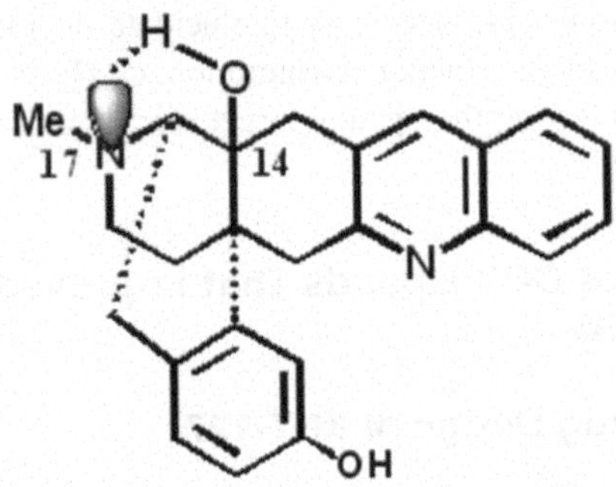

Fig. 4 Intramolecular hydrogen bond between the 14-hydroxy group and 17-nitrogen in KNT-127

hypothesis, we tried to design a compound that was a less polar derivative than SN-28 to produce an analgesic effect after systemic administration.

The basic nitrogen in SN-28 could form ammonium ion under physiological conditions, and the resulting ionized compound may be less likely to penetrate through the BBB. We postulated that the lone electron pair on the 17-nitrogen in SN-28 requires protection from forming of an ammonium ion in order to penetrate through the BBB. We already reported that the 14-hydroxyl in naltrexone could form a hydrogen bond with the lone electron pair on the 17-nitrogen to produce a less polar compound (Fig. 4). Based on the above discussion, we converted SN-28 to the novel DOR agonist, KNT-127 (Nagase et al. 2010). As we expected, KNT-127 showed higher selectivity for the DOR than SN-28 and potent agonist activity ($ED_{50} = 0.095$ nM, $ED_{50} = 0.149$ nM, respectively) when delivered by i.t. injection in the mouse acetic acid writhing test (Nagase et al. 2010). Moreover, KNT-127 showed pronounced antinociceptive effects 30 min after *s.c.* administration in the mouse acetic acid writhing test; thus, it was about 30-fold more potent than (−)TAN-67 (Nagase et al. 2010; Saitoh et al. 2011; Saitoh and Yamada 2012).

3.2 The Pharmacological Properties of KNT-127

Ours and many other studies have suggested that DOR agonists, including (±) TAN-67 and SNC80, produce potent antidepressant and anxiolytic-like effects in animal models (Saitoh and Yamada 2012). However, the DOR ligands with a diarylmethylpiperazine structure, such as SNC80, produced convulsive effects in rodents and monkeys (Comer et al. 1993; Jutkiewicz et al. 2004; Negus et al. 1994). Therefore, its clinical development has been limited. Recently, we found that KNT-127 produced no convulsive effect; thus, KNT-127 and its derivates are attracting attention as new potential treatments for depression and/or anxiety. In this section, we summarize the recent reported results of the pharmacological properties of KNT-127.

3.2.1 Antidepressant-Like Effects

We previously reported that KNT-127 produced antidepressant-like effects in a mouse forced swimming test, a screening model for antidepressants (Saitoh et al. 2011). The *s.c.* administration of KNT-127 (0.1–1.0 mg/kg) decreased the immobility time and increased the duration of swimming and climbing. These effects of KNT-127 were significantly reversed to the control level by pretreatment with NTI, suggesting that these effects were mediated by the DOR. Furthermore, the magnitude of the KNT-127 (1 mg/kg)-induced antidepressant-like effect was similar to that produced by the *s.c.* administration of imipramine (6 mg/kg), a tricyclic antidepressant. Therefore, we proposed that KNT-127 produced robust antidepressant-like effects that were mediated by DOR stimulation. In our previous studies, SNC80 produced an increase in climbing, but not swimming, in a mouse forced swimming test (Saitoh et al. 2011). These effects on the swimming and climbing behaviors following administration of KNT-127 and SNC80 were consistent with another study (Nozaki et al. 2014). As previously shown, enhanced serotonergic neurotransmission predominantly increases swimming behavior, while enhanced catecholaminergic neurotransmission increases climbing behavior in the forced swimming test (Cryan et al. 2005). This suggests that the antidepressant-like effects of KNT-127 and SNC80 may be mediated by these different neurotransmitter systems. It was reported that KNT-127 evoked the release of extracellular dopamine and glutamate levels in the rat striatum and the medial prefrontal cortex in a microdialysis study (Tanahashi et al. 2012). Although the detailed mechanisms for the antidepressant-like effects of KNT-127 have not been fully characterized, these neurotransmission may be involved in the expression of swimming and climbing behaviors.

3.2.2 Anxiolytic-Like Effects

We investigated the anxiolytic-like effects of KNT-127 using three different rat models of innate anxiety (Saitoh et al. 2013; Sugiyama et al. 2014). The *s.c.* administration of KNT-127 (0.3–3.0 mg/kg) increased the time spent in the open arm in the elevated plus-maze test in rats. These effects were significantly reversed to the control level by pretreatment with NTI, suggesting that the anxiolytic-like effects of KNT-127 were mediated by the DOR. The magnitude of the KNT-127 (3 mg/kg)-induced anxiolytic-like effects was similar to that produced by the *s.c.* administration of diazepam (1 mg/kg), a benzodiazepine anxiolytic. On the other hand, the anxiolytic-like effects of diazepam were not affected by pretreatment with NTI, indicating that these effects are not associated with the DOR. These findings were supported by results obtained from other anxiety animal models, such as from light/dark and open-field tests. Based on these results, we proposed that KNT-127 produced robust anxiolytic-like effects in rat innate anxiety models.

Amnesia, ethanol interaction, and motor coordination deficits are known as the classical side effects of benzodiazepine, and the $GABA_A$-benzodiazepine receptor pathway plays an important role in the pathophysiology of these side effects. Diazepam (1 mg/kg, *s.c.*) decreased the spontaneous alteration performance in the Y-maze test, suggesting that diazepam produced amnesia effects at the doses that caused anxiolytic-like effects, while KNT-127 (3.0 mg/kg) caused no significant performance changes in the Y-maze test (Saitoh et al. 2013). Diazepam (1 mg/kg, s. c.) also increased the ethanol sleeping time in the ethanol-induced sleeping test and foot-angle-to-walking direction in the footprint test. These results suggest that diazepam produced ethanol-interaction effects and motor coordination deficits at the doses that caused anxiolytic-like effects. Interestingly, in contrast to diazepam (1.0 mg/kg), KNT-127 (3.0 mg/kg) caused no significant performance changes in the ethanol-induced sleeping test and footprint test (Saitoh et al. 2013). Taken together, we suggested that KNT-127 did not appear to affect the $GABA_A$-benzodiazepine receptor pathway in the rat brain regions responsible for benzodiazepine side effects.

Recently, the anxiolytic effects of DOR agonists were observed in Phase II clinical studies using adult patients with anxious major depressive disorder (clinicaltrials.gov ID NCT00759395/NCT01020799). The DOR agonists are expected to be effective treatments for anxiety, without producing adverse effects associated with benzodiazepines.

3.2.3 Proconvulsant Effect

Previous studies indicated that a prototype DOR agonist SNC80 induced convulsions in about half of the rats treated with a dose of 32 mg/kg (Jutkiewicz et al. 2004). Similarly, we observed that seven of ten mice exhibited a brief convulsive event lasting 10–15 s with a catalepsy-like behavior within about 10 min of administering SNC80 (30 mg/kg). Immediately following the convulsions, these mice displayed catalepsy-like behavior for about 40 s. Interestingly, KNT-127 produced no convulsions or catalepsy, even at 30–100 times higher doses (100 mg/kg) than those required for antidepressant-like or anxiolytic-like effects in rodents. In addition, although SNC80 (10 mg/kg, *s.c.*) produced a substantial effect on the spontaneous locomotor activity in mice, KNT-127 (10 mg/kg) produced no effects. Clinically useful DOR agonists need to have minimal undesirable effects, while retaining the main medical properties. Hence, we propose that KNT-127 and its derivatives should be considered as candidate compounds for the clinical development of DOR-based novel antipsychotic drugs that lack the convulsive effects associated with other DOR agonists, which limit their therapeutic potential.

3.2.4 Potential as a Biased DOR Ligand

It has been well established that distinct agonists acting at the same G-protein-coupled receptor can engage different signaling or regulatory responses. This concept is known as biased agonism, which has important biological and therapeutic implications. Ligand-biased responses are well described in cellular models;

however, the physiological relevance of biased agonism at the behavioral level has yet to be elucidated (Violin et al. 2014).

In a previous study, Kieffer and colleague reported that DORs display differential receptor internalization properties in vivo, as SNC80 induced internalization, whereas KNT-127 did not (Nozaki et al. 2014). In contrast to SNC80, KNT-127 did not induce DOR internalization when assessed using DOR knock-in mice expressing functional fluorescent-tagged DOR (DOR-eGFP mice). While SNC80 (10 mg/kg, *i.p.*) induced receptor internalization in the striatum, hippocampus, spinal cord, and dorsal root ganglia, KNT-127 (10 mg/kg, *s.c.*) did not alter receptor distribution, as a strong fluorescent signal was detected at the cell surface in all tissues, similar to the saline control. These results suggest that KNT-127 and SNC80 induce differential signaling in the central nervous system and, therefore, have distinct behavioral consequences. Actually, KNT-127 induced an antidepressant-like effect in a biased manner, compared with SNC80. For example, repeated treatment with KNT-127 induced no tolerance to KNT-127 and/or no cross tolerance to SNC80-induced antidepressant-like effects in the forced swimming test, suggesting that the differential effects of KNT-127 and SNC80 are due to a ligand-biased agonism for the DOR-mediated tolerance effects only for SNC80 (Nozaki et al. 2014).

Interestingly, these activities are similar to that of other recently reported DOR agonists, AR-M1000390, ADL5747, and ADL5859 (Nozaki et al. 2012; Pradhan et al. 2009, 2010). Similar to KNT-127, these DOR agonists exhibited ligand-biased pharmacological effects at the DOR. These DOR agonists did not induce tolerance to an antidepressant-like effects or DOR internalization, like SNC80 does. In addition, only SNC80 and its derivatives evoked convulsions and hyperlocomotion. These findings suggested that DOR agonist-induced ligand-biased agonism possibly regulates distinct or selective intracellular signaling, neurotransmission, or long adaptation. A recent study reported that DOR agonists without convulsive effects produced decreases in both β-arrestin-2 recruitment and DOR internalization in CHO cells (Nakata et al. 2014). Furthermore, it was reported that DOR-mediated seizures were reduced in β-arrestin-2 knockout mice (Violin 2014b). More recently, Chiang et al. (2016) suggested that compared with DPDPE, SNC80 is a "super-recruiter" of β arrestin-2, whereas KNT-127 is a weak/moderate recruiter as measured in CHO cells expressing DORs. These findings suggest that a G-protein-biased ligand at the DOR may prevent DOR-mediated seizures and tolerance, while beneficial effects, such as antidepressant properties, are preserved.

4 Conclusion

The discovery of the selective DOR antagonist, NTI, led to the synthesis of (±)TAN-67, which was designed on the basis of the accessory site hypothesis and the message-address concept. After succeeding in synthesizing (±)TAN-67, many pharmacological effects induced via the DOR were identified and reported. The selective DOR antagonists, TRK-850 and TRK-851, were designed and synthesized based on the structure of NTI. The discoveries of TRK-850 and TRK-851 demonstrated the existence of the regulatory system for the cough reflex via the DOR in the central nervous system. To improve the agonistic activity of (±)TAN-67 for the DOR following systemic administration, we reexamined the accessory site of NTI, which led to the design and synthesis of SN-28. Furthermore, the low permeability of SN-28 through the BBB was improved with KNT-127, which produced high selectivity and potent agonistic activity for the DOR in vivo. Also, KNT-127 produced no convulsive effects, which is different from prototype DOR compounds, like SNC80 derivatives. The biased ligands targeting the DOR may be able to reduce the on-target seizure liability that has hindered drug discovery effects targeting selective DOR agonists. The KNT-127 type quinolinomorphinan derivatives of DOR ligands are expected to be promising candidates for the development of therapeutic DOR agonists that do not induce convulsions.

Acknowledgments We thank Prof. Tsutomu Suzuki, Prof. Junzo Kamei, Prof. Minoru Narita, (Hoshi University), Prof. Hideaki Fujii (Kitasato University), and Director Mitsuhiko Yamada (National Center of Neurology and Psychiatry) for their continuous support of research on the DOR ligand and its pharmacology.

References

Chiang T, Sansuk K, van Rijn RM (2016) Beta-arrestin 2 dependence of delta opioid receptor agonists is correlated with alcohol intake. Br J Pharmacol 173(2):332–343

Comer SD, Hoenicke EM, Sable AI, McNutt RW, Chang KJ, De Costa BR, Mosberg HI, Woods JH (1993) Convulsive effects of systemic administration of the delta opioid agonist BW373U86 in mice. J Pharmacol Exp Ther 267:888–895

Cryan JF, Valentino RJ, Lucki I (2005) Assessing substrates underlying the behavioral effects of antidepressants using the modified rat forced swimming test. Neurosci Biobehav Rev 29:547–569

Fernyhough P, Calcutt NA (2010) Abnormal calcium homeostasis in peripheral neuropathies. Cell Calcium 47:130–139

Fryer RM, Hsu AK, Eells JT, Nagase H, Gross GJ (1999) Opioid-induced second window of cardioprotection: potential role of mitochondrial KATP channels. Circ Res 84:846–851

Fryer RM, Auchampach JA, Gross GJ (2002) Therapeutic receptor targets of ischemic preconditioning. Cardiovasc Res 55:520–525

Fujii H, Takahashi T, Nagase H (2013) Non-peptidic δ opioid receptor agonists and antagonists (2000–2012). Expert Opin Ther Pat 23:1181–1208

Huh J, Gross GJ, Nagase H, Liang BT (2001) Protection of cardiac myocytes via delta(1)-opioid receptors, protein kinase C, and mitochondrial K(ATP) channels. Am J Physiol Heart Circ Physiol 280:H377–H383

Jutkiewicz EM, Eller EB, Folk JE, Rice KC, Traynor JR, Woods JH (2004) Delta-opioid agonists: differential efficacy and potency of SNC80, its 3-OH (SNC86) and 3-desoxy (SNC162) derivatives in Sprague-Dawley rats. J Pharmacol Exp Ther 309:173–181

Kamei J (1996) Role of opioidergic and serotonergic mechanisms in cough and antitussives. Pulm Pharmacol 9:349–356

Kamei J, Iwamoto Y, Misawa M, Nagase H, Kasuya Y (1994) Streptozotocin-induced diabetes selectively enhances antinociception mediated by delta 1- but not delta 2-opioid receptors. Life Sci 55:L121–L126

Kamei J, Saitoh A, Ohsawa M, Suzuki T, Misawa M, Nagase H, Kasuya Y (1995) Antinociceptive effects of the selective non-peptidic delta-opioid receptor agonist TAN-67 in diabetic mice. Eur J Pharmacol 276:131–135

Kamei J, Kawai K, Mizusuna A, Saitoh A, Morita K, Narita M, Tseng LF, Nagase H (1997) Supraspinal delta 1-opioid receptor-mediated antinociceptive properties of (-)-TAN-67 in diabetic mice. Eur J Pharmacol 322:27–30

Kamei J, Ohsawa M, Suzuki T, Saitoh A, Endoh T, Narita M, Tseng LF, Nagase H (1999) The modulatory effect of (+)-TAN-67 on the antinociceptive effects of the nociceptin/orphanin FQ in mice. Eur J Pharmacol 383:241–247

Knapp RJ, Landsman R, Waite S, Malatynska E, Varga E, Haq W, Hruby VJ, Roeske WR, Nagase H, Yamamura HI (1995) Properties of TAN-67, a nonpeptidic delta-opioid receptor agonist, at cloned human delta- and mu-opioid receptors. Eur J Pharmacol 291:129–134

Krishnan-Sarin S, Jing SL, Kurtz DL, Zweifel M, Portoghese PS, Li TK, Froehlich JC (1995a) The delta opioid receptor antagonist naltrindole attenuates both alcohol and saccharin intake in rats selectively bred for alcohol preference. Psychopharmacology (Berl) 120:177–185

Krishnan-Sarin S, Portoghese PS, Li TK, Froehlich JC (1995b) The delta 2-opioid receptor antagonist naltriben selectively attenuates alcohol intake in rats bred for alcohol preference. Pharmacol Biochem Behav 52:153–159

Lê AD, Poulos CX, Quan B, Chow S (1993) The effects of selective blockade of delta and mu opiate receptors on ethanol consumption by C57BL/6 mice in a restricted access paradigm. Brain Res 630:330–332

Levy J, Gavin JR, Sowers JR (1994) Diabetes mellitus: a disease of abnormal cellular calcium metabolism? Am J Med 96:260–273

Matsuzawa S, Suzuki T, Misawa M, Nagase H (1999) Different roles of mu-, delta- and kappa-opioid receptors in ethanol-associated place preference in rats exposed to conditioned fear stress. Eur J Pharmacol 368:9–16

Nagase H, Fujii H (2011) Opioids in preclinical and clinical trials. Top Curr Chem 299:29–62

Nagase H, Kawai K, Hayakawa J, Wakita H, Mizusuna A, Matsuura H, Tajima C, Takezawa Y, Endoh T (1998) Rational drug design and synthesis of a highly selective nonpeptide delta-opioid agonist, (4aS*,12aR*)-4a-(3-hydroxyphenyl)-2-methyl- 1,2,3,4,4a,5,12,12a-octahydropyrido[3,4-b]acridine (TAN-67). Chem Pharm Bull (Tokyo) 46:1695–1702

Nagase H, Yajima Y, Fujii H, Kawamura K, Narita M, Kamei J, Suzuki T (2001) The pharmacological profile of delta opioid receptor ligands, (+) and (-) TAN-67 on pain modulation. Life Sci 68:2227–2231

Nagase H, Osa Y, Nemoto T, Fujii H, Imai M, Nakamura T, Kanemasa T, Kato A, Gouda H, Hirono S (2009) Design and synthesis of novel delta opioid receptor agonists and their pharmacologies. Bioorg Med Chem Lett 19:2792–2795

Nagase H, Nemoto T, Matsubara A, Saito M, Yamamoto N, Osa Y, Hirayama S, Nakajima M, Nakao K, Mochizuki H, Fujii H (2010) Design and synthesis of KNT-127, a delta-opioid receptor agonist effective by systemic administration. Bioorg Med Chem Lett 20:6302–6305

Nakata E, Sakai J, Kanda T, Watanabe T, Saito D, Takahashi T, Iwai T, Hirayama S, Fujii H, Yamakawa T, Nagase H (2014) Receptor-mediated beta-arrestin signaling modulates convulsive effects. In: International narcotic research conference 2014, p 63

Negus SS, Butelman ER, Chang KJ, DeCosta B, Winger G, Woods JH (1994) Behavioral effects of the systemically active delta opioid agonist BW373U86 in rhesus monkeys. J Pharmacol Exp Ther 270:1025–1034

Nozaki C, Le Bourdonnec B, Reiss D, Windh RT, Little PJ, Dolle RE, Kieffer BL, Gavériaux-Ruff C (2012) δ-Opioid mechanisms for ADL5747 and ADL5859 effects in mice: analgesia, locomotion, and receptor internalization. J Pharmacol Exp Ther 342:799–807

Nozaki C, Nagase H, Nemoto T, Matifas A, Kieffer BL, Gaveriaux-Ruff C (2014) In vivo properties of KNT-127, a novel δ opioid receptor agonist: receptor internalization, antihyperalgesia and antidepressant effects in mice. Br J Pharmacol 171:5376–5386

Ohsawa M, Nagase H, Kamei J (1998) Role of intracellular calcium in modification of mu and delta opioid receptor-mediated antinociception by diabetes in mice. J Pharmacol Exp Ther 286:780–787

Peart JN, Patel HH, Gross GJ (2003) Delta-opioid receptor activation mimics ischemic preconditioning in the canine heart. J Cardiovasc Pharmacol 42:78–81

Portoghese PS, Sultana M, Nagase H, Takemori AE (1988) Application of the message-address concept in the design of highly potent and selective non-peptide delta opioid receptor antagonists. J Med Chem 31:281–282

Portoghese PS, Sultana M, Takemori AE (1990) Design of peptidomimetic delta opioid receptor antagonists using the message-address concept. J Med Chem 33:1714–1720

Pradhan AA, Becker JA, Scherrer G, Tryoen-Toth P, Filliol D, Matifas A, Massotte D, Gavériaux-Ruff C, Kieffer BL (2009) In vivo delta opioid receptor internalization controls behavioral effects of agonists. PLoS One 4:e5425

Pradhan AA, Walwyn W, Nozaki C, Filliol D, Erbs E, Matifas A, Evans C, Kieffer BL (2010) Ligand-directed trafficking of the δ-opioid receptor in vivo: two paths toward analgesic tolerance. J Neurosci 30:16459–16468

Roberts AJ, Gold LH, Polis I, McDonald JS, Filliol D, Kieffer BL, Koob GF (2001) Increased ethanol self-administration in delta-opioid receptor knockout mice. Alcohol Clin Exp Res 25:1249–1256

Saitoh A, Yamada M (2012) Antidepressant-like effects of delta opioid receptor agonists in animal models. Curr Neuropharmacol 10:231–238

Saitoh A, Sugiyama A, Nemoto T, Fujii H, Wada K, Oka J, Nagase H, Yamada M (2011) The novel delta opioid receptor agonist KNT-127 produces antidepressant-like and antinociceptive effects in mice without producing convulsions. Behav Brain Res 223:271–279

Saitoh A, Sugiyama A, Yamada M, Inagaki M, Oka J, Nagase H (2013) The novel δ opioid receptor agonist KNT-127 produces distinct anxiolytic-like effects in rats without producing the adverse effects associated with benzodiazepines. Neuropharmacology 67:485–493

Sakami S, Kawai K, Maeda M, Aoki T, Fujii H, Ohno H, Ito T, Saitoh A, Nakao K, Izumimoto N, Matsuura H, Endo T, Ueno S, Natsume K, Nagase H (2008) Design and synthesis of a metabolically stable and potent antitussive agent, a novel delta opioid receptor antagonist, TRK-851. Bioorg Med Chem 16:7956–7967

Schultz JE, Hsu AK, Gross GJ (1998) Ischemic preconditioning in the intact rat heart is mediated by delta1- but not mu- or kappa-opioid receptors. Circulation 97:1282–1289

Schwyzer R (1977) ACTH: a short introductory review. Ann N Y Acad Sci 297:3–26

Sugiyama A, Nagase H, Oka J, Yamada M, Saitoh A (2014) DOR(2)-selective but not DOR(1)-selective antagonist abolishes anxiolytic-like effects of the δ opioid receptor agonist KNT-127. Neuropharmacology 79:314–320

Suzuki T, Tsuji M, Mori T, Misawa M, Endoh T, Nagase H (1995) Effects of a highly selective nonpeptide delta opioid receptor agonist, TAN-67, on morphine-induced antinociception in mice. Life Sci 57:155–168

Suzuki T, Tsuji M, Mori T, Misawa M, Endoh T, Nagase H (1996) Effect of the highly selective and nonpeptide delta opioid receptor agonist TAN-67 on the morphine-induced place preference in mice. J Pharmacol Exp Ther 279:177–185

Tanahashi S, Ueda Y, Nakajima A, Yamamura S, Nagase H, Okada M (2012) Novel δ1-receptor agonist KNT-127 increases the release of dopamine and L-glutamate in the striatum, nucleus accumbens and median pre-frontal cortex. Neuropharmacology 62:2057–2067

Tseng LF, Narita M, Mizoguchi H, Kawai K, Mizusuna A, Kamei J, Suzuki T, Nagase H (1997) Delta-1 opioid receptor-mediated antinociceptive properties of a nonpeptidic delta opioid receptor agonist, (-)TAN-67, in the mouse spinal cord. J Pharmacol Exp Ther 280:600–605

van Ree JM, Gerrits MA, Vanderschuren LJ (1999) Opioids, reward and addiction: An encounter of biology, psychology, and medicine. Pharmacol Rev 51:341–396

van Rijn RM, Whistler JL (2009) The delta(1) opioid receptor is a heterodimer that opposes the actions of the delta(2) receptor on alcohol intake. Biol Psychiatry 66:777–784

van Rijn RM, Brissett DI, Whistler JL (2010) Dual efficacy of delta opioid receptor-selective ligands for ethanol drinking and anxiety. J Pharmacol Exp Ther 335:133–139
van Rijn RM, Brissett DI, Whistler JL (2012) Distinctive modulation of ethanol place preference by delta opioid receptor-selective agonists. Drug Alcohol Depend 122:156–159
Violin JD (2014) Biased ligands at mu and delta opioid receptors: targeting selective signalling to develop improved therapeutics. In: International narcotic research conference 2014, p 22
Violin JD, Crombie AL, Soergel DG, Lark MW (2014) Biased ligands at G-protein-coupled receptors: promise and progress. Trends Pharmacol Sci 35:308–316
Yajima Y, Narita M, Tsuda M, Imai S, Kamei J, Nagase H, Suzuki T (2000) Modulation of NMDA- and (+)TAN-67-induced nociception by GABA(B) receptors in the mouse spinal cord. Life Sci 68:719–725

Multifunctional Opioid Ligands

Jessica P. Anand and Deanna Montgomery

Contents

Abstract

The opioid receptor system plays a major role in the regulation of mood, reward, and pain. The opioid receptors therefore make attractive targets for the treatment of many different conditions, including pain, depression, and addiction. However, stimulation or blockade of any one opioid receptor type often leads to

J. P. Anand (✉)
Department of Pharmacology, Medical School and the Edward F. Domino Research Center, University of Michigan, Ann Arbor, MI, USA
e-mail: janand@umich.edu

D. Montgomery
Department of Medicinal Chemistry, College of Pharmacy, University of Michigan, Ann Arbor, MI, USA

E. M. Jutkiewicz (ed.), *Delta Opioid Receptor Pharmacology and Therapeutic Applications*, Handbook of Experimental Pharmacology 247, https://doi.org/10.1007/164_2018_104

on-target adverse effects that limit the clinical utility of a selective opioid agonist or antagonist. Literature precedent suggests that the opioid receptors do not act in isolation and that interactions among the opioid receptors and between the opioid receptors and other proteins may produce clinically useful targets. Multifunctional ligands have the potential to elicit desired outcomes with reduced adverse effects by allowing for the activation of specific receptor conformations and/or signaling pathways promoted as a result of receptor oligomerization or crosstalk. In this chapter, we describe several classes of multifunctional ligands that interact with at least one opioid receptor. These ligands have been designed for biochemical exploration and the treatment of a wide variety of conditions, including multiple kinds of pain, depression, anxiety, addiction, and gastrointestinal disorders. The structures, pharmacological utility, and therapeutic drawbacks of these classes of ligands are discussed.

Keywords

Anxiety · Bivalent · Depression · GPCR · Mixed efficacy · Mood · Multifunctional · Opioid · Pain · Reward

1 Introduction

Opioid agonists have long been used in the treatment of acute and chronic pain and are still widely used in the clinic today. After the discovery and cloning of the three classical opioid receptors – mu (MOR), delta (DOR), and kappa (KOR) – the search for additional and increasingly selective opioid ligands began, driven in part by the need for tools to characterize the opioid receptors. It was assumed that selective opioid agonists would be the future of opioid analgesics, and it seemed intuitive that a more specific ligand would have fewer off-target interactions and unintended effects.

Clinically relevant opioid therapeutics produce their analgesic effects through stimulation of MOR. Unfortunately, adverse effects associated with opioid analgesics such as constipation, respiratory depression, euphoria, tolerance to opioid-mediated analgesia, and physical dependence are mediated by MOR as well. Further, the development of tolerance to and dependence on opioid analgesics may contribute to the prevalence of opioid abuse (Ross and Peselow 2009; Bailey and Connor 2005; Johnston et al. 2009). The development of these undesirable side effects is problematic in many ways; not only does it complicate dosing regimens and decrease patient compliance, but it also limits the clinical utility of opioids and has been linked to increased addiction liability. As the desired analgesic effects and negative side effects are all mediated through MOR, the development of more selective MOR agonists will not address the problems associated with acute and chronic opioid analgesic use.

The stimulation of DOR or KOR has been shown to produce analgesic effects in vivo; however, there are also adverse effects associated with stimulation of each of these receptors. Stimulation of KOR produces aversive and dysphoric effects, while stimulation of DOR produces convulsions under some conditions (Lutz and Kieffer

2013). As a result, pure DOR and KOR ligands have not been pursued as therapeutic tools. However, since it has been shown that opioid receptors do not act in isolation in vitro or in vivo, the simultaneous modulation of multiple targets may generate a more desirable drug profile (Morphy et al. 2004; Morphy and Rankovic 2009; Dietis et al. 2009; Balboni et al. 2002, 2007). This strategy may allow for the activation of specific receptor conformations and/or signaling pathways promoted as a result of receptor oligomerization or crosstalk.

Multifunctional ligands that interact with multiple receptors simultaneously or with unique receptor/signaling complexes possess considerable advantages over the traditional approach of using a combination of selective drugs. Combination therapies or drug cocktails contain multiple active components with differing pharmacokinetic properties. Different drugs often need to be taken on different schedules in order to be most effective (e.g., every 6 h vs. every 12 h) due to the unique absorption and clearance rates of each chemical entity; further, the optimal absorption conditions for each drug may be different (e.g., on a full stomach, on an empty stomach, with a full glass of water). The necessary complicated dosing regimens associated with administration of multiple drugs can reduce patient compliance. In addition to these complications, coadministration of multiple chemical entities often alters metabolism and clearance rates due to off-target drug effects in the liver and kidneys. This increases the risk of patient to patient variation in efficacy and adverse drug reactions.

The multifunctional ligands reported in the literature thus far fall into two main categories: (1) bivalent or bidentate ligands, in which two separate pharmacophores are linked by a flexible spacer, and (2) multifunctional or mixed efficacy ligands, which contain a single set of binding elements that interacts with multiple targets (Fig. 1). Strategies for simultaneous modulation of multiple receptors have been

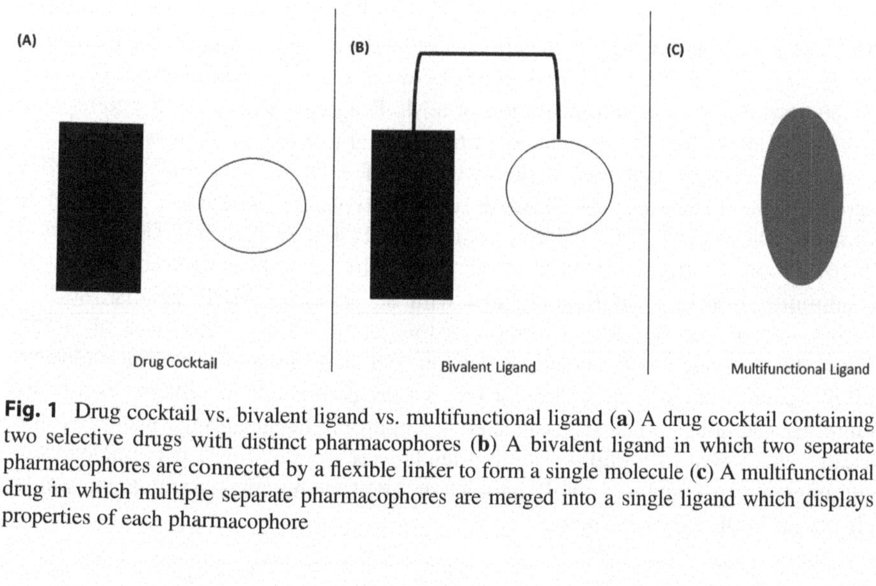

Fig. 1 Drug cocktail vs. bivalent ligand vs. multifunctional ligand (**a**) A drug cocktail containing two selective drugs with distinct pharmacophores (**b**) A bivalent ligand in which two separate pharmacophores are connected by a flexible linker to form a single molecule (**c**) A multifunctional drug in which multiple separate pharmacophores are merged into a single ligand which displays properties of each pharmacophore

developed for the treatment of pain and other disorders, a selection of which are discussed in the sections below.

2 Multifunctional Opioid/Opioid Ligands

Opioid receptors are involved in the regulation of pain, mood, and reward; as a result, opioid ligands show great therapeutic potential for treating many different sensory and mood disorders. As the different opioid receptors often exhibit opposing effects, the simultaneous modulation of multiple receptors has been proposed as a strategy to balance these therapeutic and adverse effects to elicit desired pharmacological profiles. Specific, nonselective ligands for a variety of opioid receptor combinations have been developed for a variety of applications.

2.1 DOR/MOR

It has been reported in the literature that the coadministration of a DOR agonist with a MOR agonist lessens the development of tolerance to and dependence on MOR agonists without attenuating MOR-mediated analgesia (Li et al. 2012; Lowery et al. 2011; Rozenfeld et al. 2007). Coadministration of a DOR agonist with a MOR agonist also reduces the incidence of other unwanted side effects such as stimulation of forward locomotion, which is generally interpreted as an activation of dopamine systems and has been used as an early indicator of abuse liability (Li et al. 2012). It has also been reported that small doses of DOR agonists potentiate the affinity and antinociceptive potency of MOR agonists as well as potentiating the efficacy of MOR agonists (Lowery et al. 2011; Heyman et al. 1989a, b; Horan et al. 1992; Qi et al. 1990). This suggests that the combination of a DOR agonist with a traditional MOR agonist opioid analgesic may decrease the dose necessary to produce effective analgesia, widening the therapeutic index between analgesia and adverse events and adaptations. Further, coadministration of a MOR agonist with a DOR agonist may reduce the abuse liability associated with MOR agonist opioid analgesics.

It is noteworthy that the coadministration of a DOR antagonist and a MOR agonist also decreases the adverse effects typically associated with MOR agonists (Dietis et al. 2009; Martin et al. 2000; Horan et al. 1993; Zielińska et al. 2016; Abbott and Romero 1996; Li et al. 2007). Similarly, it has been shown that the coadministration of a DOR antagonist with an addictive MOR agonist, such as heroin, can reduce self-administration of that MOR agonist (Martin et al. 2000). These data suggest that the coadministration of a DOR antagonist with a traditional MOR agonist analgesic may slow or prevent the emergence of adverse events and minimize the abuse potential associated with chronic MOR agonist use, providing a safer alternative to traditional opioid analgesics. At this time, it is unclear why administration of either a DOR agonist or DOR antagonist would have similar effects on MOR agonist effects.

2.1.1 Bifunctional DOR Agonist/MOR Agonist Ligands

Bifunctional DOR agonist/MOR agonist ligands have been developed to harness the beneficial aspects of DOR agonist/MOR agonist drug cocktails without the complicated pharmacokinetic profiles associated with administering two distinct chemical entities. These "selectively promiscuous" compounds combine two different pharmacophores in the same molecule.

The dimeric enkephalin peptide, biphalin, was one of the first reported potent DOR agonist/MOR agonist bifunctional compounds (Fig. 2). Biphalin consists of two tetrapeptides connected "tail-to-tail" by a hydrazide bridge. It produced dose-dependent antinociception in the warm water tail withdrawal assay after both peripheral administration and *intracerebroventricular* (*icv*) administration in mice. While it is expected that *icv* administration of a MOR agonist will produce antinociception, it is surprising that *intraperitoneal* (*ip*) administration of biphalin produced antinociception because unmodified peptides often do not cross the blood–brain barrier (BBB). As such, it is likely that the antinociceptive response produced by biphalin is peripherally mediated (Horan et al. 1993). Parenteral administration of biphalin significantly inhibits gastric transit (Zielińska et al. 2016). Further, when biphalin is infused directly into the brain into the lateral ventricles (*icv*), mice become physically dependent similar to that observed with morphine (*icv*), such that naloxone precipitates withdrawal signs (Abbott and Romero 1996). Overall, these findings suggest that biphalin is acting as a peripheral MOR agonist and that not all of the adverse effects associated with MOR agonists were ameliorated by this DOR agonist/MOR agonist profile (Horan et al. 1993).

More recently, a series of peptides based on endomorphin II (Tyr-Pro-Phe-Phe-NH_2), an endogenous MOR agonist, has been reported (Li et al. 2007); it was found that replacing the Tyr^1 of endomorphin II with 2′,6′-dimethyltyrosine (Dmt) increased affinity for DOR and adding moderate bulk to the 2 and 6 position of the Phe^3 produced compounds with both DOR and MOR agonist activity. A series of mixed DOR agonist/MOR agonist ligands were generated that were selective relative to KOR (Li et al. 2007). As this series is structurally similar to endogenous endomorphins, it is possible that these ligands will produce limited tolerance and dependence as compared to traditional opioid analgesics. To date, however, no in vivo data have been reported.

One of the best characterized DOR agonist/MOR agonist peptides is MMP-2200 (Tyr-D-Thr-Gly-Phe-Leu-Ser-(*O*-β-D-lactose)-NH_2), a glycosylated, bioavailable,

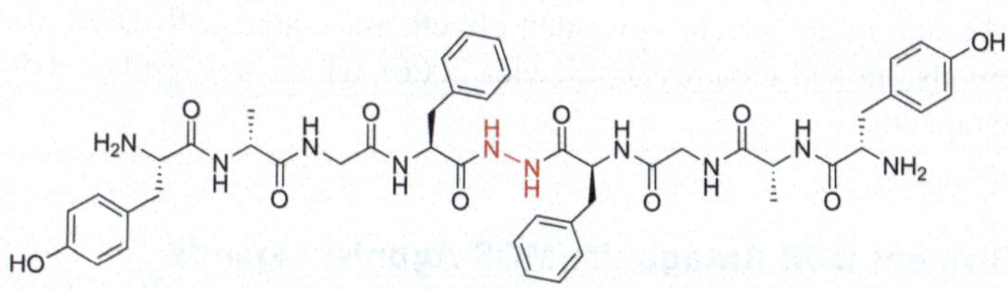

Fig. 2 Structure of biphalin ((Tyr-D-Ala-Gly-Phe-NH)$_2$), a DOR agonist/MOR agonist dimeric peptide linked by a hydrazide bridge (highlighted in red)

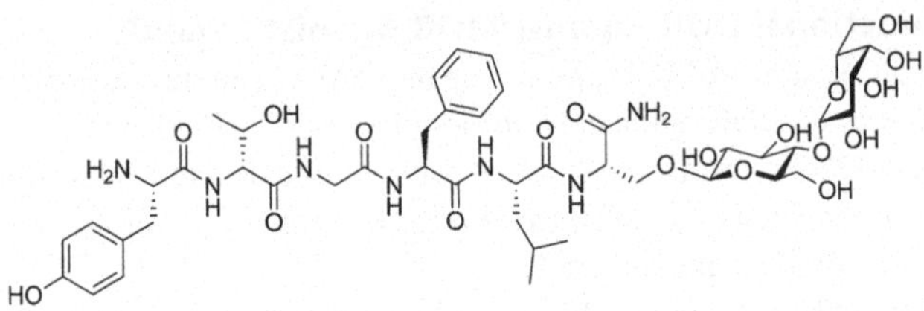

Fig. 3 Structure of MMP-2200 (Tyr-D-Thr-Gly-Phe-Leu-Ser-(O-β-D-lactose)-NH_2), a DOR agonist/MOR agonist peptide modified with a C-terminal sugar moiety to improve membrane penetration

centrally active derivative of the opioid peptide DTLES (Fig. 3). The in vivo actions of MMP-2200 have been thoroughly investigated (Li et al. 2012; Lowery et al. 2011; Polt et al. 2005). Antinociceptive tolerance was examined in the mouse warm water tail withdrawal assay. Twice daily injections of an antinociceptive dose A_{90} of MMP-2200 for 3 days produced approximately fivefold shift in ED_{50}; the same dosing regimen of morphine produced approximately 13-fold shift in ED_{50}. Mice treated twice daily with an A_{90} dose of MMP-2200 displayed significantly fewer withdrawal signs after precipitated withdrawal than mice treated twice daily for 4 days with an A_{90} dose of morphine (Lowery et al. 2011). The reinforcing effects of MMP-2200 have also been explored. Morphine produced robust self-administration in monkeys, while MMP-2200 did not. However, morphine was active in the warm water tail withdrawal assay in monkeys, while MMP-2200 did not produce antinociception in the same assay. MMP-2200 only showed antinociceptive effects in a capsaicin-induced model of allodynia in nonhuman primates (Do Carmo et al. 2008). These data suggest that the mixed efficacy DOR agonist/MOR agonist profile may reduce the negative neurochemical adaptations and addiction liability problems associated with pure MOR agonist analgesics.

There are currently very few reports of DOR/MOR dual agonist small molecules that are selective relative to KOR. One recent report describes a series of pyrrolo- and pyridomorphans that displayed full agonist activity at DOR and partial agonist activity at MOR (Kumar et al. 2014a). These compounds have not yet been explored in animal models.

While there are several DOR agonist/MOR agonist compounds that are promising leads for developing safer opioid analgesics, DOR/MOR agonist crosstalk has been pursued less vigorously than DOR antagonist/MOR agonist interactions. This may be due to the severe unwanted effects associated with DOR stimulation such as convulsions and seizures (Jutkiewicz 2006), which are significant drawbacks for any therapeutic.

2.1.2 Bivalent DOR Antagonist/MOR Agonist Ligands

The design of bivalent DOR/MOR ligands is predicated on the idea of DOR/MOR heterodimers. Literature reports suggest that using a flexible 21-atom linker between a DOR antagonist pharmacophore and a MOR agonist pharmacophore produced robust pain relief with limited development of adverse effects, while linkers of shorter or longer length showed decreased binding and/or more adverse effects (Daniels et al. 2005a; Lenard et al. 2007; Aceto et al. 2012). The Portoghese lab proposes that this 21-atom linker is the appropriate length to reach from the orthosteric binding site of DOR into the orthosteric binding site of MOR in DOR/MOR heterodimers. Additionally, they suggest that these bivalent ligands stimulate DOR/MOR heterodimers, which may be unique signaling entities, to produce antinociception without adverse effects.

MDAN-21 is one such ligand comprised of a naltrindole derivative (DOR antagonist) linked to an oxymorphone derivative (MOR agonist) by a 21-atom linker (Aceto et al. 2012) (Fig. 4). Tolerance to MDAN-21 and related bivalent compounds was tested in the radiant tail flick assay; antinociceptive dose–response curves were established before and after 3-day treatment with continuous *icv* infusion of 12 times the ED_{50}/h. Treatment with DOR antagonist/MOR agonist compounds produced less tolerance and dependence as compared with either morphine or a monovalent MOR agonist. Three-day chronic treatment with bivalent compounds with linker length in the 19–21-atom range showed no shift in ED_{50} in the radiant tail flick assay, while morphine and the monovalent MOR agonist showed approximately sixfold shift in ED_{50}. Mice receiving 3 days of continuous *icv* infusion of bivalent ligands with linker lengths in the 19–21-atom range demonstrated eight- to tenfold fewer naloxone-precipitated jumps as compared with mice treated with chronic morphine or a monovalent MOR agonist (Daniels et al. 2005a). The rewarding properties of bivalent MDAN-21 were also compared to monovalent MOR agonists with and without a monovalent DOR antagonist present. Four-day training with a monovalent MOR agonist produced significant conditioned place preference (CPP) in both the presence and absence of a monovalent DOR antagonist. In contrast, the bivalent DOR antagonist/MOR agonist ligand MDAN-21 (*iv*) did not produce significant CPP (Lenard et al. 2007). The authors suggest that by linking a DOR antagonist pharmacophore to a MOR agonist pharmacophore with a 19–21-atom linker DOR/MOR heterodimers can be activated which results in unique signaling that confers pain relief with limited development of tolerance and dependence and limited reinforcing properties. However, MDAN-21 produced inconsistent antinociceptive effects when administered peripherally suggesting that the failure of MDAN-21 to produce CPP may be due to its variable effects following peripheral administration (Aceto et al. 2012). These inconsistent results following peripheral administration may indicate variable absorption or distribution of the drug due to

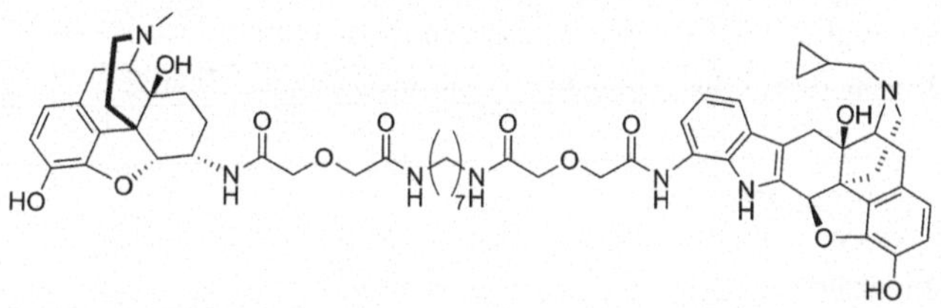

Fig. 4 Structure of MDAN-21, a bivalent DOR antagonist/MOR agonist ligand with an oxymorphone derivative linked to a naltrindole derivative by a flexible 21-atom linker

its large size or individual differences in metabolism, producing two distinct pharmacophores with their own variable pharmacokinetics.

2.1.3 Bifunctional DOR Antagonist/MOR Agonist Ligands

Small molecule DOR antagonist/MOR agonist compounds have been derived from known opioid alkaloids (Healy et al. 2013; Hiebel et al. 2007; Ananthan et al. 2012). The DOR antagonist/MOR agonist/KOR antagonist alkaloid UMB425 displayed antinociception in the mouse hot plate and tail flick assays (Fig. 5). One advantage of this small molecule scaffold is that it is active after peripheral administration, a desirable trait in a potential therapeutic. To assess the development of tolerance to the antinociceptive effects, mice were treated with an ED_{90} dose of either morphine or UMB425 twice daily for 5 days. Latencies in the hot plate and tail flick assays were determined after each dose of drug. Mice treated repeatedly with morphine developed tolerance to its antinociceptive effects on day 4 in both the hot plate and tail flick assays. Mice treated repeatedly with UMB425 developed tolerance in the hot plate assay on day 4 or day 5 and in the tail flick assay on day 5, though to a lesser degree than morphine (Healy et al. 2013); these results are consistent with a mixed efficacy DOR antagonist/MOR agonist. However, UMB425 displays selectivity for MOR (approximately 3 nM) and similar affinity for DOR and KOR (approximately 200 nM), so it is somewhat misleading to describe UMB425 as DOR antagonist/ MOR agonist. What role, if any, the KOR antagonist activity of UMB425 plays in the development of tolerance to and dependence on MOR agonists is unclear and warrants further investigation.

There have also been reports of small molecule peptidomimetic or pseudo-peptide DOR antagonist/MOR agonist compounds which are generally larger than alkaloid opioid ligands but smaller than most opioid peptides (Balboni et al. 2002; Salvadori et al. 1999; Mosberg et al. 2013; Dietis et al. 2012). Some of these ligands are based on known DOR antagonist/MOR agonist peptides (Dietis et al. 2009; Balboni et al. 2002; Healy et al. 2013; Hiebel et al. 2007; Ananthan et al. 2012; Mosberg et al. 2013). Most contain at least one amide bond but are generally more "drug-like" in size and overall physiochemical properties than peptides. Several DOR antagonist/MOR agonist pseudo-peptide ligands containing the Dimethyltyrosine-Tetrahydroisoquinoline carboxylic acid (Dmt-Tic) pharmacophore have been reported (Balboni et al. 2002; Salvadori et al. 1999; Schiller 2010). The most notable of these, UFP-505 (Dmt-Tic-Gly-NH-Bzl), is reported to be a DOR antagonist/MOR agonist compound in vitro and ex vivo (Dietis et al. 2009, 2012) and has been shown to produce less tolerance in rats as compared with morphine when given via *intrathecal* (*it*) injection (Dietis 2012)

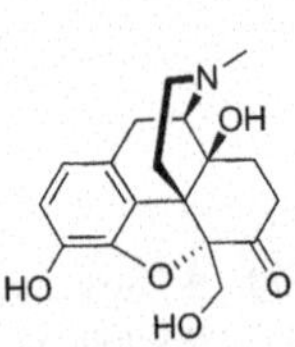

Fig. 5 Structure of UMB425, a small molecule DOR antagonist/MOR agonist/ KOR antagonist

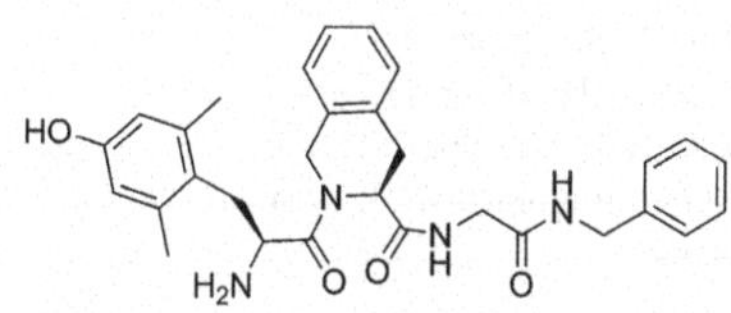

Fig. 6 Structure of UFP-505 (Dmt-Tic-Gly-NH-Bzl), a DOR antagonist/MOR agonist pseudo-peptide

(Fig. 6). However, there are no reports of any Dmt-Tic compounds which produce antinociception after peripheral administration.

A series of more constrained peptidomimetics, based on a tetrahydroquinoline scaffold and containing fewer amide bonds than the Dmt-Tic series, has also been reported (Mosberg et al. 2013; Anand et al. 2016; Bender et al. 2015; Harland et al. 2015). This series of compounds also displayed DOR antagonist/MOR agonist properties in vitro. Many of the compounds in this series displayed opioid-mediated dose-dependent antinociception after peripheral administration in mice (Mosberg et al. 2013; Anand et al. 2016; Bender et al. 2015; Harland et al. 2015), though there are no reports on the development of tolerance or physical dependence.

While traditionally thought to be less "drug-like" than small molecules, peptide ligands do possess some advantages over alkaloid and peptidomimetic opioid compounds. The larger size of peptide ligands provides many contact points between ligand and receptor. By making multiple points of contact, peptides can interact with more elements in the binding pockets of receptors and form favorable or unfavorable interactions, allowing for fine tuning of binding and efficacy profiles of multiple targets. This is a key advantage in multifunctional opioid ligands as the structural homology in the binding sites of MOR, DOR, and KOR is high; thus, it is difficult to maintain high MOR and DOR affinity without significant KOR affinity and to simultaneously produce MOR agonist activity without residual DOR or KOR agonist activity.

The Schiller peptide DIPPψNH_2 (Dmt-Tic-ψ[CH_2NH]-Phe-Phe-NH_2) is a well-characterized DOR antagonist/MOR agonist peptide (Fig. 7); it was the first reported DOR antagonist/MOR agonist compound with balanced affinity at DOR and MOR. DIPPψNH_2 produces robust antinociception in the rat tail flick assay after *icv* administration. Rats treated continuously for 4 days with a small dose of DIPPψNH_2 (*icv*) developed less tolerance in the tail flick assay than rats treated with morphine at comparable doses; however, treatment with larger doses of morphine and DIPPψNH_2 produced similar degrees of tolerance. Rats treated twice daily with DIPPψNH_2 showed fewer signs of withdrawal after treatment with antagonist than rats treated with repeated morphine, suggesting that chronic treatment with DIPPψNH_2 produces less physical dependence (Schiller et al. 1999). Unfortunately, the therapeutic potential of DIPPψNH_2 is limited by its poor BBB penetration, as expected for an unmodified peptide, and by the seizures produced after chronic administration of large doses (Schiller 2010).

The problem of BBB penetration is addressed by the cyclic glycosylated peptide, VRP26 (Dmt-c(SEtS)-[D-Cys-Aic-D-Pen]-Ser(*O*-β-D-glucose)-NH_2) (Mosberg et al. 2014) (Fig. 8). This ligand is reported to be a DOR antagonist/

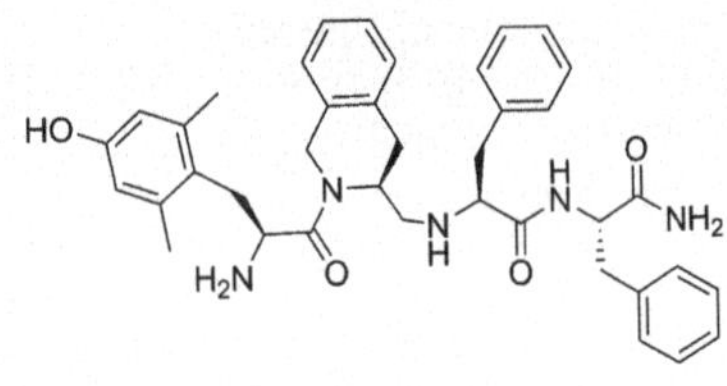

Fig. 7 Structure of DIPPψNH_2 (Dmt-Tic-ψ[CH_2NH]-Phe-Phe-NH_2), a DOR antagonist/MOR agonist peptide

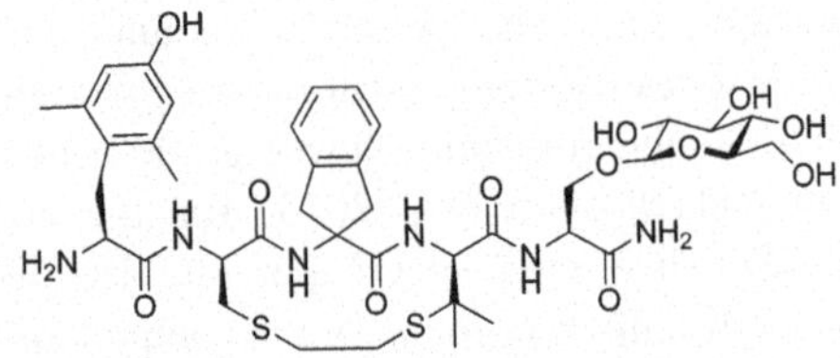

Fig. 8 Structure of VRP26 (Dmt-c(SEtS)-[D-Cys-Aic-D-Pen]-Ser(O-β-D-glucose)-NH_2), a DOR antagonist/MOR agonist peptide

MOR agonist in vitro and produced opioid-mediated antinociception in mice after peripheral administration. Further, a single bolus dose of VRP26 produced no acute tolerance in the mouse warm water tail withdrawal assay, making it a promising lead for the development of mixed efficacy opioid analgesics (Mosberg et al. 2014). Continuous infusion of VRP26 subcutaneously for 7 days did not shift the dose–response curve in the mouse warm water tail withdrawal assay, while a similar treatment with fentanyl produced a significant rightward shift in the dose–response curve, suggesting that tolerance developed to fentanyl but not VRP26. Mice treated for 7 days with a continuous infusion of fentanyl exhibited significantly more withdrawal signs after injection with naloxone than mice treated chronically with VRP26, suggesting that under these conditions physical dependence on VRP26 does not develop. In the CPP assay, fentanyl produced robust, dose-dependent increases in time spent on the drug-paired side of the apparatus as expected. While VRP26 did produce slight increases in time spent on the drug-paired side of the apparatus, the increases were not significant at any of the doses tested, which suggests that VRP26 is less rewarding than fentanyl (Anand et al. 2016). These data provide proof of concept that mixed efficacy DOR antagonist/MOR agonist compounds provide a better alternative to traditional opioid analgesics in rodent behavioral models.

2.1.4 DOR/MOR Conclusions

The mechanism(s) by which DOR ligands modulate MOR-mediated signaling are not clear. It has been suggested that DOR and MOR form functionally distinct heterodimers that signal differently than their monomeric or homomeric counterparts. In the case of DOR agonist/MOR agonist ligands, it has been proposed that these DOR/MOR heterodimers can be simultaneously occupied by both a DOR agonist and a MOR agonist and that these activated heterodimers couple to different downstream effectors, thereby producing effects different from DOR or MOR agonist stimulation alone (Gomes et al. 2000, 2011; Rios et al. 2001). Alternatively, DOR/MOR heterodimers may be desensitized, recycled, or resensitized at different rates or under different conditions than DOR or MOR alone. The desirable profile produced by multifunctional DOR/MOR ligands could also be explained without invoking heterodimers; DOR and MOR may be occupied by agonists in distinct cell populations or brain regions, in which case the confluence of these signals potentiates analgesic activity without stimulating the development of tolerance, dependence, or drug seeking behavior. Alternatively, these multifunctional ligands could simply stabilize conformations of the receptor which promote different signaling pathways which do not produce the same adverse effects as traditional opioid ligands.

Several theories have been proposed to explain how a DOR antagonist could decrease the development of tolerance to MOR agonists, many of which also involve DOR/MOR heterodimers. One theory proposes that upon DOR antagonist treatment, DOR surface expression is increased, either through blockade of basal DOR signaling such that cells traffic more DOR to the surface from intracellular stores to maintain enkephalinergic tone or through molecular chaperoning, which stabilizes the receptor and enhances trafficking to the surface of the cell from the endoplasmic reticulum (Cahill et al. 2007; Dunham and Hall 2009). MOR is co-trafficked to the plasma membrane in the form of a DOR/MOR heterodimer from the endoplasmic reticulum or vesicular stores, thereby increasing the number of active MOR binding sites available on the plasma membrane and preventing the development of tolerance through retention/increase of cell surface binding (Cvejic and Devi 1997; George et al. 2000). Another hypothesis proposes that DOR/MOR heterodimers form at the plasma membrane and that antagonist-bound DOR will remain on the cell surface and prevent internalization of agonist bound MOR through receptor/receptor dimerization; the proximity of DOR may prevent phosphorylation of MOR, thereby maintaining surface expression of active MOR (Law et al. 2005). These theories are supported by evidence which shows that DOR and MOR co-localize in the same cell in the dorsal root ganglion (Wang et al. 2010; Peng et al. 2012; Liu et al. 1995), a brain region associated with pain signaling. There also exists another set of possibilities which do not involve the dimerization of DOR and MOR. It is possible that the confluence of signals from both DOR and MOR attenuates the development of tolerance and dependence. These signals may alter the trafficking pattern of the receptors but do not necessarily do so through a direct physical interaction between DOR and MOR.

The role of DOR itself, as opposed to DOR ligands, has also been explored in the development of MOR-mediated tolerance and dependence. It has been shown that the knockdown or knockout of DOR in mice slows the development of tolerance to a MOR agonist (Kest et al. 1996; Zhu et al. 1999; Chefer and Shippenberg 2009); as DOR has a high basal signaling rate, this suggests that prevention of DOR-mediated signaling slows the development of MOR-mediated tolerance and dependence. When these data are considered with the findings from research on DOR agonist/MOR agonist and DOR antagonist/MOR agonist interactions, it becomes clear that there is a clinically significant interaction between the two receptor types. Taking advantage of this interaction by developing mixed efficacy DOR agonist/MOR agonist or DOR antagonist/MOR agonist compounds may be the key to developing a new generation of safer opioid analgesics.

2.2 DOR/KOR

The interactions of DOR and KOR have been investigated for the treatment of depression. Simultaneous administration of DOR agonist ADL5859 and KOR antagonist LY2444296 in mice resulted in synergistic antidepressant-like effects in a forced swim test, demonstrating promise for therapeutic use of a DOR agonist/KOR antagonist (Huang et al. 2016). DOR activation can also alter the effects of a KOR agonist. Initial studies showed that pretreatment with DOR agonist SNC80 blocked the KOR agonist-mediated antinociception of U50,488H in MOR knockout mice (Taylor et al. 2015). It has been proposed that DOR activation may also have the potential to reverse stress-induced addictive and depressive behaviors that result from KOR activation. Despite the demonstrated promise of these pharmacological profiles, to our knowledge, there have been no reported specific DOR/KOR bifunctional ligands at this time.

Two series of bivalent DOR/KOR ligands have been developed and used to study interactions between the two receptor types and to identify putative DOR/KOR heterodimers in vitro and in vivo. The KDAN series links the DOR antagonist naltrindole and KOR agonist ICI-199,441. KDAN-18, which joins these pharmacophores with an 18-atom spacer, exhibited antinociceptive activity in the mouse tail flick assay (Fig. 9). Based on the absence of an allosteric effect between DOR and KOR receptors that bind this ligand, the authors suggest that this compound does not interact with DOR/KOR heterodimers in which DOR and KOR are allosterically coupled but rather interacts via a bridging mechanism with DOR and KOR receptor homodimers (δ_2 and κ_1 subtypes) which are associated through a passive interface (Daniels et al. 2005b). The KDN series, on the other hand, is reported to demonstrate ligand selectivity for DOR/KOR heterodimers. KDN-21 links naltrindole and KOR antagonist 5′-guanidinonaltrindole with a 21-atom spacer (Fig. 10). This ligand displays no antinociceptive activity in the mouse tail flick assay. Based on binding studies in HEK293 cells and pharmacological studies in mice via *it* injection, the authors suggest that it bridges the two orthosteric binding sites in DOR/KOR heterodimers (Xie et al. 2005; Bhushan et al. 2004).

The localization of DOR/KOR receptor complexes to specific tissues suggests promise for the development of ligands selective for these entities for use as spinally

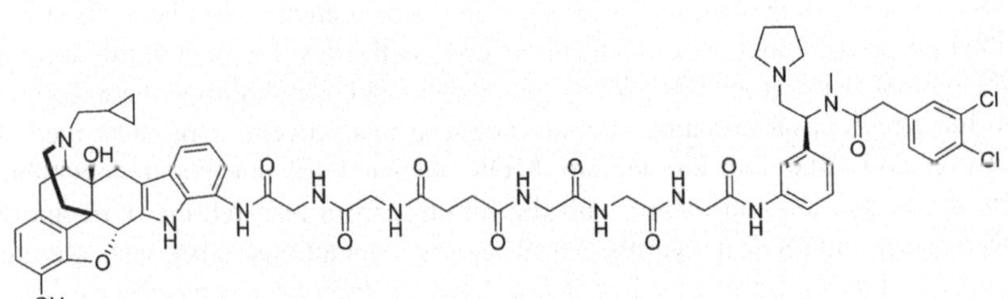

Fig. 9 Structure of KDAN-18, a bivalent DOR antagonist/KOR agonist ligand with naltrindole linked to ICI-199,441 by an 18-atom spacer

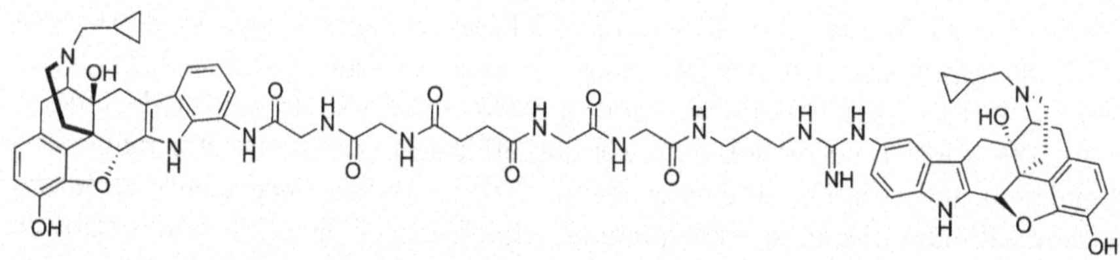

Fig. 10 Structure of KDN-21, a bivalent DOR antagonist/KOR antagonist ligand with naltrindole linked to 5′-guanidinonaltrindole by a 21-atom spacer

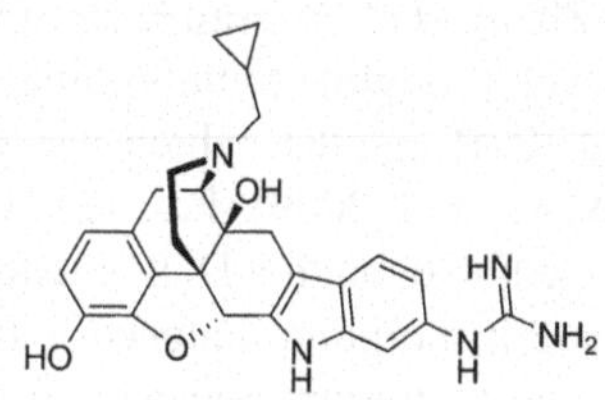

Fig. 11 Structure of 6′-guanidinonaltrindole, a small molecule agonist of DOR/KOR heterodimers

selective analgesics. 6′-guanidinonaltrindole has been reported to selectively activate DOR/KOR heterodimers but not DOR or KOR homomers and results in analgesia in the mouse tail flick assay only when the compound is administered in the spinal cord but not in the brain (Waldhoer et al. 2005) (Fig. 11).

2.3 MOR/KOR

The primary application being explored for MOR/KOR ligands is treatment of addiction to cocaine and other drugs of abuse. It has been demonstrated that KOR agonists have the potential to reduce cocaine self-administration in nonhuman primates due to their reward-modulating properties (Mello and Negus 1998; Negus et al. 1997). It has been suggested that the inhibitory effects of KOR agonists on abuse-related behaviors are a result of inhibition of dopamine release from dopaminergic neurons (Di Chiara and Imperato 1988; Maisonneuve et al. 1994). However, highly selective KOR agonists also produce severe undesirable effects such as salivation, emesis, sedation, and intense hallucinations in nonhuman primates (Mello and Negus 1998; Negus et al. 1997) and in humans (Cruz et al. 2017). It has been suggested that euphoric effects associated with weak MOR agonism may be able to balance dysphoria associated with KOR agonism, increasing the therapeutic potential of a KOR agonist. Thus, mixed MOR/KOR ligands offer potential advantages over selective KOR agonists for the treatment of drug abuse.

Orvinols are known for high affinity binding to MOR and KOR with varying efficacy and have been proposed as potential treatment for cocaine and other psychostimulant abuse, though little in vivo work has been reported on this series (Greedy et al. 2013). Nalbuphine, a mixed MOR/KOR agonist, has been shown to produce a modest attenuation of cocaine's abuse-related effects in humans

(Mello et al. 2005) (Fig. 12a). Exploration of 3-benzylaminomorphinan derivatives with full KOR agonist and partial MOR agonist properties (Neumeyer et al. 2013) led to development of MCL-101 (*butorphan*), a MOR agonist/KOR agonist (Neumeyer et al. 2000), which decreased the rewarding effects of cocaine in intracranial self-stimulation studies in rats (Provencher et al. 2013) and dose-dependently decreased cocaine self-administration with minimal side effects in rhesus monkeys (Bowen et al. 2003) (Fig. 12b). A novel quinoline derivative with MOR/KOR agonist activity, S4, was shown to inhibit naloxone-precipitated withdrawal symptoms (Deb et al. 2009) (Fig. 12c). Aminothiazolomorphinans have also been explored as mixed MOR/KOR agonists (Zhang et al. 2011a), leading to the development of (−)-3-Amino-thiazolo[5,4-b]-*N*-cyclopropylmethylmorphinan hydrochloride (ATPM), a MOR agonist/antagonist and KOR agonist, which was shown to attenuate heroin self-administration in rats (Wang et al. 2009) (Fig. 13). *ATPM* has also been shown to produce KOR- and MOR-mediated, but not DOR-mediated, antinociception in the mouse hot plate assay, to inhibit morphine-induced antinociception, and to dose-dependently attenuate tolerance to the antinociceptive effects of morphine when coadministered with morphine (Wang et al. 2009; Zhang et al. 2004). Other mixed MOR/KOR ligands have also shown some potential to elicit antinociceptive effects with limited adverse events. Endomorphin II based cyclic pentapeptides exhibiting weak MOR/KOR agonism have been reported to result in antinociceptive effects after both central and peripheral administration in mice (Perlikowska et al. 2016).

Finally, MOR/KOR ligands show promise as a treatment for gastrointestinal disorders such as irritable bowel syndrome. Quaternization with benzyl bromide of the pyridyl ring of NAP, a peripherally selective MOR ligand, resulted in BNAP, a peripherally active MOR antagonist/KOR partial agonist which resulted in inhibition

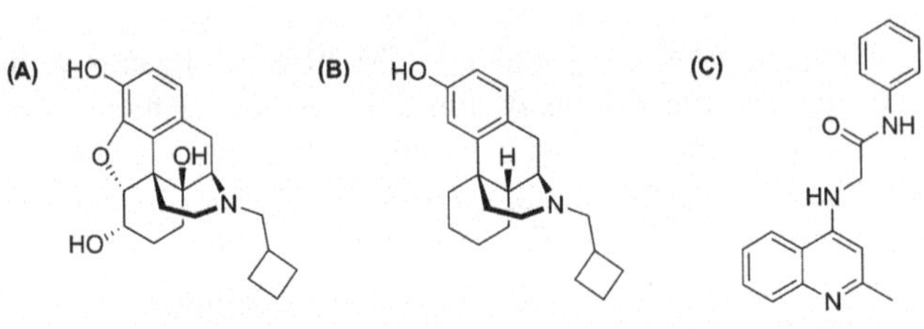

Fig. 12 Structures of (**a**) nalbuphine, (**b**) MCL-101 (butorphan), and (**c**) S4, small molecule MOR agonist/KOR agonist ligands

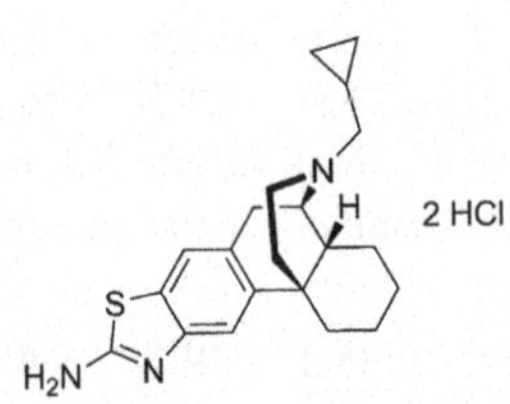

Fig. 13 Structure of ATPM, a small molecule MOR agonist/antagonist and KOR agonist

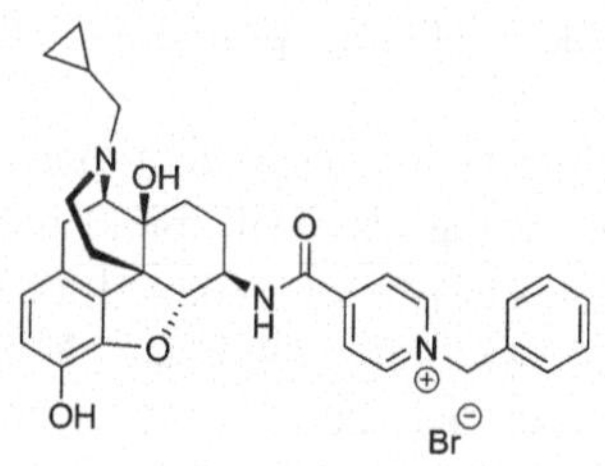

Fig. 14 Structure of BNAP, a small molecule MOR antagonist/KOR partial agonist

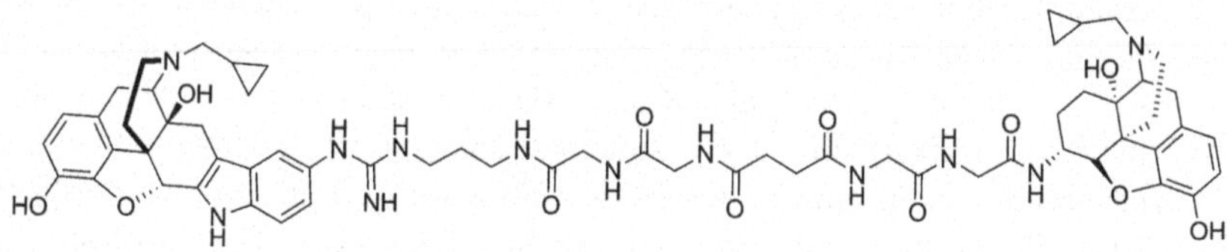

Fig. 15 Structure of KMN-21, a small molecule antagonist of MOR/KOR heterodimers

of abdominal stretching and showed analgesic activity in the acetic acid induced stretch assay in mice (Williams et al. 2016) (Fig. 14).

In addition, dimeric and bivalent MOR/KOR ligands have been developed to study receptor oligomerization. These include cyclorphan-, butorphan-, ATPM-, and other morphinan-based dimeric ligands with subnanomolar affinity for MOR and KOR which function either as MOR partial agonist/KOR full agonist or as MOR partial agonist/KOR partial agonist (Neumeyer et al. 2003; Decker et al. 2009; Zhang et al. 2011b; Peng et al. 2006). The most notable of these chemical tools is KMN-21, an antagonist of MOR/KOR heterodimers, which links MOR antagonist β-naltrexamine and KOR antagonist 5′-guanidononaltrindole with a 21-atom spacer (Zhang et al. 2009) (Fig. 15).

3 Multifunctional Opioid/Nociceptin Receptor Ligands

The nociceptin receptor (NOP) has high sequence homology with the three classical opioid receptors (MOR, DOR, and KOR) but has low affinity for standard opioid ligands, such as naloxone, due to a unique configuration of its binding site residues compared to the classical opioid receptors. As such, NOP is variably considered the fourth opioid receptor or an opioid-like receptor. NOP receptors are highly expressed in the spinal cord and many brain regions including those involved in pain, reward, drug abuse, and motor control. Its endogenous ligand, nociceptin (also known as orphinan FQ), blocks or mediates analgesic effects of opioids depending on the exposure to endogenous opioids, and NOP agonists attenuate reward properties of opioids and other drugs of abuse. However, there remain many discrepancies between studies in rodents and nonhuman primates (Toll et al. 2016; Kiguchi et al. 2016; Günther et al. 2017).

Co-immunoprecipitation studies suggest that heterodimerization may occur between NOP and members of the classical opioid receptor family (Evans et al. 2010), and nonselective opioids have been explored for NOP activity (Butour et al. 1997). While etorphine shows only moderate NOP affinity, its derivative TH-030418 shows high affinity and full agonism at NOP and all three canonical opioid receptors (Fig. 16a). TH-030418 showed potent, naloxone-reversible antinociception when administered *subcutaneously* in mice (Yu et al. 2011); however, chronic treatment resulted in dramatic tolerance development (Wen et al. 2011). Administration of buprenorphine, a mixed agonist/antagonist at classical opioid receptors with partial agonist activity at NOP, in opioid receptor knockout mice has shown that its MOR-mediated analgesia is attenuated by its NOP activation (Lufty and Cowan 2004) (Fig. 16b). Structural analogues of buprenorphine have been synthesized with the aims of increasing affinity for NOP in order to investigate the role of NOP activation in the behavioral profile of this series (Cami-Kobeci et al. 2011) and developing potential agents for relapse prevention for multiple drugs of abuse (Kumar et al. 2014b). Most notably, BU08028, a universally high affinity opioid ligand which shows full agonism at MOR, DOR, and KOR and partial agonism at NOP (Khroyan et al. 2011a), demonstrated long-lasting analgesia with reduced side effects in nonhuman primates (Ding et al. 2016) (Fig. 16c). *Intravenous* administration of MOR/DOR/KOR agonist/NOP antagonist peptides in mice resulted in antinociception without respiratory depression (Guillemyn et al. 2016). Cebranopadol (also known as GRT 6005) showed agonism at MOR, DOR, KOR, and NOP and demonstrated antinociceptive and antihypersensitive effects in rats after *iv* and oral administration with a favorable side effect profile (Linz et al. 2014) and is currently in Phase III clinical trials for several indications including cancer pain (Lambert et al. 2015) (Fig. 16d).

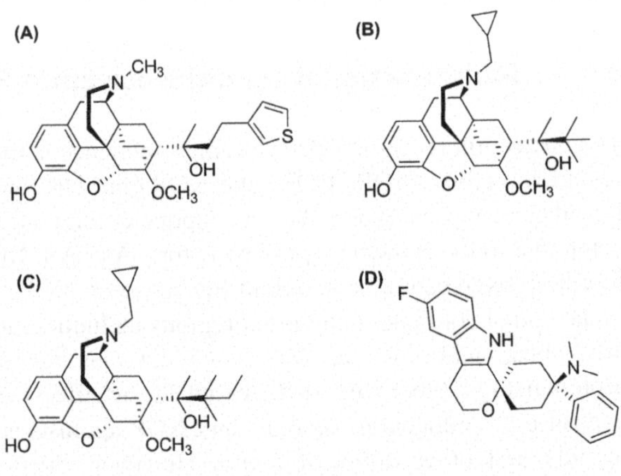

Fig. 16 Structures of (**a**) TH-030418, (**b**) buprenorphine, (**c**) BU0828, and (**d**) cebranopadol (GRT 6005), nonselective small molecule opioid agonists

3.1 DOR/NOP

DOR is also considered a therapeutic target in neuropsychiatric disorders including Parkinson's disease (Chu Sin Chung and Kieffer 2013; Pradhan et al. 2011). In rat models, DOR agonists have demonstrated antiparkinsonian changes in motor effects which are attributed to regulation of nigro-thalamic GABA neurons (Mabrouk et al. 2009). However, high doses of nonpeptidic DOR agonists have low clinical utility due to undesired side effects such as convulsions, and chronic treatment may result in tolerance to therapeutic effects (Mabrouk et al. 2014). Coadministration of DOR agonist SNC80 and NOP antagonist J-113397 in mice and rats produced synergistic antiparkinsonian effects. This observation suggests that NOP antagonism allows for reduction of dosage of DOR agonists in the treatment of Parkinson's disease with retention of full therapeutic efficacy and limited undesired effects (Mabrouk et al. 2014). However, to our knowledge, no specific DOR/NOP ligands have been reported at this time.

3.2 KOR/NOP

It has been suggested that a KOR antagonist/NOP agonist could be beneficial for preventing relapse to a variety of abused drugs (Toll et al. 2013). Recently, aryl ring analogues of buprenorphine with KOR antagonist/NOP partial agonist activity were reported and are being evaluated in vivo for this and other potential applications (Cueva et al. 2015). Nalfurafine (also known as TRK-820), a KOR agonist/NOP antagonist with MOR partial agonist activity, demonstrated antinociception without the development of dependence and adverse effects when administered *subcutaneously* in mice (Seki et al. 1999; Mizoguchi et al. 2003) and is currently in development as an antipruritic (Mustazza and Bastanzio 2011) (Fig. 17). Additionally, a series of KOR/NOP chimeric peptides, structurally based on nociceptin and dynorphin A, were prepared to delineate the functional domain of each endogenous ligand (Lapalu et al. 1997; Reinscheid et al. 1998).

3.3 MOR/NOP

MOR and NOP co-localize in many brain regions, and co-immunoprecipitation and immunofluorescence microscopy studies have shown that the two receptor types

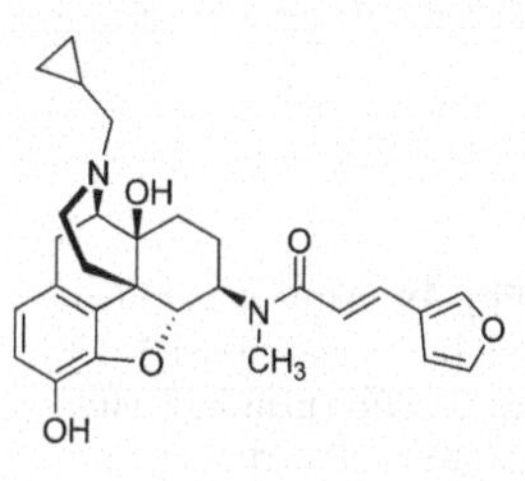

Fig. 17 Structure of nalfurafine (TRK-820), a small molecule KOR agonist/ NOP antagonist

may heterodimerize. NOP agonism has the potential to suppress opioid-induced rewarding effects without decreasing antinociceptive effects (Zaveri et al. 2013; Journigan et al. 2014). It has been suggested that a MOR/NOP agonist may have therapeutic potential as an analgesic with a wider therapeutic window and lowered addiction liability due to reduced reward and tolerance development compared to classical opioid analgesics (Spagnolo et al. 2008). It has also been proposed that a compound with sufficient NOP agonism could be used as a treatment for drug abuse (Toll et al. 2013, 2016; Kiguchi et al. 2016). Naloxone benzoylhydrazone, a MOR antagonist/KOR agonist, was shown to act as an antagonist at NOP (Fig. 18). This ligand induced antinociception without affecting locomotor activity in wild-type mice, but this effect was lost in NOP knockout mice, suggesting that NOP plays a role key in the antinociceptive effects of naloxone benzoylhydrazone (Noda et al. 1998). KGNOP1 (H-Dmt-D-Arg-Aba-β-Ala-Arg-Tyr-Tyr-Arg-Ile-Lys-NH_2), a MOR agonist/NOP antagonist pseudo-peptide, was recently reported as a candidate for dual treatment of nociceptive and neuropathic pain (Lagard et al. 2017). SR16435, a high affinity, mixed MOR/NOP partial agonist, produced antinociception with reduced development of tolerance as compared to morphine in mice (Fig. 19a). However, SR16435 also induced CPP, suggesting that partial NOP agonism is not enough to attenuate the rewarding properties associated with MOR activation (Khroyan et al. 2007; Sukhtankar et al. 2013). As a result, additional bifunctional MOR/NOP ligands with varying ratios of MOR/NOP agonist potency were developed from this scaffold in search of a ligand with a nonaddicting analgesic profile (Zaveri et al. 2013; Journigan et al. 2014). SR14150 (also known as AT-200), a MOR/NOP partial agonist, showed MOR-mediated antinociceptive and antiallodynic effects in mice (Khroyan et al. 2011b), did not induce CPP or attenuate morphine-induced CPP (Toll et al. 2009), and is a promising candidate for treatment of pain in sickle cell anemia (Vang et al. 2015) (Fig. 19b). These results suggest that a MOR/NOP partial agonist may have potential as a nonaddictive analgesic while NOP full agonism may be used to modulate opioid-induced reward (Toll et al. 2009).

Chimeric MOR/NOP ligands linking a dermorphin peptide (Tyr-D-Ala-Phe-Gly-Tyr-Pro-Ser-NH_2 or Tyr-D-Arg-Phe-β-Ala-NH_2) and a NOP peptide (Ac-Arg-Tyr-Tyr-Arg-Ile-Lys-NH_2) were developed as tools to study putative MOR/NOP

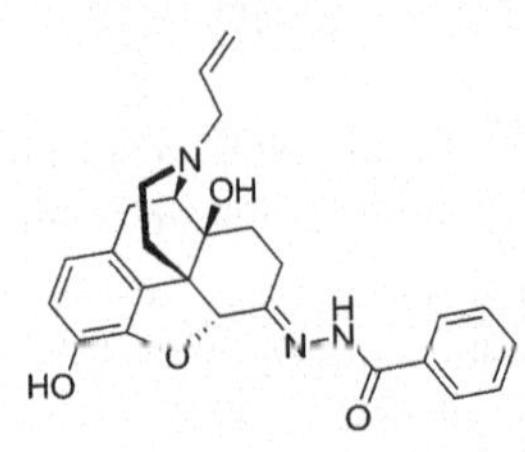

Fig. 18 Structure of naloxone benzoylhydrazone, a small molecule MOR antagonist/KOR agonist/NOP antagonist

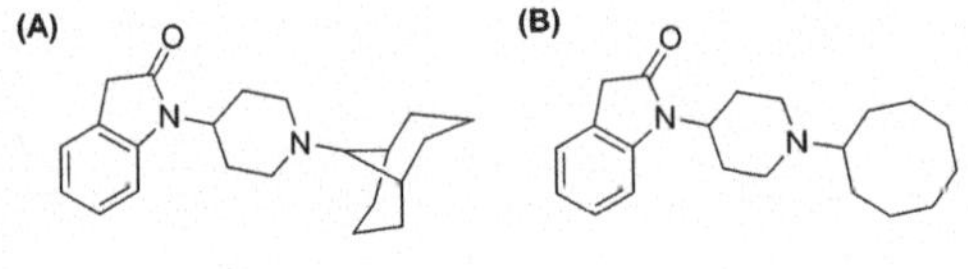

Fig. 19 Structures of (**a**) SR16435 and (**b**) SR14150 (AT-200), small molecule MOR/NOP partial agonists

heterodimers (Kawano et al. 2006) and demonstrated potent antinociceptive activity in the mouse tail flick assay following *it* administration but low activity following *icv* administration (Kawano et al. 2007). The bivalent MOR/NOP agonist, DeNo, is a combination of dermorphin and nociceptin which displays only weak antinociceptive properties but may be useful as a tool for investigating simultaneous activation of MOR and NOP (Bird et al. 2016).

4 Multifunctional Opioid/Non-opioid Ligands

Opioid receptors are known to interact with other GPCRs, and the existence of various heterodimers remains controversial (Rios et al. 2001). Ligands have been developed to explore the existence of constitutive heterodimers and to selectively target these complexes. In addition, induction of non-endogenous opioid receptor heteromers through the use of bivalent ligands has been proposed as a way to promote unique pharmacology (Portoghese et al. 2017). Many bivalent ligands have been developed to explore crosstalk between opioid receptors and other systems, and these compounds are being explored for a variety of therapeutic uses.

4.1 Opioid/Cannabinoid

Cannabinoid receptors are found primarily in brain and neuronal tissue, and agonists of these receptors have been linked to many behavioral effects including analgesia and regulation of mood and appetite. Several endogenous agonists for the cannabinoid receptors have been identified, including eicosanoids; however, the primary endogenous agonists of the cannabinoid receptors are uncertain. There are two known cannabinoid receptor types (CB_1 and CB_2), and some evidence suggests the existence of additional types or subtypes (Howlett et al. 2002). Activation of cannabinoid and opioid receptors results in similar behavioral effects including antinociception and regulation of mood, and both types of receptor are expressed in brain regions associated with antinociception. There is evidence which suggests that the cannabinoid 1 receptor (CB_1) heterodimerizes with each of the classical opioid receptors (Bushlin et al. 2010; Rios et al. 2009). Rimonabant (also known as SR141716), a CB_1 antagonist, has been shown to bind to MOR and inhibit signaling in mouse cortex and MOR-CHO membranes (Cinar and Szucs 2009) and to inhibit DOR function at micromolar concentrations (Zádor et al. 2014). It has also been shown that coadministration of opioid and cannabinoid receptor agonists may have a synergistic antinociceptive effect (Grenald et al. 2017), and simultaneous activation of both opioid and cannabinoid receptors results in highly effective analgesia in neuropathic pain animal models (Kleczkowska et al. 2013). Preclinical coadministration studies suggest promise for the development of multifunctional opioid/cannabinoid ligands as analgesics which can be dosed at lower concentrations than opioids alone (Nielsen et al. 2017).

Bivalent opioid/CB_1 ligands were also developed from high affinity DOR/MOR peptide Tyr-D-Ala-Gly-Phe-NH_2 and rimonabant as tools for investigating crosstalk and synergistic effects (Mollica et al. 2017). It has been proposed that a bivalent MOR/CB_1 ligand which activates one receptor and blocks activity at the other may be useful as an analgesic since association between the two receptors leads to an antagonistic response (Bushlin et al. 2010). However, bivalent ligands based on MOR agonist α-oxymorphamine and rimonabant showed no reduced tolerance development compared to coadministration of the monovalent ligands, suggesting that MOR/CB_1 is not an important target for reduction of opioid tolerance (Le Naour et al. 2013). MOR antagonist/CB_1 antagonist bivalent ligands were developed from the opioid agonist fentanyl and rimonabant (Fernández-Fernández et al. 2014). Coadministration studies have shown such ligands to have potential therapeutic applications including reduction of pruritic response induced by rimonabant and regulation of alcohol intake and feeding behavior (Rowland et al. 2002; Tallett et al. 2009; Wright and Rodgers 2013).

4.2 Opioid/Neurokinin-1

Neurokinin-1 receptors (NK1R) are widely expressed throughout the central nervous system and often co-localize with the three classical opioid receptors (Pinto et al. 2008). The endogenous agonist for NK1R, Substance P (SP), is released in primary afferents in response to pain and other noxious stimuli (Besson 1999). The resulting stimulation of NK1R produces inflammation and signals of stress and pain (Xiao et al. 2016). In other words, NK1R functions to oppose the opioid receptors; stimulation of NK1R is nociceptive, while stimulation of the opioid receptors is antinociceptive. Interestingly, stimulation of the opioid receptors can inhibit SP release, and conversely, stimulation of NK1R modulates opioid receptor function and the development of adverse effects associated with chronic opioid analgesic use (Xiao et al. 2016).

Due to this intertwined relationship, a series of multifunctional opioid agonist/NK1R antagonist peptides/peptidomimetics that combine opioid and NK1R peptide sequences has been developed (Yamamoto et al. 2007; Nair et al. 2013, 2015). While many of these compounds produced antinociception in vivo, sometimes more potently than morphine, repeated administration produced tolerance to the antinociceptive effects of these compounds. Other mixed opioid agonist/NK1R antagonist peptides have also been developed (Betti et al. 2015; Dyniewicz et al. 2017). Surprisingly, some of these compounds do not produce cross-tolerance with morphine (Betti et al. 2015) and therefore may provide a novel class of compounds for treating pain.

4.3 MOR/CCR5

Opioid agonists have been shown to induce the expression of C-C chemokine receptor type 5 (CCR5), which assists in the entry of the AIDS virus into immune cells. Opioid receptors and CCR5 are closely situated on the cell membrane (Suzuki et al. 2002), and heterodimerization between MOR and CCR5 has been proposed (Chen et al. 2004). A bivalent MOR agonist/CCR5 antagonist, MCC22, produced potent antinociception in the mouse tail flick assay and is a candidate for use both in treatment of chronic pain and in blocking penetration of HIV into the central nervous system (Akgun et al. 2015) (Fig. 20). Bivalent MOR antagonist/CCR5 antagonist ligands have been developed from MOR antagonist naltrexone and CCR5 antagonist maraviroc as probes to study putative MOR/CCR5 heterodimerization during progression of opioid-enhanced NeuroAIDS. Maraviroc alone does not effectively inhibit HIV infection in primary human astrocytes in the presence of morphine while a bivalent MOR/CCR5 antagonist ligand inhibits HIV invasion in both the presence and absence of morphine (Yuan et al. 2012, 2013; Arnatt et al. 2016).

4.4 MOR/mGluR5

The metabotropic glutamate-5 receptor (mGluR5) is widely distributed in the central nervous system, including in the dorsal horn and the glia of the spinal cord. This receptor modulates synaptic transmission, neuronal excitability, and plasticity (Akgun et al. 2013). Allosteric modulation of mGluR5 by antagonist MPEP has been shown to enhance MOR-mediated antinociception and suppress the development of morphine tolerance and dependence (Schröder et al. 2009). Heteromerization of MOR and mGluR5 has also been proposed; to target these putative heteromers, bivalent ligand MMG22, containing MOR agonist oxymorphone and mGluR5 antagonist *m*-methoxy-MPEP pharmacophores, was developed (Fig. 21). This ligand showed potent, long-lasting antinociception in a mouse model of bone cancer pain and is an

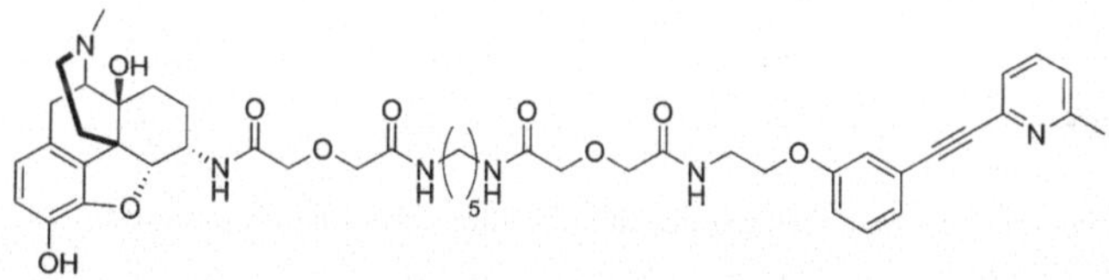

Fig. 20 Structure of MCC22, a bivalent MOR agonist/CCR5 antagonist ligand with a 22-atom linker

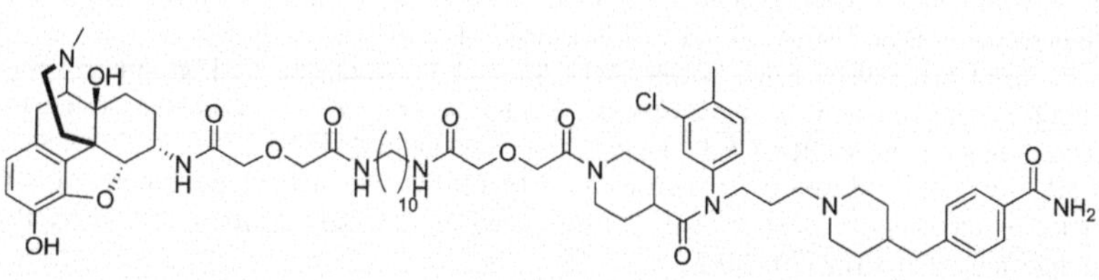

Fig. 21 Structure of MMG22, a bivalent MOR agonist/mGluR5 antagonist ligand with oxymorphone linked to an MPEP derivative by a 22-atom linker

excellent candidate for the treatment of chronic, intractable pain via spinal administration (Akgun et al. 2013; Smeester et al. 2014).

5 Conclusions

Opioid receptors play a major role in the regulation of pain, mood, and reward; however, selective ligands for MOR, DOR, and KOR all have on-target adverse effects which complicate their use in treating pain, mood disorders, and addiction (Lutz and Kieffer 2013). As a result, the development of multifunctional ligands which display activity at multiple opioid receptors or activity at both opioid receptors and other receptors has been the focus of a great deal of research. The move away from selective ligands has been seen within and outside the opioid field, and many researchers in the GPCR community are developing multifunctional ligands as both tools and therapeutics.

Some of the ligands described in the sections above have been used as tools to elucidate the mechanisms of crosstalk between receptors as well as the organization of receptors into oligomers, which may provide a more nuanced view of the physiological role of receptor/receptor interactions and signaling. The information gathered from these studies is an important step in the rational design of novel therapeutics that display multifunctional activity. In fact, many of the ligands discussed in this chapter show improvement over selective opioid ligands with regard to therapeutic efficacy or reduction in the development of adverse effects in animal models and show promise as novel therapeutics for the treatment of pain, addiction, and mood disorders.

References

Abbott NJ, Romero IA (1996) Transporting therapeutics across the blood-brain barrier. Mol Med Today 2(3):106–113

Aceto MD, Harris LS, Negus SS, Banks ML, Hughes LD, Akgun E, Portoghese PS (2012) MDAN-21: a bivalent opioid ligand containing mu-agonist and delta-antagonist pharmacophores and its effects in rhesus monkeys. Int J Med Chem 2012:1–6

Akgun E, Javed MI, Lunzer MM, Smeester BA, Beitz AJ, Portoghese PS (2013) Ligands that interact with putative MOR-mGluR5 heteromer in mice with inflammatory pain produce potent antinociception. Proc Natl Acad Sci 110(28):11595–11599

Akgun E, Javed MI, Lunzer MM, Powers MD, Sham YY, Watanabe Y, Portoghese PS (2015) Inhibition of inflammatory and neuropathic pain by targeting a mu opioid receptor/chemokine receptor5 heteromer (MOR-CCR5). J Med Chem 58(21):8647–8657

Anand JP, Boyer BT, Mosberg HI, Jutkiewicz EM (2016) The behavioral effects of a mixed efficacy antinociceptive peptide, VRP26, following chronic administration in mice. Psychopharmacology 233(13):2479–2487

Ananthan S, Saini SK, Dersch CM, Xu H, McGlinchey N, Giuvelis D, Bilsky EJ, Rothman RB (2012) 14-Alkoxy-and 14-acyloxypyridomorphinans: μ agonist/δ antagonist opioid analgesics with diminished tolerance and dependence side effects. J Med Chem 55(19):8350–8363

Arnatt CK, Falls BA, Yuan Y, Raborg TJ, Masvekar RR, El-Hage N, Selley DE, Nicola AV, Knapp PE, Hauser KF et al (2016) Exploration of bivalent ligands targeting putative mu opioid receptor and chemokine receptor CCR5 dimerization. Bioorg Med Chem 24(22):5969–5987

Bailey CP, Connor M (2005) Opioids: cellular mechanisms of tolerance and physical dependence. Curr Opin Pharmacol 5:60–68

Balboni G, Guerrini R, Salvadori S, Bianchi C, Rizzi D, Bryant SD, Lazarus LH (2002) Evaluation of the Dmt-Tic pharmacophore: conversion of a potent delta-opioid receptor antagonist into a potent delta agonist and ligands with mixed properties. J Med Chem 45(3):713–720

Balboni G, Onnis V, Congiu C, Zotti M, Sasaki Y, Ambo A, Bryant SD, Jinsmaa Y, Lazarus LH, Lazzari I et al (2007) Further studies on the effect of lysine at the C-terminus of the Dmt-Tic opioid pharmacophore. Bioorg Med Chem 15(9):3143–3151

Bender AM, Griggs NW, Anand JP, Traynor JR, Jutkiewicz EM, Mosberg HI (2015) Asymmetric synthesis and in vitro and in vivo activity of tetrahydroquinolines featuring a diverse set of polar substitutions at the 6 position as mixed-efficacy mu opioid receptor/delta opioid receptor ligands. ACS Chem Neurosci 6(8):1428–1435

Besson JM (1999) The neurobiology of pain. Lancet 353:1610–1615

Betti C, Starnowska J, Mika J, Dyniewicz J, Frankiewicz L, Novoa A, Bochynska M, Keresztes A, Kosson P, Makuch W et al (2015) Dual alleviation of acute and neuropathic pain by fused opioid agonist-neurokinin 1 antagonist peptidomimetics. ACS Med Chem Lett 6(12):1209–1214

Bhushan RG, Sharma SK, Xie Z, Daniels DJ, Portoghese PS (2004) A bivalent ligand (KDN-21) reveals spinal delta and kappa opioid receptors are organized as heterodimers that give rise to delta1 and kappa2 phenotypes. Selective targeting of delta-kappa heterodimers. J Med Chem 47(12):2969–2972

Bird MF, Camilla Cerlesi M, Brown M, Malfacini D, Vezzi V, Molinari P, Micheli L, Di Cesare Mannelli L, Ghelardini C, Guerrini R et al (2016) Characterisation of the novel mixed mu-NOP peptide ligand dermorphin-N/OFQ (DeNo). PLoS One 11(6):1–22

Bowen CA, Stevens Negus S, Zong R, Neumeyer JL, Bidlack JM, Mello NK (2003) Effects of mixed-action kappa/mu opioids on cocaine self-administration and cocaine discrimination by rhesus monkeys. Neuropsychopharmacology 28:1125–1139

Bushlin I, Rozenfeld R, Devi LA (2010) Cannabinoid-opioid interactions during neuropathic pain and analgesia. Curr Opin Pharmacol 10(1):80–86

Butour JL, Moisand C, Mazarguil H, Mollereau C, Meunier JC (1997) Recognition and activation of the opioid receptor-like ORL1 receptor by nociceptin, nociceptin analogs and opioids. Eur J Pharmacol 321(1):97–103

Cahill CM, Holdridge SV, Morinville A (2007) Trafficking of delta-opioid receptors and other g-protein-coupled receptors: implications for pain and analgesia. Trends Pharmacol Sci 28(1):23–31

Cami-Kobeci G, Polgar WE, Khroyan TV, Toll L, Husbands SM (2011) Structural determinants of opioid and NOP receptor activity in derivatives of buprenorphine. J Med Chem 54(19):6531–6537

Chefer VI, Shippenberg TS (2009) Augmentation of morphine-induced sensitization but reduction in morphine tolerance and reward in delta-opioid receptor knockout mice. Neuropsychopharmacology 34(4):887–898

Chen C, Li J, Bot G, Szabo I, Rogers TJ, Liu-Chen LY (2004) Heterodimerization and cross-desensitization between the mu-opioid receptor and the chemokine CCR5 receptor. Eur J Pharmacol 483:175–186

Chu Sin Chung P, Kieffer BL (2013) Delta opioid receptors in brain function and diseases. Pharmacol Ther 140(1):112–120

Cinar R, Szucs M (2009) CB1 receptor-independent actions of SR141716 on G-protein signaling: coapplication with the mu-opioid agonist Tyr-D-Ala-Gly-(NMe)Phe-Gly-Ol unmasks novel, pertussis toxin-insensitive opioid signaling in mu-opioid receptor-Chinese hamster ovary cells. J Pharmacol Exp Ther 330(2):567–574

Cruz A, Domingos S, Gallardo E, Martinho A (2017) A unique natural selective kappa-opioid receptor agonist, Salvinorin A, and its roles in human therapeutics. Phytochemistry 137:9–14

Cueva JP, Roche C, Ostovar M, Kumar V, Clark MJ, Hillhouse TM, Lewis JW, Traynor JR, Husbands SM (2015) C7β-methyl analogues of the orvinols: the discovery of kappa opioid antagonists with nociceptin/orphanin FQ peptide (NOP) receptor partial agonism and low, or zero, efficacy at mu opioid receptors. J Med Chem 58(10):4242–4249

Cvejic S, Devi LA (1997) Dimerization of the delta opioid receptor. J Biol Chem 272(43):26959–26964

Daniels DJ, Lenard NR, Etienne CL, Law PY, Roerig SC, Portoghese PS (2005a) Opioid-induced tolerance and dependence in mice is modulated by the distance between pharmacophores in a bivalent ligand series. Proc Natl Acad Sci 102(52):19208–19213

Daniels DJ, Kulkarni A, Xie Z, Bhushan RG, Portoghese PS (2005b) A bivalent ligand (KDAN-18) containing delta-antagonist and kappa-agonist pharmacophores bridges delta2 and kappa1 opioid receptor phenotypes. J Med Chem 48(6):1713–1716

Deb I, Paira P, Hazra A, Banerjee S, Dutta PK, Mondal NB, Das S (2009) Synthesis and characterizations of novel quinoline derivatives having mixed ligand activities at the kappa and mu receptors: potential therapeutic efficacy against morphine dependence. Bioorg Med Chem 17(16):5782–5790

Decker M, Fulton BS, Zhang B, Knapp BI, Bidlack JM, Neumeyer JL (2009) Univalent and bivalent ligands of butorphan: characteristics of the linking chain determine the affinity and potency of such opioid ligands. J Med Chem 52(23):7389–7396

Di Chiara G, Imperato A (1988) Drugs abused by humans preferentially increase synaptic dopamine concentrations in the mesolimbic system of freely moving rats. Proc Natl Acad Sci 85:5274–5278

Dietis N (2012) Strategies to reduce morphine tolerance in cancer: evaluation of the bifunctional opioid UFP-505. University of Leicester

Dietis N, Guerrini R, Calo G, Salvadori S, Rowbotham DJ, Lambert DG (2009) Simultaneous targeting of multiple opioid receptors: a strategy to improve side-effect profile. Br J Anaesth 103(1):38–49

Dietis N, McDonald J, Molinari S, Calo G, Guerrini R, Rowbotham DJ, Lambert DG (2012) Pharmacological characterization of the bifunctional opioid ligand H-Dmt-Tic-Gly-NH-Bzl (UFP-505). Br J Anaesth 108(2):262–270

Ding H, Czoty PW, Kiguchi N, Cami-Kobeci G, Sukhtankar DD, Nader MA, Husbands SM, Ko M-C (2016) A novel orvinol analog, BU08028, as a safe opioid analgesic without abuse liability in primates. Proc Natl Acad Sci 113:E5511–E5518

Do Carmo GP, Polt R, Bilsky EJ, Rice KC, Negus SS (2008) Behavioral pharmacology of the mu/delta opioid glycopeptide MMP2200 in rhesus monkeys. J Pharmacol Exp Ther 326(3):939–948

Dunham JH, Hall RA (2009) Enhancement of the surface expression of G protein-coupled receptors. Trends Biotechnol 27(9):541–545

Dyniewicz J, Lipinski PF, Kosson P, Lesniak A, Bochynska-Czyz M, Muchowska A, Tourwe D, Ballet S, Misicka A, Lipkowski AW (2017) Hydrazone linker as a useful tool for preparing chimeric peptide/nonpeptide bifunctional compounds. ACS Med Chem Lett 8(1):73–77

Evans RM, You H, Hameed S, Altier C, Mezghrani A, Bourinet E, Zamponi GW (2010) Heterodimerization of ORL1 and opioid receptors and its consequences for *N*-type calcium channel regulation. J Biol Chem 285(2):1032–1040

Fernández-Fernández C, Callado LF, Girón R, Sánchez E, Erdozain AM, López-Moreno JA, Morales P, de Fonseca FR, Fernández-Ruiz J, Goya P et al (2014) Combining rimonabant and fentanyl in a single entity: preparation and pharmacological results. Drug Des Devel Ther 8:263–277

George SR, Fan T, Xie Z, Tse R, Tam V, Varghese G, O'Dowd BF (2000) Oligomerization of mu- and delta-opioid receptors: generation of novel functional properties. J Biol Chem 275(34):26128–26135

Gomes I, Jordan BA, Gupta A, Trapaidze N, Nagy V, Devi LA (2000) Heterodimerization of mu and delta opioid receptors: a role in opiate synergy. J Neurosci 20:1–5

Gomes I, IJzerman AP, Ye K, Maillet EL, Devi LA (2011) G protein-coupled receptor heteromerization: a role in allosteric modulation of ligand binding. Mol Pharmacol 79(6):1044–1052

Greedy BM, Bradbury F, Thomas MP, Grivas K, Cami-Kobeci G, Archambeau A, Bosse K, Clark MJ, Aceto M, Lewis JW et al (2013) Orvinols with mixed kappa/mu opioid receptor agonist activity. J Med Chem 56(8):3207–3216

Grenald SA, Young MA, Wang Y, Ossipov MH, Ibrahim MM, Largent-Milnes TM, Vanderah TW (2017) Synergistic attenuation of chronic pain using mu opioid and cannabinoid receptor 2 agonists. Neuropharmacology 116:59–70

Guillemyn K, Starnowska J, Lagard C, Dyniewicz J, Rojewska E, Mika J, Chung NN, Utard V, Kosson P, Lipkowski AW et al (2016) Bifunctional peptide-based opioid agonist-nociceptin antagonist ligands for dual treatment of acute and neuropathic pain. J Med Chem 59(8):3777–3792

Günther T, Dasgupta P, Mann A, Miess E, Kliewer A, Fritzwanker S, Steinborn R, Schulz S (2017) Targeting multiple opioid receptors – improved analgesics with reduced side effects? Br J Pharmacol.

Harland AA, Yeomans L, Griggs NW, Anand JP, Pogozheva ID, Jutkiewicz EM, Traynor JR, Mosberg HI (2015) Further optimization and evaluation of bioavailable, mixed-efficacy mu-opioid receptor (MOR) agonists/delta-opioid receptor (DOR) antagonists: balancing MOR and DOR affinities. J Med Chem 58(22):8952–8969

Healy JR, Bezawada P, Shim J, Jones JW, Kane MA, MacKerell AD, Coop A, Matsumoto RR (2013) Synthesis, modeling, and pharmacological evaluation of UMB 425, a mixed μ agonist/δ antagonist opioid analgesic with reduced tolerance liabilities. ACS Chem Neurosci 4(9):1256–1266

Heyman JS, Vaught JL, Mosberg HI, Haaseth RC, Porreca F (1989a) Modulation of mu-mediated antinociception by delta agonists in the mouse: selective potentiation of morphine and normorphine by [D-Pen2, D-Pen5]enkephalin. Eur J Pharmacol 165(1):1–10

Heyman JS, Jiang Q, Rothman RB, Mosberg HI, Porreca F (1989b) Modulation of mu-mediated antinociception by delta agonists: characterization with antagonists. Eur J Pharmacol 169:43–52

Hiebel AC, Yong SL, Bilsky E, Giuvelis D, Deschamps JR, Parrish DA, Aceto MD, May EL, Harris LS, Coop A et al (2007) Probes for narcotic receptor mediated phenomena. 34. Synthesis and structure – activity relationships of a potent μ-agonist δ-antagonist and an exceedingly potent antinociceptive in the enantiomeric C9-substituted 5-(3-Hydroxyphenyl)-*N*-phenylethylmorphan series. J Med Chem 50(16):3765–3776

Horan P, Tallarida RJ, Haaseth RC, Matsunaga TO, Hruby VJ, Porreca F (1992) Antinociceptive interactions of opioid delta receptor agonists with morphine in mice: supra- and sub-additivity. Life Sci 50:1535–1541

Horan PJ, Mattia A, Bilsky EJ, Weber S, Davis TP, Yamamura HI, Malatynska E, Appleyard SM, Slaninova J, Misicka A (1993) Antinociceptive profile of biphalin, a dimeric enkephalin analog. J Pharmacol Exp Ther 265:1446–1454

Howlett AC, Barth F, Bonner TI, Cabral G, Casellas P, Devane WA, Felder CC, Herkenham M, Mackie K, Martin BR, International Union of Pharmacology et al (2002) XXVII. Classification of cannabinoid receptors. Pharmacol Rev 54(2):161–202

Huang P, Tunis J, Parry C, Tallarida R, Liu-Chen LY (2016) Synergistic antidepressant-like effects between a kappa opioid antagonist (LY2444296) and a delta opioid agonist (ADL5859) in the mouse forced swim test. Eur J Pharmacol 781:53–59

Johnston LD, O'Malley PM, Bachman JG, Schulenberg JE (2009) Monitoring the future: national survey results on drug use, 1975–2008, vol 1. NIH Publication No. 09–7402, pp 1–721

Journigan VB, Polgar WE, Khroyan TV, Zaveri NT (2014) Designing bifunctional nop receptor-mu opioid receptor ligands from NOP-receptor selective scaffolds. Part II. Bioorg Med Chem 22(8):2508–2516

Jutkiewicz EM (2006) The antidepressant-like effects of delta-opioid receptor agonists. Mol Interv 6(3):162–169

Kawano S, Ambo A, Sasaki Y (2006) Synthesis and receptor binding properties of chimeric peptides containing a mu-opioid receptor ligand and nociceptin/orphanin FQ receptor ligand Ac-RYYRIK-amide. Bioorg Med Chem Lett 16(18):4839–4841

Kawano S, Ito R, Nishiyama M, Kubo M, Matsushima T, Minamisawa M, Ambo A, Sasaki Y (2007) Receptor binding properties and antinociceptive effects of chimeric peptides consisting of a mu-opioid receptor agonist and an ORL1 receptor antagonist. Biol Pharm Bull 30(7):1260–1264

Kest B, Lee CE, McLemore GL, Inturrisi CE (1996) An antisense oligodeoxynucleotide to the delta opioid receptor (DOR-1) inhibits morphine tolerance and acute dependence in mice. Brain Res Bull 39(3):185–188

Khroyan TV, Zaveri NT, Polgar WE, Orduna J, Olsen C, Jiang F, Toll L (2007) SR 16435 [1-(1-(bicyclo[3.3.1]nonan-9-Yl)piperidin-4-Yl)indolin-2-one], a novel mixed nociceptin/orphanin FQ/mu-opioid receptor partial agonist: analgesic and rewarding properties in mice. J Pharmacol Exp Ther 320(2):934–943

Khroyan TV, Polgar WE, Cami-kobeci G, Husbands SM, Zaveri NT, Toll L (2011a) The first universal opioid ligand, (2S)-2-[(5R,6R,7R,14S)-*N*-cyclopropylmethyl-4,5-epoxy-6,14-ethano-3-hydroxy-6-methoxymorphinan-7-Yl]-3,3-dimethylpentan-2-Ol (BU08028): characterization of the in vitro profile and in vivo behavioral effects in mouse mode. J Pharmacol Exp Ther 336(3):952–961

Khroyan TV, Polgar WE, Orduna J, Montenegro J, Jiang F, Zaveri NT, Toll L (2011b) Differential effects of nociceptin/orphanin FQ (NOP) receptor agonists in acute versus chronic pain: studies with bifunctional NOP/mu receptor agonists in the sciatic nerve ligation chronic pain model in mice. J Pharmacol Exp Ther 339(2):687–693

Kiguchi N, Ding H, Ko MC (2016) Central N/OFQ-NOP receptor system in pain modulation. In: Advances in pharmacology, vol 75. Elsevier, Amsterdam, pp 217–243

Kleczkowska P, Lipkowski AW, Tourwé D, Ballet S (2013) Hybrid opioid/non-opioid ligands in pain research. Curr Pharm Des 19:7435–7450

Kumar V, Clark MJ, Traynor JR, Lewis JW, Husbands SM (2014a) Pyrrolo- and pyridomorphinans: non-selective opioid antagonists and delta opioid agonists/mu opioid partial agonists. Bioorg Med Chem 22(15):4067–4072

Kumar V, Ridzwan IE, Grivas K, Lewis JW, Clark MJ, Meurice C, Jimenez-Gomez C, Pogozheva I, Mosberg H, Traynor JR et al (2014b) Selectively promiscuous opioid ligands: discovery of high affinity/low efficacy opioid ligands with substantial nociceptin opioid peptide receptor affinity. J Med Chem 57(10):4049–4057

Lagard C, Chevillard L, Guillemyn K, Risede P, Laplanche J-L, Spetea M, Ballet S, Megarbane B (2017) Bifunctional peptide-based opioid agonist/nociceptin antagonist ligand for dual treatment of nociceptive and neuropathic pain. Pain 158(3):505–515

Lambert DG, Bird MF, Rowbotham DJ (2015) Cebranopadol: a first in-class example of a nociceptin/orphanin FQ receptor and opioid receptor agonist. Br J Anaesth 114(3):364–366

Lapalu S, Moisand C, Mazarguil H, Cambois G, Mollereau C, Meunier JC (1997) Comparison of the structure-activity relationships of nociceptin and dynorphin a using chimeric peptides. FEBS Lett 417(3):333–336

Law PY, Erickson-Herbrandson LJ, Zha QQ, Solberg J, Chu J, Sarre A, Loh HH (2005) Heterodimerization of mu- and delta-opioid receptors occurs at the cell surface only and requires receptor-G protein interactions. J Biol Chem 280(12):11152–11164

Le Naour M, Akgun E, Yekkirala A, Lunzer MM, Powers MD, Kalyuzhny AE, Portoghese PS (2013) Bivalent ligands that target μ opioid (MOP) and cannabinoid1 (CB1) receptors are potent analgesics devoid of tolerance. J Med Chem 56:5505–5513

Lenard NR, Daniels DJ, Portoghese PS, Roerig SC (2007) Absence of conditioned place preference or reinstatement with bivalent ligands containing mu-opioid receptor agonist and delta-opioid receptor antagonist pharmacophores. Eur J Pharmacol 566:75–82

Li T, Shiotani K, Miyazaki A, Tsuda Y, Ambo A, Sasaki Y, Jinsmaa Y, Marczak E, Bryant SD, Lazarus LH et al (2007) Bifunctional [2′,6′-dimethyl-L-tyrosine1]endomorphin-2 analogues substituted at position 3 with alkylated phenylalanine derivatives yield potent mixed mu-agonist/delta-antagonist and dual mu-agonist/delta-agonist opioid ligands. J Med Chem 50(12):2753–2766

Li Y, Lefever MR, Muthu D, Bidlack JM, Bilsky EJ, Polt R (2012) Opioid glycopeptide analgesics derived from endogenous enkephalins and endorphins. Future Med Chem 4(2):205–226

Linz K, Christoph T, Tzschentke TM, Koch T, Schiene K, Gautrois M, Schröder W, Kögel BY, Beier H, Englberger W et al (2014) Cebranopadol: a novel potent analgesic nociceptin/orphanin FQ peptide and opioid receptor agonist. J Pharmacol Exp Ther 349(3):535–548

Liu NJ, Xu T, Xu C, Li CQ, Yu YX, Kang HG, Han JS (1995) Cholecystokinin octapeptide reverses mu-opioid-receptor-mediated inhibition of calcium current in rat dorsal root ganglion neurons. J Pharmacol Exp Ther 275(3):1293–1299

Lowery JJ, Raymond TJ, Giuvelis D, Bidlack JM, Polt R, Bilsky EJ (2011) In vivo characterization of MMP-2200, a mixed delta/mu opioid agonist, in mice. J Pharmacol Exp Ther 336(3):767–778

Lufty K, Cowan A (2004) Buprenorphine : a unique drug with complex pharmacology. Curr Neuropharmacol 2(4):395–402

Lutz PE, Kieffer BL (2013) Opioid receptors: distinct roles in mood disorders. Trends Neurosci 36(3):195–206

Mabrouk OS, Marti M, Salvadori S, Morari M (2009) The novel delta opioid receptor agonist UFP-512 dually modulates motor activity in hemiparkinsonian rats via control of the nigro-thalamic pathway. Neuroscience 164(2):360–369

Mabrouk OS, Viaro R, Volta M, Ledonne A, Mercuri N, Morari M (2014) Stimulation of delta opioid receptor and blockade of nociceptin/orphanin FQ receptor synergistically attenuate parkinsonism. J Neurosci 34(39):12953–12962

Maisonneuve IM, Archer S, Glick SD (1994) U50,488, a kappa opioid receptor agonist, attenuates cocaine-induced increases in extracellular dopamine in the nucleus accumbens of rats. Neurosci Lett 181:57–60

Martin TJ, Kim SA, Cannon DG, Sizemore GM, Bian D, Porreca F, Smith JE (2000) Antagonism of delta(2) opioid receptors by Naltrindole-5′-isothiocyanate attenuates heroin self-administration but not antinociception in rats. J Pharmacol Exp Ther 294(3):975–982

Mello NK, Negus SS (1998) Effects of kappa opioid agonists on cocaine- and food-maintained responding by rhesus monkeys. J Pharmacol Exp Ther 286(2):812–824

Mello NK, Mendelson JH, Sholar MB, Jaszyna-Gasior M, Goletiani N, Siegel AJ (2005) Effects of the mixed mu/kappa opioid nalbuphine on cocaine-induced changes in subjective and cardio-vascular responses in men. Neuropsychopharmacology 30(3):618–632

Mizoguchi H, Hung KC, Leitermann R, Narita M, Nagase H, Suzuki T, Tseng LF (2003) Blockade of mu-opioid receptor-mediated G-protein activation and antinociception by TRK-820 in mice. Eur J Pharmacol 461(1):35–39

Mollica A, Pelliccia S, Famiglini V, Stefanucci A, Macedonio G, Chiavaroli A, Orlando G, Brunetti L, Ferrante C, Pieretti S et al (2017) Exploring the first rimonabant analog-opioid peptide hybrid compound, as bivalent ligand for CB1 and opioid receptors. J Enzyme Inhib Med Chem 32(1):444–451

Morphy R, Rankovic Z (2009) Designing multiple ligands – medicinal chemistry strategies and challenges. Curr Pharm Des 15(6):587–600

Morphy R, Kay C, Rankovic Z (2004) From magic bullets to designed multiple ligands. Res Focus Rev 9(15):641–652

Mosberg HI, Yeomans L, Harland AA, Bender AM, Sobczyk-kojiro K, Anand JP, Clark MJ, Jutkiewicz EM, Traynor JR (2013) Opioid peptidomimetics: leads for the design of bioavailable mixed efficacy mu opioid receptor (MOR) agonist/delta opioid receptor (DOR) antagonist ligands. J Med Chem 56:2139–2149

Mosberg HI, Yeomans L, Anand JP, Porter V, Sobczyk-Kojiro K, Traynor JR, Jutkiewicz EM (2014) Development of a bioavailable μ opioid receptor (MOPr) agonist, δ opioid receptor

(DOPr) antagonist peptide that evokes antinociception without development of acute tolerance. J Med Chem 57(7):3148–3153
Mustazza C, Bastanzio G (2011) Development of nociceptin receptor (NOP) agonists and antagonists. Med Res Rev 31(4):605–648
Nair P, Yamamoto T, Largent-Milnes TM, Cowell S, Kulkarni V, Moye S, Navratilova E, Davis P, Ma SW, Vanderah TW et al (2013) Truncation of the peptide sequence in bifunctional ligands with mu and delta opioid receptor agonist and neurokinin 1 receptor antagonist activities. Bioorg Med Chem Lett 23(17):4975–4978
Nair P, Yamamoto T, Cowell S, Kulkarni V, Moye S, Navratilova E, Davis P, Ma SW, Vanderah TW, Lai J et al (2015) Discovery of tripeptide-derived multifunctional ligands possessing delta/mu opioid receptor agonist and neurokinin 1 receptor antagonist activities. Bioorg Med Chem Lett 25(17):3716–3720
Negus SS, Mello NK, Portoghese PS, Lin CE (1997) Effects of kappa opioids on cocaine self-administration by rhesus monkeys. J Pharmacol Exp Ther 282(1):44–55
Neumeyer JL, Bidlack JM, Zong R, Bakthavachalam V, Gao P, Cohen DJ, Negus SS, Mello NK (2000) Synthesis and opioid receptor affinity of morphinan and benzomorphan derivatives: mixed kappa agonists and mu agonists/antagonists as potential pharmacotherapeutics for cocaine dependence. J Med Chem 43(1):114–122
Neumeyer JL, Zhang A, Xiong W, Gu X-H, Hilbert JE, Knapp BI, Negus SS, Mello NK, Bidlack JM (2003) Design and synthesis of novel dimeric morphinan ligands for kappa and mu opioid receptors. J Med Chem 46(24):5162–5170
Neumeyer JL, Zhang B, Zhang T, Sromek AW, Knapp BI, Cohen DJ, Bidlack JM (2013) Synthesis, binding affinity, and functional in vitro activity of 3-benzylaminomorphinan and 3-benzylaminomorphine ligands at opioid receptors. J Med Chem 55(8):3878–3890
Nielsen S, Sabioni P, Trigo JM, Ware MA, Betz-Stablein BD, Murnion B, Lintzeris N, Khor KE, Farrell M, Smith A et al (2017) Opioid-sparing effect of cannabinoids: a systematic review and meta-analysis. Neuropsychopharmacology 42:1752–1765
Noda Y, Mamiya T, Nabeshima T, Nishi M, Higashioka M, Takeshima H (1998) Loss of antinociception induced by naloxone benzoylhydrazone in nociceptin receptor-knockout mice. J Biol Chem 273(29):18047–18051
Peng X, Knapp BI, Bidlack JM, Neumeyer JL (2006) Synthesis and preliminary in vitro investigation of bivalent ligands containing homo- and heterodimeric pharmacophores at mu, delta, and kappa opioid receptors. J Med Chem 49(1):256–262
Peng J, Sarkar S, Chang SL (2012) Opioid receptor expression in human brain and peripheral tissues using absolute quantitative real-time RT-PCR. Drug Alcohol Depend 124:223–228
Perlikowska R, Piekielna J, Gentilucci L, De Marco R, Cerlesi MC, Calo G, Artali R, Tömböly C, Kluczyk A, Janecka A (2016) Synthesis of mixed MOR/KOR efficacy cyclic opioid peptide analogs with antinociceptive activity after systemic administration. Eur J Med Chem 109:276–286
Pinto M, Sousa M, Lima D, Tavares I (2008) Participation of mu-opioid, GABA(B), and NK1 receptors of major pain control medullary areas in pathways targeting the rat spinal cord: implications for descending modulation of nociceptive transmission. J Comp Neurol 510(2):175–187
Polt R, Dhanasekaran M, Keyari CM (2005) Glycosylated neuropeptides: a new vista for neuropsychopharmacology? Med Res Rev 25(5):557–585
Portoghese PS, Akgun E, Lunzer MM (2017) Heteromer induction: an approach to unique pharmacology? ACS Chem Neurosci 8(3):426–428
Pradhan AA, Befort K, Nozaki C, Gavériaux-Ruff C, Kieffer BL (2011) The delta opioid receptor: an evolving target for the treatment of brain disorders. Trends Pharmacol Sci 32(10):581–590
Provencher BA, Sromek AW, Li W, Russell S, Chartoff E, Knapp BI, Bidlack JM, Neumeyer JL (2013) Synthesis and pharmacological evaluation of aminothiazolomorphinans at the mu and kappa opioid receptors. J Med Chem 56(21):8872–8878

Qi JN, Mosberg HI, Porreca F (1990) Modulation of the potency and efficacy of mu-mediated antinociception by delta agonist in the mouse. J Pharmacol Exp Ther 254(2):683–689
Reinscheid RK, Higelin J, Henningsen RA, Monsma FJ, Civelli O (1998) Structures that delineate orphanin FQ and dynorphin a pharmacological selectivities. J Biol Chem 273(3):1490–1495
Rios CD, Jordan BA, Gomes I, Devi LA (2001) G-protein-coupled receptor dimerization: modulation of receptor function. Pharmacol Ther 92:71–87
Rios C, Gomes I, Devi LA (2009) μ opioid and CB1 cannabinoid receptor interactions: reciprocal inhibition of receptor signaling and neuritogenesis. Br J Pharmacol 148(4):387–395
Ross S, Peselow E (2009) The neurobiology of addictive disorders. Clin Neuropharmacol 32(5):269–276
Rowland NE, Mukherjee M, Robertson K (2002) Effects of the cannabinoid receptor antagonist SR 141716, alone and in combination with dexfenfluramine or naloxone, on food intake in rats. Psychopharmacology 159(1):111–116
Rozenfeld R, Abul-Husn NS, Gomez I, Devi LA (2007) An emerging role for the delta opioid receptor in the regulation of mu opioid receptor function. Sci World J 7(S2):64–73
Salvadori S, Guerrini R, Balboni G, Bianchi C, Bryant SD, Cooper PS, Lazarus LH (1999) Further studies on the Dmt-Tic pharmacophore: hydrophobic substituents at the C-terminus endow delta antagonists to manifest mu agonism or mu antagonism. J Med Chem 42(24):5010–5019
Schiller PW (2010) Bi- or multifunctional opioid peptide drugs. Life Sci 86:598–603
Schiller PW, Fundytus ME, Merovitz L, Weltrowska G, Nguten TM-D, Lemieux C, Chung NN, Coderre TJ (1999) The opioid mu agonist/delta antagonist DIPP-NH2(psi) produces a potent analgesic effect, no physical dependence and less tolerance than morphine in rats. J Med Chem 42(18):3520–3526
Schröder H, Wu DF, Seifert A, Rankovic M, Schulz S, Höllt V, Koch T (2009) Allosteric modulation of metabotropic glutamate receptor 5 affects phosphorylation, internalization, and desensitization of the mu-opioid receptor. Neuropharmacology 56(4):768–778
Seki T, Awamura S, Kimura C, Ide S, Sakano K, Minami M, Nagase H, Satoh M (1999) Pharmacological properties of TRK-820 on cloned mu-, delta- and kappa-opioid receptors and nociceptin receptor. Eur J Pharmacol 376:159–167
Smeester BA, Lunzer MM, Akgun E, Beitz AJ, Portoghese PS (2014) Targeting putative mu opioid/metabotropic glutamate receptor-5 heteromers produces potent antinociception in a chronic murine bone cancer model. Eur J Pharmacol 743:48–52
Spagnolo B, Calo G, Polgar WE, Jiang F, Olsen CM, Berzetei-Gurske I, Khroyan TV, Husbands SM, Lewis JW, Toll L et al (2008) Activities of mixed NOP and mu-opioid receptor ligands. Br J Pharmacol 153(3):609–619
Sukhtankar DD, Zaveri NT, Husbands SM, Ko M-C (2013) Effects of spinally administered bifunctional nociceptin/orphanin FQ peptide receptor/mu-opioid receptor ligands in mouse models of neuropathic and inflammatory pain. J Pharmacol Exp Ther 346:11–22
Suzuki S, Chuang LF, Yau P, Doi RH, Chuang RY (2002) Interactions of opioid and chemokine receptors: oligomerization of mu, kappa, and delta with CCR5 on immune cells. Exp Cell Res 280(2):192–200
Tallett AJ, Blundell JE, Rodgers RJ (2009) Effects of acute low-dose combined treatment with naloxone and AM 251 on food intake, feeding behaviour and weight gain in rats. Pharmacol Biochem Behav 91(3):358–366
Taylor AMW, Roberts KW, Pradhan AA, Akbari HA, Walwyn W, Lutfy K, Carroll FI, Cahill CM, Evans CJ (2015) Anti-nociception mediated by a κ opioid receptor agonist is blocked by a δ receptor agonist. Br J Pharmacol 172(2):691–703
Toll L, Khroyan TV, Polgar WE, Jiang F, Olsen C, Zaveri NT (2009) Comparison of the antinociceptive and antirewarding profiles of novel bifunctional nociceptin receptor/mu-opioid receptor ligands: implications for therapeutic applications. J Pharmacol Exp Ther 331(3):954–964

Toll L, Khroyan TV, Polgar WE, Husbands SM, Zaveri NT, Lucie PS, International, S. R. I, Ave R, Park M, Ba B et al (2013) Pharmacology of mixed NOP/mu ligands. In: Research and development of opioid-related ligands. American Chemical Society, Washington, pp 369–391

Toll L, Bruchas MR, Calo' G, Cox BM, Zaveri NT (2016) Nociceptin/orphanin FQ receptor structure, signaling, ligands, functions, and interactions with opioid systems. Pharmacol Rev 68(2):419–457

Vang D, Paul JA, Nguyen J, Tran H, Vincent L, Yasuda D, Zaveri NT, Gupta K (2015) Small-molecule nociceptin receptor agonist ameliorates mast cell activation and pain in sickle mice. Haematologica 100(12):1517–1525

Waldhoer M, Fong J, Jones RM, Lunzer MM, Sharma SK, Kostenis E, Portoghese PS, Whistler JL (2005) A heterodimer-selective agonist shows in vivo relevance of G protein-coupled receptor dimers. Proc Natl Acad Sci 102(25):9050–9055

Wang Y, Tao Y, Li F, Wang Y, Xu X, Chen J, Cao Y, Chi Z, Neumeyer JL, Zhang A et al (2009) Pharmacological characterization of ATPM [(−)-3-amino-thiazolo[5,4-B]-*N*-cyclopropylmethylmorphianan hydrochloride], a novel mixed kappa-agonist and mu-agonist/–antagonist that attenuates morphine antinociceptive tolerance and heroin self-administration behavior. J Pharmacol Exp Ther 329(1):306–313

Wang H-B, Zhao B, Zhong Y-Q, Li K-C, Li Z-Y, Wang Q, Lu Y-J, Zhang Z-N, He S-Q, Zheng H-C et al (2010) Coexpression of δ- and μ-opioid receptors in nociceptive sensory neurons. Proc Natl Acad Sci 107(29):13117–13122

Wen Q, Yu G, Li YL, Yan LD, Gong ZH (2011) Pharmacological mechanisms underlying the antinociceptive and tolerance effects of the 6,14-bridged oripavine compound 030418. Acta Pharmacol Sin 32(10):1215–1224

Williams DA, Zheng Y, David BG, Yuan Y, Zaidi SA, Stevens DL, Scoggins KL, Selley DE, Dewey WL, Akbarali HI et al (2016) 6β-*N*-heterocyclic substituted naltrexamine derivative BNAP: a peripherally selective mixed MOR/KOR ligand. ACS Chem Neurosci 7(8):1120–1129

Wright FL, Rodgers RJ (2013) Low dose naloxone attenuates the pruritic but not anorectic response to rimonabant in male rats. Psychopharmacology 226:415–431

Xiao J, Zeng S, Wang X, Babazada H, Li Z, Liu R, Yu W (2016) Neurokinin 1 and opioid receptors: relationships and interactions in nervous system. Transl Perioper Pain Med 19(3):11–21

Xie Z, Xie Z, Bhushan RG, Bhushan RG, Daniels DJ, Daniels DJ, Portoghese PS, Portoghese PS (2005) Interaction of bivalent ligand KDN21 with heterodimeric delta-kappa opioid receptors in human embryonic kidney 293 cells. Mol Pharmacol 68(4):1079–1086

Yamamoto T, Nair P, Davis P, Ma S, Navratilova E, Moye S, Tumati S, Lai J, Vanderah TW, Yamamura HI et al (2007) Design, synthesis, and biological evaluation of novel bifunctional C-terminal-modified peptides for delta/mu opioid receptor agonists and neurokinin-1 receptor antagonists. J Med Chem 50(12):2779–2786

Yu G, Di Yan L, Li YL, Wen Q, Dong HJ, Gong ZH (2011) TH-030418: a potent long-acting opioid analgesic with low dependence liability. Naunyn Schmiedeberg's Arch Pharmacol 384(2):125–131

Yuan Y, Arnatt CK, Li G, Haney KM, Ding D, Jacob JC, Selley DE, Zhang Y (2012) Design and synthesis of a bivalent ligand to explore the putative heterodimerization of the mu opioid receptor and the chemokine receptor CCR5. Org Biomol Chem 10(13):2633

Yuan Y, Arnatt CK, El-Hage N, Dever SM, Jacob JC, Selley DE, Hauser KF, Zhang Y (2013) A bivalent ligand targeting the putative mu opioid receptor and chemokine receptor CCR5 heterodimers: binding affinity versus functional activities. Med Chem Commun 4(5):847–851

Zádor F, Kocsis D, Borsodi A, Benyhe S (2014) Micromolar concentrations of rimonabant directly inhibits delta opioid receptor specific ligand binding and agonist-induced G-protein activity. Neurochem Int 67(1):14–22

Zaveri NT, Jiang F, Olsen C, Polgar WE, Toll L (2013) Designing bifunctional NOP receptor-mu opioid receptor ligands from NOP receptor-selective scaffolds. Part I. Bioorg Med Chem Lett 23(11):3308–3313

Zhang A, Xiong W, Hilbert JE, DeVita EK, Bidlack JM, Neumeyer JL (2004) 2-aminothiazole-derived opioids. Bioisosteric replacement of phenols. J Med Chem 47(8):1886–1888
Zhang S, Yekkirala A, Tang Y, Portoghese PS (2009) A bivalent ligand (KMN-21) antagonist for mu/kappa heterodimeric opioid receptors. Bioorg Med Chem Lett 19(24):6978–6980
Zhang T, Yan Z, Sromek A, Knapp BI, Scrimale T, Bidlack JM, Neumeyer JL (2011a) Aminothiazolomorphinans with mixed κ and μ opioid activity. J Med Chem 54(6):1903–1913
Zhang B, Zhang T, Sromek AW, Scrimale T, Bidlack JM, Neumeyer JL (2011b) Synthesis and binding affinity of novel mono- and bivalent morphinan ligands for kappa, mu, and delta opioid receptors. Bioorg Med Chem 19(9):2808–2816
Zhu Y, King MA, Schuller AGP, Nitsche JF, Reidl M, Elde RP, Unterwald E, Pasternak GW, Pintar JE (1999) Retention of supraspinal delta-like analgesia and loss of morphine tolerance in delta opioid receptor knockout mice. Neuron 24:243–252
Zielińska M, Jarmuz A, Sałaga M, Lipkowski AW, Fichna J (2016) Mixed MOP/DOP agonist biphalin elicits anti-transit effect in mouse models mimicking diarrhea-predominant irritable bowel syndrome symptoms. Pharmacol Rep 68(1):32–36

Contribution of Delta-Opioid Receptors to Pathophysiological Events Explored by Endogenous Enkephalins

Bernard P. Roques

Contents

Abstract

Very few discoveries in the neurosciences have triggered clinical speculation and experimentation regarding the etiology of psychiatric illness to the same extent as that following identification of the opiate receptor(s) and subsequent isolation of endogenous morphine-like peptides. There is overwhelming evidence in animals and in human that opioids are involved in behaviorally relevant issues such as the modulation of pain, the response to stress, motivation, addiction, sexuality, food intake, etc., but our knowledge on the possible relation between opioids and mental illness is still very limited.

These responses could be explored eitheir by using higlhy selective delta agonist or by emphasizing the effects of phasically secreted endogenous opioid peptides, enkephalin. Both approaches were investigated in particular through protection of enkephalin degradation by dual enkephalinase ihibitors DENKIs such as RB101, PL37 or PL265.

B.P. Roques (✉)
Membre de l'Académie des Sciences (France et Europe), U1022 INSERM/UMR 8258 CNRS, Université Paris-Descartes (Paris V), CSO Pharmaleads SAS, 4 Avenue de l'Observatoire, 75006 Paris, France
e-mail: bernard.roques@parisdescartes.fr; bernard.roques@pharmaleads.com

E. M. Jutkiewicz (ed.), *Delta Opioid Receptor Pharmacology and Therapeutic Applications*, Handbook of Experimental Pharmacology 247, DOI 10.1007/164_2016_17

Keywords
Synthetic opioids • Endogenous opioids peptides • Enkephalin • Enkephalinases • Aminopeptidase inhibition • DENKIs, Dual enkephalinase inhibitor • Dopamine • Depression

The apparent influence of endogenous opioids on the regulation of behavior has led to considerable efforts to associate alterations in opioidergic systems with psychiatric symptoms. The hypothesized associations have been investigated basically by three different methods. First, the effect of a selective opioid agonist or its antagonist on behavior is observed in normal volunteers as well as in psychiatric patients. The substance is usually administered intravenously or orally. The second type of study involves the determination of opioid concentrations in the CSF or serum of psychotic patients. The third method was to investigate the behavioral effects of increasing the level of endogenous opioids secreted by a stimuli in a particular brain structure though their physiological protection of degradation by specific degrading enzymes.

The biological effects of endogenous opioid peptides are mediated through three classes of naloxone-sensitive opioid receptors: mu (μ), kappa (k), and delta (∂). Each receptor is stimulated preferentially by specific endogenous peptides (Met-enkephalin, Leu-enkephalin, and β-endorphin) for the μ- and ∂-receptors. These endorphins are generated from maturation of long gene precursor called preproenkephalin for Met- and Leu-Enks, pre-proopiomelanocortin for β-endorphin, whereas dynorphin, the agonist of k-opioid receptor, is issued from preprodynorphin.

Several lines of evidence have suggested a role for opioid receptor systems in depression, including early studies investigating the potential antidepressant therapy of endogenous opioid peptides in humans. For example, it was shown that basal serum β-endorphin (an endogenous opioid peptide that binds to μ- and ∂-opioid receptors) levels were significantly elevated in patients with depression after antidepressant treatment (Darko et al. 1992; Djurovic et al. 1999). It was also reported that plasma β-endorphin levels were found to be elevated in patients after electroconvulsive shocks (ECS) for the treatment of depression, suggesting that endogenous μ- and/or ∂-opioid receptors were involved, at least in part, in the mechanisms of the ECS antidepressant activity (Emrich et al. 1979; Inturrisi et al. 1982). Consistently β-endorphin was reported to produce rapidly an antidepressant action in depressed patients (Kline et al. 1977). Some clinical reports also demonstrated the effectiveness of the μ-opioid receptor agonists oxycodone, oxymorphone, and buprenorphine in patients with refractory major depression (Bodkin et al. 1995; Stoll and Rueter 1999). Conversely, it was indicated that the nonselective opioid receptor antagonist naltrexone induced self-reported mental depression to volunteer subjects, in a placebo-controlled open study (Hollister et al. 1981). Taken together, these findings suggest that the endogenous opioid

Table 1 Affinity and selectivity for μ- and ∂-opioid binding sites of sterically constrained cyclic and linear enkephalins

Compounds	KI(∂) nM [^{3}H]DSTBULET	KI(μ) nM [^{3}H]DAGO	μ/∂
Tyr-D-Thr-Gly-Phe-Leu-Thr, DTLET	1.61	25	16
Tyr-D-Pen-Gly-Phe-D-Pen, DPDPE	8.85	993	110
Tyr-D-Ser(OtBu)-Gly-Phe-Leu-Thr, DSTBULET	2.81	374	130
Tyr-D-Ser(OtBu)-Gly-Phe-Leu-Thr(OtBu), BUBU	1.69	480	280
Tyr-D-Cys(OtBu)-Gly-Phe-Leu-Thr(OtBu), BUBUC	2.90	2,980	1,020
Tyr-D-Met-Phe-His-Leu-Met-Asp-NH_2, Deltorphin	2.40[a]	1,630[a]	679
Diallyl-Tyr-Aib-Aib-Phe-Leu, ICI 174,864	311	29,200	94
Naltrindole, NTI	1.30	74	57

[a]Using different radioligands

systems have important roles in depression, but clinical demonstration using a well-characterized exogenous opiate is still expected.

The ∂-opioid receptor was cloned in the 1990s (Evans et al. 1992; Kieffer et al. 1992). It was reported that these ∂-opioid receptors were located in the olfactory bulb, cerebral cortex, striatum, amygdala, hippocampus, brainstem nuclei, and spinal cord in rodents (Erbs et al. 2015; Pradhan et al. 2011; Le Merrer et al. 2009; Delay-Goyet et al. 1990). Madar et al. (1996) suggest that the distribution pattern of ∂-opioid receptors in the human brain using [^{11}C]-methyl-naltrindole was partially consistent with the location of the major regions involved in the modulation of mood and emotion (Madar et al. 1996). Interestingly, Filliol et al. (2000) found that ∂-opioid receptor knockout mice exhibited increases in the immobility times in the forced swimming test. This finding also suggests that the endogenous ∂-opioid receptor systems significantly contribute to the regulation of mood and emotion confirming the early results obtained in rodent using well-adapted behavioral tests and the first designed selective ∂-agonists (Table 1) (review in Roques et al. 1993).

Moreover, there are two different strategies to modulate the responses elicited by a neuropeptide. The classical approach is to use exogenous agonists that ubiquitously stimulate the peptide receptor(s), a strategy that is thought to be associated with serious drawbacks related to overstimulation of receptors (e.g., morphine). A more physiological approach is to modulate the extracellular concentrations of endogenous peptide effectors by inhibiting their metabolizing enzymes (Roques et al. 2012).

The disadvantages of peptidase inhibitors compared with exogenous agonists or antagonists could be their lower pharmacological potencies. However, this disadvantage is compensated for by their more physiological effects that correlate with the phasic release of their peptide substrates in brain structures recruited by a particular stimulus (e.g., pain, stress, or emotion) and the absence or small change in either the secretion of the peptide or expression of its targets (e.g., metabolizing enzymes and receptors).

We have privileged this approach (review in Roques et al. 2012) for the main endogenous opioid peptides enkephalins in designing potent inhibitors of the two degrading peptidases involved in the interruption of the message conveyed by the two peptides. These two enzymes, neutral endopeptidase (NEP or neprilysin) and aminopeptidase N (APN), belong to the class of zinc metallopeptidases (Roques et al. 2012).

1 Structure, Biological, and Pharmacological Properties of Dual Inhibitors of Enkephalinase (DENKIs)

Taking into account the substantial similarities in the active sites of zinc metallopeptidases (Fournie-Zaluski et al. 2009; Oefner et al. 2004; Marie-Claire et al. 2000; Tiraboschi et al. 1999), the rational design of potent selective or dual inhibitors of NEP and APN (Roques et al. 1993; Noble and Roques 2007; Roques 2000; Fournie-Zaluski and Roques 2002) has led to the selection of molecules that contain a strong metal-coordinating group (e.g., a thiol, carboxyl, hydroxamate, or phosphinic group) and are able to satisfy all possible energetically favorable interactions with at least one of the S1–S2′ subsites surrounding the catalytic site, as evidenced by inhibitor co-crystallization (Fournie-Zaluski et al. 2009; Oefner et al. 2004) (reviewed in references Noble and Roques 2007; Thanawala et al. 2008; Roques 2000; Mina-Osorio 2008; Fournié-Zaluski and Roques 2002).

The first DENK inhibitors (Fig. 1) were designed in 1984 (Fournie-Zaluski et al. 1984) using the hydroxamate group as a zinc-chelating moiety, assuming that the strength of its coordination to the metal should counterbalance a "less-than-perfect" fit of the inhibitor side chains to the active sites of the two metallopeptidases (Roques et al. 1993) that are obviously not identical (Fournie-Zaluski et al. 2009; Oefner et al. 2004). Accordingly, kelatorphan strongly inhibits NEP ($IC_{50} = 1.8$ nM) and less efficiently inhibits APN ($IC_{50} = 380$ nM).

Kelatorphan was the first compound that completely inhibited enkephalin catabolism (Bourgoin et al. 1986). It had antinociceptive effects in numerous acute nociceptive animal models (Fournie-Zaluski et al. 1984, 1985), and after intrathecal administration, it induced longer-lasting analgesia in patients with cancer (M.C. Fournié-Zaluski and J. Meynadier, unpublished observations) than the combination of both bestatin and thiorphan (Meynadier et al. 1988). Kelatorphan was also active in complete Freund's adjuvant-induced arthritis in rats, a widely used model of chronic pain (Kayser et al. 1989; Perrot et al. 1993), and it reduced nociception by 60% in mononeuropathic rats (Kayser et al. 1989; Lee et al. 1994). The entrance of kelatorphan into the brain is very limited, and therefore, the analgesic effects observed in arthritic rats are assumed to be due to a peripheral effect at the level of injured tissues (Maldonado et al. 1994).

Due to its very weak ability to enter the brain, kelatorphan is not active in antidepressant tests. This is also the case with other class of DENKIs in which the chelating group is a phosphinic moiety (Fig. 1). Thus, these compounds such as in

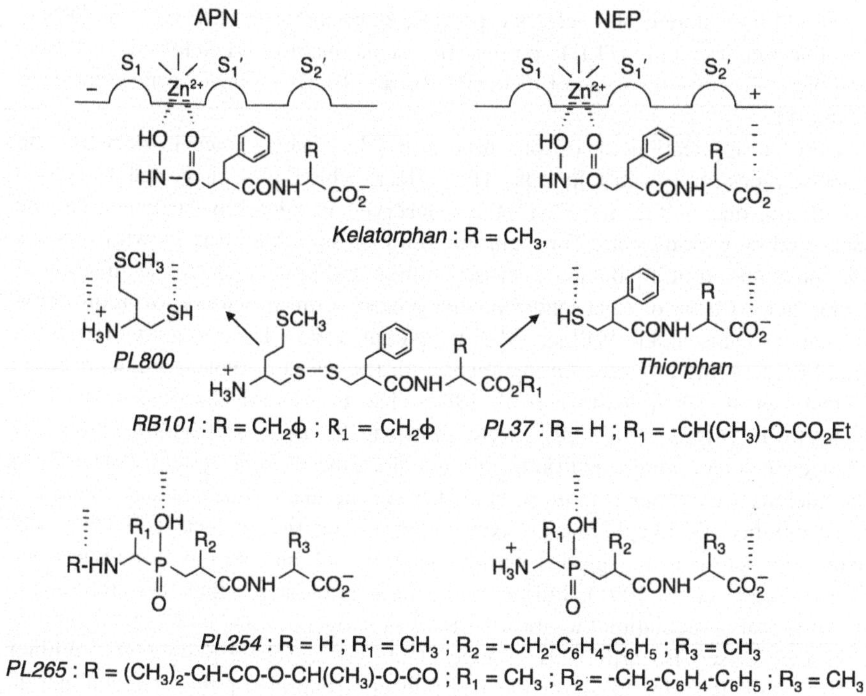

Fig. 1 Schematic inhibition of the two zinc metallopeptidases: aminopeptidase N (APN) and neprilysin (NEP) by the three different families of inhibitors

PL264 which have nanomolar NEP and APN inhibitory potency were shown to be active against neuropathic pain at the peripheral level (Poras et al. 2014).

Nevertheless, even after introduction of hydrophobic protecting group as in PL265, these phosphinic DENKIs have a very weak tendency to cross the blood-brain barrier and could not be used as putative antidepressants. In contrast, as easily shown with RB101, the DENKIs releasing in the brain (Fournie-Zaluski et al. 1992) the two highly selective APN and NEP inhibitors are able to diffuse easily in brain structures as demonstrated by autoradiographic studies (Jardinaud et al. 2004) and are the most interesting dual inhibitors for studying the role of ENKs in the control of mood. Therefore, the orally active DENKI PL37 (Fig. 1) which is in clinical trials (at this time phase II) to reduce neuropathic pain and which is about 10 times more potent by *i.v.* route than RB101 (Benoist et al. 2002) warrants an investigation in depressive disorders.

Several types of dual enkephalinase inhibitors (DENKIs) have been synthesized (Fig. 1) depending on the structure of the zinc-chelating moiety. Two families are able to recognize and inhibit directly NEP and APN with nanomolar affinity (i.e., kelatorphan or phosphinic inhibitors), while another family is able to form to two highly selective and potent inhibitors of NEP and APN issued from the cleavage of the disulfide bond (Roques et al. 2012).

In addition, potent and selective peptidic delta agonists (DSLET, Tyr-D-Ser-Gly-Phe-Leu-Thr and DTLET) derived from enkephalins and endowed with good oral bioavailability for some of them (BUBU and BUBUC) have been synthesized (Table 1).

Chronic application of various mild stress has been shown to decrease the responsiveness to reward in rats. This effect, which was suggested to mimic anhedonia, one of the main symptoms observed in depressive patients, can be measured by various tests. Thus, chronic mild stress (CMS) was shown to reduce the consumption of a palatable sucrose solution and to decrease the acquisition of preferences for a distinct environment paired with a variety of reinforcing substances (Muscat and Willner 1992; Papp et al. 1993). These negative responses could be prevented by chronic treatment with tricyclic or atypical antidepressants (Smadja et al. 1995; Willner et al. 1987). The behavioral changes, induced by exposure to chronic mild stress, were shown to be associated with a number of changes in dopaminergic neurotransmission in the mesolimbic system, especially in the nucleus accumbens (Stamford et al. 1991). The nucleus accumbens contains a large number of enkephalinergic cell bodies giving rise to local collaterals and axons projecting to the globus pallidus-ventral pallidum region (for review see Groenewegen et al. 1991). Furthermore, there is evidence that this structure is instrumental in mediating the reward effects of exogenous and endogenous opioids (for reviews see Bozarth 1991; Dauge et al. 1992; Scheel-Krüger and Willner 1991). With the aim to study the possible contribution of the enkephalinergic system in the anhedonia-like state induced by chronic mild stress, microdialysis was used to measure the extracellular levels of [Met]enkephalin-like material in the rostral part of the nucleus accumbens (N-Acc) of freely moving rats exposed or not to chronically mild stress (Bertrand et al. 1997). In both groups, the basal levels of [Met]enkephalin-like material (Met-LI) were found to be similar. Exposure of the two groups to a newly introduced rat (a stressful situation) leads to increased extracellular levels of [Met]enkephalin in the controls but not in chronic mild-stressed rats. A likely explanation for the lack of effect of the fearful social confrontation on the extracellular levels of Met-LI in CMS rats could be that the stress procedure rendered the enkephalinergic system, located in the N-Acc, unable to adapt to a threatening stimulus.

It has been shown that mice with a disruption of the preproenkephalin gene displayed a reduction in locomotor activity and anxious and aggressive behavior (Konig et al. 1996). Furthermore, numerous studies, using inhibitors of enkephalins catabolism, have shown that endogenous enkephalins induce antidepressant-like effects in animal models of "depression" such as the learned helplessness, the conditioned suppression of motility test, and the Porsolt assay (Baamonde et al. 1992; Tejedor-Real et al. 1995). Thus, RB101 has also shown anxiolytic effects mainly through DOR stimulation (Nieto et al. 2005; Jardinaud et al. 2005; McNally 2005) as they remain present in MOR KO mice (Noble and Roques 2007; Nieto et al 2005). Consistent with these results, PENK KO mice exhibit anxiogenic responses, increased aggressiveness (Konig et al. 1996; Ragnauth et al. 2001), stronger anxiety, and depressive posttraumatic stress disorder (Kung et al. 2010).

A role of endogenous enkephalins in the reward process is supported by the efficiency of naloxone (injected *i.p.* or in the N-Acc.) to suppress the preferential consumption of sucrose solution by rats and to block the reinforcing effect of a sucrose solution in a place preference paradigm (Agmo et al. 1995). However, in the CMS paradigm, chronic morphine administration was reported to reverse anhedonia the first and second weeks of stress procedure, while chronic treatment with the dual inhibitor of enkephalin-degrading enzymes, RB101, failed to reproduce this effect (Smadja et al. 1995). A plausible explanation could be that the CMS procedure induces a subsensitivity of the opioid receptors rendering them unable to respond to the weak increase in enkephalin levels produced by the inhibitor. This assumption is compatible with a dysfunctioning of the dopaminergic system observed in the CMS rats. In the N-Acc, D_2 receptors are mainly located on enkephalinergic neurons, and dopamine was reported to reduce their functioning (Le Moine and Bloch 1995). Therefore, the persistent enhancement of dopamine release in the mesolimbic system and the subsensitivity of D_2 receptors (Papp et al. 1993; Stamford et al. 1991), induced by the CMS, would decrease the enkephalin-controlled reward.

However, this could also explain the lack of increased level of enkephalins in the N-Acc in animal CMS faced with a threatening situation. It is interesting to observe that the inability to respond to emotional stimuli is one of the syndromes observed in depressive states.

Most studies on the etiology of depression have been devoted to the role of catecholamines, whereas the possible involvement of peptides, especially opioids, has been poorly investigated, partly because of the lack of appropriate tools. The study with CMS of Bertrand et al. (1997) shows that the association of microdialysis techniques and pharmacological experiments should allow this new direction to be explored in particular by using the numerous recently synthesized systematically active non-peptide selective delta agonists (Saitoh and Yamada 2012).

2 Mutual Role of Dopaminergic and Endogenous Opioid Signaling in Behavioral Control

High densities of both opioid receptors and NEP are found in the nucleus accumbens and the caudate nucleus (Waksman et al. 1986; Kieffer et al. 1992), forebrain structures that are involved in emotional, cognitive, and motor functions and receive a rich innervation from dopaminergic neurons located in the VTA and substantia nigra. Various pharmacological and biochemical studies have also shown that morphine and enkephalins are involved in the control of behavior such as arousal, locomotion, self-administration, self-stimulation, learning, and memory functions through modulation of the motor (nigrostriatal) and limbic cortical (mesocorticolimbic) dopaminergic systems.

Chronic administration of antidepressant drugs increases Met-enkephalin-like immune reactivity in the striatum and N-Acc of rat brain (De Felipe et al. 1985). A similar increase has also been observed after electroconvulsive shock or following chronic administration of lithium (Staunton et al. 1982). Moreover, chronic haloperidol treatment increases the levels of both the D2 receptor and preproenkephalin mRNA (Le Moine et al. 1991).

Several pharmacological studies have been carried out to clarify the role of opioids on dopaminergic systems. Moreover, stimulation of ∂-opioid receptors in the VTA by local injection of the selective agonists DSTBULET, DTLET, BUBU, or an inhibitor of enkephalin catabolism (kelatorphan) induced hyperactivity in familiar (home cage) or unfamiliar (open-field and four-hole box) environments. These effects were suppressed by a ∂-selective antagonist. The μ-agonist DAMGO also increased locomotion in the actimeter but decreased the activity in the open-field and four-hole box tests, possibly reflecting an increase in emotion and fear (Calenco-Choukroun et al. 1991a). The differences in the responses induced by kelatorphan or ∂-agonists with those produced by DAMGO suggest that μ- and ∂-receptors are involved in different neuronal pathways in the VTA. This is supported both by the association of only some enkephalinergic terminals with tyrosine hydroxylase-containing neurons in the rat VTA (Sesack and Pickel 1992) and by a study in which 6-hydroxydopamine-induced lesions of the rat mesoaccumbens pathway were found to abolish the effects of kelatorphan or BUBU in the VTA but not those elicited by the μ-agonist DAMGO (Calenco-Choukroun et al. 1991b). Taken together these results show that the endogenous enkephalins preferentially bind to ∂-receptors to induce hyperactivity. The conditions under which they could activate μ-receptors remain an open question.

The complexity of the interactions between dopaminergic and opioidergic systems in the N-Acc is illustrated by the results of 6-hydroxydopamine-induced lesions of the dopaminergic neurons of the VTA and chronic neuroleptic treatment, both of which potentiate the behavioral effects of exogenous opioids (Stinus et al. 1985, 1986; Kalivas and Bronson 1985) or kelatorphan (Maldonado et al. 1990b) infused into the nucleus accumbens. In agreement with the presence of D2 receptors on the forebrain enkephalin neurons, D2-dopamine receptor antagonists, such as sulpiride, or the mixed D1–D2 antagonist haloperidol, but not the D1 antagonist SCH23390, have been found to facilitate the opioid-related behavioral effects induced by kelatorphan (Maldonado et al. 1990).

Because both the enkephalinergic and dopaminergic systems in the N-Acc appear to function in parallel to increase locomotor activity (Kalivas et al. 1983), the supersensitivity of the opioid system after long-term dopaminergic blockade could be interpreted as a homeostatic mechanism to maintain normal locomotor activity. Moreover, the behavioral supersensitivity to endogenous opioids protected by kelatorphan in the N-Acc appears to be maximal after 2–3 weeks. This delay corresponds to the first appearance of the antipsychotic effects of neuroleptics, suggesting that alterations in the opioidergic system, very likely through its interrelations with the dopaminergic pathway, could be taking place in a neuronal system critically involved in the control of mood (Roques et al. 1985; MacLennan and Maier 1983; Kennedy et al. 2006). Kelatorphan also produces the same pattern of increase in brain stimulation or locomotor activity by i.c.v. administered amphetamine and ∂-opioid agonists (Heidbreder et al. 1988).

A link between opioidergic and dopaminergic systems has also been demonstrated by the clear antidepressant-like effects observed in the forced swimming and suppression of mobility test following i.v. administration of the systemically active mixed inhibitor RB101 in mice. These effects, which were shown to be related to ∂-receptor and D1 receptor activation, produced an increase in dopamine turnover in the striatum (Baamonde et al. 1992).

Injection of DTLET or kelatorphan into the rat striatum increased locomotor activity (Roques et al. 1985; Dauge et al. 1988). This effect was reversed by the dopamine antagonist thioproperazine and could be related to a specific D2-induced increase in the spontaneous and K+ -induced release of newly synthesized striatal dopamine (Petit et al. 1986; Daugé et al. 1989).

The tonic inhibition of the striatal opioid neurons by the nigrostriatal dopaminergic input suggests that under normal conditions, dopamine release is under the control of ∂-receptors, tonically stimulated by endogenous enkephalins (Petit et al. 1986). Because haloperidol was shown to increase the expression of the striatal D2 receptors located on enkephalins neurons (Le Moine et al. 1991), the tardive dyskinesia syndrome induced by long-term treatment with neuroleptics might be, at least partially, due to excessive μ (akinesia) and ∂ (tremor) effects induced by disinhibition of the enkephalinergic neurons normally negatively controlled by the dopaminergic input. The molecular events that control the relationships between these interactions, however, are still largely unknown, and it could be interesting to study the effects of dual NEP/APN inhibitors.

The interrelationships between the opioidergic and dopaminergic systems in the mesocorticolimbic and nigrostriatal pathways provide strong support for a crucial role of endogenous opioids in the control of mood. Amphetamine enhances the release of central dopamine, and chronic use of this drug results in psychotic symptoms resembling schizophrenia. Several recent studies have shown a cross-sensitization among amphetamine, stressful stimuli, and kelatorphan. This suggests that a hypersecretion of endogenous opioid peptides in the mesocorticolimbic pathway could induce an exaggerated behavioral response to stressful environmental stimuli, whose repetition could induce psychotic symptoms (MacLennan and Maier 1983). Conversely, depression might result from a deficiency in enkephalin release, minimizing their rewarding and euphorogenic effects. Likewise, drug abuse could be caused by a deficiency in the internal opioid-controlled rewarding system.

According to their mechanism of action, the DENKIs increasing the extra-synaptic concentration of endogenous enkephalins by protecting them from degradation increase the pharmacological responses generated by the amounts of non-protected peptides release under the action of specific stimuli. This has been put to use to design DENKIs leading very efficient analgesic properties to relieve pain induced by very strong nociceptive stimuli, for example, by excess of nociception and various types of neuropathic pain (review in Roques et al. 2012).

However, another interest of this physiological manipulation of the opioidergic system is the demonstration of its very efficient association with compounds acting on the same diseases but through a different mechanism of action. These synergetic analgesic responses against cancer pain as well as various neuropathic pain were obtained by the association of various DENKIs such as PL37 and PL265 with clinically used antiepileptics such as gabapentinoids (Menendez et al. 2008; Gonzalez-Rodriguez et al. 2009; Bonnard et al. 2015) or cannabinoids agonists (Gonzalez-Rodriguez et al. 2015, in preparation).

Another interesting observation has been the very strong synergistic increase of morphine analgesia induced by its association with the DENKIs RB101 (Mas-Nieto et al. 2001) allowing to reduce by a factor 7 the efficient dose of morphine. All these results drove us to investigate the interest of associating a classical dopaminergic antidepressant such as amisulpride and the dual NEP/APN inhibitor RB101 (Cordonnier et al. 2005).

3 The Mutual Influence of Exogenous and Endogenous Opioids on the Dopaminergic Functions (Dauge et al. 1992; Di Chiara and Imperato 1988; Wood et al. 1980)

The interaction between both systems was also demonstrated by administration of more or less selective dopamine antagonists. Chronic administration of haloperidol increased both the synthesis of endogenous enkephalins in the rat striatum and, to a lesser extent, in the nucleus accumbens (Hong et al. 1978) and the preproenkephalin mRNA expression in the rat striatum and pituitary (Jaber et al. 1994; Normand et al. 1987, 1988). Whereas many studies investigated the interaction between the two systems using exogenous opioid ligands, few studies have been devoted to the regulation of the endogenous opioid system following chronic treatment with neuroleptics. As previously discussed, the best way to achieve this aim is to protect the endogenous opioids from enkephalinase-controlled inactivation. This process occurs after synaptic release and in structures only recruited by specific stimuli. Amisulpride, an antipsychotic, was selected because of its low propensity to induce extrapyramidal side effects, avoiding motor disturbances that could induce bias in behavioral tests in particular following chronic treatments.

This study has been focused on behaviors in which RB101 had already been proved to be efficient in mice, i.e., analgesia in the hot-plate test (Noble et al. 1992) and antidepressant-like effects in the forced swim test and locomotor activity (Baamonde et al 1992).

A strong potentiation of RB101-induced hyperlocomotor effects was observed after 3 weeks of treatment with amisulpride. This result reinforces previous studies showing that chronic administration of neuroleptics sensitized the enkephalinergic system, potentiating the locomotor activity induced by the dual inhibitor of enkephalin catabolism, kelatorphan (Maldonado et al. 1990). However, whereas in this previous study using sulpiride or haloperidol it was clearly demonstrated that

3 weeks of treatment was necessary to achieve these potentiating effects on opioid responses, we observed that 5 days of treatment with amisulpride were sufficient. This suggests a more rapid onset of action of this latter compound compared with other neuroleptics. This duration of treatment was retained for the remainder of the study. It is interesting to notice that the effect of a 5-day treatment could still be seen even after 3 days of withdrawal suggesting that this treatment has durable effects.

Whereas no potentiation of the analgesic effects of RB101 by the amisulpride 5-day treatment was found in the hot-plate test, its hyperlocomotor and antidepressant-like effects were potentiated as observed in locomotor activity recording and forced swim test, respectively. Altogether these results suggest that the effects of the association of amisulpride and RB101 could take place more specifically in the nigrostriatal system and the limbic system.

SNC80 (2.5 mg/kg i.p.) alone induced hyperlocomotor and antidepressant-like effects, which were potentiated by a treatment of 5 days with amisulpride. NTI (5 mg/kg s.c.) totally blocked both the effects of SNC80 and SNC80 + amisulpride, demonstrating the specific involvement of delta-opioid receptors in the observed responses.

These results indicate that a chronic treatment with amisulpride potentiates the action of RB101. This effect seems to be restricted to behavioral responses induced by opioids acting on delta-opioid receptors. This could be due to an increase in the density of the delta-opioid receptors or even a sensitization of these receptors. The presence of a high concentration of delta-opioid receptors in brain areas involved in motor and motivational control suggests the existence of selective interactions between opioids and D2 receptors in certain brain regions because of the behaviorally selective effects of the amisulpride + RB101 combination. Thus, a high density of delta-opioid receptors has been found within the limbic system which is known to control emotional responses and reward behavior (Mansour et al. 1988), and amisulpride has also a preferential action on the limbic system (Moller 2003). Another hypothesis could be an enhancement in the synthesis of preproenkephalins and an increase in the release of enkephalins, in agreement with previous studies showing an increase of preproenkephalins in mice lacking the dopaminergic D2 receptor (Baik et al. 1995; Maldonado et al. 1997), or after a chronic blockade of dopaminergic neurotransmission with antagonists.

Whatever the explanation, the results suggest an interesting new therapeutic approach in CNS disorders, for example, in the treatment of depression, owing to the implication of the delta receptor in mood regulation.

An extension of the facilitation by chronic treatment with the D2 dopamine antagonist amisulpride (Cordonnier et al. 2005) was used to investigate whether a blockade of the dopaminergic system could lead to a more physiological "opioid substitution" compared with exogenous opioid agonists such as methadone and buprenorphine which are currently used as substitutes in pharmacotherapy of opioid addiction. With these pharmacotherapies, the level of relapse unfortunately remains very high.

The goal was therefore to act on both dopaminergic and opioid systems and, particularly, to investigate whether a blockade of the dopaminergic system could potentiate the endogenous opioid system, leading to a more physiological "opiate substitution" compared with exogenous opioid agonists.

Expression of morphine-induced locomotor sensitization was abolished after combined treatment with amisulpride (20 mg kg^{-1}, i.p.) and RB101 (80 mg kg^{-1}, i.p.), whereas these drugs were not effective when used alone. These results were compared with the effects of amisulpride combined with buprenorphine (0.1 mg kg^{-1}, i.p.) or methadone (2.5 mg kg^{-1}, i.p.) upon morphine-induced behavioral sensitization. Whereas the combination of amisulpride and buprenorphine partially blocked the expression of morphine sensitization, amisulpride + methadone was not effective in this paradigm.

The combination of amisulpride + RB101 appears to be very efficient in blocking the expression of morphine-induced behavioral sensitization. This could reflect a reinstatement of a balance between the function of the dopamine and opioid systems and could represent a new approach in maintenance treatments for opiate addiction.

It is important to notice that the synergetic approach in improving the antidepressant actions of dopaminergic agents by endogenous enkephalins occurs by a concomitant increase in the levels of endogenous enkephalins protected by a DENKI and recruiting selectively delta receptors in structures where the state and syndromes of depression release the opioid peptides.

Finally, recent studies using delta-opioid receptors expressed in GABAergic forebrain neurons (DIx-DOR) (Chu Sin Chung et al. 2015) yielded curiously opposite behavioral responses since both low anxiety was found for Dlx-DOR mice which contrast with the well-known increase in anxiety produced in mice by native DOR knockout (Chu Sin Chung et al. 2015) and DOR antagonist. The dual anxiolytic and anxiogenic roles for DORs open novel perspectives in the area of DOR function and anxiety disorders and warrant investigation of their physiological function by using endogenous enkephalins protected by DENKIs.

In conclusion the development of (1) dual orally active ENK inhibitors with strong analgesic properties and immediate antidepressant effects (Noble and Roques 2007) and (2) delta agonists devoid of side effects may lead to significant improvements in the treatment of depression and mood disorders.

References

Agmo A, Galvan A, Talamantes B (1995) Reward and reinforcement produced by drinking sucrose: two processes that may depend on different neurotransmitters. Pharmacol Biochem Behav 52:403–414

Baamonde A, Dauge V, Ruiz-Gayo M, Fulga IG, Turcaud S, Fournie-Zaluski MC, Roques BP (1992) Antidepressant-type effects of endogenous enkephalins protected by systemic RB 101 are mediated by opioid delta and dopamine D1 receptor stimulation. Eur J Pharmacol 216:157–166

Baik JH, Picetti R, Saiardi A, Thiriet G, Dierich A, Depaulis A, Le Meur M, Borrelli E (1995) Parkinsonian-like locomotor impairment in mice lacking dopamine D2 receptors. Nature 377:424–428

Benoist JM, Keime F, Montagne J, Noble F, Fournie-Zaluski MC, Roques BP, Willer JC, Le Bars D (2002) Depressant effect on a C-fibre reflex in the rat, of RB101, a dual inhibitor of enkephalin-degrading enzymes. Eur J Pharmacol 445:201–210

Bertrand E, Smadja C, Mauborgne A, Roques BP, Daugé V (1997) Social interaction increases the extracellular levels of [Met]enkephalin in the nucleus accumbens of control but not of chronic mild stressed rats. Neuroscience 80:17–20

Bodkin JA, Zornberg GL, Lukas SE, Cole JO (1995) Buprenorphine treatment of refractory depression. J Clin Psychopharmacol 15:49–57

Bonnard E, Poras H, Nadal X, Maldonado R, Fournie-Zaluski MC, Roques BP (2015) Long-lasting oral analgesic effects of N-protected aminophosphinic dual ENKephalinase inhibitors (DENKIs) in peripherally controlled pain. Pharmacol Res Perspect 3:e00116

Bourgoin S, Le Bars D, Artaud F, Clot AM, Bouboutou R, Fournie-Zaluski MC, Roques BP, Hamon M, Cesselin F (1986) Effects of kelatorphan and other peptidase inhibitors on the in vitro and in vivo release of methionine-enkephalin-like material from the rat spinal cord. J Pharmacol Exp Ther 238:360–366

Bozarth MA (1991) The mesolimbic dopamine system as a model reward system. In: Willner P, Scheel-Krüger J (eds) The mesolimbic dopamine system : from motivation to action. Wiley, New York, pp 301–330

Calenco-Choukroun G, Dauge V, Gacel G, Feger J, Roques BP (1991a) Opioid delta agonists and endogenous enkephalins induce different emotional reactivity than mu agonists after injection in the rat ventral tegmental area. Psychopharmacology (Berl) 103:493–502

Calenco-Choukroun G, Dauge V, Gacel G, Roques BP (1991b) Lesion of dopamine mesolimbic neurons blocks behavioral effects induced by the endogenous enkephalins but not by a mu-opioid receptor agonist. Eur J Pharmacol 209:267–271

Chu Sin Chung P, Keyworth HL, Martin-Garcia E, Charbogne P, Darcq E, Bailey A, Filliol D, Matifas A, Scherrer G, Ouagazzal AM, Gaveriaux-Ruff C, Befort K, Maldonado R, Kitchen I, Kieffer BL (2015) A novel anxiogenic role for the delta opioid receptor expressed in GABAergic forebrain neurons. Biol Psychiatry 77:404–415

Cordonnier L, Sanchez M, Roques BP, Noble F (2005) Facilitation of enkephalins-induced delta-opioid behavioral responses by chronic amisulpride treatment. Neuroscience 135:1–10

Darko DF, Risch SC, Gillin JC, Golshan S (1992) Association of beta-endorphin with specific clinical symptoms of depression. Am J Psychiatry 149:1162–1167

Dauge V, Rossignol P, Roques BP (1988) Comparison of the behavioural effects induced by administration in rat nucleus accumbens or nucleus caudatus of selective mu and delta opioid peptides or kelatorphan an inhibitor of enkephalin-degrading-enzymes. Psychopharmacology (Berl) 96:343–352

Daugé V, Rossignol P, Roques BP (1989) Blockade of dopamine receptors reverses the behavioral effects of endogenous enkephalins in the nucleus caudatus but not in the nucleus accumbens : involvement of ∂ and μ opioid receptors. Psychopharmacology (Berl) 99:168–175

Dauge V, Kalivas PW, Duffy T, Roques BP (1992) Effect of inhibiting enkephalin catabolism in the VTA on motor activity and extracellular dopamine. Brain Res 599:209–214

De Felipe MC, De Ceballos ML, Gil C, Fuentes JA (1985) Chronic antidepressant treatment increases enkephalin levels in n. accumbens and striatum of the rat. Eur J Pharmacol 112:119–122

Delay-Goyet P, Zajac JM, Roques BP (1990) Improved quantitative radioautography of rat brain ∂-opioid binding sites using [(3)H]DSTBULET, a new highly potent and selective linear enkephalin analogue. Neurochem Int 16:341–368

Di Chiara G, Imperato A (1988) Drugs abused by humans preferentially increase synaptic dopamine concentrations in the mesolimbic system of freely moving rats. Proc Natl Acad Sci U S A 85:5274–5278

Djurovic D, Milic-Askrabic J, Majkic-Singh N (1999) Serum beta-endorphin level in patients with depression on fluvoxamine. Farmaco 54:130–133

Emrich HM, Hollt V, Kissling W, Fischler M, Laspe H, Heinemann H, von Zerssen D, Herz A (1979) beta-Endorphin-like immunoreactivity in cerebrospinal fluid and plasma of patients with schizophrenia and other neuropsychiatric disorders. Pharmakopsychiatr Neuropsychopharmakol 12:269–276

Erbs E, Paget L, Scherrer G, Matifas A, Filliol D, Vonesch JL, Koch M, Kessler P, Hentsch D, Birling MC, Koutsourakis M, Vasseur, Veinante P, Kieffeter BL, Massotte D (2015) A mu-delta opioid receptor brain atlas reveals neuronal co-occurrence in subcortical networks. Brain Struct Funct 220:677–702

Evans CJ, Keith DE Jr, Morrison H, Magendzo K, Edwards RH (1992) Cloning of a delta opioid receptor by functional expression. Science 258:1952–1955

Filliol D, Ghozland S, Chluba J, Martin M, Matthes HW, Simonin F, Befort K, Gaveriaux-Ruff C, Dierich A, LeMeur M, Valverde O, Maldonado R, Kieffer BL (2000) Mice deficient for delta- and mu-opioid receptors exhibit opposing alterations of emotional responses. Nat Genet 25:195–200

Fournie-Zaluski MC, Roques BP (2002) New selective aminopeptidase N inhibitors and potential therapeutic. In: Langner J, Snyder A (eds) Ectopeptidases : CD13/Aminopeptidase N and CD26/Dipeptidyl peptidase IV in medicine and biology. Kluwer Academic/Plenum Publishers, New York, pp 51–94

Fournie-Zaluski MC, Chaillet P, Bouboutou R, Coulaud A, Cherot P, Waksman G, Costentin J, Roques BP (1984) Analgesic effects of kelatorphan, a new highly potent inhibitor of multiple enkephalin degrading enzymes. Eur J Pharmacol 102:525–528

Fournie-Zaluski MC, Coulaud A, Bouboutou R, Chaillet P, Devin J, Waksman G, Costentin J, Roques BP (1985) New bidentates as full inhibitors of enkephalin-degrading enzymes: synthesis and analgesic properties. J Med Chem 28:1158–1169

Fournie-Zaluski MC, Coric P, Turcaud S, Lucas E, Noble F, Maldonado R, Roques BP (1992) “Mixed inhibitor-prodrug” as a new approach toward systemically active inhibitors of enkephalin-degrading enzymes. J Med Chem 35:2473–2481

Fournie-Zaluski MC, Poras H, Roques BP, Nakajima Y, Ito K, Yoshimoto T (2009) Structure of aminopeptidase N from Escherichia coli complexed with the transition-state analogue aminophosphinic inhibitor PL250. Acta Crystallogr D Biol Crystallogr 65:814–822

Gonzalez-Rodriguez S, Pevida M, Roques BP, Fournie-Zaluski MC, Hidalgo A, Menendez L, Baamonde A (2009) Involvement of enkephalins in the inhibition of osteosarcoma-induced thermal hyperalgesia evoked by the blockade of peripheral P2X3 receptors. Neurosci Lett 465:285–289

Gonzalez-Rodriguez S, Menendez L, Lastra A, Baamonde A, Roques BP, Fournié-Zaluski MC, Poras H (2015) Antihyperalgesic effects of PL265, an orally active dual enkephalinase inhibitor (DENKI) in a mice model of tibial bone cancer and synergistic responses when combined with gabapentin, the purinergic antagonist □-317491 or a cannabinoid CB2 agonist. In preparation

Groenewegen HJ, Berendse HW, Meredith GE, Haber SN, Voorn P, Woters JG, Lohman AHM (1991) Functional anatomy of the ventral, limbic system-innervated striatum. In: Willner P, Scheel-Krüger J (eds) The mesolimbic dopamine system: from motivation to action. Wiley, New York, pp 19–59

Heidbreder C, Roques B, Vanderhaeghen JJ, De Witte P (1988) Kelatorphan, a potent enkephalinases inhibitor, presents opposite properties when injected intracerebroventricularly or into the nucleus accumbens on intracranial self-stimulation. Neurochem Int 12:347–350

Hollister LE, Johnson K, Boukhabza D, Gillespie HK (1981) Aversive effects of naltrexone in subjects not dependent on opiates. Drug Alcohol Depend 8:37–41

Hong JS, Yang HY, Fratta W, Costa E (1978) Rat striatal methionine-enkephalin content after chronic treatment with cataleptogenic and noncataleptogenic antischizophrenic drugs. J Pharmacol Exp Ther 205:141–147

Inturrisi CE, Alexopoulos G, Lipman R, Foley K, Rossier J (1982) beta-Endorphin immunoreactivity in the plasma of psychiatric patients receiving electroconvulsive treatment. Ann N Y Acad Sci 398:413–423

Jaber M, Tison F, Fournier MC, Bloch B (1994) Differential influence of haloperidol and sulpiride on dopamine receptors and peptide mRNA levels in the rat striatum and pituitary. Brain Res Mol Brain Res 23:14–20

Jardinaud F, Banisadr G, Noble F, Melik-Parsadaniantz S, Chen H, Dugave C, Laplace H, Rostene W, Fournie-Zaluski MC, Roques BP, Popovici T (2004) Ontogenic and adult whole body distribution of aminopeptidase N in rat investigated by in vitro autoradiography. Biochimie 86:105–113

Jardinaud F, Crete D, Canestrelli C, Ledent C, Roques BP, Noble F (2005) CB1 receptor knockout mice show similar behavioral modifications to wild-type mice when enkephalin catabolism is inhibited. Brain Res 1063:77–83

Kalivas PW, Bronson M (1985) Mesolimbic dopamine lesions produce an augmented behavioral response to enkephalin. Neuropharmacology 24:931–936

Kalivas PW, Widerlov E, Stanley D, Breese G, Prange AJ Jr (1983) Enkephalin action on the mesolimbic system: a dopamine-dependent and a dopamine-independent increase in locomotor activity. J Pharmacol Exp Ther 227:229–237

Kayser V, Fournie-Zaluski MC, Guilbaud G, Roques BP (1989) Potent antinociceptive effects of kelatorphan (a highly efficient inhibitor of multiple enkephalin-degrading enzymes) systemically administered in normal and arthritic rats. Brain Res 497:94–101

Kennedy SE, Koeppe RA, Young EA, Zubieta JK (2006) Dysregulation of endogenous opioid emotion regulation circuitry in major depression in women. Arch Gen Psychiatry 63:1199–1208

Kieffer BL, Befort K, Gaveriaux-Ruff C, Hirth CG (1992) The delta-opioid receptor: isolation of a cDNA by expression cloning and pharmacological characterization. Proc Natl Acad Sci U S A 89:12048–12052

Kline NS, Li CH, Lehmann HE, Lajtha A, Laski E, Cooper T (1977) Beta-endorphin--induced changes in schizophrenic and depressed patients. Arch Gen Psychiatry 34:1111–1113

Konig M, Zimmer AM, Steiner H, Holmes PV, Crawley JN, Brownstein MJ, Zimmer A (1996) Pain responses, anxiety and aggression in mice deficient in pre-proenkephalin. Nature 383:535–538

Kung JC, Chen TC, Shyu BC, Hsiao S, Huang AC (2010) Anxiety- and depressive-like responses and c-fos activity in preproenkephalin knockout mice: oversensitivity hypothesis of enkephalin deficit-induced posttraumatic stress disorder. J Biomed Sci 17:29

Le Merrer J, Becker JA, Befort K, Kieffer BL (2009) Reward processing by the opioid system in the brain. Physiol Rev 89:1379–1412

Le Moine C, Bloch B (1995) D1 and D2 dopamine receptor gene expression in the rat striatum: sensitive mRNA probes demonstrate prominent segregation of D1 and D2 mRNAs in distinct neuronal populations of the dorsal and ventral striatum. J Comp Neurol 355:426–481

Le Moine C, Normand E, Bloch B (1991) Phenotypical characterization of the rat striatal neurons expressing the D1 dopamine receptor gene. Proc Natl Acad Sci U S A 88:4206–4209

Lee SH, Kayser V, Guilbaud G (1994) Antinociceptive effect of systemic kelatorphan, in mononeuropathic rats, involves different opioid receptor types. Eur J Pharmacol 264:61–67

MacLennan AJ, Maier SF (1983) Coping and the stress-induced potentiation of stimulant stereotypy in the rat. Science 219:1091–1093

Madar I, Lever JR, Kinter CM, Scheffel U, Ravert HT, Musachio JL, Mathews WB, Dannals RF, Frost JJ (1996) Imaging of delta opioid receptors in human brain by N1'-([11C]methyl) naltrindole and PET. Synapse 24:19–28

Maldonado R, Dauge V, Feger J, Roques BP (1990) Chronic blockade of D2 but not D1 dopamine receptors facilitates behavioural responses to endogenous enkephalins, protected by kelatorphan, administered in the accumbens in rats. Neuropharmacology 29:215–223

Maldonado R, Valverde O, Turcaud S, Fournie-Zaluski MC, Roques BP (1994) Antinociceptive response induced by mixed inhibitors of enkephalin catabolism in peripheral inflammation. Pain 58:77–83

Maldonado R, Saiardi A, Valverde O, Samad TA, Roques BP, Borrelli E (1997) Absence of opiate rewarding effects in mice lacking dopamine D2 receptors. Nature 388:586–589

Mansour A, Khachaturian H, Lewis ME, Akil H, Watson SJ (1988) Anatomy of CNS opioid receptors. Trends Neurosci 11:308–314

Marie-Claire C, Tiraboschi G, Ruffet E, Inguimbert N, Fournie-Zaluski MC, Roques BP (2000) Exploration of the S(')(1) subsite of neprilysin: a joined molecular modeling and site-directed mutagenesis study. Proteins 39:365–371

Mas-Nieto M, Pommier B, Tzavara ET, Caneparo A, Da Nascimento S, Le Fur G, Roques BP, Noble F (2001) Reduction of opioid dependence by the CB(1) antagonist SR141716A in mice: evaluation of the interest in pharmacotherapy of opioid addiction. Br J Pharmacol 132:1809–1816

McNally GP (2005) Facilitation of fear extinction by midbrain periaqueductal gray infusions of RB101(S), an inhibitor of enkephalin-degrading enzymes. Behav Neurosci 119:1672–1677

Menendez L, Hidalgo A, Meana A, Poras H, Fournie-Zaluski MC, Roques BP, Baamonde A (2008) Inhibition of osteosarcoma-induced thermal hyperalgesia in mice by the orally active dual enkephalinase inhibitor PL37. Potentiation by gabapentin. Eur J Pharmacol 596:50–55

Meynadier J, Dalmas S, Lecomte JM, Gros C, Schwartz JC (1988) Potent analgesic effects of inhibitors of enkephalin metabolism administered intrathecally to cancer patients. Pain Clin 2:201–206

Mina-Osorio P (2008) The moonlighting enzyme CD13: old and new functions to target. Trends Mol Med 14:361–371

Moller HJ (2003) Amisulpride: limbic specificity and the mechanism of antipsychotic atypicality. Prog Neuropsychopharmacol Biol Psychiatry 27:1101–1111

Muscat R, Willner P (1992) Suppression of sucrose drinking by chronic mild unpredictable stress: a methodological analysis. Neurosci Biobehav Rev 16:507–517

Nieto MM, Guen SL, Kieffer BL, Roques BP, Noble F (2005) Physiological control of emotion-related behaviors by endogenous enkephalins involves essentially the delta opioid receptors. Neuroscience 135:305–313

Noble F, Roques BP (2007) Protection of endogenous enkephalin catabolism as natural approach to novel analgesic and antidepressant drugs. Expert Opin Ther Targets 11:145–159

Noble F, Soleilhac JM, Soroca-Lucas E, Turcaud S, Fournie-Zaluski MC, Roques BP (1992) Inhibition of the enkephalin-metabolizing enzymes by the first systemically active mixed inhibitor prodrug RB 101 induces potent analgesic responses in mice and rats. J Pharmacol Exp Ther 261:181–190

Normand E, Popovici T, Fellmann D, Bloch B (1987) Anatomical study of enkephalin gene expression in the rat forebrain following haloperidol treatment. Neurosci Lett 83:232–236

Normand E, Popovici T, Onteniente B, Fellmann D, Piatier-Tonneau D, Auffray C, Bloch B (1988) Dopaminergic neurons of the substantia nigra modulate preproenkephalin A gene expression in rat striatal neurons. Brain Res 439:39–46

Oefner C, Roques BP, Fournie-Zaluski MC, Dale GE (2004) Structural analysis of neprilysin with various specific and potent inhibitors. Acta Crystallogr D Biol Crystallogr 60:392–396

Papp M, Muscat R, Willner P (1993) Subsensitivity to rewarding and locomotor stimulant effects of a dopamine agonist following chronic mild stress. Psychopharmacology (Berl) 110:152–158

Perrot S, Kayser V, Fournie-Zaluski MC, Roques BP, Guilbaud G (1993) Antinociceptive effect of systemic PC 12, a prodrug mixed inhibitor of enkephalin-degrading enzymes, in normal and arthritic rats. Eur J Pharmacol 241:129–133

Petit F, Hamon M, Fournie-Zaluski MC, Roques BP, Glowinski J (1986) Further evidence for a role of delta-opiate receptors in the presynaptic regulation of newly synthesized dopamine release. Eur J Pharmacol 126:1–9

Poras H, Bonnard E, Dange E, Fournié-Zaluski MC, Roques BP (2014) New orally active dual ENKephalinase inhibitors (DENKIs) for central and peripheral pain treatment. J Med Chem 57:5748–5763

Pradhan AA, Befort K, Nozaki C, Gaveriaux-Ruff C, Kieffer BL (2011) The delta opioid receptor: an evolving target for the treatment of brain disorders. Trends Pharmacol Sci 32:581–590

Ragnauth A, Schuller A, Morgan M, Chan J, Ogawa S, Pintar J, Bodnar RJ, Pfaff DW (2001) Female preproenkephalin-knockout mice display altered emotional responses. Proc Natl Acad Sci U S A 98:1958–1963

Roques BP (2000) Novel approaches to targeting neuropeptide systems. Trends Pharmacol Sci 21:475–483

Roques BP, Dauge V, Gacel G, Fournie-Zaluski MC (1985) Selective agonists and antagonists of delta-opioid receptors and inhibitors of enkephalins metabolism. Potential use in mental illness. In: Shagass C, Josiassen RC, Bridger WH, Weiss KJ, Stoff D, Simpon GM (eds) Biological psychiatry, developments in psychiatry, vol. 7. Elsevier, New York

Roques BP, Noble F, Dauge V, Fournie-Zaluski MC, Beaumont A (1993) Neutral endopeptidase 24.11: structure, inhibition, and experimental and clinical pharmacology. Pharmacol Rev 45:87–146

Roques BP, Fournie-Zaluski MC, Wurm M (2012) Inhibiting the breakdown of endogenous opioids and cannabinoids to alleviate pain. Nat Rev Drug Discov 11(4):292–310

Saitoh A, Yamada M (2012) Antidepressant-like effects of ∂ opioid receptor agonists in animal models. Curr Neuropharmacol 10:231–238

Scheel-Krüger J, Willner P (1991) The mesolimbic system: principles of operation. In: Willner P, Scheel-Krüger J (eds) The mesolimbic dopamine system: from motivation to action. Wiley, New York, pp 559–597

Sesack SR, Pickel VM (1992) Dual ultrastructural localization of enkephalin and tyrosine hydroxylase immunoreactivity in the rat ventral tegmental area: multiple substrates for opiate-dopamine interactions. J Neurosci 12:1335–1350

Smadja C, Ruiz-Taron F, Turcaud S, Roques BP, Maldonado R (1995) Effects induced by endogenous and exogenous opioids in different animal models of depression: modulation by the CCK system. In: Traynor J (ed) Analgesia, vol 1(4–6). INRC Proceedings 1995, pp 742–745

Stamford JA, Muscat R, O'Connor JJ, Patel J, Trout SJ, Wieczorek WJ, Kruk ZL, Willner P (1991) Voltammetric evidence that subsensitivity to reward following chronic mild stress is associated with increased release of mesolimbic dopamine. Psychopharmacology (Berl) 105:275–282

Staunton DA, Deyo SN, Shoemaker WJ, Ettenberg A, Bloom FE (1982) Effects of chronic lithium on enkephalin systems and pain responsiveness. Life Sci 31:1837–1840

Stinus L, Winnock M, Kelley AE (1985) Chronic neuroleptic treatment and mesolimbic dopamine denervation induce behavioural supersensitivity to opiates. Psychopharmacology (Berl) 85:323–328

Stinus L, Nadaud D, Jauregui J, Kelley AE (1986) Chronic treatment with five different neuroleptics elicits behavioral supersensitivity to opiate infusion into the nucleus accumbens. Biol Psychiatry 21:34–48

Stoll AL, Rueter S (1999) Treatment augmentation with opiates in severe and refractory major depression. Am J Psychiatry 156:2017

Tejedor-Real P, Mico JA, Maldonado R, Roques BP, Gibert-Rahola J (1995) Implication of endogenous opioid system in the learned helplessness model of depression. Pharmacol Biochem Behav 52:145–152

Thanawala V, Kadam VJ, Ghosh R (2008) Enkephalinase inhibitors: potential agents for the management of pain. Curr Drug Targets 9:887–894

Tiraboschi G, Jullian N, Thery V, Antonczak S, Fournie-Zaluski MC, Roques BP (1999) A three-dimensional construction of the active site (region 507-749) of human neutral endopeptidase (EC.3.4.24.11). Protein Eng 12:141–149

Waksman G, Hamel E, Fournie-Zaluski MC, Roques BP (1986) Autoradiographic comparison of the distribution of the neutral endopeptidase "enkephalinase" and of mu and delta opioid receptors in rat brain. Proc Natl Acad Sci U S A 83:1523–1527

Willner P, Towell A, Sampson D, Sophokleous S, Muscat R (1987) Reduction of sucrose preference by chronic unpredictable mild stress, and its restoration by a tricyclic antidepressant. Psychopharmacology (Berl) 93:358–364

Wood PL, Stotland M, Richard JW, Rackham A (1980) Actions of mu, kappa, sigma, delta and agonist/antagonist opiates on striatal dopaminergic function. J Pharmacol Exp Ther 215:697–703

Part II

DOPR Signaling

Ligand-Directed Signaling at the Delta Opioid Receptor

Ana Vicente-Sanchez and Amynah A. Pradhan

Contents

Abstract

Delta opioid receptors (δORs) regulate a number of physiological functions, and agonists for this receptor are being pursued for the treatment of mood disorders, chronic pain, and migraine. A major challenge to the development of these compounds is that, like many G-protein coupled receptors (GPCRs), agonists at the δOR can induce very different signaling and receptor trafficking events. This concept, known as ligand-directed signaling, functional selectivity, or biased agonism, can result in different agonists producing highly distinct behavioral consequences. In this chapter, we highlight the in vitro and in vivo evidence for ligand-directed signaling and trafficking at the δOR. A number of biological implications of agonist-directed signaling at the δOR have been demonstrated. Importantly, ligand-specific effects can impact both acute behavioral effects of delta agonists, as well as the long-term adaptations induced by chronic drug treatment. A better understanding of the specific signaling cascades that regulate these differential behavioral effects would

A. Vicente-Sanchez
Department of Psychiatry, University of Illinois at Chicago, Chicago, IL, USA

A.A. Pradhan (✉)
Department of Psychiatry, University of Illinois at Chicago, Chicago, IL, USA

Department of Psychiatry, UIC, 1601 W Taylor St (MC 912), Chicago, IL 60612, USA
e-mail: apradhan@psych.uic.edu

E. M. Jutkiewicz (ed.), *Delta Opioid Receptor Pharmacology and Therapeutic Applications*, Handbook of Experimental Pharmacology 247, DOI 10.1007/164_2017_39

help to guide rational drug design, ultimately resulting in δOR agonists with fewer adverse effects.

Keywords
Pain • Receptor trafficking • Tolerance

1 Introduction

Delta opioid receptors (δORs) are G-protein coupled receptors (GPCRs) that primarily couple to the inhibitory Gαi/o family of G-proteins. Classically, it was thought that GPCRs could exist in two different states, an "on" and an "off" state. According to this model, all agonists increased the active conformation of the receptor (on state), which resulted in an increase in all signaling cascades associated with this active form (for review, see Kenakin 2004). Today, it is widely recognized that GPCRs can actually exist in multiple conformations, and ligands can stabilize different active states (Lagerstrom and Schioth 2008; Costa-Neto et al. 2016). Each receptor conformation produces distinct receptor–effector complexes, which triggers differing signaling pathways and receptor trafficking events (Kenakin 2012; Reiter et al. 2012). This concept is referred to as ligand- or agonist-directed signaling, functional selectivity, or biased agonism (5, 6).

Divergent functional responses from ligand-bound GPCRs can be modulated at a number of different levels. Traditionally, it was thought that responses derived from the activation of GPCRs were primarily due to G-protein-dependent signaling, which was initiated at the cell membrane. Despite the importance of this pathway in GPCR-mediated effects, it is currently known that these receptors can also signal through G-protein-independent cascades, initiated at the cell membrane and other subcellular compartments (Costa-Neto et al. 2016). The best characterized signaling proteins in this category are arrestins, which are involved in GPCR internalization and desensitization (Pierce et al. 2002; Reiter et al. 2012). Arrestins also act as scaffolds to other signaling proteins that are activated after ligand binding initiating a new wave of intracellular signaling events (Sorkin and von Zastrow 2009; Rajagopal et al. 2010). Thus, functional selectivity may be observed by ligands promoting G-protein-dependent or -independent signaling or both. Further, different signaling cascades within each of these categories may or may not be initiated by a given ligand, therefore conferring an added degree of complexity to GPCR signaling. The promise of biased agonism lies in the possibility of developing ligands that preferentially initiate signaling cascades which correlate with a desired biological effect, while avoiding signaling events that evoke less desirable or adverse events.

In recent years, the δOR has attracted increasing attention for its therapeutic potential. Although δOR agonists are poor analgesics for the treatment of acute pain (Gallantine and Meert 2005), they are highly effective in animal models of pain associated with chronic inflammation, neuropathy, cancer, and diabetes (for review, see Pradhan et al. 2011; Vicente-Sanchez et al. 2016). In addition, genetic deletion of the δOR in mice revealed its role in emotional processing (Filliol et al. 2000), and pharmacological studies confirmed the anxiolytic and antidepressant effects of δ agonists (Broom et al. 2002a; Saitoh et al. 2004, 2005; Perrine et al. 2006). Further, δORs have also recently been established as novel targets for the treatment of migraine (Pradhan et al. 2014; Charles and Pradhan 2016; Rice et al. 2016). Delta agonists are thus being developed for the treatment of chronic pain, migraine, and anxiety, and depression. Moreover, it has been shown that delta agonists bear neuroprotective and motor control properties, and they are currently under investigation for the treatment of hypoxic/ischemic stress (Husain et al. 2012; He et al. 2013; Maslov et al. 2013), and Parkinson disease (Hudzik et al. 2000). Importantly, unlike mu opioid receptor agonists, δ agonists are not self-administered and do not appear to possess substantial rewarding properties on their own (Negus et al. 1994, 1998; Brandt et al. 2001; Do Carmo et al. 2009). However, some δOR agonists can produce convulsions (Comer et al. 1993; Negus et al. 1994; Jutkiewicz et al. 2006) which has limited the development of these compounds. This adverse effect is an example of ligand-directed signaling as only some agonists produce δOR-dependent convulsions, while others do not (Pradhan et al. 2012; Chu Sin Chung et al. 2014). A better understanding of the receptor conformation states and signaling cascades that elicit this adverse effect would open new avenues for the therapeutic potential of delta agonist.

2 In Vitro Evidence for Ligand-Directed Signaling at the Delta Opioid Receptor

Cell-free assays have been used to show that different δOR agonists can produce distinct conformational changes upon binding to the δOR. Plasmon-waveguide resonance (PWR) spectroscopy assays have established that ligands of different efficacies impose distinct structural constraints on purified human δOR incorporated into lipid bilayers. In these studies, the peptide ligands DPDPE and deltorphin II produced different PWR spectral changes compared to non-peptide ligands such as TAN67 and SNC80 (Salamon et al. 2000, 2002; Alves et al. 2004). This type of spectroscopy has also been used to show that the different conformational states induced by agonists can stabilize distinct δOR-Gαi/o subunits interactions (Alves et al. 2003, 2004; Hruby et al. 2010).

Activation of δORs can initiate both G-protein-dependent and -independent signaling pathways. In terms of G-protein coupling, results obtained in cellular systems are consistent with the notion that agonists can produce selective engagement between the δOR and different Gα subunits. In HEK293 cells expressing human δOR-Gαi or Gαo fusion proteins, DADLE preferentially activated Gαi (Moon et al. 2001). Further, in SK-N-BE cells which endogenously express δOR, etorphine evoked coupling with Gαi2, Gαi1/3, and one PTX-insensitive Gα subunit; while the peptides DPDPE and deltorphin I preferentially stimulated coupling with Gαo2 and Gαi2 (Allouche et al. 1999). These results indicate that ligand-induced receptor conformations differentially evoked interaction of the receptor with individual G-proteins which could subsequently effect downstream signaling cascades.

Several studies using bioluminescence resonance energy transfer (BRET) indicate that δORs are in preengaged complexes with different signaling molecules, which can be distinctly modulated by ligands. For example, a large panel of opioid ligands were shown to have differing abilities to activate G-proteins and recruit arrestins at the δOR (Molinari et al. 2010). Notably, most ligands showed low efficacy for arrestin 3 (β-arrestin 2) despite considerable efficacy at recruiting G-proteins. In addition, those ligands that induced strong G-protein coupling, but showed weak or null efficacy for arrestin 3-δOR interactions, also acted as competitive antagonists for arrestin 3 binding. Thus, these ligands were agonists at one end point (G-protein coupling), and partial agonists at another (arrestin recruitment) (Molinari et al. 2010). Ligand-specific conformational rearrangements that occur at the level of Gα and Gβγ subunits have also been observed. DPDPE, SNC80, and morphine binding to the δOR were all found to decrease the distance between position 60 of the Gαi1 protein and the N terminus of Gγ; while the ligand TICP drew these same regions apart (Audet et al. 2008). These conformational differences likely account for the finding that the former three ligands inhibit cAMP activity and promote ERK phosphorylation, while TICP is an inverse agonist at adenylate cyclase, and still an agonist in the ERK pathway (Audet et al. 2005, 2008). Moreover, the recent crystal structure of the δOR revealed that the sodium allosteric site in the receptor coordinated residues to form an efficacy switch regulating biased signaling. Mutations of this sodium site, which disrupted ion binding, increased agonist-induced arrestin recruitment, and converted antagonists/weak partial agonists into potent arrestin-biased ligands (Fenalti et al. 2014).

Moreover, real-time analysis of second messenger levels with BRET-based biosensors also shows that ligand-specific conformational changes at the receptor level are transduced to downstream effectors. Studies examining the effect of different δ ligands on cAMP production showed that the ligand efficacy of δOR agonists to activate G-protein signaling and their ability to induce δOR internalization modulated the early stages of the cyclase inhibition. However, internalization was not predictive of the pattern of the cyclase response after sustained stimulation of the δOR (Tudashki et al. 2014). Additionally, studies in HEK293 cells with the label-free dynamic mass redistribution technology have revealed that a large number of opioid ligands exhibit pathway-biased agonism at the δOR (Morse et al. 2013). Further, constitutive association between δOR, Gβ1γ2, and the G-protein-gated inwardly rectifying K+ channel, Kir3.1–3.2 was detected using BRET (Richard-Lalonde et al. 2013). In response to different agonists, Gβγ subunits adopted distinct conformations which, in turn, modulated Kir3.1–3.2 channel activity in an agonist-specific manner (Richard-Lalonde et al. 2013).

As for many GPCRs, agonist-induced activation of the δOR often leads to receptor internalization and trafficking. δOR internalization has been observed following binding of endogenous opioids (leu- and met-enkephalin), peptides (DPDPE, deltorphin I and II), and small molecules (SNC80 and BW373U86) (Pradhan et al. 2009; Bradbury et al. 2009). Compared to the mu opioid receptor which is rapidly recycled back to the cell surface, the δOR is predominantly targeted for degradation through the Endosomal Sorting Complex Required for Transport (ESCRT) machinery (Henry et al. 2011). However, depending on the ligand and the conditions under which it is tested, δ agonists have been shown to evoke a number of different trafficking events. For example, in the neuroblastoma cell line SK-N-BE, SNC80, DPDPE, and deltorphin I were shown to induce δOR internalization and sorting to lysosomes; while leu- and met-enkephalin, and etorphine induced receptor recycling following internalization (Marie et al. 2003; Lecoq et al. 2004). In another experimental model, DPDPE was found to promote δOR internalization and recycling in cortical neuronal cultures and HEK293 cells, while SNC80 induced receptor internalization and degradation (Audet et al. 2012). Differences in ligand-induced trafficking events could be due to agonists producing: distinct receptor conformation states, differences in receptor phosphorylation, and/or recruitment of divergent signaling/trafficking molecules. It is important to keep in mind that findings from different cellular models may be due to differences in signaling and trafficking machinery which can vary across cell types. Nevertheless, these in vitro studies indicate that different δ agonists can produce distinct receptor conformations, coupling, signaling, and trafficking events consistent with ligand-directed signaling at this receptor.

3 In Vivo Evidence for Ligand-Directed Signaling at the Delta Opioid Receptor

The concept of ligand-directed signaling has profound behavioral implications, both in terms of understanding the complexity of GPCR pharmacology and for facilitating drug development (Bosier and Hermans 2007; Galandrin et al. 2007; Pradhan et al. 2012; Charfi et al. 2015; Costa-Neto et al. 2016). However, the evidence for this phenomenon is primarily based on in vitro experiments using recombinant cell systems. A major challenge in GPCR research, today, is to demonstrate the physiological relevance of agonist-biased signaling and regulation.

Early experiments using antisense techniques revealed in vivo differences in δ agonist-induced G-protein coupling. For example, knockdown of brain Gαo subunits significantly reduced the potency of DPDPE but not deltorphin II-induced antinociception, while the opposite was true following Gαq knockdown (Sanchez-Blazquez and Garzon 1998).

Deeper insight into the behavioral consequences of ligand-specific effects was made possible with the development of a knock-in mouse model expressing fluorescent δORs (DOR-eGFP) (Scherrer et al. 2006). In these animals, the endogenous δOR is replaced by a fluorescent-tagged δOR which allows direct visualization of the receptor in tissue. Studies with primary cultures from the central and peripheral nervous system of these animals showed that SNC80 induced rapid and robust DOR-eGFP internalization (Scherrer et al. 2006; Pradhan et al. 2009; Poole et al. 2011), whereas the δOR agonist, ARM390, did not produce detectable receptor sequestration (Pradhan et al. 2009). This difference in trafficking occurred despite these two ligands sharing similar binding and G-protein activation profiles at the δOR. In vivo studies showed that this difference in internalization had important biological consequences (Pradhan et al. 2009, 2010, 2016). In acute tolerance studies, mice were treated twice with equipotent doses of the high-internalizing agonist SNC80; and the low-internalizing agonist, ARM390. Initial injections of each drug produced similar anti-hyperalgesic effects in a Complete Freund's Adjuvant (CFA)-model of inflammatory pain. However, repeated injection of SNC80 produced a complete acute behavioral tolerance, which was correlated with uncoupling of the receptor from G-proteins and profound receptor internalization. In contrast, ARM390 continued to be effective following the second injection, which corresponded with intact receptor coupling and expression on the cell surface. Notably, the difference in internalization dynamics between SNC80 and ARM390 could be due to phosphorylation at the critical serine 363 site of δOR, which was only observed after treatment with SNC80 (Pradhan et al. 2009). This study was the first to demonstrate that internalization of δOR directly controlled behavioral effects of receptor activation.

Importantly, chronic treatment with SNC80 or ARM390 also produced agonist-selective adaptations (Pradhan et al. 2010) (Fig. 1). In this study, animals were treated daily for 5 days with SNC80 or ARM390. Chronic treatment with either agonist produced a complete tolerance to its anti-hyperalgesic effects in the CFA model of inflammatory pain. However, the mechanism of tolerance was different between the two agonists. Chronic SNC80 induced a generalized tolerance to all δOR-mediated behaviors, including pain-relief, locomotor stimulation, and anxiolysis. Under these circumstances, significant downregulation of δORs were detected throughout the peripheral and central nervous system, indicating widespread degradation of the receptor. In contrast, chronic ARM390 resulted only in analgesic tolerance; and δOR agonists continued to be effective at inducing centrally mediated behaviors (locomotion and anxiolysis). Interestingly, the number, coupling, and cell surface localization of the δOR was maintained. Tolerance to the pain-relieving effects of ARM390 appeared to be due to uncoupling of the δOR from second messenger signaling cascades (voltage dependent Ca^{2+} channels) within the dorsal root ganglia (Pradhan et al. 2010). Together, these results show that the distinct internalization properties of SNC80 and ARM390 have important behavioral implications following both acute and chronic treatment, and that non-internalizing agonists may be better for the development of centrally mediated pathologies.

More recent evidence using arrestin knockout mice indicates that high- and low-internalizing δOR agonists may also differentially recruit arrestin isoforms (Table 1). Arrestins are major mediators of receptor trafficking and intracellular signaling cascades and are a significant source of ligand bias (Reiter et al. 2012). The high-internalizing δ agonist, SNC80, appears to preferentially recruit arrestin 2 vs. arrestin 3, resulting in δOR desensitization and acute tolerance. Thus, knockout of arrestin 2 results in increased potency and decreased acute tolerance to high-internalizing δ agonists (Mittal et al. 2013; Pradhan et al. 2016). In contrast, low-internalizing δ agonists, such as ARM390 and JNJ20788560, preferentially recruit arrestin 3, which facilitates the rate of receptor resensitization thus encouraging subsequent receptor reactivity. In this case, knockout of arrestin 3 resulted in increased tolerance to low-internalizing δ agonists, which corresponded with reduced rates of δOR resensitization. Further, BRET analysis revealed that the δOR was in preengaged complexes with arrestin 3 at the cell membrane, and low-internalizing agonists preferentially encouraged increased interaction between these molecules at the cell surface. Therefore, in the absence of arrestin 3,

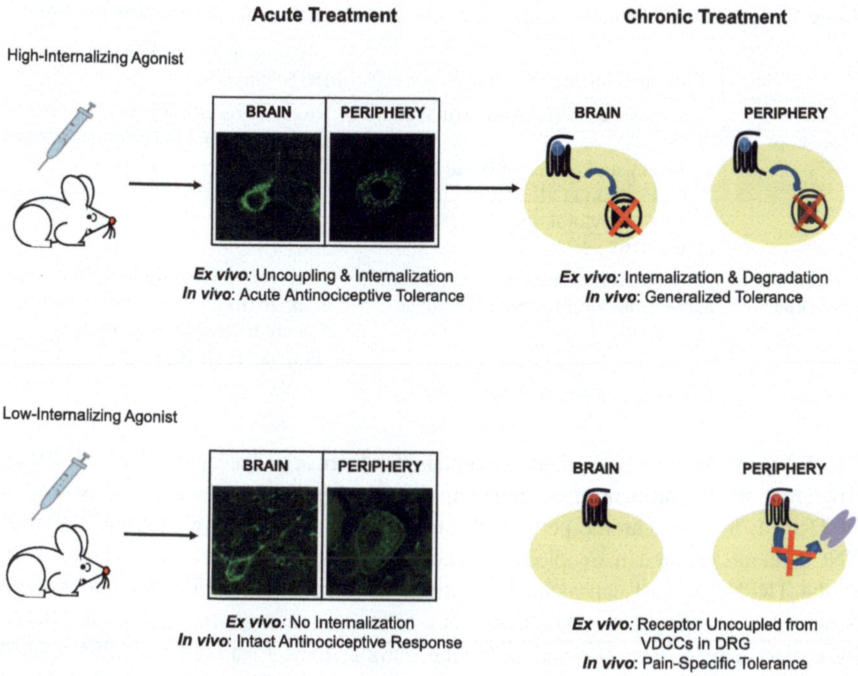

Fig. 1 Ligand-directed trafficking at the delta opioid receptor (δOR) has important behavioral implications. High-internalizing agonists, such as SNC80, produce robust internalization of DOR-eGFP following systemic injection, as shown in representative images from the brain (hippocampus) and periphery (dorsal root ganglia, DRG) of DOR-eGFP knock-in mice. Acutely, internalization of δOR correlates with receptor uncoupling and acute tolerance to the pain-relieving effects of SNC80. Chronic treatment with a high-internalizing agonist also produces δOR downregulation, resulting in tolerance to all delta agonist-mediated behaviors. In contrast, low-internalizing agonists like ARM390 do not produce detectable DOR-eGFP receptor trafficking, which correlates with δOR–G-protein coupling, and no acute tolerance. Chronic treatment with a low-internalizing agonist only results in a pain-specific tolerance, induced by uncoupling of δOR from voltage dependent Ca^{2+} channels (VDCCs) at the level of the DRGs. δOR-mediated behaviors regulated at the level of the brain, such as anxiolysis and locomotor stimulation, remain intact following chronic treatment with low-internalizing agonist (Pradhan et al. 2009, 2010)

low-internalizing agonist-bound receptor could now undergo a longer term desensitization, resulting in less receptor resensitization (Cahill et al. 2016; Pradhan et al. 2016). These results show how different ligands, which have similar behavioral effects (pain-relieving, anxiolytic, antimigraine), can regulate the δOR in vivo in completely different ways.

The ability of SNC80 and ARM390 to produce differential interactions between the δOR and arrestins has also been shown to have other downstream signaling consequences (Rowan et al. 2014). In rat trigeminal ganglia neurons, activation of the δOR by SNC80 increased the recruitment of arrestin 3 to the receptor, and away from the transient receptor potential vanilloid receptor type 1 (TRPV1). The association of

Table 1 Agonists differentially regulate interactions between δ opioid receptor and arrestins

	High-internalizing (SNC80)	Low-internalizing (ARM390 and JNJ20788560)
Arrestin 2 Knockout	↓ receptor desensitization (Mittal et al. 2013) ↑ efficacy and potency (Mittal et al. 2013; Pradhan et al. 2016) ↓ acute behavioral tolerance (Pradhan et al. 2016)	No effect on efficacy/potency or acute behavioral tolerance (Pradhan et al. 2016)
Arrestin 3 Knockout	No effect on receptor desensitization or acute behavioral tolerance (Pradhan et al. 2016)	↓ receptor resensitization (Pradhan et al. 2016) ↑ acute behavioral tolerance (Pradhan et al. 2016)

Arrestin 2 = β-arrestin 1; Arrestin 3 = β-arrestin 2

TRPV1 with arrestin 3 is required to maintain the desensitized receptor (Por et al. 2012). Therefore, the impartment of this interaction induced by SNC80 led to the sensitization of TRPV1. In vivo, chronic peripherally restricted (in the hindpaw) administration of SNC80 in rats resulted in an opioid-induced hyperalgesic state, and enhanced sensitivity to the TRPV1 agonist capsaicin. In contrast, the association of TRPV1 and arrestin 3 was not attenuated by ARM390, and behavioral symptoms of opioid-induced hyperalgesia were not observed following chronic local treatment with this agonist (Rowan et al. 2014). The lack of interaction between ARM390 and arrestin 3 observed in this study is in contrast to the knockout mouse studies described above (Pradhan et al. 2016), and this difference could be due to differences in species, anatomical location, and/or route of drug administration (systemic vs. local). Nevertheless, this study again highlights the differences between δ agonists to regulate behavioral outcomes.

Ligand-specific effects for δOR agonists have also been observed in other behavioral paradigms. Recent publications have shown that ethanol withdrawal results in anxiety that is differentially alleviated by the δ agonists SNC80 and TAN67 (van Rijn et al. 2010; van Rijn et al. 2012). Both SNC80 and TAN67 significantly reduced alcohol withdrawal-induced anxiety in mice. However, SNC80 also increased overall ethanol consumption, an effect that was arrestin 3 dependent (Chiang et al. 2015), while TAN67 decreased ethanol consumption (van Rijn et al. 2010). Moreover, these two drugs had opposite effects in the induction of ethanol place preference; where SNC80 prevented the expression of ethanol place preference, TAN67 increased this preference (van Rijn et al. 2012). In view of these results, TAN67 has been postulated to augment the rewarding effects of lower doses of ethanol thus diminishing its consumption, while SNC80 might enhance the rewarding effects of alcohol resulting in increased alcohol consumption (Alongkronrusmee et al. 2016). This phenomenon indicates that agonist-specific activation of different signaling pathways results in a distinct modulation of ethanol-induced behaviors, and suggests that agonists like TAN67 would be more effective for the treatment of alcohol abuse disorders (Alongkronrusmee et al. 2016). Taken together, the in vivo data strongly indicate that differences in δ agonist-induced signaling and trafficking can have profound effects on behavioral responses.

4 Conclusions and Future Directions

Like many GPCRs, ligands for the δOR can produce highly distinct effects. This agonist-directed trafficking and signaling can have important behavioral consequences that need to be considered for the therapeutic development of δ agonists. For instance, behavioral studies looking at long-term tolerance would suggest that non-internalizing δ agonists may be better for the treatment of anxiety and depression, since central δOR function remains intact following chronic treatment. In addition, agonists that do not recruit arrestin 3 may be more desirable for alcohol use disorders. Furthermore, a major caveat to the development of δ agonists is that some, but not all, agonists also produce convulsions, an effect that is dependent on the activation of the δOR (Broom et al. 2002b; Chung et al. 2015). A better understanding of this ligand-specific effect would increase the likelihood of developing a successful δOR pharmacotherapy, and some preliminary studies have started to elucidate the mechanism of this agonist-selective effect. In conditional knockout mice lacking δORs in forebrain GABAergic neurons, the convulsant activity of SNC80 was abolished (Chung et al. 2015). These findings shed some light on the neuroanatomical site and neurotransmitter system involved in δOR-induced convulsions. Further, the finding that drugs such as SNC80 specifically recruit Kir3.1–3.2 channels also points to a potential mechanism of pro-convulsant δ agonists (Nagi et al. 2015). Future studies expanding upon this work will provide important insight on how different δ agonists produce profoundly different behavioral responses, potentially leading to in vitro screening tools to screen novel drug candidates.

Acknowledgements This work was supported by NIH-NIDA grants DA031243 and DA040688.

References

Allouche S, Polastron J, Hasbi A, Homburger V, Jauzac P (1999) Differential G-protein activation by alkaloid and peptide opioid agonists in the human neuroblastoma cell line SK-N-BE. Biochem J 342(Pt 1):71–78

Alongkronrusmee D, Chiang T, van Rijn RM (2016) Delta opioid pharmacology in relation to alcohol behaviors. In: Handbook of experimental pharmacology. doi:10.1007/164_2016_30

Alves ID, Salamon Z, Varga E, Yamamura HI, Tollin G, Hruby VJ (2003) Direct observation of G-protein binding to the human delta-opioid receptor using plasmon-waveguide resonance spectroscopy. J Biol Chem 278(49):48890–48897. doi:10.1074/jbc.M306866200

Alves ID, Ciano KA, Boguslavski V, Varga E, Salamon Z, Yamamura HI, Hruby VJ, Tollin G (2004) Selectivity, cooperativity, and reciprocity in the interactions between the delta-opioid receptor, its ligands, and G-proteins. J Biol Chem 279(43):44673–44682. doi:10.1074/jbc.M404713200

Audet N, Paquin-Gobeil M, Landry-Paquet O, Schiller PW, Pineyro G (2005) Internalization and Src activity regulate the time course of ERK activation by delta opioid receptor ligands. J Biol Chem 280(9):7808–7816. doi:10.1074/jbc.M411695200

Audet N, Gales C, Archer-Lahlou E, Vallieres M, Schiller PW, Bouvier M, Pineyro G (2008) Bioluminescence resonance energy transfer assays reveal ligand-specific conformational changes within preformed signaling complexes containing delta-opioid receptors and heterotrimeric G proteins. J Biol Chem 283(22):15078–15088

Audet N, Charfi I, Mnie-Filali O, Amraei M, Chabot-Dore AJ, Millecamps M, Stone LS, Pineyro G (2012) Differential association of receptor-Gbetagamma complexes with beta-arrestin2 determines recycling bias and potential for tolerance of delta opioid receptor agonists. J Neurosci 32(14):4827–4840. doi:10.1523/jneurosci.3734-11.2012

Bosier B, Hermans E (2007) Versatility of GPCR recognition by drugs: from biological implications to therapeutic relevance. Trends Pharmacol Sci 28(8):438–446

Bradbury FA, Zelnik JC, Traynor JR (2009) G protein independent phosphorylation and internalization of the delta opioid receptor. J Neurochem 109(5):1526–1535

Brandt MR, Furness MS, Rice KC, Fischer BD, Negus SS (2001) Studies of tolerance and dependence with the delta-opioid agonist SNC80 in rhesus monkeys responding under a schedule of food presentation. J Pharmacol Exp Ther 299(2):629–637

Broom DC, Jutkiewicz EM, Folk JE, Traynor JR, Rice KC, Woods JH (2002a) Convulsant activity of a non-peptidic delta-opioid receptor agonist is not required for its antidepressant-like effects in Sprague-Dawley rats. Psychopharmacology (Berl) 164(1):42–48

Broom DC, Nitsche JF, Pintar JE, Rice KC, Woods JH, Traynor JR (2002b) Comparison of receptor mechanisms and efficacy requirements for delta-agonist-induced convulsive activity and antinociception in mice. J Pharmacol Exp Ther 303(2):723–729

Cahill CM, Walwyn W, Taylor AM, Pradhan AA, Evans CJ (2016) Allostatic mechanisms of opioid tolerance beyond desensitization and downregulation. Trends Pharmacol Sci 37(11):963–976. doi:10.1016/j.tips.2016.08.002

Charfi I, Audet N, Bagheri Tudashki H, Pineyro G (2015) Identifying ligand-specific signalling within biased responses: focus on delta opioid receptor ligands. Br J Pharmacol 172(2):435–448. doi:10.1111/bph.12705

Charles A, Pradhan AA (2016) Delta-opioid receptors as targets for migraine therapy. Curr Opin Neurol 29(3):314–319. doi:10.1097/wco.0000000000000311

Chiang T, Sansuk K, van Rijn RM (2015) Beta-arrestin 2 dependence of delta opioid receptor agonists is correlated with alcohol intake. Br J Pharmacol. doi:10.1111/bph.13374

Chu Sin Chung P, Boehrer A, Stephan A, Matifas A, Scherrer G, Darcq E, Befort K, Kieffer BL (2014) Delta opioid receptors expressed in forebrain GABAergic neurons are responsible for SNC80-induced seizures. Behav Brain Res 278C:429–434. doi:10.1016/j.bbr.2014.10.029

Chung PC, Boehrer A, Stephan A, Matifas A, Scherrer G, Darcq E, Befort K, Kieffer BL (2015) Delta opioid receptors expressed in forebrain GABAergic neurons are responsible for SNC80-induced seizures. Behav Brain Res 278:429–434. doi:10.1016/j.bbr.2014.10.029

Comer SD, Hoenicke EM, Sable AI, McNutt RW, Chang KJ, De Costa BR, Mosberg HI, Woods JH (1993) Convulsive effects of systemic administration of the delta opioid agonist BW373U86 in mice. J Pharmacol Exp Ther 267(2):888–895

Costa-Neto CM, Parreiras ESLT, Bouvier M (2016) A pluridimensional view of biased agonism. Mol Pharmacol 90(5):587–595. doi:10.1124/mol.116.105940

Do Carmo GP, Folk JE, Rice KC, Chartoff E, Carlezon WA Jr, Negus SS (2009) The selective non-peptidic delta opioid agonist SNC80 does not facilitate intracranial self-stimulation in rats. Eur J Pharmacol 604(1–3):58–65

Fenalti G, Giguere PM, Katritch V, Huang XP, Thompson AA, Cherezov V, Roth BL, Stevens RC (2014) Molecular control of delta-opioid receptor signalling. Nature 506(7487):191–196. doi:10.1038/nature12944

Filliol D, Ghozland S, Chluba J, Martin M, Matthes HW, Simonin F, Befort K, Gaveriaux-Ruff C, Dierich A, LeMeur M, Valverde O, Maldonado R, Kieffer BL (2000) Mice deficient for delta- and mu-opioid receptors exhibit opposing alterations of emotional responses. Nat Genet 25 (2):195–200

Galandrin S, Oligny-Longpre G, Bouvier M (2007) The evasive nature of drug efficacy: implications for drug discovery. Trends Pharmacol Sci 28(8):423–430

Gallantine EL, Meert TF (2005) A comparison of the antinociceptive and adverse effects of the mu-opioid agonist morphine and the delta-opioid agonist SNC80. Basic Clin Pharmacol Toxicol 97(1):39–51

He X, Sandhu HK, Yang Y, Hua F, Belser N, Kim DH, Xia Y (2013) Neuroprotection against hypoxia/ischemia: delta-opioid receptor mediated cellular/molecular events. Cell Mol Life Sci 70(13):2291–2303. doi:10.1007/s00018-012-1167-2
Henry AG, White IJ, Marsh M, von Zastrow M, Hislop JN (2011) The role of ubiquitination in lysosomal trafficking of delta-opioid receptors. Traffic 12(2):170–184
Hruby VJ, Alves I, Cowell S, Salamon Z, Tollin G (2010) Use of plasmon waveguide resonance (PWR) spectroscopy for examining binding, signaling and lipid domain partitioning of membrane proteins. Life Sci 86(15–16):569–574. doi:10.1016/j.lfs.2009.02.027
Hudzik TJ, Howell A, Payza K, Cross AJ (2000) Antiparkinson potential of delta-opioid receptor agonists. Eur J Pharmacol 396(2–3):101–107
Husain S, Abdul Y, Potter DE (2012) Non-analgesic effects of opioids: neuroprotection in the retina. Curr Pharm Des 18(37):6101–6108
Jutkiewicz EM, Baladi MG, Folk JE, Rice KC, Woods JH (2006) The convulsive and electroencephalographic changes produced by nonpeptidic delta-opioid agonists in rats: comparison with pentylenetetrazol. J Pharmacol Exp Ther 317(3):1337–1348
Kenakin T (2004) Principles: receptor theory in pharmacology. Trends Pharmacol Sci 25(4):186–192
Kenakin TP (2012) Biased signalling and allosteric machines: new vistas and challenges for drug discovery. Br J Pharmacol 165(6):1659–1669
Lagerstrom MC, Schioth HB (2008) Structural diversity of G protein-coupled receptors and significance for drug discovery. Nat Rev Drug Discov 7(4):339–357
Lecoq I, Marie N, Jauzac P, Allouche S (2004) Different regulation of human delta-opioid receptors by SNC-80 [(+)-4-[(alphaR)-alpha-((2S,5R)-4-allyl-2,5-dimethyl-1-piperazinyl)-3-methoxybenzyl]-N,N-diethylbenzamide] and endogenous enkephalins. J Pharmacol Exp Ther 310(2):666–677
Marie N, Lecoq I, Jauzac P, Allouche S (2003) Differential sorting of human delta-opioid receptors after internalization by peptide and alkaloid agonists. J Biol Chem 278(25):22795–22804
Maslov LN, Naryzhnaia NV, Tsibulnikov SY, Kolar F, Zhang Y, Wang H, Gusakova AM, Lishmanov YB (2013) Role of endogenous opioid peptides in the infarct size-limiting effect of adaptation to chronic continuous hypoxia. Life Sci 93(9–11):373–379. doi:10.1016/j.lfs.2013.07.018
Mittal N, Roberts K, Pal K, Bentolila LA, Fultz E, Minasyan A, Cahill C, Pradhan A, Conner D, DeFea K, Evans C, Walwyn W (2013) Select G-protein-coupled receptors modulate agonist-induced signaling via a ROCK, LIMK, and beta-arrestin 1 pathway. Cell Rep 5(4):1010–1021. doi:10.1016/j.celrep.2013.10.015
Molinari P, Vezzi V, Sbraccia M, Gro C, Riitano D, Ambrosio C, Casella I, Costa T (2010) Morphine-like opiates selectively antagonize receptor-arrestin interactions. J Biol Chem 285 (17):12522–12535
Moon HE, Cavalli A, Bahia DS, Hoffmann M, Massotte D, Milligan G (2001) The human delta opioid receptor activates G(i1)alpha more efficiently than G(o1)alpha. J Neurochem 76(6):1805–1813
Morse M, Sun H, Tran E, Levenson R, Fang Y (2013) Label-free integrative pharmacology on-target of opioid ligands at the opioid receptor family. BMC Pharmacol Toxicol 14:17. doi:10.1186/2050-6511-14-17
Nagi K, Charfi I, Pineyro G (2015) Kir3 channels undergo arrestin-dependant internalization following delta opioid receptor activation. Cell Mol Life Sci. doi:10.1007/s00018-015-1899-x
Negus SS, Butelman ER, Chang KJ, DeCosta B, Winger G, Woods JH (1994) Behavioral effects of the systemically active delta opioid agonist BW373U86 in rhesus monkeys. J Pharmacol Exp Ther 270(3):1025–1034
Negus SS, Gatch MB, Mello NK, Zhang X, Rice K (1998) Behavioral effects of the delta-selective opioid agonist SNC80 and related compounds in rhesus monkeys. J Pharmacol Exp Ther 286 (1):362–375
Perrine SA, Hoshaw BA, Unterwald EM (2006) Delta opioid receptor ligands modulate anxiety-like behaviors in the rat. Br J Pharmacol 147(8):864–872
Pierce KL, Premont RT, Lefkowitz RJ (2002) Seven-transmembrane receptors. Nat Rev Mol Cell Biol 3(9):639–650

Poole DP, Pelayo JC, Scherrer G, Evans CJ, Kieffer BL, Bunnett NW (2011) Localization and regulation of fluorescently labeled delta opioid receptor, expressed in enteric neurons of mice. Gastroenterology 141(3):982–991

Por ED, Bierbower SM, Berg KA, Gomez R, Akopian AN, Wetsel WC, Jeske NA (2012) Beta-Arrestin-2 desensitizes the transient receptor potential vanilloid 1 (TRPV1) channel. J Biol Chem 287(44):37552–37563. doi:10.1074/jbc.M112.391847

Pradhan AA, Becker JA, Scherrer G, Tryoen-Toth P, Filliol D, Matifas A, Massotte D, Gaveriaux-Ruff C, Kieffer BL (2009) In vivo delta opioid receptor internalization controls behavioral effects of agonists. PLoS One 4(5):e5425

Pradhan AA, Walwyn W, Nozaki C, Filliol D, Erbs E, Matifas A, Evans C, Kieffer BL (2010) Ligand-directed trafficking of the delta-opioid receptor in vivo: two paths toward analgesic tolerance. J Neurosci 30(49):16459–16468

Pradhan AA, Befort K, Nozaki C, Gaveriaux-Ruff C, Kieffer BL (2011) The delta opioid receptor: an evolving target for the treatment of brain disorders. Trends Pharmacol Sci 32(10):581–590

Pradhan AA, Smith ML, Kieffer BL, Evans CJ (2012) Ligand-directed signalling within the opioid receptor family. Br J Pharmacol 167(5):960–969

Pradhan AA, Smith ML, Zyuzin J, Charles A (2014) Delta-opioid receptor agonists inhibit migraine-related hyperalgesia, aversive state and cortical spreading depression in mice. Br J Pharmacol 171 (9):2375–2384. doi:10.1111/bph.12591

Pradhan AA, Perroy J, Walwyn WM, Smith ML, Vicente-Sanchez A, Segura L, Bana A, Kieffer BL, Evans CJ (2016) Agonist-specific recruitment of arrestin isoforms differentially modify delta opioid receptor function. J Neurosci 36(12):3541–3551. doi:10.1523/jneurosci.4124-15.2016

Rajagopal S, Rajagopal K, Lefkowitz RJ (2010) Teaching old receptors new tricks: biasing seven-transmembrane receptors. Nat Rev Drug Discov 9(5):373–386

Reiter E, Ahn S, Shukla AK, Lefkowitz RJ (2012) Molecular mechanism of beta-arrestin-biased agonism at seven-transmembrane receptors. Annu Rev Pharmacol Toxicol 52:179–197

Rice FL, Xie JY, Albrecht PJ, Acker E, Bourgeois J, Navratilova E, Dodick DW, Porreca F (2016) Anatomy and immunochemical characterization of the non-arterial peptidergic diffuse dural innervation of the rat and Rhesus monkey: implications for functional regulation and treatment in migraine. Cephalalgia. doi:10.1177/0333102416677051

Richard-Lalonde M, Nagi K, Audet N, Sleno R, Amraei M, Hogue M, Balboni G, Schiller PW, Bouvier M, Hebert TE, Pineyro G (2013) Conformational dynamics of Kir3.1/Kir3.2 channel activation via delta-opioid receptors. Mol Pharmacol 83(2):416–428. doi:10.1124/mol.112.081950

Rowan MP, Szteyn K, Doyle AP, Gomez R, Henry MA, Jeske NA (2014) Beta-arrestin-2-biased agonism of delta opioid receptors sensitizes transient receptor potential vanilloid type 1 (TRPV1) in primary sensory neurons. Mol Pain 10:50. doi:10.1186/1744-8069-10-50

Saitoh A, Kimura Y, Suzuki T, Kawai K, Nagase H, Kamei J (2004) Potential anxiolytic and antidepressant-like activities of SNC80, a selective delta-opioid agonist, in behavioral models in rodents. J Pharmacol Sci 95(3):374–380

Saitoh A, Yoshikawa Y, Onodera K, Kamei J (2005) Role of delta-opioid receptor subtypes in anxiety-related behaviors in the elevated plus-maze in rats. Psychopharmacology (Berl) 182 (3):327–334

Salamon Z, Cowell S, Varga E, Yamamura HI, Hruby VJ, Tollin G (2000) Plasmon resonance studies of agonist/antagonist binding to the human delta-opioid receptor: new structural insights into receptor-ligand interactions. Biophys J 79(5):2463–2474. doi:10.1016/s0006-3495(00)76489-7

Salamon Z, Hruby VJ, Tollin G, Cowell S (2002) Binding of agonists, antagonists and inverse agonists to the human delta-opioid receptor produces distinctly different conformational states distinguishable by plasmon-waveguide resonance spectroscopy. J Pept Res 60(6):322–328

Sanchez-Blazquez P, Garzon J (1998) Delta opioid receptor subtypes activate inositol-signaling pathways in the production of antinociception. J Pharmacol Exp Ther 285(2):820–827

Scherrer G, Tryoen-Toth P, Filliol D, Matifas A, Laustriat D, Cao YQ, Basbaum AI, Dierich A, Vonesh JL, Gaveriaux-Ruff C, Kieffer BL (2006) Knockin mice expressing fluorescent delta-opioid receptors uncover G protein-coupled receptor dynamics in vivo. Proc Natl Acad Sci U S A 103(25):9691–9696

Sorkin A, von Zastrow M (2009) Endocytosis and signalling: intertwining molecular networks. Nat Rev Mol Cell Biol 10(9):609–622

Tudashki HB, Robertson DN, Schiller PW, Pineyro G (2014) Endocytic profiles of delta-opioid receptor ligands determine the duration of rapid but not sustained cAMP responses. Mol Pharmacol 85(1):148–161. doi:10.1124/mol.113.089003

van Rijn RM, Brissett DI, Whistler JL (2010) Dual efficacy of delta opioid receptor-selective ligands for ethanol drinking and anxiety. J Pharmacol Exp Ther 335(1):133–139

van Rijn RM, Brissett DI, Whistler JL (2012) Distinctive modulation of ethanol place preference by delta opioid receptor-selective agonists. Drug Alcohol Depend 122(1–2):156–159. doi:10.1016/j.drugalcdep.2011.09.024

Vicente-Sanchez A, Segura L, Pradhan AA (2016) The delta opioid receptor tool box. Neuroscience 338:145–159. doi:10.1016/j.neuroscience.2016.06.028

Delta Opioid Receptor Expression and Function in Primary Afferent Somatosensory Neurons

Amaury François and Grégory Scherrer

Contents

Abstract

The functional diversity of primary afferent neurons of the dorsal root ganglia (DRG) generates a variety of qualitatively and quantitatively distinct somatosensory experiences, from shooting pain to pleasant touch. In recent years, the identification of dozens of genetic markers specifically expressed by subpopulations of DRG neurons has dramatically improved our understanding of this diversity and provided

A. François • G. Scherrer (✉)
Department of Anesthesiology, Perioperative and Pain Medicine, Stanford Neurosciences Institute, Stanford University School of Medicine, Palo Alto, CA, USA

Department of Molecular and Cellular Physiology, Stanford Neurosciences Institute, Stanford University School of Medicine, Palo Alto, CA, USA

Department of Neurosurgery, Stanford Neurosciences Institute, Stanford University School of Medicine, Palo Alto, CA, USA
e-mail: gs25@stanford.edu

E. M. Jutkiewicz (ed.), *Delta Opioid Receptor Pharmacology and Therapeutic Applications*, Handbook of Experimental Pharmacology 247, https://doi.org/10.1007/164_2017_58

the tools to manipulate their activity and uncover their molecular identity and function. Opioid receptors have long been known to be expressed by discrete populations of DRG neurons, in which they regulate cell excitability and neurotransmitter release. We review recent insights into the identity of the DRG neurons that express the delta opioid receptor (DOR) and the ion channel mechanisms that DOR engages in these cells to regulate sensory input. We highlight recent findings derived from DORGFP reporter mice and from in situ hybridization and RNA sequencing studies in wild-type mice that revealed DOR presence in cutaneous mechanosensory afferents eliciting touch and implicated in tactile allodynia. Mechanistically, we describe how DOR modulates opening of voltage-gated calcium channels (VGCCs) to control glutamatergic neurotransmission between somatosensory neurons and postsynaptic neurons in the spinal cord dorsal horn. We additionally discuss other potential signaling mechanisms, including those involving potassium channels, which DOR may engage to fine tune somatosensation. We conclude by discussing how this knowledge may explain the analgesic properties of DOR agonists against mechanical pain and uncovers an unanticipated specialized function for DOR in cutaneous mechanosensation.

Keywords

Delta opioid receptor • Excitability • Ion channels • Mechanosensation • Neuroanatomy • Neurotransmitter release • Pain • Primary afferent dorsal root ganglion neurons • Touch

1 Introduction: Diversity of Primary Afferent Somatosensory Neurons

Primary afferent somatosensory neurons detect and transmit information eliciting perception of temperature, touch, itch, pain, and positioning of body parts. Primary afferent somatosensory neurons are pseudounipolar neurons; their peripheral axons innervate organs in the periphery (e.g., skin, viscera, muscles, bones), and their central axons project onto neurons of the spinal cord dorsal horn and, in rare cases, brainstem dorsal column nuclei. Their somata are collected in the dorsal root ganglia (DRG, cervical to sacral segmental levels) and trigeminal ganglia (TG, head). Primary afferent somatosensory neurons shape the perception of our external and internal environments and are essential for our survival. The diversity of somatosensory stimuli translates into the outstanding functional and molecular diversity of primary afferent neurons. While primary afferent sensory neurons were categorized based on the size of their cell bodies (small, medium, large), myelination (unmyelinated, thinly, thickly myelinated), conduction velocity (slow C, fast Aδ, and Aβ fibers), activation threshold (low [touch] and high [pain]), and sensitivity (mechanical, thermal, chemical), genetic engineering techniques in mice have revealed in recent years dozens of classes, with specific molecular identities and contributions to somatosensation (Abraira and Ginty 2013; Delmas et al. 2011;

Basbaum et al. 2009; Lumpkin and Bautista 2005; Lewin and Moshourab 2004; Le Pichon and Chesler 2014).

These studies have transformed our understanding of somatosensation by identifying genetic markers to label and manipulate discrete neural populations. For example, small to medium DRG neurons were considered to encode pain and medium to large neurons to encode touch or proprioception. We now know that small unmyelinated neurons include mechanosensory neurons that express TH, VGLUT3, and TAFA4 and are essential to touch (Delfini et al. 2013; Li et al. 2011; Seal et al. 2009), and pruritoreceptors that express MrgA3 and are essential to itch (Liu et al. 2009), but not to pain. Similarly, it is clear today that large myelinated neurons are not limited to mechanosensory neurons encoding touch, but include multiple classes of nociceptors (Woodbury et al. 2008; Luo et al. 2007; Ghitani et al. 2017; Arcourt et al. 2017). This has important practical implications for experimental design and data interpretation. First, labeling or recording DRG/TG neurons does not mean that pain is studied; the results are just as likely to be relevant to itch, touch, or proprioception. Second, studying unidentified neurons means pooling data from neurons that are almost certainly functionally heterogeneous, even if their cell bodies have similar sizes. This is also exemplified by the TRP channels TRPV1 and TRPM8, which are markers expressed by two populations of small-diameter neurons that specifically respond to noxious heat and capsaicin versus cool and menthol, respectively (Basbaum et al. 2009; Peier et al. 2002).

Opioid receptors, including DOR, have long been known to be expressed in DRG and TG across species (Fields et al. 1980; Buzas and Cox 1997; Zhu et al. 1998). Since our understanding of the functional organization of primary afferent somatosensory neurons is far more advanced for mouse DRG neurons, we describe in this chapter newly acquired knowledge regarding the identity of DOR-expressing DRG neurons in rodents and DOR function in these cells.

2 DOR Distribution in Dorsal Root Ganglia

It is accepted within the field that several classes of DRG neurons express DOR; however, the precise identity of these neurons is disputed (see 2.3.). Because the expression pattern of a gene within DRG neuron subtypes defines its contribution to somatosensation, resolving the identity of DOR-expressing neurons is of critical importance to defining the therapeutic potential of peripheral DOR analgesics (see Conclusion). We first focus on novel knowledge gained from the use of recently identified genetic markers for subpopulations of cutaneous mechanosensitive neurons in DORGFP mice. We next discuss these results in the context of the pre-existing literature on DOR distribution in DRG.

2.1 DOR-Expressing DRG Neurons Are Predominantly A LTMRs

The characteristic feature of DOR distribution in DRG, compared to that of the mu opioid receptor (MOR), is DOR enrichment in neurons with large-diameter cell bodies and myelinated axons (Fig. 1). DRG neurons with myelinated axons express neurofilament 200 (NF200) and are born earlier than unmyelinated nociceptors during embryonic development (Abraira and Ginty 2013; Liu and Ma 2011; Lallemend and Ernfors 2012). In situ hybridization studies in rodents established

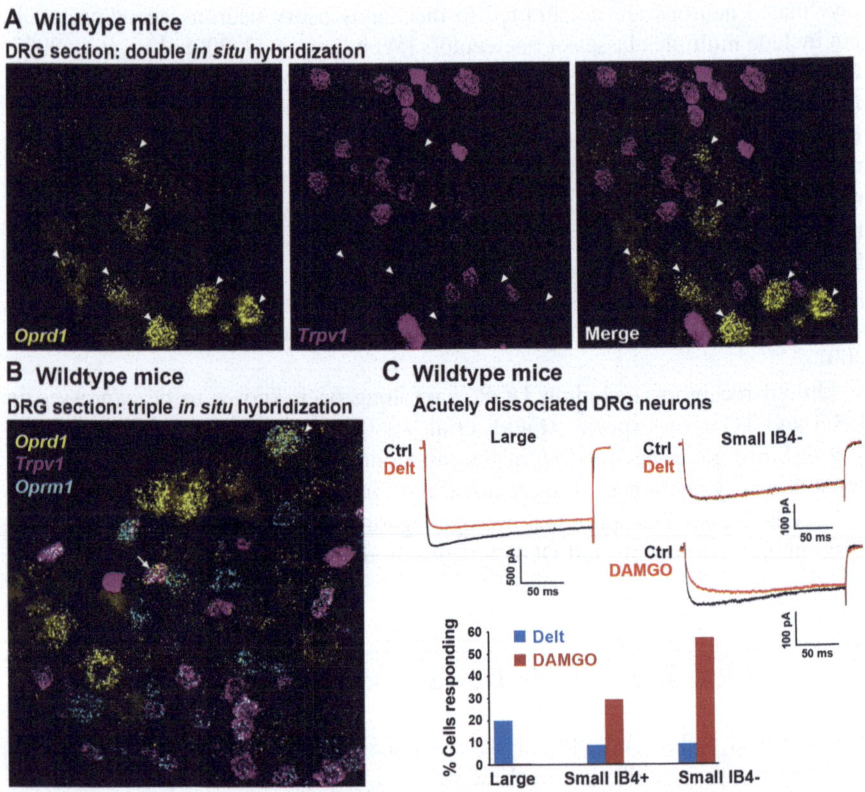

Fig. 1 DOR is predominantly expressed by large-diameter DRG neurons, not by TRPV1+ and MOR+ small-diameter C nociceptors. (**a**) Single mRNA molecule labeling in DRG sections from wild-type mice reveals that DOR (*Oprd1* gene) and TRPV1 are expressed by different populations of DRG neurons. Arrowheads indicate DOR+ neurons. (**b**) The great majority of TRPV1+ peptidergic C nociceptors co-express MOR (*Oprm1* gene). MOR is also found in other DRG neuron types, including large-diameter neurons in which it occasionally co-occurs with DOR (*arrowhead*). The *arrow* shows a rare example of a C nociceptor co-expressing MOR, TRPV1, and DOR. (**c**) Consistent with histological data, electrophysiological recordings show that the DOR agonist deltorphin II (Delt), in contrast to the MOR agonist DAMGO, predominantly inhibits calcium currents in large-diameter DRG neurons from wild-type mice (modified from Bardoni et al. 2014). Images in **a** and **b** were provided by Dong Wang, Scherrer laboratory

the presence and dense labeling of *Oprd1* mRNA in DRG early during development in large and NF200-immunoreactive DRG neurons (Zhu et al. 1998; Mennicken et al. 2003; Bardoni et al. 2014).

More recently, the development of tools including DORGFP knockin mice (Scherrer et al. 2006) and RNA sequencing (Usoskin et al. 2015; Wang et al. 2009; Kolodziejczyk et al. 2015) has enabled the identification and categorization of DOR-expressing (DOR+) large-diameter NF200+ DRG neurons. Co-immunolabeling studies using DORGFP reporter mice revealed that the majority of DORGFP+ NF200+ neurons express the neurotrophin receptors Ret and/or TrkC (~60%, (Bardoni et al. 2014)), which identify several classes of low-threshold mechanosensitive A fibers (A low-threshold mechanoreceptors, A LTMRs).

Among these, DORGFP is predominantly expressed by Ret+ TrkC+ A LTMRs that innervate hair. These neurons project to the skin where their DORGFP+ axons arborize densely and form circumferential endings around numerous hair follicles (Fig. 2a, c). A recent study uncovered the anatomical and physiological properties of this class of A LTMRs also known as Aβ Field receptors (Bai et al. 2015), showing that they are activated by light stroking of the skin and skin indentation (including in the noxious range). Note also that Aβ Field receptor afferents are among the largest cells in mammals. Their central axons not only synapse in the spinal cord dorsal horn but also extend collaterals that ascend the dorsal columns and terminate in the gracile and cuneate nuclei in the brainstem (Bai et al. 2015). At the lumbar level, their peripheral axon can innervate the distal aspects of the limbs. Injection of the B fragment of cholera toxin (CTB), a retrograde tracer, in the brainstem dorsal column nuclei of DORGFP mice results in the backlabeling of numerous DORGFP+ DRG neurons, confirming DOR expression in this class of DRG neuron (Bardoni et al. 2014).

Other classes of A LTMRs that frequently express DORGFP (Bardoni et al. 2014) are Merkel cell afferents (Lumpkin and Bautista 2005) and Meissner corpuscle afferents (Pare et al. 2001). Merkel cell afferents are Aβ LTMRs that express TrkC (Bai et al. 2015), respond to skin indentation, and adapt slowly to mechanical stimulation (SA class of A LTMRs), while Meissner corpuscle afferents are rapidly adapting (RA) Ret+ Aβ LTMRs (Luo et al. 2009) and are low frequency vibration detectors (Fig. 3).

Importantly, DOR is not restricted to A LTMRs, but is also expressed by high-threshold mechanosensitive A fibers (i.e., A HTMRs, myelinated nociceptors), most of which can be identified by co-expression of NF200 and the nociceptor markers CGRP or TrkA (the receptor for NGF). Thus, approximately 36% of DORGFP+ NF200+ DRG neurons co-express TrkA and CGRP. Interestingly, this population of DRG neurons frequently co-expresses DOR and MOR (Bardoni et al. 2014; Joseph and Levine 2010), as MOR is expressed by virtually all C and A peptidergic nociceptors. NF200+ TrkA/CGRP+ neurons include cutaneous Aδ mechanonociceptors forming epidermal free nerve endings (Lawson et al. 2008), but also myelinated nociceptors innervating other tissues such as muscle afferents (Jankowski et al. 2013; Alvarez et al. 1992). Thus far the molecular identity of DOR+ myelinated DRG neurons suggests that proprioceptors, which innervate

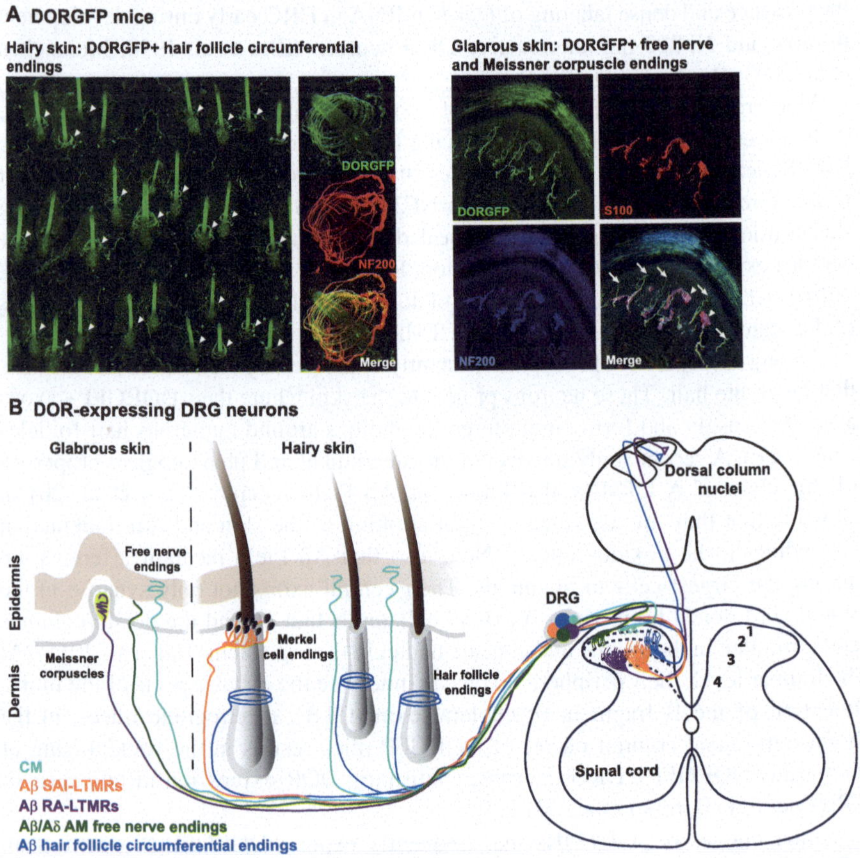

Fig. 2 DOR is expressed by several classes of cutaneous mechanosensitive neurons. (**a**) DORGFP mice reveal DOR expression at the peripheral terminals of A low-threshold mechanoreceptors (LTMRs) forming circumferential endings around hair follicles (arrowheads in *left panel*). DOR is also expressed by DRG neurons innervating the glabrous skin, particularly by C nonpeptidergic nociceptors that co-express MrgprD and end as free nerve endings in the stratum granulosum (*arrows* in *right panel*) and by A LTMRs forming Meissner corpuscles (*arrowheads* in *right panel*). Modified from Bardoni et al. (2014). (**b**) Schematic summarizing the peripheral and central projection patterns of DOR+ cutaneous DRG neurons. *CM* C mechanonociceptors, *SA* slowly adapting, *RA* rapidly adapting

muscle spindles and tendons and encode positioning of body parts in space, do not express DOR (Bardoni et al. 2014) as they lack co-expression of parvalbumin and Runx3 (de Nooij et al. 2013).

This characterization of DOR+ DRG neurons in DORGFP reporter mice was recently confirmed in wild-type mice by an unbiased large-scale single-cell RNA sequencing study (Usoskin et al. 2015) and functional assays in spinal cord slices or acutely dissociated DRG neurons (Bardoni et al. 2014). Note that the single-cell RNA sequencing dataset from Usoskin et al. (2015) is publicly available at http://

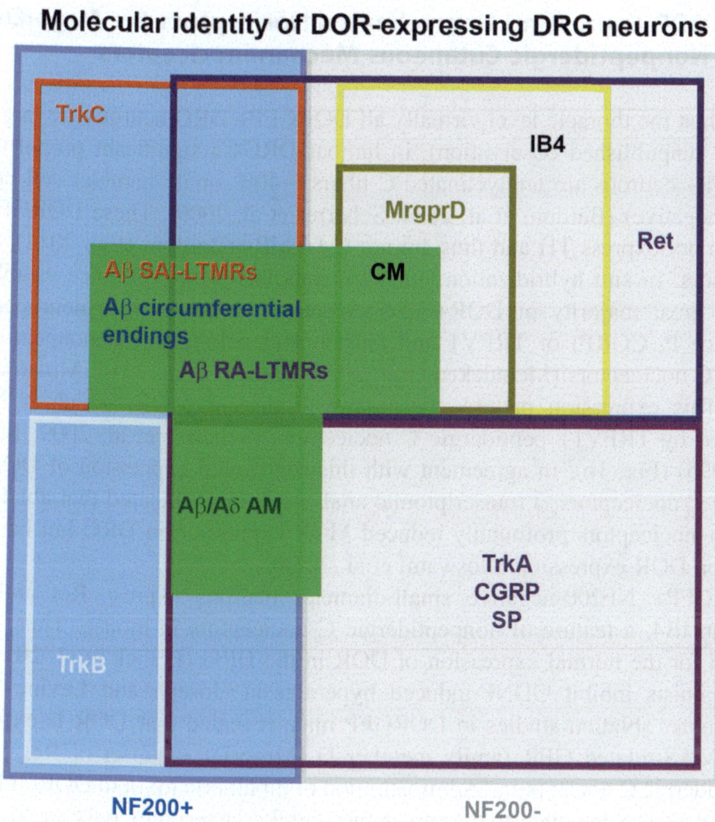

Fig. 3 Functional and genetic diversity of A LTMRs and nociceptors expressing DOR. Multiple classes of molecularly heterogeneous DRG neurons express DOR. These different populations include myelinated nociceptors that express TrkA, as well as A LTMRs expressing other neurotrophin receptors such as Ret and TrkC. Ret is also expressed by the C nociceptors that express DOR, which correspond to a subpopulation of IB4-binding MrgprD+ mechanosensitive afferents

linnarssonlab.org/drg/. This webpage from the Linnarsson laboratory website contains a search engine that displays a scatter plot of expression in DRG neuron subpopulations for any gene of interest. Thus, searches for *Oprd1*, *Oprm1*, and *Trpv1* genes show that *Oprm1* and *Trpv1* transcripts are enriched in neurofilament-negative DRG neurons that express peptides such as CGRP and substance P (i.e., unmyelinated peptidergic nociceptors), while *Oprd1* transcripts are found in neurofilament-positive mechanosensory DRG neurons, consistent with neuroanatomical and electrophysiological studies in DORGFP and wild-type mice.

2.2 DOR-Expressing Unmyelinated Nociceptors Are MrgprD+ Nonpeptidergic Cutaneous Mechanonociceptors

Although at the thoracic level virtually all DORGFP+ DRG neurons are large and NF200+ (unpublished observation), in lumbar DRGs a significant proportion of DORGFP+ neurons are unmyelinated C fibers (~40% small-diameter cell bodies, NF200-negative) (Bardoni et al. 2014; Scherrer et al. 2009). These DORGFP+ C fibers do not express TH and thus are not C LTMRs (Bardoni et al. 2014), but C nociceptors. In situ hybridization studies in rats and wild-type mice established that the great majority of DOR+ C nociceptors do not express neuropeptides (substance P, CGRP) or TRPV1 and thus mostly belong to the nonpeptidergic class of C nociceptors (Mennicken et al. 2003; Bardoni et al. 2014; Minami et al. 1995). This expression pattern contrasts with that of MOR, which is highly expressed by TRPV1+ peptidergic C nociceptors (Scherrer et al. 2009; Minami et al. 1995) (Fig. 1b). In agreement with this segregated expression of DOR and MOR in C nociceptors, a transcriptomic analysis in rats indicated that ablation of TRPV1+ nociceptors profoundly reduced MOR expression in DRG but had little impact on DOR expression (Goswami et al. 2014).

DORGFP+ NF200-negative small-diameter neurons express Ret and bind the lectin IB4, a feature of nonpeptidergic C nociceptors in mouse. Thus, Ret is essential for the normal expression of DOR in the DRG (Franck et al. 2011) and DOR agonists inhibit GDNF-induced hyperalgesia (Joseph and Levine 2010). Furthermore, ablation studies in DORGFP mice revealed that DOR is expressed by the MAS-related GPR family member D (MrgprD) + subset of Ret + IB4+ nonpeptidergic C nociceptors. Administration of diphtheria toxin to DORGFP mice crossed with mutant mice that express the diphtheria receptor only in MrdprD+ neurons resulted in an almost complete loss of DOR+ small-diameter DRG neurons (Bardoni et al. 2014). MrdprD+ DRG neurons are cutaneous C nociceptors that respond to punctate mechanical stimuli and are critical to acute mechanical pain (Zylka et al. 2005; Dussor et al. 2008; Cavanaugh et al. 2009; Rau et al. 2009). Consistent with a function in cutaneous mechanonociception, DORGFP+ MrgprD+ nociceptors densely innervate the glabrous skin, forming epidermal free nerve endings that reach the stratum granulosum (Fig. 2b), in contrast to peptidergic nociceptors that terminate less superficially. Studies of visceral innervation thus far indicate that in contrast to TRPV1+ MOR+ peptidergic C nociceptors, which densely innervate viscera (Cavanaugh et al. 2011; De Schepper et al. 2008; Jones et al. 2005), DORGFP+ afferent terminals are very rare in visceral tissue (Scherrer et al. 2009).

2.3 Controversy Regarding DOR Distribution in DRG Neurons: Historical Considerations

The controversy regarding the expression pattern of DOR in DRG stems from three major sources. First, there is disagreement regarding which methods most accurately identify DOR-expressing neurons (antibodies, in situ hybridization, radioligand binding, RNA sequencing, DORGFP reporter mouse). Second, there are significant differences in the functional organization of DRG neurons in mice and rats. Third, there is disagreement regarding the interpretation of functional studies that used intrathecal opioid ligands or knockout mice to probe DOR function in pain at the DRG/spinal level. We focus our discussion below on the first point.

Following cloning of the *Oprd1* gene in 1992 (Kieffer et al. 1992; Evans et al. 1992) and the determination of the DOR amino acid sequence, anti-DOR antibodies could be generated. Immunohistochemistry became the method of choice to investigate DOR distribution in tissues. In 1993, Dado, Elde, and colleagues reported the production of a rabbit antiserum generated by injection of a synthetic peptide consisting of the amino acids 3–17 of the DOR sequence (i.e., amino-terminal, extracellular). During the following decade, the immunoreactivity generated by this antibody (Ab^{3-17}-ir) was used extensively to analyze DOR distribution in DRG and its subcellular localization in primary afferent neurons (Dado et al. 1993; Zhang et al. 1998; Riedl et al. 2009; Overland et al. 2009; Bao et al. 2003; Guan et al. 2005). Early studies reported that Ab^{3-17}-ir was mostly associated with small-diameter DRG neurons co-expressing substance P and CGRP (Dado et al. 1993; Zhang et al. 1998). These reports strongly influenced research directions in the following decade and led to the idea that DOR would be predominantly expressed by peptidergic C nociceptors. Ab^{3-17} was also used in studies that proposed that DOR agonists can promote the insertion of DOR into the DRG neuron plasma membrane, via a direct substance P-DOR interaction in large dense-core vesicles [(Bao et al. 2003; Guan et al. 2005), reviewed by (Gendron et al. 2015)]. At the time, *Oprd1* knockout mice were not available (Filliol et al. 2000; Zhu et al. 1999), and the pre-adsorption test was instead used to test Ab^{3-17}-ir specificity. Preincubation of the Ab^{3-17} with the 3–17 synthetic peptide eliminated Ab^{3-17}-ir, confirming the high affinity of the synthetic peptide for the antibody. It is clear, however, that the pre-adsorption test does not demonstrate that the immunoreactivity pattern generated by Ab^{3-17} in tissues results from recognition of DOR (Holmseth et al. 2012).

An important misconception is that the controversy originates from observations made using the DORGFP reporter mouse. This view is factually incorrect: it was apparent that Ab^{3-17}-ir pattern in DRG and spinal cord did not match DOR distribution defined with other techniques, well before the generation of DORGFP mice in 2006. In 1995, Minami et al. using double in situ hybridization studies established that mRNAs encoding DOR and substance P were localized in distinct DRG neurons (Minami et al. 1995), consistent with Mennicken et al. later observation that *Oprd1* mRNA was preferentially found in NF200+ and large-diameter DRG neurons (Mennicken et al. 2003). Ab^{3-17}-ir pattern in the CNS also did not match the known distribution of *Oprd1* mRNA or the binding pattern of DOR radioligands (Arvidsson et al. 1995; Mansour et al. 1987; Kitchen et al. 1997). In the spinal cord, Ab^{3-17}-ir was restricted to the terminals of peptidergic DRG neurons in the superficial dorsal horn, while *Oprd1* mRNA (Mennicken et al.

2003; Cahill et al. 2001) and DOR radioligand binding (Mennicken et al. 2003) are present throughout the dorsal and ventral horns, consistent with electrophysiological recordings documenting the existence of DOR-expressing spinal neurons (Eckert and Light 2002). Note also that numerous other anti-DOR antibodies were produced and generated different i.r. patterns compared to Ab^{3-17}-ir. Some of these antibodies may have recognized DOR based on the observation that their i.r. pattern resembled DOR radioligand binding pattern (e.g., see Cahill et al. 2001). In the absence of specificity control in knockout mice for each serum and lot used, antibody specificity remains difficult to estimate.

The generation of DORGFP mice drew attention to and revived the pre-existing controversy, because the DORGFP expression pattern does not match the Ab^{3-17}-ir pattern, but is consistent with *Oprd1* mRNA distribution or DOR radioligand binding pattern in wild-type mice throughout the nervous system (Bardoni et al. 2014; Scherrer et al. 2006; Scherrer et al. 2009). DOR KO mice allowed examination of these inconsistencies in the 2000s and showed that binding of radioligand was lost in knockout mice (Bardoni et al. 2014; Scherrer et al. 2009; Goody et al. 2002), validating the specificity of this approach. By contrast, Ab^{3-17}-ir was intact in DOR knockout mice (Scherrer et al. 2009), indicating that this antibody recognizes a molecule other than DOR. Other studies reported decreased Ab^{3-17}-ir in DOR knockout mice (Overland et al. 2009; Wang et al. 2010), a puzzling result given the absence of *Oprd1* mRNA and DOR binding sites in the great majority of Ab^{3-17}-immunoreactive neurons and regions in the nervous system in wild-type mice.

As mentioned initially, species differences have also contributed to the controversy regrading DOR expression pattern (and co-expression with MOR) in DRG. TRPV1+ CGRP+ DRG neurons and IB4-binding nonpeptidergic nociceptors are almost completely nonoverlapping populations in mice, but substantially overlap in rats (Price and Flores 2007). Regarding functional assays, intrathecal injections of DOR ligands may also activate receptors expressed by spinal neurons, or present on brainstem descending axons, preventing definitive conclusions about DOR in DRG neurons. Virtually all DOR agonists can to some extent bind and activate MOR, particularly at high doses and given broader MOR expression in DRG neurons. Often, naltrindole is used to provide evidence of DOR involvement; however, naltrindole can also block MOR-mediated responses. For example, naltrindole co-injection can block the increase in latency for tail withdrawal caused by intrathecal DAMGO (unpublished observation). Data on DOR expression and function in cultured DRG neurons can also be difficult to interpret. DRG somatosensory neurons are tuned to respond to their extracellular environment and injuries. Axotomy (i.e., occurring during dissection) and the composition of the culture medium (e.g., serum, growth factors, neurotrophins such as NGF) profoundly alter gene expression (e.g., ion channels and GPCRs defining excitability and responsiveness) and transform DRG neuron molecular identity. To our knowledge, no study has rigorously characterized how the culturing process and the factors present in the medium impact each class of DRG neuron, including opioid receptor expression and signaling, compared to the in vivo physiologic condition. While cultured DRG neurons can be very useful to study molecular mechanisms such as gating of ion channels (i.e., TRPV1 gating by capsaicin) or ligand binding to GPCRs, whether they properly model physiological receptor expression and function in DRG neurons in vivo is uncertain. To address this aim, it may be preferable to use acutely dissociated DRG neurons within hours of dissection. It is clear that all methods used to study DOR distribution have limitations, including DORGFP mice. While the knockin approach is likely to faithfully report *Oprd1* promoter activity and identify DOR-expressing cells, the insertion/addition of the GFP sequence/tag may modify certain aspects of mRNA processing and protein function. The use of multiple approaches will likely be necessary to address this controversy. Specifically, the emergence of novel techniques for quantifying gene expression with unprecedented sensitivity in wild-type mouse or rat DRG, by RNA sequencing and combinatorial detection of single mRNA molecules for multiple genes (Fig. 1a, b), will be particularly useful, given that protein function in a cell implies mRNA presence in its cell body.

3 Molecular and Physiological Consequences of DOR Activation in DRG Neurons

Very few studies have investigated DOR signaling in DRG neurons in vivo or in acutely dissociated DRG neurons that endogenously express DOR, such as A LTMRs. We review these studies here, along with those using cultured DRG neurons and DRG neurons transfected to express DOR. We focus on ion channel mechanisms relevant for DOR control of DRG neuron excitability and neurotransmitter release (Fig. 4). We recommend other excellent review articles and book chapters describing general mechanisms of DOR signaling via G protein and adenylate cyclase or β–arrestin pathways that have been described in other cell types and might also occur in somatosensory neurons (Gendron et al. 2015; Dang and Christie 2012; Pradhan et al. 2011; Lamberts and Traynor 2013; Zaki et al. 1996; Cahill et al. 2016; Gendron et al. 2016; Charfi et al. 2015; Fujita et al. 2014; Pradhan et al. 2012; Georgoussi 2015; Al-Hasani and Bruchas 2011; Gaveriaux-Ruff and Kieffer 2011; Zollner and Stein 2007).

3.1 Inhibition of Voltage-Gated Calcium Channels

Voltage-gated calcium channels (VGCCs) are activated by depolarization of the plasma membrane, resulting in calcium flow into the cell (i.e., inward current). While this increase in intracellular calcium can contribute to action potential (AP) generation (Fatt and Katz 1953), its main purpose in mammals is to couple electrical excitation to calcium-dependent intracellular mechanisms, including synaptic vesicle fusion for neurotransmitter release. VGCCs are composed of a main α subunit that forms the pore of the channel. This protein is comprised of four repeated domains that each contain six transmembrane segments (TM) and several

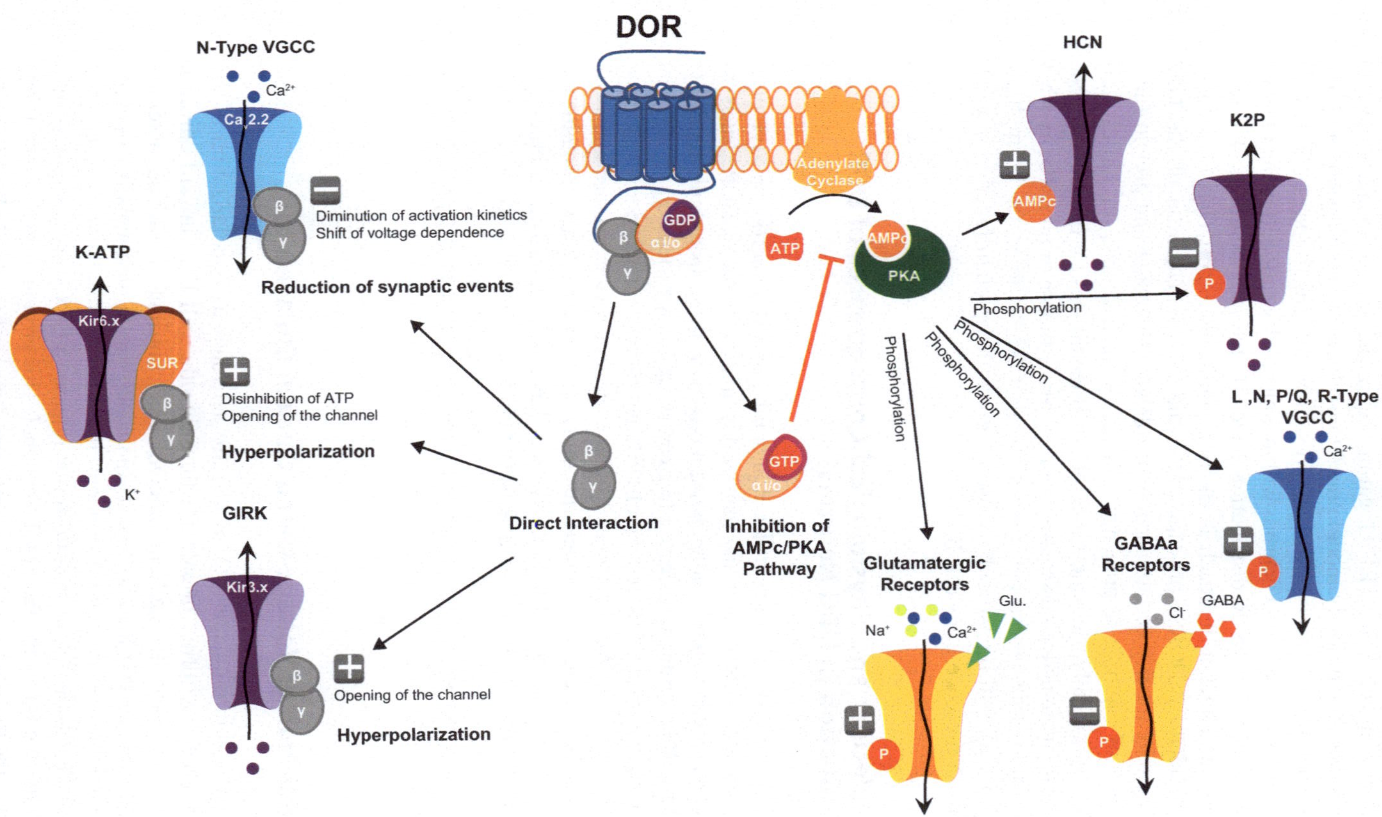

Fig. 4 Putative ion channel mechanisms engaged by DOR in DRG neurons. Binding of endogenous or exogenous DOR agonists activates G proteins. Subsequent separation of the G protein subunits triggers distinct signaling cascades: on the *left*, direct interaction of Gβγ with N-Type VGCCs leads to reduced calcium current amplitude and decreased neurotransmitter release, while direct interaction with GIRK and/or KATP channels increases potassium conductance, resulting in membrane hyperpolarization. On the right, inhibition of adenylate cyclase by Gαi/o reduces cAMP production and PKA activation, which alters the phosphorylation states of different receptors and ion channels and leads to decreased excitability

intracellular loops (IL) with residues for protein interactions and posttranslational modifications (Tanabe et al. 1987). The α subunit requires an additional intracellular β subunit and an extracellular α2δ subunit to form a functional VGCC (except for T-type VGCCs, which lack the β subunit). β and α2δ subunits modulate membrane expression, opening probability, and voltage dependence of the α subunit (Arikkath and Campbell 2003). Ten genes code for the α subunit and form three families of VGCCs (types) with distinct distributions and functions: L-type VGCCs (Cav1.1, Cav1.2, Cav1.3, Cav1.4); P/Q, N, and R-type VGCCs (Cav2.1, Cav2.2, and Cav2.3); and T-type VGCCs (Cav3.1, Cav3.2, Cav3.3) (Catterall 2011; Simms and Zamponi 2014; Zamponi 2016).

L-type, T-type, and most importantly, P/Q and N-type, VGCCs have been functionally described in DRG neurons of various species (Evans et al. 1996; Heinke et al. 2004; Murakami et al. 2001; Nowycky et al. 1985). Cav2.1 (P/Q) and Cav2.2 (N-type) appear to be particularly important for synaptic transmission between primary afferent somatosensory neurons and spinal cord neurons (Simms and Zamponi 2014). Cav2.1 is highly expressed in large-diameter DRG neurons, while Cav2.2 is preferentially found in small-diameter DRG neurons (Bell et al. 2004; Murali et al. 2015; Saegusa et al. 2001; Yusaf et al. 2001). Both Cav2.1 and Cav2.2 are localized in axon presynaptic terminals. Action potentials cause the opening of the channel and calcium influx at close proximity of the SNAP-SNARE complex. This local increase in calcium concentration triggers the fusion of synaptic vesicles with the plasma membrane for neurotransmitter release (Sudhof 2004). Approximately 45% and 15% of the total DRG VGCC current is thought to be conducted through Cav2.2 and Cav2.1, respectively (Wilson et al. 2000). The Cav2.2 α subunit possesses a crucial sequence for opioid receptor-mediated inhibition, in the intracellular loop (IL1) between the TM I and II. Following opioid receptor activation, dissociation, and binding of the βγ subunits of Gi/o proteins (Gβγ) to this site causes a shift in the voltage dependence of calcium channel activation to more positive membrane potentials and slows down the activation kinetics of the Cav2.2 current, resulting in reduced neurotransmitter release (Herlitze et al. 1997; Ikeda 1996). For opioid receptor-mediated inhibition to occur, opioid receptors and VGCCs need to be close to each other. Additionally, opioid receptor-mediated modulation of VGCCs is voltage dependent: depolarization (e.g., during a train of action potentials) causes Gβγ to dissociate from the channel and drastically reduces inhibition. Because IL1 is also important for interactions between Cav2.2 α and β subunits, Gβγ binding may displace Cav2.2 β due to steric hindrance, which also contributes to Gβγ inhibition of VGCCs (Buraei and Yang 2010). Note that other GPCRs, coupled to Gq or Gs proteins, can positively regulate VGCCs, including through the recruitment of kinases that phosphorylate the channel and counteract opioids' inhibition of calcium influx (Zamponi et al. 1997). Cav2.2 is important not only for glutamate release but also for the release of calcitonin gene-related peptide (CGRP) and substance P, two major pro-nociceptive neuropeptides (Evans et al. 1996; Brittain et al. 2011; Westenbroek et al. 1998).

While a wealth of studies has demonstrated MOR-meditated inhibition of VGCCs in DRG neurons (Schroeder et al. 1991; Moises et al. 1994; Wu et al. 2004; Schroeder and McCleskey 1993; Raingo et al. 2007; Jiang et al. 2013; Pan et al. 2008) particularly in small-diameter nociceptors, DOR coupling to VGCCs remains controversial. Thus, several studies reported that DOR agonists only minimally impact calcium current amplitude and kinetics in DRG neurons from naïve rodents (i.e., uninjured) (Walwyn et al. 2005; Pradhan et al. 2013; Brackley et al. 2016; Acosta and Lopez 1999; Wu et al. 2008). Another interpretation of these data, however, is that most of the cells patched in these studies did not belong to the class of DRG neurons that normally express DOR. Indeed, blind recording approaches (i.e., unidentified cells) are unlikely to randomly select DOR-expressing neurons in the cultures used. First, DOR is expressed by fewer neurons in DRG, compared to MOR, for example. Second, the methods used to culture DRG neurons generally select for small-diameter neurons, in particular peptidergic, when NGF, the TrkA agonist that promotes the survival and growth of this class of DRG neurons, is present in the culture medium. Because they are more sensitive to enzymatic and mechanical dissociation, and because factors necessary for their survival (GDNF, neurotrophin 3, neurotrophin 4) are rarely included in the medium, large-diameter neurons are most often rare in these cultures and not recorded. Consistent with this interpretation, Bardoni et al. reported, in acutely dissociated DRG neurons from wild-type mice, that the DOR agonist deltorphin II inhibits VGCCs both in a subset of large-diameter and small-diameter IB4-binding DRG neurons, but very rarely in IB4-negative (presumably peptidergic) small-diameter neurons. Furthermore, DOR agonists reduce the amplitude of EPSCs evoked in dorsal horn neurons by stimulation of primary afferents, in particular A fiber-mediated EPSCs in laminae III-V (Bardoni et al. 2014; Glaum et al. 1994; Fan et al. 1993; Rogers and Henderson 1990). Because the reduction in amplitude was accompanied by an increase in paired-pulse ratio, this result establishes that DOR inhibits neurotransmitter release from A fibers, presumably through inhibition of VGCCs at their terminals (Bardoni et al. 2014). Similarly, release of the endogenous DOR agonist enkephalin in the spinal cord reduces A fiber input, and suppression of enkephalin release induces mechanical pain (François et al. 2017), suggesting that endogenous DOR signaling also inhibits VGCCs presynaptically to gate mechanical pain. DORGFP fusion receptor, either endogenously expressed by acutely dissociated DRG neurons (Bardoni et al. 2014) or heterogeneously expressed in cultured neurons (Pettinger et al. 2013), also inhibited VGCCs following application of DOR agonist. Finally, studies that demonstrated that DOR agonists reduce CGRP release (Overland et al. 2009; Patwardhan et al. 2006) also support the idea that DOR inhibits VGCCs, regardless of the origin of this effect (i.e., small- versus large-diameter CGRP+ DRG neurons). Data reporting a decrease in substance P release by DOR agonists (Kouchek et al. 2013; Normandin et al. 2013; Zachariou and Goldstein 1996) are more difficult to interpret given that substance P, contrary to CGRP, is also expressed by spinal cord neurons (Gutierrez-Mecinas et al. 2014). DOR function can be upregulated in a variety of regions and conditions (see Gendron et al. 2015 for review). Of particular interest are the studies that showed increased DOR function and coupling to VGCCs in DRG neurons. Thus, inflammation and bradykinin can augment the efficacy of DOR agonists to inhibit VGCCs (Walwyn et al. 2005; Pradhan et al. 2013; Brackley et al. 2016; Mittal et al. 2013) or the proportion of DRG neurons in which this inhibitory effect is seen (Pettinger et al. 2013). The mechanisms underlying this gain in function are actively investigated, and recent studies suggest that receptor export and signaling molecules controlling receptor desensitization and trafficking such as GRK2 and β−arrestin 1 are involved (Brackley et al. 2016; Pettinger et al. 2013; Mittal et al. 2013; Gendron et al. 2006).

3.2 Activation of Inward-Rectifier Potassium Channels

Potassium channels represent a family of more than 80 genes that can be divided in six major classes: voltage-gated with six TM (Kv), inwardly rectifier with two TM (Kir), tandem pore domain with 4 TM (K2P), and calcium-activated BK, Sk, and IK channels (Yu and Catterall 2004). Among them, Kir potassium channels are particularly important for opioid modulation of cell excitability. Kirs are composed of four main α subunits that form the pore of the channel. Each α subunit comprises 2 TMs, and intracellular N- and C-terminals. Opening of Kir channels leads to potassium flowing out of the cell (i.e., outward current), which hyperpolarizes the membrane, reducing excitability and synaptic transmission (Hille 1992). In DRG neurons, two main families of Kir have been described: ATP-sensitive (KATP) and G protein-coupled inwardly rectifying K^+ (GIRK) channels.

KATP channels are gated by ATP (Kir6.x family) and are composed of four Kir6.1 and/or Kir6.2 subunits, supplemented by four SUR 1 and/or two (sulfonylurea receptor) subunits. The SUR subunits are 17 TM proteins, containing two nucleotide-binding sites that act as ATP sensors. At cellular physiological concentration, ATP binding to these sites results in inhibition of Kir6.x channel opening (Aguilar-Bryan et al. 1998). Gβγ proteins can also interact with these sites. Following activation of G proteins, Gβγ binds to SUR and decreases the inhibition by ATP, resulting in facilitation of Kir6.x channels opening (Wada et al. 2000). Immunohistochemical studies suggest that Kir6.2, SUR1, and SUR2 are expressed in DRG neurons in rats (Kawano et al. 2009a; Zoga et al. 2010). Consistent with this, functional evidence suggests that KATP currents can inhibit substance P release from rat DRG neurons (Ohkubo and Shibata 1995; Sarantopoulos et al. 2003). Furthermore, KATP activation has been shown to reduce mechanical and thermal nociception in naïve mice, and following inflammatory and neuropathic pain induced by bradykinin and axotomy, respectively (Zoga et al. 2010; Du et al. 2011; Kawano et al. 2009b; Pacheco and Duarte 2005). Several studies demonstrated that peripheral morphine antinociception involves KATP activation, linking opioid receptors to these channels (Afify et al. 2013; Cunha et al. 2010; Rodrigues and Duarte 2000). In a similar way, it appears that antinociception caused by DOR agonists can derive from KATP activation (Gutierrez et al. 2012; Pena-dos-Santos et al. 2009; Saloman et al. 2011). Whether these observations result from direct binding of Gβγ dissociated from DOR following its activation remains unclear. To note, most of these studies were performed in rats, it is unclear that opioid receptor signaling involving KATP channels is conserved in DRG from other species, particularly mice and humans.

GIRK channels are directly regulated by Gi/o proteins (Kir3.x family) (Ocana et al. 2004). GIRK channels are comprised of four α subunits, each with two 2 TMs and intracellular C- and N-terminals, which interact with Gβγ. In the absence of G protein interaction, the C- and N-terminus maintain the channel in a closed conformation. Following G protein activation, binding of Gβγ to the intracellular domains switches the channel to an open conformation (Mark and Herlitze 2000). Four genes encode different α subunits: Kir3.1 to Kir3.4 (also named GIRK1 to GIRK 4). To form a functional pore, α subunits are assembled into homo- or hetero-tetramers. The combination of these four subunits influences the properties of the channel. GIRK1 needs to be associated with another class of GIRK α subunit to form a functional potassium current (i.e., obligatory heteromer) (Jelacic et al. 2000; Kennedy et al. 1996; Lesage et al. 1995). GIRK channels are widely distributed in the CNS and represent major effectors of opioid receptors in neurons, including for DOR. In DRG neurons, on the other hand, the expression of GIRK channels remains debated. An RT-PCR and electrophysiological analysis suggested that GIRK 1 through 4 are expressed in rat DRG (Gao et al. 2007). Furthermore, a recent study not only confirmed that GIRK channels are present in DRG but also showed that GIRK2 critically contributes to peripheral morphine analgesia in rats and humans (Nockemann et al. 2013). In this article, the authors also provide evidence that GIRK channels are not expressed in mouse DRG. If GIRK channels are present in DOR-expressing DRG neurons, they likely are modulated by DOR. Very few studies have explored this possibility. Regarding DOR modulation of GIRK, DOR agonists can activate GIRK 1 and 2 in rat trigeminal ganglia (Chung et al. 2014).

3.3 Indirect Modulation of Ion Channels

DOR may also indirectly control ion channel opening and neuron excitability, through Gi/o, Gs or Gq modulation of phosphorylation of ion channels by PKA, PKC, or phosphatidylinositol-4,5-bisphosphate (PtdIns(4,5)P2), and diacylglycerol (DAG) abundance.

Opioid receptors regulate the activity of adenylate cyclase (AC) (Nestler 2004; Gilman 1987). AC produces cAMP, a second messenger that activates downstream effectors, particularly protein kinase A (PKA). PKA phosphorylates proteins and modifies their properties, including ion channels and receptors that regulate neuronal excitability. Early studies in cell lines established that opioid receptors, including DOR, inhibit AC via Gαi/o (Prather et al. 1994; Roerig et al. 1992; Wong et al. 1992; Wong et al. 1991). This effect, however, seems dependent on the neuronal context. Thus, while activation of DOR or MOR results in AC inhibition in neurons in numerous brain regions (Buzas et al. 1994; Chneiweiss et al. 1988; Eybalin et al. 1987; Izenwasser et al. 1993; Law et al. 1981), an increase in AC was reported in the olfactory bulb, spinal cord explant, and DRG (Makman et al. 1988; Olianas and Onali 1995; Shen and Crain 1989). In DRG, these studies revealed that activation of opioid receptors, including DOR, can lead to a dual modulation, i.e., inhibitory via Gi/o or excitatory via Gs, of AC, depending on the dose of agonists used (Fan et al. 1993; Crain and Shen 1990; Tang et al. 1995). More recently, AC activation by morphine has been proposed to contribute to morphine tolerance and withdrawal, via a Gs-mediated increase of TRPV1 activity and CGRP release (Spahn et al. 2013; Tumati et al. 2011; Tumati et al. 2010; Yue et al. 2008). Recent research on the topic is lacking, and no study has specifically examined DOR coupling to AC in DRG neurons that endogenously express this opioid receptor. DOR activation in DRG neurons can potentially modulate all targets downstream of cAMP and PKA, including receptors or ion channels such as hyperpolarization-activated cyclic nucleotide-gated (HCN) and two pore domain (K2P or K+ leak) channels (see below). Note that in addition to the classical Gβγ inhibition, VGCC activity can be modulated by PKA phosphorylation, especially for L-Type channels (Altier and Zamponi 2008; Sculptoreanu et al. 1993). L-Type calcium channels are expressed in small- and large-diameter DRG neurons (Scroggs and Fox 1992) and may be indirectly regulated by DOR.

HCN channels carry a Na^{+} K^{+} inward current and are activated by membrane hyperpolarization (−60 to −90 mV). Opening of HCN channels elicits membrane depolarization toward the threshold for action potential generation. In neurons, HCN channels are notably responsible for rhythmicity in pacemaker cells and contribute to rebound activity and resting potential (Biel et al. 2009). Importantly, HCN channel opening is facilitated by binding of cAMP (i.e., in a PKA-independent manner) and is indirectly modulated by GPCRs controlling AC activity. In DRG, HCN 1 and 3 are thought to be expressed by Aβ and Aδ fibers, whereas HCN2 is predominantly expressed by C nociceptors and reportedly contributes to inflammatory and neuropathic pain (Emery et al. 2011; Momin et al. 2008; Weng et al. 2012). It follows that if DOR is coupled to AC and alters cAMP concentration in DRG neurons, DOR agonist antinociceptive activity may involve modulation of HCN channel activity. Among K2P channels, TREK-1, TREK-2, TRAAK, and TRESK channels are of particular interest. These channels determine neuron resting potential and, in DRG neurons, sensitivity to mechanical and thermal stimuli (Alloui et al. 2006; Brohawn et al. 2014; Honore 2007; Noel et al. 2009). All types of DRG neurons are thought to express some isoform of TREK and/or TRESK (Alloui et al. 2006; Kang and Kim 2006). TREK channels are inhibited by PKA and PKC phosphorylation, hydrolysis of PtdIns(4, 5)P2, DAG (Honore 2007), potentially linking TREK function to DOR signaling in DRG neurons. Consistent with this idea, a recent study described a direct link between MOR activation and TREK-1 activation (Devilliers et al. 2013).

4 Conclusion: Implications for DOR Function in Somatosensation and Pain Control

The characterization of DOR+ DRG neurons has so far revealed that many of these cells have the molecular identity and anatomical properties of cutaneous mechanosensory neurons known to initiate the perception of touch and mechanical pain. The mechanosensitivity of DOR+ DRG neurons was directly demonstrated by electrophysiological recording using an ex vivo somatosensory system preparation in which the skin, cutaneous nerve, DRGs, and spinal cord are dissected in continuity (Koerber and Woodbury 2002). DORGFP+ DRG neurons did not respond to noxious heat applied to the skin, but were very sensitive and fired vigorously in response to stimulation with a mechanical probe (Bardoni et al. 2014). These studies also revealed the existence of several types of DOR+ myelinated mechanosensory afferents, with distinct conduction velocities (covering Aδ and Aβ range) and thresholds (firing in response to innocuous and/or noxious mechanical stimulation).

The signaling and ion channel mechanisms engaged by DOR in DRG neurons generally result in a reduction in cell excitability and/or neurotransmitter release. In other words, DOR activation in DRG is expected to reduce the perception of the somatosensory modalities encoded by the DRG neurons in which the receptor is predominantly expressed: touch and mechanical pain. A large number of studies have thus reported that DOR agonists reduce cutaneous mechanical sensitivity, both acutely (Scherrer et al. 2009; Normandin et al. 2013; Pacheco and Duarte 2005; Pacheco Dda et al. 2012; Fuchs et al. 1999; Cao et al. 2001) and in the context of hypersensitivity (i.e., allodynia) induced by nerve or tissue injuries (i.e., chronic inflammatory or neuropathic pain) (Scherrer et al. 2009; Joseph and Levine 2010; Pradhan et al. 2013; Kouchek et al. 2013; Saloman et al. 2011; Cao et al. 2001; Sluka et al. 2002; Desmeules et al. 1993; Gaveriaux-Ruff et al. 2011; Stewart and Hammond 1994; Kabli and Cahill 2007; Obara et al. 2009; Nozaki et al. 2012; Gaveriaux-Ruff et al. 2008; Stein et al. 1989; Zhou et al. 1998; Hervera et al. 2010; Otis et al. 2011; Pradhan et al. 2009; Le Bourdonnec et al. 2009). The antinociceptive properties of DOR agonists in rodent pain models led researchers to propose that DOR agonists were particularly useful against chronic pain compared to acute pain (Gaveriaux-Ruff and Kieffer 2011; Pradhan et al. 2013; Stewart and Hammond 1994; Kabli and Cahill 2007; Cahill et al. 2007; Vanderah 2010; Hurley and Hammond 2000). The emergence of this concept results, in part, from the fact that cutaneous mechanical sensitivity has long been evaluated almost exclusively in animals with neuropathic or inflammatory hypersensitivity, but very rarely acutely in healthy uninjured animals. By contrast, in most studies before the mid-2000s ((re)emergence of the concept of pain modalities and peripheral nociceptive labelled lines (Ma 2010; Craig 2003; Pereira and Alves 2011; Emery et al. 2016)), acute sensitivity in healthy animals was measured using heat as a noxious stimulus. It is clear today that evaluating cutaneous mechanical sensitivity in the setting of injury probes the function of A LTMRs that normally encode touch and not only nociceptors. Following injury, loss of inhibition in dorsal horn circuits enables polysynaptic neurotransmission between A LTMRs and nociceptive neurons, resulting in innocuous mechanical stimulation being perceived as painful (i.e., mechanical allodynia) (Basbaum et al. 2009; Sandkuhler 2009). It is very likely that the anti-allodynic properties of DOR agonists result from the beneficial reduction in neurotransmitter release from A LTMRs. DOR activation in A LTMRs detecting movements across the skin, and in A LTMRs/A HTMRs/ MrgprD+ C nociceptors responding to skin indentation, could alleviate dynamic and static allodynia, respectively. Note that DOR expression in A LTMRs suggests that DOR may also modulate mechanosensation in the absence of injury; however, evaluating touch perception in rodents remains challenging. To our knowledge no studies have yet evaluated the impact of DOR activation or knockout on touch modalities encoded by the different A LTMRs expressing DOR (e.g., movement across the skin, vibrations, texture coding).

Much work is still needed to determine if DOR is expressed by other types of afferents, innervating other tissues (e.g., muscle or bone), or coding other somatosensory modalities (e.g. itch), and how tissue and nerve injury impact DOR expression and signaling in DRG during inflammatory and neuropathic chronic pain. Similarly, analyzing DOR distribution in trigeminal ganglia and DOR signaling at the peripheral terminals might uncover new therapeutic indications for DOR agonists. Finally, DOR is expressed by a variety of neurons in brain regions processing somatosensory information, such as the amygdala and prefrontal/cingulate cortices. Future studies will resolve DOR function in these regions and establish its contribution to shaping pain experience, including attribution of negative emotional valence (e.g., pain unpleasantness), execution of motivated behaviors (e.g., avoidance), and cognitive and psychological maladaptations (e.g., pain catastrophizing, mood disorders).

Acknowledgments G.S. was supported by NIH/NIDA grant DA031777 and a Rita Allen Foundation and American Pain Society award. A.F. was supported by a Stanford Dean's fellowship. We thank Dr. Dong Wang for providing unpublished in situ hybridization images (Fig. 1a, b), and Elizabeth Sypek and Dr. Kristen Hymel Scherrer for their helpful comments on the manuscript.

References

Abraira VE, Ginty DD (2013) The sensory neurons of touch. Neuron 79:618–639

Acosta CG, Lopez HS (1999) Delta opioid receptor modulation of several voltage-dependent Ca(2+) currents in rat sensory neurons. J Neurosci 19:8337–8348

Afify EA, Khedr MM, Omar AG, Nasser SA (2013) The involvement of K(ATP) channels in morphine-induced antinociception and hepatic oxidative stress in acute and inflammatory pain in rats. Fundam Clin Pharmacol 27:623–631

Aguilar-Bryan L et al (1998) Toward understanding the assembly and structure of KATP channels. Physiol Rev 78:227–245

Al-Hasani R, Bruchas MR (2011) Molecular mechanisms of opioid receptor-dependent signaling and behavior. Anesthesiology 115:1363–1381

Alloui A et al (2006) TREK-1, a K+ channel involved in polymodal pain perception. EMBO J 25:2368–2376

Altier C, Zamponi GW (2008) Signaling complexes of voltage-gated calcium channels and G protein-coupled receptors. J Recept Signal Transduct Res 28:71–81

Alvarez FJ, Kavookjian AM, Light AR (1992) Synaptic-interactions between gaba-immuno-reactive profiles and the terminals of functionally defined myelinated nociceptors in the monkey and cat spinal-cord. J Neurosci 12:2901–2917

Arcourt A et al (2017) Touch receptor-derived sensory information alleviates acute pain signaling and fine-tunes nociceptive reflex coordination. Neuron 93:179–193

Arikkath J, Campbell KP (2003) Auxiliary subunits: essential components of the voltage-gated calcium channel complex. Curr Opin Neurobiol 13:298–307

Arvidsson U et al (1995) Delta-opioid receptor immunoreactivity: distribution in brainstem and spinal cord, and relationship to biogenic amines and enkephalin. J Neurosci 15:1215–1235

Bai L et al (2015) Genetic identification of an expansive mechanoreceptor sensitive to skin stroking. Cell 163:1783–1795

Bao L et al (2003) Activation of delta opioid receptors induces receptor insertion and neuropeptide secretion. Neuron 37:121–133

Bardoni R et al (2014) Delta opioid receptors presynaptically regulate cutaneous mechanosensory neuron input to the spinal cord dorsal horn. Neuron 81:1312–1327

Basbaum AI, Bautista DM, Scherrer G, Julius D (2009) Cellular and molecular mechanisms of pain. Cell 139:267–284

Bell TJ, Thaler C, Castiglioni AJ, Helton TD, Lipscombe D (2004) Cell-specific alternative splicing increases calcium channel current density in the pain pathway. Neuron 41:127–138

Biel M, Wahl-Schott C, Michalakis S, Zong X (2009) Hyperpolarization-activated cation channels: from genes to function. Physiol Rev 89:847–885

Brackley AD, Gomez R, Akopian AN, Henry MA, Jeske NA (2016) GRK2 constitutively governs peripheral delta opioid receptor activity. Cell Rep 16:2686–2698

Brittain JM et al (2011) Suppression of inflammatory and neuropathic pain by uncoupling CRMP-2 from the presynaptic Ca(2)(+) channel complex. Nat Med 17:822–829

Brohawn SG, Su Z, MacKinnon R (2014) Mechanosensitivity is mediated directly by the lipid membrane in TRAAK and TREK1 K+ channels. Proc Natl Acad Sci U S A 111:3614–3619

Buraei Z, Yang J (2010) The ss subunit of voltage-gated Ca2+ channels. Physiol Rev 90:1461–1506

Buzas B, Cox BM (1997) Quantitative analysis of mu and delta opioid receptor gene expression in rat brain and peripheral ganglia using competitive polymerase chain reaction. Neuroscience 76:479–489

Buzas B, Izenwasser S, Portoghese PS, Cox BM (1994) Evidence for delta opioid receptor subtypes regulating adenylyl cyclase activity in rat brain. Life Sci 54:PL101–PL106

Cahill CM et al (2001) Immunohistochemical distribution of delta opioid receptors in the rat central nervous system: evidence for somatodendritic labeling and antigen-specific cellular compartmentalization. J Comp Neurol 440:65–84

Cahill CM, Holdridge SV, Morinville A (2007) Trafficking of delta-opioid receptors and other G-protein-coupled receptors: implications for pain and analgesia. Trends Pharmacol Sci 28:23–31

Cahill CM, Walwyn W, Taylor AMW, Pradhan AAA, Evans CJ (2016) Allostatic mechanisms of opioid tolerance beyond desensitization and down regulation. Trends Pharmacol Sci 37:963–976

Cao CQ, Hong Y, Dray A, Perkins M (2001) Spinal delta-opioid receptors mediate suppression of systemic SNC80 on excitability of the flexor reflex in normal and inflamed rat. Eur J Pharmacol 418:79–87

Catterall WA (2011) Voltage-gated calcium channels. Cold Spring Harb Perspect Biol 3:a003947

Cavanaugh DJ et al (2009) Distinct subsets of unmyelinated primary sensory fibers mediate behavioral responses to noxious thermal and mechanical stimuli. Proc Natl Acad Sci U S A 106:9075–9080

Cavanaugh DJ et al (2011) Restriction of transient receptor potential vanilloid-1 to the peptidergic subset of primary afferent neurons follows its developmental downregulation in nonpeptidergic neurons. J Neurosci 31:10119–10127
Charfi I, Audet N, Tudashki HB, Pineyro G (2015) Identifying ligand-specific signalling within biased responses: focus on delta opioid receptor ligands. Br J Pharmacol 172:435–448
Chneiweiss H, Glowinski J, Premont J (1988) Mu and delta opiate receptors coupled negatively to adenylate cyclase on embryonic neurons from the mouse striatum in primary cultures. J Neurosci 8:3376–3382
Chung MK et al (2014) Peripheral G protein-coupled inwardly rectifying potassium channels are involved in delta-opioid receptor-mediated anti-hyperalgesia in rat masseter muscle. Eur J Pain 18:29–38
Craig AD (2003) Pain mechanisms: labeled lines versus convergence in central processing. Annu Rev Neurosci 26:1–30
Crain SM, Shen KF (1990) Opioids can evoke direct receptor-mediated excitatory effects on sensory neurons. Trends Pharmacol Sci 11:77–81
Cunha TM et al (2010) Morphine peripheral analgesia depends on activation of the PI3Kgamma/AKT/nNOS/NO/KATP signaling pathway. Proc Natl Acad Sci U S A 107:4442–4447
Dado RJ, Law PY, Hoh HH, Elde R (1993) Immunofluorescent indentification of a delta ([delta])-opioid receptor on primary afferent nerve terminals. Neuroreport 5:341–344
Dang VC, Christie MJ (2012) Mechanisms of rapid opioid receptor desensitization, resensitization and tolerance in brain neurons. Br J Pharmacol 165:1704–1716
De Schepper HU et al (2008) TRPV1 receptor signaling mediates afferent nerve sensitization during colitis-induced motility disorders in rats. Am J Physiol Gastrointest Liver Physiol 294: G245–G253
Delfini MC et al (2013) TAFA4, a chemokine-like protein, modulates injury-induced mechanical and chemical pain hypersensitivity in mice. Cell Rep 5:378–388
Delmas P, Hao J, Rodat-Despoix L (2011) Molecular mechanisms of mechanotransduction in mammalian sensory neurons. Nat Rev Neurosci 12:139–153
Desmeules JA, Kayser V, Guilbaud G (1993) Selective opioid receptor agonists modulate mechanical allodynia in an animal model of neuropathic pain. Pain 53:277–285
Devilliers M et al (2013) Activation of TREK-1 by morphine results in analgesia without adverse side effects. Nat Commun 4:2941
de Nooij JC, Doobar S, Jessell TM (2013) Etv1 inactivation reveals proprioceptor subclasses that reflect the level of NT3 expression in muscle targets. Neuron 77:1055–1068
Du X, Wang C, Zhang H (2011) Activation of ATP-sensitive potassium channels antagonize nociceptive behavior and hyperexcitability of DRG neurons from rats. Mol Pain 7:35
Dussor G, Zylka MJ, Anderson DJ, McCleskey EW (2008) Cutaneous sensory neurons expressing the Mrgprd receptor sense extracellular ATP and are putative nociceptors. J Neurophysiol 99:1581–1589
Eckert WA III, Light AR (2002) Hyperpolarization of substantia gelatinosa neurons evoked by mu-, kappa-, delta 1-, and delta 2-selective opioids. J Pain 3:115–125
Emery EC, Young GT, Berrocoso EM, Chen L, McNaughton PA (2011) HCN2 ion channels play a central role in inflammatory and neuropathic pain. Science 333:1462–1466
Emery EC et al (2016) In vivo characterization of distinct modality-specific subsets of somatosensory neurons using GCaMP. Sci Adv 2:e1600990
Evans CJ, Keith DE Jr, Morrison H, Magendzo K, Edwards RH (1992) Cloning of a delta opioid receptor by functional expression. Science 258:1952–1955
Evans AR, Nicol GD, Vasko MR (1996) Differential regulation of evoked peptide release by voltage-sensitive calcium channels in rat sensory neurons. Brain Res 712:265–273
Eybalin M, Pujol R, Bockaert J (1987) Opioid receptors inhibit the adenylate cyclase in guinea pig cochleas. Brain Res 421:336–342

Fan SF, Shen KF, Crain SM (1993) Mu and delta-opioid agonists at low concentrations decrease voltage-dependent K+ currents in F11 neuroblastoma X drug neuron hybrid-cells via cholera toxin-sensitive receptors. Brain Res 605:214–220

Fatt P, Katz B (1953) The electrical properties of crustacean muscle fibres. J Physiol 120:171–204

Fields HL, Emson PC, Leigh BK, Gilbert RF, Iversen LL (1980) Multiple opiate receptor sites on primary afferent fibres. Nature 284:351–353

Filliol D et al (2000) Mice deficient for delta- and gamma-opioid receptors exhibit opposing alterations of emotional responses. Nat Genet 25:195–200

Franck MC et al (2011) Essential role of ret for defining non-peptidergic nociceptor phenotypes and functions in the adult mouse. Eur J Neurosci 33:1385–1400

François A et al (2017) A brainstem-spinal cord inhibitory circuit for mechanical pain modulation by GABA and enkephalins. Neuron 93(4):822–839.e6

Fuchs PN, Roza C, Sora I, Uhl G, Raja SN (1999) Characterization of mechanical withdrawal responses and effects of mu-, delta- and kappa-opioid agonists in normal and mu-opioid receptor knockout mice. Brain Res 821:480–486

Fujita W, Gomes I, Devi LA (2014) Revolution in GPCR signalling: opioid receptor heteromers as novel therapeutic targets: IUPHAR review 10. Br J Pharmacol 171:4155–4176

Gao XF, Zhang HL, You ZD, Lu CL, He C (2007) G protein-coupled inwardly rectifying potassium channels in dorsal root ganglion neurons. Acta Pharmacol Sin 28:185–190

Gaveriaux-Ruff C, Kieffer BL (2011) Delta opioid receptor analgesia: recent contributions from pharmacology and molecular approaches. Behav Pharmacol 22:405–414

Gaveriaux-Ruff C, Karchewski LA, Hever X, Matifas A, Kieffer BL (2008) Inflammatory pain is enhanced in delta opioid receptor-knockout mice. Eur J Neurosci 27:2558–2567

Gaveriaux-Ruff C et al (2011) Genetic ablation of delta opioid receptors in nociceptive sensory neurons increases chronic pain and abolishes opioid analgesia. Pain 152:1238–1248

Gendron L et al (2006) Morphine and pain-related stimuli enhance cell surface availability of somatic delta-opioid receptors in rat dorsal root ganglia. J Neurosci 26:953–962

Gendron L, Mittal N, Beaudry H, Walwyn W (2015) Recent advances on the delta opioid receptor: from trafficking to function. Br J Pharmacol 172:403–419

Gendron L, Cahill CM, von Zastrow M, Schiller PW, Pineyro G (2016) Molecular pharmacology of delta-opioid receptors. Pharmacol Rev 68:631–700

Georgoussi Z (2015) The other side of opioid receptor signaling: regulation by protein-protein interaction. Spring 4:L21

Ghitani N et al (2017) Specialized mechanosensory nociceptors mediating rapid responses to hair pull. Neuron 95:944–954.e944

Gilman AG (1987) G proteins: transducers of receptor-generated signals. Annu Rev Biochem 56:615–649

Glaum SR, Miller RJ, Hammond DL (1994) Inhibitory actions of delta 1-, delta 2-, and mu-opioid receptor agonists on excitatory transmission in lamina II neurons of adult rat spinal cord. J Neurosci 14:4965–4971

Goody RJ, Oakley SM, Filliol D, Kieffer BL, Kitchen I (2002) Quantitative autoradiographic mapping of opioid receptors in the brain of delta-opioid receptor gene knockout mice. Brain Res 945:9–19

Goswami SC et al (2014) Molecular signatures of mouse TRPV1-lineage neurons revealed by RNA-Seq transcriptome analysis. J Pain 15:1338–1359

Guan JS et al (2005) Interaction with vesicle luminal protachykinin regulates surface expression of delta-opioid receptors and opioid analgesia. Cell 122:619–631

Gutierrez VP et al (2012) The peripheral L-arginine-nitric oxide-cyclic GMP pathway and ATP-sensitive K(+) channels are involved in the antinociceptive effect of crotalphine on neuropathic pain in rats. Behav Pharmacol 23:14–24

Gutierrez-Mecinas M, Watanabe M, Todd AJ (2014) Expression of gastrin-releasing peptide by excitatory interneurons in the mouse superficial dorsal horn. Mol Pain 10:79

Heinke B, Balzer E, Sandkuhler J (2004) Pre- and postsynaptic contributions of voltage-dependent Ca2+ channels to nociceptive transmission in rat spinal lamina I neurons. Eur J Neurosci 19:103–111

Herlitze S, Hockerman GH, Scheuer T, Catterall WA (1997) Molecular determinants of inactivation and G protein modulation in the intracellular loop connecting domains I and II of the calcium channel alpha1A subunit. Proc Natl Acad Sci U S A 94:1512–1516

Hervera A, Negrete R, Leanez S, Martin-Campos J, Pol O (2010) The role of nitric oxide in the local antiallodynic and antihyperalgesic effects and expression of delta-opioid and cannabinoid-2-receptors during neuropathic pain in mice. J Pharmacol Exp Ther 334:887–896

Hille B (1992) Ionic channels of excitable membranes, 2nd edn. Sinauer Associates, Inc., Sunderland, MA

Holmseth S et al (2012) Specificity controls for immunocytochemistry: the antigen preadsorption test can lead to inaccurate assessment of antibody specificity. J Histochem Cytochem 60:174–187

Honore E (2007) The neuronal background K2P channels: focus on TREK1. Nat Rev Neurosci 8:251–261

Hurley RW, Hammond DL (2000) The analgesic effects of supraspinal mu and delta opioid receptor agonists are potentiated during persistent inflammation. J Neurosci 20:1249–1259

Ikeda SR (1996) Voltage-dependent modulation of N-type calcium channels by G-protein beta gamma subunits. Nature 380:255–258

Izenwasser S, Buzas B, Cox BM (1993) Differential regulation of adenylyl cyclase activity by mu and delta opioids in rat caudate putamen and nucleus accumbens. J Pharmacol Exp Ther 267:145–152

Jankowski MP, Rau KK, Ekmann KM, Anderson CE, Koerber HR (2013) Comprehensive phenotyping of group III and IV muscle afferents in mouse. J Neurophysiol 109:2374–2381

Jelacic TM, Kennedy ME, Wickman K, Clapham DE (2000) Functional and biochemical evidence for G-protein-gated inwardly rectifying K+ (GIRK) channels composed of GIRK2 and GIRK3. J Biol Chem 275:36211–36216

Jiang YQ, Andrade A, Lipscombe D (2013) Spinal morphine but not ziconotide or gabapentin analgesia is affected by alternative splicing of voltage-gated calcium channel CaV2.2 pre-mRNA. Mol Pain 9:67

Jones RC III, Xu L, Gebhart GF (2005) The mechanosensitivity of mouse colon afferent fibers and their sensitization by inflammatory mediators require transient receptor potential vanilloid 1 and acid-sensing ion channel 3. J Neurosci 25:10981–10989

Joseph EK, Levine JD (2010) Mu and delta opioid receptors on nociceptors attenuate mechanical hyperalgesia in rat. Neuroscience 171:344–350

Kabli N, Cahill CM (2007) Anti-allodynic effects of peripheral delta opioid receptors in neuropathic pain. Pain 127:84–93

Kang D, Kim D (2006) TREK-2 (K2P10.1) and TRESK (K2P18.1) are major background K+ channels in dorsal root ganglion neurons. Am J Phys Cell Physiol 291:C138–C146

Kawano T et al (2009a) ATP-sensitive potassium currents in rat primary afferent neurons: biophysical, pharmacological properties, and alterations by painful nerve injury. Neuroscience 162:431–443

Kawano T et al (2009b) Suppressed Ca2+/CaM/CaMKII-dependent K(ATP) channel activity in primary afferent neurons mediates hyperalgesia after axotomy. Proc Natl Acad Sci U S A 106:8725–8730

Kennedy ME, Nemec J, Clapham DE (1996) Localization and interaction of epitope-tagged GIRK1 and CIR inward rectifier K+ channel subunits. Neuropharmacology 35:831–839

Kieffer BL, Befort K, Gaveriaux-Ruff C, Hirth CG (1992) The delta-opioid receptor: isolation of a cDNA by expression cloning and pharmacological characterization. Proc Natl Acad Sci U S A 89:12048–12052

Kitchen I, Slowe SJ, Matthes HW, Kieffer B (1997) Quantitative autoradiographic mapping of mu-, delta- and kappa-opioid receptors in knockout mice lacking the mu-opioid receptor gene. Brain Res 778:73–88
Koerber HR, Woodbury CJ (2002) Comprehensive phenotyping of sensory neurons using an ex vivo somatosensory system. Physiol Behav 77:589–594
Kolodziejczyk AA, Kim JK, Svensson V, Marioni JC, Teichmann SA (2015) The technology and biology of single-cell RNA sequencing. Mol Cell 58:610–620
Kouchek M, Takasusuki T, Terashima T, Yaksh TL, Xu Q (2013) Effects of intrathecal SNC80, a delta receptor ligand, on nociceptive threshold and dorsal horn substance p release. J Pharmacol Exp Ther 347:258–264
Lallemend F, Ernfors P (2012) Molecular interactions underlying the specification of sensory neurons. Trends Neurosci 35:373–381
Lamberts JT, Traynor JR (2013) Opioid receptor interacting proteins and the control of opioid signaling. Curr Pharm Design 19:7333–7347
Law PY, Wu J, Koehler JE, Loh HH (1981) Demonstration and characterization of opiate inhibition of the striatal adenylate cyclase. J Neurochem 36:1834–1846
Lawson JJ, McIlwrath SL, Woodbury CJ, Davis BM, Koerber HR (2008) TRPV1 unlike TRPV2 is restricted to a subset of mechanically insensitive cutaneous nociceptors responding to heat. J Pain 9:298–308
Le Bourdonnec B et al (2009) Spirocyclic delta opioid receptor agonists for the treatment of pain: discovery of N,N-diethyl-3-hydroxy-4-(spiro[chromene-2,4′-piperidine]-4-yl) benzamide (ADL5747). J Med Chem 52:5685–5702
Le Pichon CE, Chesler AT (2014) The functional and anatomical dissection of somatosensory subpopulations using mouse genetics. Front Neuroanat 8:21
Lesage F et al (1995) Molecular properties of neuronal G-protein-activated inwardly rectifying K+ channels. J Biol Chem 270:28660–28667
Lewin GR, Moshourab R (2004) Mechanosensation and pain. J Neurobiol 61:30–44
Li L et al (2011) The functional organization of cutaneous low-threshold mechanosensory neurons. Cell 147:1615–1627
Liu Y, Ma Q (2011) Generation of somatic sensory neuron diversity and implications on sensory coding. Curr Opin Neurobiol 21:52–60
Liu Q et al (2009) Sensory neuron-specific GPCR Mrgprs are itch receptors mediating chloroquine-induced pruritus. Cell 139:1353–1365
Lumpkin EA, Bautista DM (2005) Feeling the pressure in mammalian somatosensation. Curr Opin Neurobiol 15:382–388
Luo W et al (2007) A hierarchical NGF signaling cascade controls ret-dependent and ret-independent events during development of nonpeptidergic DRG neurons. Neuron 54:739–754
Luo W, Enomoto H, Rice FL, Milbrandt J, Ginty DD (2009) Molecular identification of rapidly adapting mechanoreceptors and their developmental dependence on ret signaling. Neuron 64:841–856
Ma Q (2010) Labeled lines meet and talk: population coding of somatic sensations. J Clin Invest 120:3773–3778
Makman MH, Dvorkin B, Crain SM (1988) Modulation of adenylate cyclase activity of mouse spinal cord-ganglion explants by opioids, serotonin and pertussis toxin. Brain Res 445:303–313
Mansour A, Khachaturian H, Lewis ME, Akil H, Watson SJ (1987) Autoradiographic differentiation of mu, delta, and kappa opioid receptors in the rat forebrain and midbrain. J Neurosci 7:2445–2464
Mark MD, Herlitze S (2000) G-protein mediated gating of inward-rectifier K+ channels. Eur J Biochem 267:5830–5836
Mennicken F et al (2003) Phylogenetic changes in the expression of delta opioid receptors in spinal cord and dorsal root ganglia. J Comp Neurol 465:349–360

Minami M, Maekawa K, Yabuuchi K, Satoh M (1995) Double in-situ hybridization study on coexistence of mu-opioid, delta-opioid and kappa-opioid receptor messenger-rnas with preprotachykinin-a messenger-rna in the rat dorsal-root ganglia. Mol Brain Res 30:203–210

Mittal N et al (2013) Select G-protein-coupled receptors modulate agonist-induced signaling via a ROCK, LIMK, and beta-arrestin 1 pathway. Cell Rep 5:1010–1021

Moises HC, Rusin KI, Macdonald RL (1994) Mu- and kappa-opioid receptors selectively reduce the same transient components of high-threshold calcium current in rat dorsal root ganglion sensory neurons. J Neurosci 14:5903–5916

Momin A, Cadiou H, Mason A, McNaughton PA (2008) Role of the hyperpolarization-activated current Ih in somatosensory neurons. J Physiol 586:5911–5929

Murakami M et al (2001) Distribution of various calcium channel alpha(1) subunits in murine DRG neurons and antinociceptive effect of omega-conotoxin SVIB in mice. Brain Res 903:231–236

Murali SS et al (2015) High-voltage-activated calcium current subtypes in mouse DRG neurons adapt in a subpopulation-specific manner after nerve injury. J Neurophysiol 113:1511–1519

Nestler EJ (2004) Historical review: molecular and cellular mechanisms of opiate and cocaine addiction. Trends Pharmacol Sci 25:210–218

Nockemann D et al (2013) The K(+) channel GIRK2 is both necessary and sufficient for peripheral opioid-mediated analgesia. EMBO Mol Med 5:1263–1277

Noel J et al (2009) The mechano-activated K+ channels TRAAK and TREK-1 control both warm and cold perception. EMBO J 28:1308–1318

Normandin A, Luccarini P, Molat JL, Gendron L, Dallel R (2013) Spinal mu and delta opioids inhibit both thermal and mechanical pain in rats. J Neurosci 33:11703–11714

Nowycky MC, Fox AP, Tsien RW (1985) Three types of neuronal calcium channel with different calcium agonist sensitivity. Nature 316:440–443

Nozaki C et al (2012) Delta-opioid mechanisms for ADL5747 and ADL5859 effects in mice: analgesia, locomotion, and receptor internalization. J Pharmacol Exp Ther 342:799–807

Obara I et al (2009) Local peripheral opioid effects and expression of opioid genes in the spinal cord and dorsal root ganglia in neuropathic and inflammatory pain. Pain 141:283–291

Ocana M, Cendan CM, Cobos EJ, Entrena JM, Baeyens JM (2004) Potassium channels and pain: present realities and future opportunities. Eur J Pharmacol 500:203–219

Ohkubo T, Shibata M (1995) ATP-sensitive K+ channels mediate regulation of substance P release via the prejunctional histamine H3 receptor. Eur J Pharmacol 277:45–49

Olianas MC, Onali P (1995) Participation of delta opioid receptor subtypes in the stimulation of adenylyl cyclase activity in rat olfactory bulb. J Pharmacol Exp Ther 275:1560–1567

Otis V, Sarret P, Gendron L (2011) Spinal activation of delta opioid receptors alleviates cancer-related bone pain. Neuroscience 183:221–229

Overland AC et al (2009) Protein kinase C mediates the synergistic interaction between agonists acting at alpha2-adrenergic and delta-opioid receptors in spinal cord. J Neurosci 29:13264–13273

Pacheco Dda F, Pacheco CM, Duarte ID (2012) Peripheral antinociception induced by delta-opioid receptors activation, but not mu- or kappa-, is mediated by Ca(2)(+)-activated cl(−) channels. Eur J Pharmacol 674:255–259

Pacheco DF, Duarte ID (2005) Delta-opioid receptor agonist SNC80 induces peripheral antinociception via activation of ATP-sensitive K+ channels. Eur J Pharmacol 512:23–28

Pan HL et al (2008) Modulation of pain transmission by G-protein-coupled receptors. Pharmacol Ther 117:141–161

Pare M, Elde R, Mazurkiewicz JE, Smith AM, Rice FL (2001) The Meissner corpuscle revised: a multiafferented mechanoreceptor with nociceptor immunochemical properties. J Neurosci 21:7236–7246

Patwardhan AM et al (2006) PAR-2 agonists activate trigeminal nociceptors and induce functional competence in the delta opioid receptor. Pain 125:114–124

Peier AM et al (2002) A TRP channel that senses cold stimuli and menthol. Cell 108:705–715

Pena-dos-Santos DR et al (2009) Activation of peripheral kappa/Delta opioid receptors mediates 15-deoxy-(Delta12,14)-prostaglandin J2 induced-antinociception in rat temporomandibular joint. Neuroscience 163:1211–1219

Pereira JC Jr, Alves RC (2011) The labelled-lines principle of the somatosensory physiology might explain the phantom limb phenomenon. Med Hypotheses 77:853–856

Pettinger L, Gigout S, Linley JE, Gamper N (2013) Bradykinin controls pool size of sensory neurons expressing functional delta-opioid receptors. J Neurosci 33:10762–10771

Pradhan AA et al (2009) In vivo delta opioid receptor internalization controls behavioral effects of agonists. PLoS One 4:e5425

Pradhan AA, Befort K, Nozaki C, Gaveriaux-Ruff C, Kieffer BL (2011) The delta opioid receptor: an evolving target for the treatment of brain disorders. Trends Pharmacol Sci 32:581–590

Pradhan AA, Smith ML, Kieffer BL, Evans CJ (2012) Ligand-directed signalling within the opioid receptor family. Br J Pharmacol 167:960–969

Pradhan A, Smith M, McGuire B, Evans C, Walwyn W (2013) Chronic inflammatory injury results in increased coupling of delta opioid receptors to voltage-gated Ca2+ channels. Mol Pain 9:8

Prather PL, Loh HH, Law PY (1994) Interaction of delta-opioid receptors with multiple G proteins: a non-relationship between agonist potency to inhibit adenylyl cyclase and to activate G proteins. Mol Pharmacol 45:997–1003

Price TJ, Flores CM (2007) Critical evaluation of the colocalization between calcitonin gene-related peptide, substance P, transient receptor potential vanilloid subfamily type 1 immunoreactivities, and isolectin B4 binding in primary afferent neurons of the rat and mouse. J Pain 8:263–272

Raingo J, Castiglioni AJ, Lipscombe D (2007) Alternative splicing controls G protein-dependent inhibition of N-type calcium channels in nociceptors. Nat Neurosci 10:285–292

Rau KK et al (2009) Mrgprd enhances excitability in specific populations of cutaneous murine polymodal nociceptors. J Neurosci 29:8612–8619

Riedl MS et al (2009) Coexpression of alpha 2A-adrenergic and delta-opioid receptors in substance P-containing terminals in rat dorsal horn. J Comp Neurol 513:385–398

Rodrigues AR, Duarte ID (2000) The peripheral antinociceptive effect induced by morphine is associated with ATP-sensitive K(+) channels. Br J Pharmacol 129:110–114

Roerig SC, Loh HH, Law PY (1992) Identification of three separate guanine nucleotide-binding proteins that interact with the delta-opioid receptor in NG108-15 neuroblastoma x glioma hybrid cells. Mol Pharmacol 41:822–831

Rogers H, Henderson G (1990) Activation of mu- and delta-opioid receptors present on the same nerve terminals depresses transmitter release in the mouse hypogastric ganglion. Br J Pharmacol 101:505–512

Saegusa H et al (2001) Suppression of inflammatory and neuropathic pain symptoms in mice lacking the N-type Ca2+ channel. EMBO J 20:2349–2356

Saloman JL, Niu KY, Ro JY (2011) Activation of peripheral delta-opioid receptors leads to anti-hyperalgesic responses in the masseter muscle of male and female rats. Neuroscience 190:379–385

Sandkuhler J (2009) Models and mechanisms of hyperalgesia and allodynia. Physiol Rev 89:707–758

Sarantopoulos C, McCallum B, Sapunar D, Kwok WM, Hogan Q (2003) ATP-sensitive potassium channels in rat primary afferent neurons: the effect of neuropathic injury and gabapentin. Neurosci Lett 343:185–189

Scherrer G et al (2006) Knockin mice expressing fluorescent delta-opioid receptors uncover G protein-coupled receptor dynamics in vivo. Proc Natl Acad Sci U S A 103:9691–9696

Scherrer G et al (2009) Dissociation of the opioid receptor mechanisms that control mechanical and heat pain. Cell 137:1148–1159

Schroeder JE, McCleskey EW (1993) Inhibition of Ca2+ currents by a mu-opioid in a defined subset of rat sensory neurons. J Neurosci 13:867–873

Schroeder JE, Fischbach PS, Zheng D, McCleskey EW (1991) Activation of mu opioid receptors inhibits transient high- and low-threshold Ca2+ currents, but spares a sustained current. Neuron 6:13–20

Scroggs RS, Fox AP (1992) Multiple Ca2+ currents elicited by action potential waveforms in acutely isolated adult rat dorsal root ganglion neurons. J Neurosci 12:1789–1801

Sculptoreanu A, Scheuer T, Catterall WA (1993) Voltage-dependent potentiation of L-type Ca2+ channels due to phosphorylation by cAMP-dependent protein kinase. Nature 364:240–243

Seal RP et al (2009) Injury-induced mechanical hypersensitivity requires C-low threshold mechanoreceptors. Nature 462:651–655

Shen KF, Crain SM (1989) Dual opioid modulation of the action potential duration of mouse dorsal root ganglion neurons in culture. Brain Res 491:227–242

Simms BA, Zamponi GW (2014) Neuronal voltage-gated calcium channels: structure, function, and dysfunction. Neuron 82:24–45

Sluka KA, Rohlwing JJ, Bussey RA, Eikenberry SA, Wilken JM (2002) Chronic muscle pain induced by repeated acid injection is reversed by spinally administered mu- and delta-, but not kappa-, opioid receptor agonists. J Pharmacol Exp Ther 302:1146–1150

Spahn V et al (2013) Opioid withdrawal increases transient receptor potential vanilloid 1 activity in a protein kinase A-dependent manner. Pain 154:598–608

Stein C, Millan MJ, Shippenberg TS, Peter K, Herz A (1989) Peripheral opioid receptors mediating antinociception in inflammation - evidence for involvement of mu-receptors, delta-receptors and kappa-receptors. J Pharmacol Exp Ther 248:1269–1275

Stewart PE, Hammond DL (1994) Activation of spinal delta-1 or delta-2 opioid receptors reduces carrageenan-induced hyperalgesia in the rat. J Pharmacol Exp Ther 268:701–708

Sudhof TC (2004) The synaptic vesicle cycle. Annu Rev Neurosci 27:509–547

Tanabe T et al (1987) Primary structure of the receptor for calcium channel blockers from skeletal muscle. Nature 328:313–318

Tang T, Kiang JG, Cote T, Cox BM (1995) Opioid-induced increase in [Ca2+]i in ND8-47 neuroblastoma x dorsal root ganglion hybrid cells is mediated through G protein-coupled delta-opioid receptors and desensitized by chronic exposure to opioid. J Neurochem 65:1612–1621

Tumati S, Roeske WR, Vanderah TW, Varga EV (2010) Sustained morphine treatment augments prostaglandin E2-evoked calcitonin gene-related peptide release from primary sensory neurons in a PKA-dependent manner. Eur J Pharmacol 648:95–101

Tumati S, Roeske WR, Largent-Milnes TM, Vanderah TW, Varga EV (2011) Intrathecal PKA-selective siRNA treatment blocks sustained morphine-mediated pain sensitization and antinociceptive tolerance in rats. J Neurosci Methods 199:62–68

Usoskin D et al (2015) Unbiased classification of sensory neuron types by large-scale single-cell RNA sequencing. Nat Neurosci 18:145–153

Vanderah TW (2010) Delta and kappa opioid receptors as suitable drug targets for pain. Clin J Pain 26:S10–S15

Wada Y et al (2000) A region of the sulfonylurea receptor critical for a modulation of ATP-sensitive K(+) channels by G-protein betagamma-subunits. EMBO J 19:4915–4925

Walwyn W et al (2005) Induction of delta opioid receptor function by up-regulation of membrane receptors in mouse primary afferent neurons. Mol Pharmacol 68:1688–1698

Wang Z, Gerstein M, Snyder M (2009) RNA-Seq: a revolutionary tool for transcriptomics. Nat Rev Genet 10:57–63

Wang HB et al (2010) Coexpression of delta- and mu-opioid receptors in nociceptive sensory neurons. Proc Natl Acad Sci U S A 107:13117–13122

Weng X, Smith T, Sathish J, Djouhri L (2012) Chronic inflammatory pain is associated with increased excitability and hyperpolarization-activated current (Ih) in C- but not Adelta-nociceptors. Pain 153:900–914

Westenbroek RE, Hoskins L, Catterall WA (1998) Localization of Ca2+ channel subtypes on rat spinal motor neurons, interneurons, and nerve terminals. J Neurosci 18:6319–6330

Wilson SM et al (2000) The status of voltage-dependent calcium channels in alpha 1E knock-out mice. J Neurosci 20:8566–8571
Wong YH et al (1991) Mutant alpha subunits of Gi2 inhibit cyclic AMP accumulation. Nature 351:63–65
Wong YH, Conklin BR, Bourne HR (1992) Gz-mediated hormonal inhibition of cyclic AMP accumulation. Science 255:339–342
Woodbury CJ, Kullmann FA, McIlwrath SL, Koerber HR (2008) Identity of myelinated cutaneous sensory neurons projecting to nocireceptive laminae following nerve injury in adult mice. J Comp Neurol 508:500–509
Wu ZZ, Chen SR, Pan HL (2004) Differential sensitivity of N- and P/Q-type Ca2+ channel currents to a mu opioid in isolectin B4-positive and -negative dorsal root ganglion neurons. J Pharmacol Exp Ther 311:939–947
Wu ZZ, Chen SR, Pan HL (2008) Distinct inhibition of voltage-activated Ca2+ channels by delta-opioid agonists in dorsal root ganglion neurons devoid of functional T-type Ca2+ currents. Neuroscience 153:1256–1267
Yu FH, Catterall WA (2004) The VGL-chanome: a protein superfamily specialized for electrical signaling and ionic homeostasis. Sci STKE 2004:re15
Yue X et al (2008) Sustained morphine treatment augments basal CGRP release from cultured primary sensory neurons in a Raf-1 dependent manner. Eur J Pharmacol 584:272–277
Yusaf SP, Goodman J, Pinnock RD, Dixon AK, Lee K (2001) Expression of voltage-gated calcium channel subunits in rat dorsal root ganglion neurons. Neurosci Lett 311:137–141
Zachariou V, Goldstein BD (1996) Delta-opioid receptor modulation of the release of substance P-like immunoreactivity in the dorsal horn of the rat following mechanical or thermal noxious stimulation. Brain Res 736:305–314
Zaki PA et al (1996) Opioid receptor types and subtypes: the delta receptor as a model. Annu Rev Pharmacol Toxicol 36:379–401
Zamponi GW (2016) Targeting voltage-gated calcium channels in neurological and psychiatric diseases. Nat Rev Drug Discov 15:19–34
Zamponi GW, Bourinet E, Nelson D, Nargeot J, Snutch TP (1997) Crosstalk between G proteins and protein kinase C mediated by the calcium channel alpha1 subunit. Nature 385:442–446
Zhang X, Bao L, Arvidsson U, Elde R, Hokfelt T (1998) Localization and regulation of the delta-opioid receptor in dorsal root ganglia and spinal cord of the rat and monkey: evidence for association with the membrane of large dense-core vesicles. Neuroscience 82:1225–1242
Zhou L, Zhang Q, Stein C, Schafer M (1998) Contribution of opioid receptors on primary afferent versus sympathetic neurons to peripheral opioid analgesia. J Pharmacol Exp Ther 286:1000–1006
Zhu YX, Hsu MS, Pintar JE (1998) Developmental expression of the mu, kappa, and delta opioid receptor mRNAs in mouse. J Neurosci 18:2538–2549
Zhu Y et al (1999) Retention of supraspinal delta-like analgesia and loss of morphine tolerance in delta opioid receptor knockout mice. Neuron 24:243–252
Zoga V et al (2010) KATP channel subunits in rat dorsal root ganglia: alterations by painful axotomy. Mol Pain 6:6
Zollner C, Stein C (2007) Opioids. Handb Exp Pharmacol 177:31–63
Zylka MJ, Rice FL, Anderson DJ (2005) Topographically distinct epidermal nociceptive circuits revealed by axonal tracers targeted to Mrgprd. Neuron 45:17–25

Evidence and Function Relevance of Native DOR–MOR Heteromers

Catherine M. Cahill and Edmund Ong

Contents

Abstract

Opioid receptors are the sites of action for morphine and most other clinically used opioid drugs. Abundant evidence now demonstrates that different opioid receptor types can physically associate to form heteromers. Owing to their constituent monomers' involvement in analgesia, mu/delta opioid receptor (M/DOR) heteromers have been a particular focus of attention. Understandings of the physiological relevance and indisputable proof of M/DOR formation in vivo are still evolving. This aspect of the field has been slow to progress in large part by the limitations of most available experimental models; recently however, promising progress is being made. As a result, the long-repeated promise of opioid receptor heteromers as selective therapeutic targets is now being realized.

C. M. Cahill (✉)
Department of Psychiatry and Biobehavioral Sciences, Hatos Center for Neuropharmacology, Semel Institute for Neuroscience and Human Behavior, University of California Los Angeles, Los Angeles, CA, USA
e-mail: cmcahill@ucla.edu

E. Ong
Department of Anesthesiology, Duke University, Durham, NC, USA

E. M. Jutkiewicz (ed.), *Delta Opioid Receptor Pharmacology and Therapeutic Applications*, Handbook of Experimental Pharmacology 247, https://doi.org/10.1007/164_2018_112

Keywords
Analgesia · Antinociception · Delta · Dimer · Heterodimer · Heteromer · Mu · Opiate · Opioid · Pain · Receptor interaction · Receptor trafficking

1 Introduction

There is increasing convincing evidence that G-protein coupled receptors can form functional complexes as dimers or heteromers and that such complexes extend to the family of opioid receptors. Existent opioid receptor heteromers were first described by Jordan and Devi (1999) where functional and physical interaction was demonstrated between kappa (KOR) and delta (DOR) opioid receptors. Although, the concept of an opioid receptor complex was proposed earlier following the observation of noncompetitive binding interactions between mu opioid receptor (MOR) and DOR ligands (Rothman et al. 1992). Subsequently, many studies have provided further evidence for the existence of opioid receptor heteromers using various experimental approaches including co-immunoprecipitation, immunocytochemistry with novel heteromer antibodies, bioluminescence and Foerster resonance energy transfer, and electrophysiology. The proximity and interaction assay research supporting the existence of opioid heteromers has previously been thoroughly reviewed elsewhere (Costantino et al. 2012; Stockton and Devi 2012). However, there has been much debate of the physiological significance of such complexes and, initially, whether they truly existed in vivo. A preponderance of early evidence for their existence relied upon the use of heterologous expression systems in immortalized cell lines. These tools provide unparalleled experimental control. They permit the generation of precise conditions with maximum favorability for the detection of opioid receptor heteromers and a wealth of approaches to intricately dissect their functionality. A great deal of information about the mu–delta opioid receptor (M/DOR) heteromer has been gained using these models. Until recently, the deficit lay in the uncertain physiological relevance of those precisely engineered conditions. That is, opioid receptors are typically expressed in limited quantities in neuronal tissue with each opioid receptor type under tight and differential translational and trafficking control, notwithstanding the lack of clear subcellular co-localization. Meanwhile, these models often expressed these receptors in very large quantities (e.g., using CMV promoters) in HEK293 or CHO cells lacking the same control mechanisms. While these contrived models were very useful in providing information about the potential for MOR–DOR interactions and how M/DOR as a distinct receptor species behaved, the degree to which these interactions and behaviors occur in normal, physiological systems was a matter of some debate. This review will focus on recent in vivo and ex vivo research demonstrating the cellular localization, function, and unique signalling of MOR and DOR (M/DOR) heteromers. Three criteria have been recently proposed for demonstrating heteromers in native tissue: (1) physical proximity via direct interaction or allosteric interaction, (2) unique pharmacology of the heteromer from the individual receptor type, and (3) disruption of the heteromer leads to loss of the heteromer-specific properties (Gomes et al. 2016). To date, only a few heteromer complexes fit all three of these criteria, including the M/DOR heteromer (Gomes et al. 2016).

2 Historical Physiological Interactions

Indications for MOR–DOR interactions first arose from seemingly paradoxical findings of two related, but distinct research avenues: (1) DOR analgesia as a therapeutic target, and (2) the mechanisms of tolerance to MOR-mediated analgesia.

Delta Opioid Receptor Analgesia There has been interest in developing DOR agonists as novel therapeutics for treating pain (in particular, selective small molecule drugs). The DOR system is upregulated in various models of chronic pain and selective ligands show encouraging analgesic profiles in reducing pain hypersensitivities associated with tissue and nerve injury (Cahill et al. 2007). The attractiveness of this target emanated from studies showing that activation of DOR had: (1) minimal or no rewarding properties that may trigger addiction liability, (2) no life-threatening effects of respiratory depression, and (3) did not alter nociceptive responses in the absence of injury or pathology (Gendron et al. 2016; Spahn and Stein 2017).

Mu Opioid Receptor Analgesic Tolerance Perplexingly, many studies also support the development of DOR antagonists as novel analgesic therapeutics based on research demonstrating that this same receptor limits MOR function, possibly through the formation of MOR–DOR complexes.

A concerted effort has been made to understand the mechanisms of opioid analgesic tolerance because that tolerance limits effectiveness of pain treatment. This analgesic tolerance also jeopardizes compliance, because tolerance to other pharmacological effects such as constipation does not develop at the same rate. Opioid tolerant states have also been reported in certain chronic pain states, even though subjects may be "opioid-naïve." For example, neuropathic pain, defined as pain caused by damage or dysfunction of the nervous system, is a challenge to treat as it is often refractory to many pharmacotherapies, including opioid analgesics (Gilron et al. 2006).

Understandably, most studies have focused on understanding functional changes in the MOR as the majority of opioid analgesics target this receptor type. The first inclination that DOR restricted MOR activity was with pharmacological in vivo studies. Morphine analgesic tolerance was attenuated by co-administration of naltrindole, a DOR antagonist (Abdelhamid et al. 1991; Fundytus et al. 1995), and similar effects were evident by reducing expression of DORs by antisense knockdown (Kest et al. 1996), or constitutive DOR knockout mice (Chefer and Shippenberg 2009; Zhu et al. 1999) or disruption of cyclin-dependent kinase 5, which is required for phosphorylation of Thr-161 and trafficking of DOR to plasma membranes (Xie et al. 2009).

This research was surprising given the previous reports showing analgesic synergy between DOR and MOR agonists in both naïve and morphine tolerant mice (Porreca et al. 1987). Interestingly, even ultralow dose DOR antagonists suppressed morphine-induced analgesic tolerance (Abul-Husn et al. 2007). This suggested that the effect might not be driven solely by the absence/presence or blockade/activation of DOR, but perhaps an interaction between the two receptors. In this case, DOR ligands would act allosterically; this would explain the matching effects of DOR agonism and DOR antagonism.

Evidence that opioid receptors, like other GPCRs, could form heteromeric complexes was subsequently demonstrated by multiple research groups. The existence of MOR and DOR (M/DOR) heteromers in native tissue was first identified by Devi and colleagues using novel antibodies for the heteromer created by immunization subtraction methods (Gupta et al. 2010). Importantly, this study demonstrated that M/DOR heteromers abundance was augmented in animals following chronic morphine treatment. The existence of M/DOR heteromers was confirmed using an innovative approach to insert a TAT domain peptide into the membrane in the correct orientation where it could interrupt the formation of the M/DOR complex. TAT fusion-interfering peptide corresponding to the second intracellular loop of the DOR (Tat-DOR-2L) reduced cell surface expression of DOR and disrupted the formation of M/DOR heteromers (Xie et al. 2009) as well as reduced the development of morphine tolerance in a model of inflammatory pain (Chen et al. 2012). Further, systemic administration of MOR^{TM1}-TAT, which corresponds to the first transmembrane domain of the MOR, but not MOR^{TM3}-TAT, disrupted the formation of the M/DOR heteromer, and consequently increased morphine antinociception and attenuated the development of morphine analgesic tolerance (He et al. 2011).

One caveat to consider for experiments evaluating morphine tolerance is the influence of memory with repeated testing. This is an experimental confound which manifests as a gradual reduction in pain threshold with repeated testing. It is perhaps not surprising that rodents may learn that they will be removed from an environment associated with a noxious stimulus with repeated testing. The use of appropriate saline-injected controls may not capture changes in baseline nociceptive threshold due to a floor effect in that most tests have calibrated instrumentation to produce short latencies because they are predicting analgesic effects following morphine administration. Together, this change in pain thresholds and this methodological limitation may overstate analgesic tolerance. There are reports in the literature of this behavioral sensitization/tolerance. Behavioral tolerance was reported after exposing rats to a nonfunctional hot plate that involved habituation to the novel distractive stimuli (Bardo and Hughes 1979). Another study examined the differential effects of weekly compared to daily exposure of a rat to the hotplate test. In this study, a sensitization phenomenon was evident, where nociceptive thresholds decreased with weekly testing (Espejo and Mir 1994). More relevant was the finding that morphine can facilitate memory, which was proposed to contribute to associative learning in antinociceptive tolerance to morphine. Thus, repeated administration of morphine in the same or different environments or when animals were moved to a different context showed that morphine antinociceptive tolerance was significantly reversed by the change in context (Nakama-Kitamura and Doe 2003). These findings indicate that morphine develops associative and nonassociative antinociceptive tolerance, indicating that antinociceptive tolerance to morphine has contextual specificity. This is relevant to the conclusion that DOR contributes to morphine tolerance because DOR is necessary for hippocampal learning. DOR knockout mice or administration of the DOR antagonist naltrindole impaired hippocampal-dependent novel object recognition learning, demonstrating that DOR activity modulates learning and memory performance (Le Merrer et al. 2013). DOR antagonism (pharmacological or functional) may inhibit morphine-induced effects on memory.

3 Physical Evidence

Many immunohistochemical studies identified MOR and DOR co-localization. These studies have come under intense criticism due to purported lack of DOR antibody specificity, where immunolabeling remained present in constitutive knock-out mice (Gendron et al. 2016). The generation of an MOR-mcherry knockin mouse (Erbs et al. 2015; Gardon et al. 2014) allowed for breeding with the DOR-eGFP (Scherrer et al. 2006) to create a double knockin mouse. This mouse was used for extensive mapping of MOR and DOR throughout the peripheral and central nervous systems (Erbs et al. 2015), where receptors could be visualized with subcellular resolution. MOR and DOR were often co-expressed with high density in many brain regions and also identified to be co-localized within large dorsal root ganglia neurons (Erbs et al. 2015), in contrast to previous findings using immunohistochemical techniques in DOR-eGFP mice (Scherrer et al. 2009). Conversely, this same mouse [DOR-eGFP] was used to show MOR and DOR co-localization within enteric neurons of the myenteric plexus (Poole et al. 2011), which may account for the ability of DOR to inhibit gastrointestinal secretion and motility. Other electrophysiological (Egan and North 1981) and pharmacologic (Fox-Threlkeld et al. 1994) studies support DOR/MOR co-expression by enteric neurons.

Although the existence of GPCR heteromers was proposed almost two decades ago, there remains some scepticism of the existence of such receptor complexes in vivo, due to the general lack of tools available for detection of such complexes. There have been major advances in this tool kit that provide validation of M/DOR existence and the capacity of producing physiological effects. The crystal structures of MOR and DOR support the possibility of direct interaction between the two receptor types. Using an unbiased coarse-grained molecular dynamics simulation of freely diffusing opioid receptors in an explicit lipid–water environment, Provasi and colleagues identified the formation of M/DOR heteromers during the simulation. Importantly, once formed, the complex did not dissociate (Provasi et al. 2015). Further, in this latter study the minimum distance between each crystal structure within the heteromer was identified as 10 Å. This finding complements research showing that functional activity of bivalent ligands with linked mu agonist and delta antagonist pharmacophores have the greatest activity with a linkage spacer length of 22 Å (Lenard et al. 2007; Daniels et al. 2005; Yekkirala et al. 2013).

Further evidence that M/DOR exist in vivo used co-immunoprecipitation techniques similar to previous studies with various heterologous cell systems. For example, co-immunoprecipitation of spinal cord tissue revealed the existence of constitutively expressed M/DOR heteromers (Xie et al. 2009; Gomes et al. 2004; He et al. 2011). Because of the questionable specificity of DOR antibodies required for such studies, a novel approach of subtraction immunization was taken to produce an M/DOR specific antibody (Gomes et al. 2014). Using this antibody, in vivo expression of M/DOR was visualized in various brain structures (Gupta et al. 2010). The subcellular co-localization together with co-immunoprecipitation studies strengthens the existence of M/DOR heteromers, especially in subcortical networks involved in eating, sexual behavior, and response to aversive stimuli (Erbs et al. 2015).

It is not clear if M/DOR are synthesized within intracellular compartments and are trafficked to the membrane as a functional unit or formed at the plasma membrane. It is generally well accepted that the majority of DORs leaving the endoplasmic reticulum do not mature or traffic to the plasma membrane. This results in low expression of functional DORs on the cell surface. Rather, DORs are primarily degraded in lysosomal pathways. The formation of M/DOR may be one mechanism to enhance DOR maturation and trafficking. A Golgi chaperone, receptor transport protein 4 (RTP4), was shown to regulate the expression and cell surface trafficking of M/DOR heteromers (Décaillot et al. 2008). Chaperoning resulted in an increase in the cellular signalling of these receptors. A recent elegant review is available on molecular and pharmacological chaperones for GPCRs (Williams and Devi 2010). Cell surface trafficking of DOR is also evident in models of chronic inflammatory pain (Cahill et al. 2003; Morinville et al. 2004a; Gendron et al. 2007) or after prolonged morphine treatment (Cahill et al. 2001; Hack et al. 2005; Lucido et al. 2005; Morinville et al. 2004b). Prolonged morphine treatment also increases the abundance of M/DOR in various brain regions as detected by heteromeric antibodies (Gupta et al. 2010). Subsequently, it was proposed that morphine acts as a pharmacochaperone bringing the M/DOR heteromer to the cell surface (Costantino et al. 2012). In contrast, other studies provide evidence that M/DOR heteromers form at the cell surface (Law et al. 2005). This alternative is supported by studies showing that DOR and MOR can interact via transmembrane domains in coarse-grained molecular dynamics simulations (Provasi et al. 2015). Since data support both formation of heteromers within the receptor maturation process and their formation at the cell surface, it would not be unreasonable to suggest that both processes may occur depending on the physiological processes that engage formation of the heteromer.

4 Functional Evidence: *Pharmacological Subtypes – Bias Ligand Signalling or Heteromers?*

Pharmacological studies have proposed DOR-1 and DOR-2 subtypes. [D-Pen^2,D-Pen^5]enkephalin (DPDPE, DOR-1 agonist) and [D-Ala^2,Glu^4]deltorphin (Deltorphin II, DOR-2 agonist) both elicit antinociception in various pain models but repeated intracerebroventricular administration of either ligand was shown not to produce cross-tolerance to the other agonist (Mattia et al. 1991). Moreover, opposite effects on ethanol consumption were produced using delta subtype (DOR-1 and DOR-2) selective ligands (van Rijn and Whistler 2009).

The existence of DOR subtypes does not easily comport with molecular studies where only one transcript for DOR has been identified and splice variants for the DOR have not been described. However, there are many mechanisms that one could envisage to create alternative behaviors of different DOR ligands. One explanation for pharmacological subtypes is the existence and functional activity of heteromers (van Rijn and Whistler 2009), where the M/DOR heteromer was proposed to account for the DOR-1 subtype (van Rijn and Whistler 2009). Other studies suggest that DOR-2 subtype accounts for heteromers, where antagonism of DOR-2, but not DOR-1, reduced the development of morphine tolerance following chronic morphine treatment in a model of inflammatory pain (Beaudry et al. 2015). Using electrophysiological techniques on a slice preparation of the ventral tegmental area, DPDPE and Deltorphin II were shown to elicit opposing depolarization or hyperpolarization effects in the same neuron, which was not predicted by MOR agonist-induced effects, topographical localization, or whether it was positive for tyrosine hydroxylase or not (Margolis et al. 2017). While these data may argue against M/DOR heteromers explaining the DOR subtype phenomenon in this midbrain structure, this latter study identified that: (1) MOR agonist-induced effects could be augmented by a DOR antagonist and vice versa, (2) DOR agonist effects could be augmented with MOR selective antagonist CTAP, and finally (3) most VTA neurons expressed both DOR and MOR (Margolis et al. 2017). Together, these data support previous findings that DOR antagonists increase the potency and intrinsic efficacy of MOR agonists in cells co-expressing both receptors (Gomes et al. 2000, 2004). MOR ligands are capable of allosterically enhancing DOR radioligand binding and vice versa, which suggests strong positive cooperativity between the two receptor units. These data support the concept that DOR ligands (including antagonists) will allosterically enhance MOR ligand binding leading to the potentiation of MOR-mediated effects including antinociception.

In cultured cells, M/DOR heteromers have unique signalling properties compared to either MOR or DOR alone: signalling switched from a G-protein dependent (monomeric) to an independent (heteromeric) pathway (Rozenfeld and Devi 2007). MOR or DOR monomeric receptor activation couples to G-protein signalling cascades, and there has been a concerted effort to develop ligands that only couple through this G-protein signalling rather than β-arrestin, as the latter is proposed to account for unwanted pharmacological effects such as respiratory depression (Siuda et al. 2017). In contrast, the M/DOR heteromer led to a constitutive recruitment of β-arrestin-2 to the receptor complex resulting in changes in the spatiotemporal regulation of ERK1/2 signalling. However, treatment with an MOR or DOR ligand switched signalling to a non-β-arrestin-2-mediated signalling. Thus, the heteromer and the bias lines of drug development are trying to achieve the same fate – less β-arrestin signalling.

The identification that M/DOR heteromers primarily signal via β-arrestins led to the development of MOR agonists with DOR antagonist properties that were devoid of β-arrestin-2 recruitment activity. These compounds promised to have a unique pharmacology that would produce less respiratory depression, less GI dysfunction, and lower propensity to induce tolerance and dependence compared to morphine. Such compounds were synthesized based on endomorphin structure (Cai et al. 2014) or drug library screening for β-arrestin recruitment (Gomes et al. 2013). CYM51010 was identified through the latter method. The involvement of M/DOR heteromers in CYM51010-induced antinociception following spinal administration was confirmed by co-administration of a heteromeric antibody that acts as a functional antagonist at the receptor complex (Gomes et al. 2013). Importantly, this chemical elicited antinociception but reduced tolerance and physical dependence compared to morphine. Others took the approach to identify M/DOR selective ligands with the hypothesis that the heteromer would produce analgesia but be devoid of many side effects (Pinello et al. 2010). For example, 6′-guanidinonaltrindole was reported to produce analgesia following spinal administration (but not into the brain) via the unique property of selectively activating only M/DOR heteromers but not either MOR or DOR alone (Waldhoer et al. 2005). Chemists also synthesized bivalent ligands with MOR agonist and DOR antagonist pharmacophores, which with specific spacers (21 atoms) allowed for potent analgesic activity but devoid of tolerance and dependence (Daniels et al. 2005). Small molecule chemicals with similar pharmacology of MOR agonist and DOR antagonist properties were also reported to produce analgesia with less analgesic tolerance and dependence (Ananthan et al. 2012). The possibility of those drugs with MOR agonist and DOR antagonist properties have less side effect profile led to the development of eluxadoline (Breslin et al. 2012), which is now FDA approved for treatment of diarrhea associated with irritable bowel syndrome (Levio and Cash 2017). Eluxadoline-induced reductions in gastrointestinal transit were reduced in constitutive DOR knockout mice (Fujita et al. 2014). Using M/DOR heteromer antibodies as functional antagonists, Fujita and colleagues showed that eluxadoline-mediated signalling could be partially blocked (Fujita et al. 2014). Together, these data suggest that eluxadoline effects on gut motility are mediated, in part, by M/DOR heteromers. Figure 1 depicts a cartoon comparing DOR and MOR monomeric and M/DOR heteromeric formation, trafficking, signalling, and pharmacological effects.

5 Conclusions

In this review, we provide concordant and compelling evidence of the existence and functionality of M/DOR heteromers in endogenous tissues. Through the refinement and execution of physiologically relevant experimental tools, there is now an advancement of M/DOR heteromer understanding beyond the confines of earlier, more contrived model systems while also reinforcing and complementing those preceding findings. Attention has now shifted from the mere existence of heteromers towards a more determined effort to understand the processes by which they are

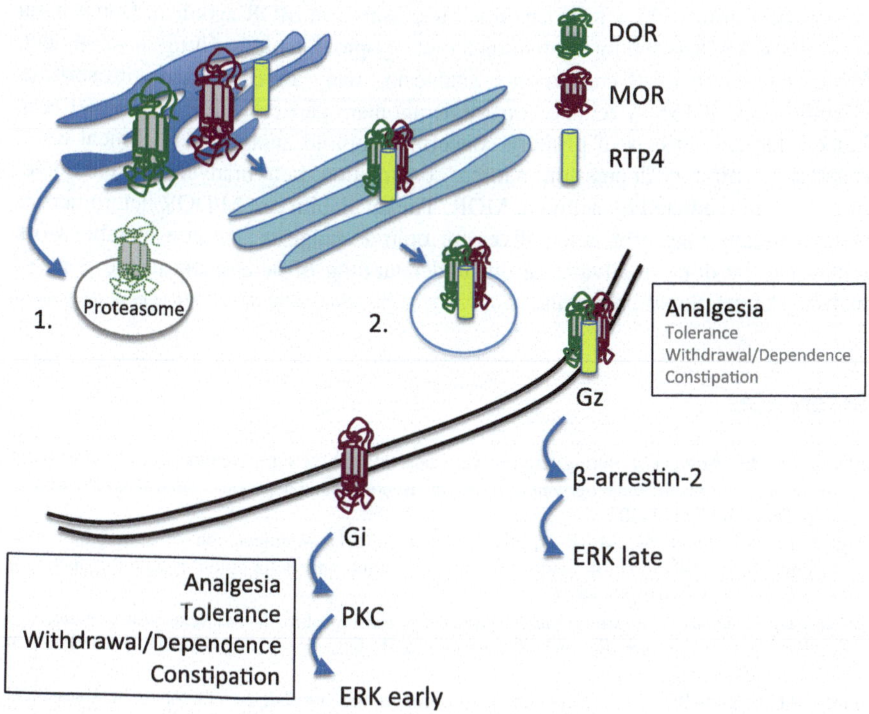

Fig. 1 Proposed model of mu/delta opioid receptor (M/DOR) heteromeric formation and signaling to plasma membranes. 1. DORs in the endoplasmic reticulum are often misfolded and targeted to degradation pathways. A high portion of MOR mature through the endoplasmic reticulum and Golgi network which allows for high cell surface expression. 2. M/DOR heteromers are formed in the endoplasmic reticulum and stabilized by RTP4 chaperone to allow for translocation to plasma membranes. Once at the plasma membrane, MOR signal predominantly through G-protein Gi/o coupling, which in turn generates PKC phosphorylation and early activation of mitogen activated protein kinase ERK activation. In contrast, M/DOR heteromers predominantly activate β-arrestin pathways although have been shown to activate Gz proteins in cell culture models. The heteromer also activates mitogen activated protein kinase ERK activation but the temporal and spatial activation is different than MOR alone, where activation of ERK is minutes later. Following agonist stimulation, MOR produces many pharmacological effects including analgesia, and prolonged treatment produces cellular adaptations or allostasis that contributes to the development of constipation, physical dependence, and analgesic tolerance, whereas activation of the heteromer also produces robust analgesia but appears to induce fewer negative effects caused by allostatic adaptations

formed and regulated as well as their behavior as receptors. Thus, although functional interactions between MOR and DORs may arise, such as competition for downstream effector systems, the research highlighted above confirms that physical interaction exists in the formation of heteromeric complexes.

Understanding M/DOR heteromers as distinct opioid receptor species naturally raises the prospect of these heteromers as therapeutic targets. The relevant literature certainly make this assertion, and justifiably so. Clinical opioid pharmacology has always been limited by a reliance, albeit necessary, on MOR agonism. Under basal conditions, MOR is the most obvious target of opioid analgesic drug development. While actions on DOR can produce analgesia, many are associated with seizures (Chung et al. 2015). A reliance on MOR agonism carries with it adverse effects. Indeed, the side effects of primary concern for opioid analgesics in clinical use – sedation, respiratory depression, nausea, constipation, itch, bradycardia, and addiction – are all mediated by action at MOR. The availability of M/DOR heteromers as distinct targets may offer alternatives for opioid analgesia, but considerable work remains to be done in advancing our understanding of heteromers to the point of realizing translational potentials.

References

Abdelhamid EE, Sultana M, Portoghese PS, Takemori AE (1991) Selective blockage of delta opioid receptors prevents the development of morphine tolerance and dependence in mice. J Pharmacol Exp Ther 258(1):299–303

Abul-Husn NS, Sutak M, Milne B, Jhamandas K (2007) Augmentation of spinal morphine analgesia and inhibition of tolerance by low doses of mu- and delta-opioid receptor antagonists. Br J Pharmacol 151(6):877–887

Ananthan S et al (2012) 14-Alkoxy- and 14-acyloxypyridomorphinans: mu agonist/delta antagonist opioid analgesics with diminished tolerance and dependence side effects. J Med Chem 55(19):8350–8363

Bardo MT, Hughes RA (1979) Exposure to a nonfunctional hot plate as a factor in the assessment of morphine-induced analgesia and analgesic tolerance in rats. Pharmacol Biochem Behav 10(4):481–485

Beaudry H, Gendron L, Moron JA (2015) Implication of delta opioid receptor subtype 2 but not delta opioid receptor subtype 1 in the development of morphine analgesic tolerance in a rat model of chronic inflammatory pain. Eur J Neurosci 41(7):901–907

Breslin HJ et al (2012) Identification of a dual delta OR antagonist/mu OR agonist as a potential therapeutic for diarrhea-predominant Irritable Bowel Syndrome (IBS-d). Bioorg Med Chem Lett 22(14):4869–4872

Cahill CM et al (2001) Prolonged morphine treatment targets delta opioid receptors to neuronal plasma membranes and enhances delta-mediated antinociception. J Neurosci 21(19):7598–7607

Cahill CM, Morinville A, Hoffert C, O'Donnell D, Beaudet A (2003) Up-regulation and trafficking of delta opioid receptor in a model of chronic inflammation: implications for pain control. Pain 101:199–208

Cahill CM, Holdridge SV, Morinville A (2007) Trafficking of delta-opioid receptors and other G-protein-coupled receptors: implications for pain and analgesia. Trends Pharmacol Sci 28(1):23–31

Cai J et al (2014) Endomorphin analogues with mixed mu-opioid (MOP) receptor agonism/delta-opioid (DOP) receptor antagonism and lacking beta-arrestin2 recruitment activity. Bioorg Med Chem 22(7):2208–2219

Chefer VI, Shippenberg TS (2009) Augmentation of morphine-induced sensitization but reduction in morphine tolerance and reward in delta-opioid receptor knockout mice. Neuropsychopharmacology 34(4):887–898

Chen H-J et al (2012) Disruption of delta-opioid receptor phosphorylation at threonine 161 attenuates morphine tolerance in rats with CFA-induced inflammatory hypersensitivity. Neurosci Bull 28(2):182–192

Chung PCS et al (2015) Delta opioid receptors expressed in forebrain GABAergic neurons are responsible for SNC80-induced seizures. Behav Brain Res 278:429–434

Costantino CM, Gomes I, Stockton SD, Lim MP, Devi LA (2012) Opioid receptor heteromers in analgesia. Expert Rev Mol Med 14(April):e9
Daniels DJ et al (2005) Opioid-induced tolerance and dependence in mice is modulated by the distance between pharmacophores in a bivalent ligand series. Proc Natl Acad Sci U S A 102(52):19208–19213
Décaillot FM, Rozenfeld R, Gupta A, Devi LA (2008) Cell surface targeting of mu-delta opioid receptor heterodimers by RTP4. Proc Natl Acad Sci U S A 105(41):16045–16050
Egan TM, North RA (1981) Both mu and delta opiate receptors exist on the same neuron. Science 214(4523):923–924
Erbs E et al (2015) A mu-delta opioid receptor brain atlas reveals neuronal co-occurrence in subcortical networks. Brain Struct Funct 220(2):677–702
Espejo EF, Mir D (1994) Differential effects of weekly and daily exposure to the hot plate on the rat's behavior. Physiol Behav 55(6):1157–1162
Fox-Threlkeld JE et al (1994) Identification of mechanisms and sites of actions of mu and delta opioid receptor activation in the canine intestine. J Pharmacol Exp Ther 268(2):689–700
Fujita W et al (2014) Molecular characterization of eluxadoline as a potential ligand targeting mu-delta opioid receptor heteromers. Biochem Pharmacol 92(3):448–456
Fundytus ME, Schiller PW, Shapiro M, Weltrowska G, Coderre TJ (1995) Attenuation of morphine tolerance and dependence with the highly selective delta-opioid receptor antagonist TIPP[psi]. Eur J Pharmacol 286(1):105–108
Gardon O et al (2014) Expression of mu opioid receptor in dorsal diencephalic conduction system: new insights for the medial habenula. Neuroscience 277:595–609
Gendron L, Pintar JE, Chavkin C (2007) Essential role of mu opioid receptor in the regulation of delta opioid receptor-mediated antihyperalgesia. Neuroscience 150(4):807–817
Gendron L, Cahill CM, Von Zastrow M, Schiller PW, Pineyro G (2016) Molecular pharmacology of δ-opioid receptors. Pharmacol Rev 68:631–700
Gilron I, Watson CPN, Cahill CM, Moulin DE (2006) Neuropathic pain: a practical guide for the clinician. CMAJ 175(3):265–275
Gomes I et al (2000) Heterodimerization of mu and delta opioid receptors: a role in opiate synergy. J Neurosci 20(22):RC110
Gomes I et al (2004) A role for heterodimerization of mu and delta opiate receptors in enhancing morphine analgesia. Proc Natl Acad Sci U S A 101(14):5135–5139
Gomes I et al (2013) Identification of a μ-δ opioid receptor heteromer-biased agonist with antinociceptive activity. Proc Natl Acad Sci U S A 110(29):12072–12077
Gomes I, Gupta A, Bushlin I, Devi LA (2014) Antibodies to probe endogenous G protein-coupled receptor heteromer expression, regulation, and function. Front Pharmacol 5:268
Gomes I et al (2016) G protein-coupled receptor heteromers. Annu Rev Pharmacol Toxicol 56:403–425
Gupta A et al (2010) Increased abundance of opioid receptor heteromers after chronic morphine administration. Sci Signal 3(131):ra54
Hack SP, Bagley EE, Chieng BCH, Christie MJ (2005) Induction of delta-opioid receptor function in the midbrain after chronic morphine treatment. J Neurosci 25(12):3192–3198
He S-Q et al (2011) Facilitation of mu-opioid receptor activity by preventing delta-opioid receptor-mediated codegradation. Neuron 69(1):120–131
Jordan BA, Devi LA (1999) G-protein-coupled receptor heterodimerization modulates receptor function. Nature 399(6737):697–700
Kest B, Lee CE, McLemore GL, Inturrisi CE (1996) An antisense oligodeoxynucleotide to the delta opioid receptor (DOR-1) inhibits morphine tolerance and acute dependence in mice. Brain Res Bull 39(3):185–188
Law P-Y et al (2005) Heterodimerization of mu- and delta-opioid receptors occurs at the cell surface only and requires receptor-G protein interactions. J Biol Chem 280(12):11152–11164

Le Merrer J, Rezai X, Scherrer G, Becker JAJ, Kieffer BL (2013) Impaired hippocampus-dependent and facilitated striatum-dependent behaviors in mice lacking the delta opioid receptor. Neuropsychopharmacology 38(6):1050–1059

Lenard NR, Daniels DJ, Portoghese PS, Roerig SC (2007) Absence of conditioned place preference or reinstatement with bivalent ligands containing mu-opioid receptor agonist and delta-opioid receptor antagonist pharmacophores. Eur J Pharmacol 566(1–3):75–82

Levio S, Cash BD (2017) The place of eluxadoline in the management of irritable bowel syndrome with diarrhea. Ther Adv Gastroenterol 10(9):715–725

Lucido AL, Morinville A, Gendron L, Stroh T, Beaudet A (2005) Prolonged morphine treatment selectively increases membrane recruitment of delta-opioid receptors in mouse basal ganglia. J Mol Neurosci 25(3):207–214

Margolis EB, Fujita W, Devi LA, Fields HL (2017) Two delta opioid receptor subtypes are functional in single ventral tegmental area neurons, and can interact with the mu opioid receptor. Neuropharmacology 123:420–432

Mattia A, Vanderah T, Mosberg HI, Porreca F (1991) Lack of antinociceptive cross-tolerance between [D-Pen2, D-Pen5]enkephalin and [D-Ala2]deltorphin II in mice: evidence for delta receptor subtypes. J Pharmacol Exp Ther 258(2):583–587

Morinville A, Cahill CM, Kieffer B, Collier B, Beaudet A (2004a) Mu-opioid receptor knockout prevents changes in delta-opioid receptor trafficking induced by chronic inflammatory pain. Pain 109(3):266–273

Morinville A et al (2004b) Morphine-induced changes in delta opioid receptor trafficking are linked to somatosensory processing in the rat spinal cord. J Neurosci 24:5549–5559

Nakama-Kitamura M, Doe N (2003) The influence of contextual cue on antinociceptive tolerance and facilitation of memory with morphine. J Pharmacol Sci 92(3):237–244

Pinello C et al (2010) Characterization of an agonist probe for opioid receptor mu 1 (OPRM1)-opioid receptor delta 1 (OPRD1) heterodimerization. National Center for Biotechnology Information, Bethesda

Poole DP et al (2011) Localization and regulation of fluorescently labeled delta opioid receptor, expressed in enteric neurons of mice. Gastroenterology 141(3):982–991.e18

Porreca F, Heyman JS, Mosberg HI, Omnaas JR, Vaught JL (1987) Role of mu and delta receptors in the supraspinal and spinal analgesic effects of [D-Pen2, D-Pen5]enkephalin in the mouse. J Pharmacol Exp Ther 241(2):393–400

Provasi D, Boz MB, Johnston JM, Filizola M (2015) Preferred supramolecular organization and dimer interfaces of opioid receptors from simulated self-association. PLoS Comput Biol 11(3): e1004148

Rothman RB et al (1992) Probing the opioid receptor complex with (+)-trans-SUPERFIT. II. Evidence that mu ligands are noncompetitive inhibitors of the delta cx opioid peptide binding site. Peptides 13(6):1137–1143

Rozenfeld R, Devi LA (2007) Receptor heterodimerization leads to a switch in signaling: beta-arrestin2-mediated ERK activation by mu-delta opioid receptor heterodimers. FASEB J 21(10):2455–2465

Scherrer G et al (2006) Knockin mice expressing fluorescent delta-opioid receptors uncover G protein-coupled receptor dynamics in vivo. Proc Natl Acad Sci U S A 103(25):9691–9696

Scherrer G et al (2009) Dissociation of the opioid receptor mechanisms that control mechanical and heat pain. Cell 137(6):1148–1159

Siuda ER, Carr R 3rd, Rominger DH, Violin JD (2017) Biased mu-opioid receptor ligands: a promising new generation of pain therapeutics. Curr Opin Pharmacol 32:77–84

Spahn V, Stein C (2017) Targeting delta opioid receptors for pain treatment: drugs in phase I and II clinical development. Expert Opin Investig Drugs 26(2):155–160

Stockton SD, Devi LA (2012) Functional relevance of μ-δ opioid receptor heteromerization: a role in novel signaling and implications for the treatment of addiction disorders: from a symposium on new concepts in mu-opioid pharmacology. Drug Alcohol Depend 121(3):167–172

van Rijn RM, Whistler JL (2009) The delta(1) opioid receptor is a heterodimer that opposes the actions of the delta(2) receptor on alcohol intake. Biol Psychiatry 66(8):777–784
Waldhoer M et al (2005) A heterodimer-selective agonist shows in vivo relevance of G protein-coupled receptor dimers. Proc Natl Acad Sci U S A 102(25):9050–9055
Williams D, Devi LA (2010) Escorts take the lead molecular chaperones as therapeutic targets. Prog Mol Biol Transl Sci 91:121–149
Xie W-Y et al (2009) Disruption of Cdk5-associated phosphorylation of residue threonine-161 of the delta-opioid receptor: impaired receptor function and attenuated morphine antinociceptive tolerance. J Neurosci 29(11):3551–3564
Yekkirala AS, Kalyuzhny AE, Portoghese PS (2013) An immunocytochemical-derived correlate for evaluating the bridging of heteromeric mu-delta opioid protomers by bivalent ligands. ACS Chem Biol 8(7):1412–1416
Zhu Y et al (1999) Retention of supraspinal delta-like analgesia and loss of morphine tolerance in delta opioid receptor knockout mice. Neuron 24(1):243–252

Part III

DOPR and Disease

OPRD1 Genetic Variation and Human Disease

Richard C. Crist and Toni-Kim Clarke

Contents

Abstract

The *OPRD1* gene encodes the delta-opioid receptor, which has multiple functions including regulating reward pathways. The gene contains more than 2,000 verified genetic variants but only 2 currently have evidence for specific functions: rs1042114 disrupts maturation of the receptor and rs569356 affects *OPRD1* expression. These polymorphisms and others in the gene have been found to be associated with human diseases. The most reproducible data are associations between opioid addiction and three variants in intron 1 (rs2236861, rs2236857, and rs3766951), which have been described in a number of independent populations. Several publications also point toward an association between anorexia and a haplotype block containing rs569356 and rs533123. Unfortunately the mechanisms underlying these two effects are currently unknown. In contrast, rs1042114 has been linked to Alzheimer's disease through an increasingly well-defined mechanism by which the variant allele reduces

R.C. Crist (✉)
Center for Neurobiology and Behavior, Department of Psychiatry, University of Pennsylvania School of Medicine, 125 South 31st Street, Room 2207, Philadelphia, PA 19104, USA
e-mail: crist@mail.med.upenn.edu

T.-K. Clarke
Division of Psychiatry, University of Edinburgh, Royal Edinburgh Hospital, Edinburgh, UK

E. M. Jutkiewicz (ed.), *Delta Opioid Receptor Pharmacology and Therapeutic Applications*,
Handbook of Experimental Pharmacology 247, DOI 10.1007/164_2016_112

production of the beta-amyloid plaques associated with the disease. Additional studies of *OPRD1* variants are necessary to replicate current findings and to delineate the functional roles of relevant polymorphisms.

Keywords
Addiction • Alzheimer's disease • Anorexia • Delta opioid receptor • OPRD1

1 The Delta-Opioid Receptor Gene

The delta-opioid receptor (DOR) is encoded by the delta-opioid receptor gene (*OPRD1*). The human *OPRD1* gene consists of three known exons located on chromosome 1 between 28812142 and 28863696 bp (UCSC Genome Browser, Build GRCh38/hg38, http://genome.ucsc.edu). *OPRD1* is ~51 kb in length and the bulk of the gene is made up of the first intron, which is more than 45 kb in length. The majority of human genes encode two or more splice variants, whereby the transcribed messenger RNA is alternatively processed to encode different proteins. However, there is little information on alternative splicing in *OPRD1*. One or two potential splice variants have been identified in transcript datasets but there is currently no evidence that these transcripts are translated in vivo (UCSC Genome Browser, Build GRCh38/hg38, http://genome.ucsc.edu). Given the substantial amount of translated splice variants known to be produced by another opioid receptor gene, *OPRM1*, it is possible that alternative splicing in *OPRD1* still occurs but perhaps features significant temporal and spatial regulation.

2 Genetic Diversity in *OPRD1*

The 1000 Genomes Project has identified 2,240 polymorphisms within *OPRD1* and the flanking 5 kb regions (http://www.1000genomes.org). The vast majority of these variants occur within the large first intron (Fig. 1). Although almost 100 coding (exonic) variants have been identified, only 5 have a minor allele frequency greater than 1%. Three of these coding variants primarily occur in specific ethnic groups and are virtually unstudied: rs77585854 (South American and Hispanic populations), rs111622802 (African populations), and rs118175398 (Asian populations). These SNPs may be worth investigating in more targeted studies focusing on those specific populations. The remaining two common coding variants, rs1042114 and rs2234918, are present across a range of ethnicities.

A number of studies originally focused on rs1042114, which causes a phenylalanine amino acid to be replaced by a cysteine residue at position 27 in the N-terminus of the DOR protein sequence. This variant could hypothetically alter ligand binding but no differences were observed in the affinities of the wild-type and variant receptors for many known agonists and antagonists (Leskela et al. 2009). However, the rs1042114 polymorphism has still been shown to affect DOR function

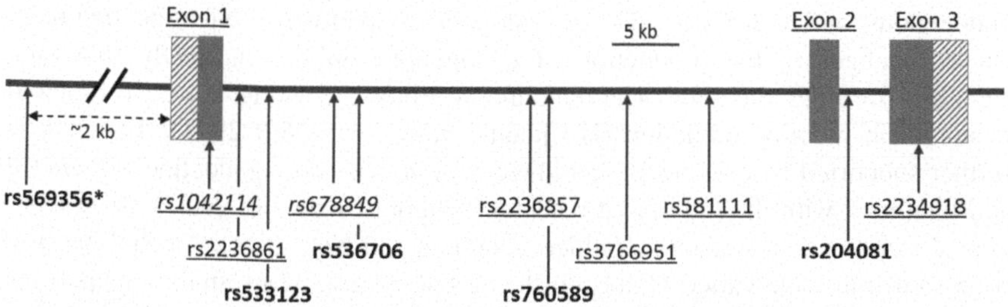

Fig. 1 Diagram of the *OPRD1* gene including the locations of polymorphisms significantly or nominally associated with human disease. *Gray boxes* indicate exons and *boxes* with *diagonal lines* indicate untranslated regions. Indications of associated disease are as follows: *bold* = anorexia nervosa; *asterisk* = obesity; *underlined* = addiction; *italics* = Alzheimer's disease or related phenotypes. SNP and exon locations taken from December 2013 (GRCh38/hg38) build of the human genome in UCSC Genome Browser (http://genome.ucsc.edu)

by affecting the amount of receptors present on the cell surface (Leskela et al. 2009, 2012). The Cys-27 protein accumulates in intracellular compartments due to impaired maturation of the protein to the cell surface and an increased rate of internalization compared to Phe-27 receptors (Leskela et al. 2009). Heterodimers can also be formed between the Phe-27 and Cys-27 forms of the protein, resulting in a dominant-negative effect in which transport of the mature Phe-27 receptor to the surface is also impaired (Leskela et al. 2012). The only other *OPRD1* polymorphism with functional evidence is rs569356, which is located in the promoter region ~2 kb upstream of the transcription start site. Luciferase assays in HEK293 cells found an association between the genotype at the rs569356 and transcription levels (Zhang et al. 2010). Constructs carrying the G allele had increased transcription compared to those with the A allele (Zhang et al. 2010). It is important to note that rs569356 is in linkage disequilibrium (LD) with rs1042114 in populations of both European and African descent ($r^2 = 1.0$), meaning that the polymorphisms are inherited together and the allele at one locus can be used to predict the allele at the other. Although the in vitro findings described above studied each SNP in isolation, the LD between the variants means that in association studies of certain ethnic groups it is impossible to determine which of the SNPs is causative.

3 Drug Abuse and Addiction

Opioid receptors are the targets of a large number of drugs, including heroin and many analgesics, and are therefore intimately involved in opioid addiction. Activation of the mu-opioid receptor or DOR by exogenous opioids results in downstream signaling through reward pathways. Although the opioid receptors are not directly bound by other drugs of abuse, the receptors are still involved in mediating the physiological response to these drugs via other mechanisms. For example, endogenous and exogenous opioids have similar effects at opioid receptors and the endogenous molecules can also activate reward pathways. Nicotine treatment causes changes in the production of endogenous opioids indirectly following signaling through nicotinic acetylcholine receptors, creating a role for opioid receptors in nicotine addiction (Hadjiconstantinou and Neff 2011). This link is further supported by changes in opioid receptor activity during nicotine withdrawal and treatment with the mu-opioid receptor antagonist naloxone causes some withdrawal symptoms in smokers (Hadjiconstantinou and Neff 2011). Alcohol has also been shown to have varied effects on the opioid system. Many studies support the release of opioid ligands following alcohol consumption, although these findings may vary by brain region, the amount of alcohol, and the species being studied (Herz 1997). Opioid antagonists are also able to reduce alcohol intake in rodent models and this effect has been observed with the DOR-specific antagonist naltrindole (Herz 1997). Increases in the endogenous DOR ligand enkephalin have the opposite effect on alcohol consumption, providing additional evidence for the role of the receptor in alcoholism-like behaviors (Herz 1997). It is possible that many of these effects are specific to certain brain regions. Activity at some types of opioid receptors is altered in cocaine addicts and this change occurs in distinct areas of the brain (Yoo et al. 2012).

Addictions are complex phenotypes that are affected by many environmental factors like socioeconomic status and peer groups. Like most human phenotypes, however, there is also a genetic component to substance-use disorders. For example, alcoholism is known to run in families and twin studies have consistently found genetics to account for ~60% of susceptibility to the disease (Edenberg 2011). Similar studies have found cocaine and opioid addiction to be 30–50% hereditary (Kendler et al. 2003; Merikangas et al. 1998). The genetic component of addiction is likely the result of polymorphisms that alter the expression or function of proteins involved in the disorder. Given the known links between the opioid receptors and addiction to an array of substances, it is possible that SNPs in those genes will be associated with susceptibility to addictive behavior.

The earliest case-control studies of *OPRD1* polymorphisms focused primarily on heroin addiction. Despite heroin and abused opioid analgesics primarily targeting the mu-opioid receptor, many of these drugs also have some affinity for DOR and evidence supports a role for DOR in signaling following opioid use (Janecka et al. 2004). Mayer et al. (1997) initially analyzed rs2234918 in a small number of German heroin addicts and found the minor allele of the synonymous SNP to be significantly more common in cases compared to controls (Table 1). However, a follow-up study in another German cohort failed to replicate this finding and the effect was not observed in Chinese heroin addicts either (Franke et al. 1999; Xu et al. 2002). The latter study could have conceivably been the result of ethnic differences between European and Asian populations but the lack of significance in the larger German replication sample suggests that the initial finding may have been a false positive.

Subsequent experiments began to expand their focus beyond opioid addiction and to increase the number of analyzed polymorphisms. An analysis of 18 *OPRD* SNPs in European-American families selected for alcoholism identified no significant

Table 1 *OPRD1* polymorphisms associated with human disease

SNP ID	Location	Variant type	Phenotype	Population	Reference
rs569356	chr1:29136686	Upstream	Anorexia nervosa	European-American	Brown et al. (2007)
			Obesity	European	Kvaloy et al. (2013)
rs1042114	chr1:29138975	Non-synonymous coding	Opioid addiction	European-American	Zhang et al. (2010)
			Beta-amyloid production	N/A	Leskela et al. (2009) and Sarajarvi et al. (2011)
rs2236861	chr1:29139756	Intronic	Opioid addiction	European descent (American/Israeli)	Levran et al. (2008)
				European descent (Australian)	Nelson et al. (2012)
				European	Beer et al. (2013)
rs533123	chr1:29141155	Intronic	Anorexia nervosa	European descent (American/European)	Wang et al. (2011)
rs678849	chr1:29145188	Intronic	Cocaine addiction	African-American	Crist et al. (2013a)
			Opioid addiction treatment	European-American	Crist et al. (2013b)
			Tau/beta-amyloid ratio	European descent (American/Australian)	Roussotte et al. (2014)
rs536706	chr1:29147338	Intronic	Anorexia nervosa	European descent (American/European)	Bergen et al. (2003)
rs2236857	chr1:29161609	Intronic	Opioid addiction	European descent (American/Israeli)	Levran et al. (2008)
				European descent (Australian)	Nelson et al. (2012)
rs760589	chr1:29162465	Intronic	Anorexia nervosa	European descent (American/European)	Bergen et al. (2003)
rs3766951	chr1:29169559	Intronic	Opioid addiction	European descent (American/Israeli)	Levran et al. (2008)
				European descent (Australian)	Nelson et al. (2012)

(continued)

Table 1 (continued)

SNP ID	Location	Variant type	Phenotype	Population	Reference
rs581111	chr1:29175373	Intronic	Opioid addiction treatment	European-American	Clarke et al. (2014)
rs204081	chr1:29186719	Intronic	Anorexia nervosa	European descent (American/European)	Bergen et al. (2003)
rs2234918	chr1:29189597	Synonymous coding	Opioid addiction	European	Mayer et al. (1997)

associations between genotypes and either alcohol or opioid dependence, although the final number of people meeting DSM criteria for opioid dependence was only 83 (Xuei et al. 2007). The analysis in the opioid addict subpopulation was likely lacking in statistical power given the small sample size and would not have been able to detect associations with small or moderate effect sizes. A lack of association between alcoholism and *OPRD1* polymorphisms was also observed in small sample of Taiwanese patients and a larger study examining 11 variants in the gene in European-American drug addicts (Loh el et al. 2004; Zhang et al. 2008). However, the latter analysis did find rs1042114, the non-synonymous SNP in exon 1, to be associated with opioid addiction (Zhang et al. 2008). This study also failed to find any associations with cocaine addiction or overall drug dependence in the European-American population, findings that have since been supported by other published data (Crist et al. 2013a; Maher et al. 2011; Zhang et al. 2008). Although no association has been observed between *OPRD1* variants and cocaine addiction in European-Americans, rs678849 genotype is associated with cocaine addiction in African-Americans (Crist et al. 2013a). As with many of the significant results described above, however, the validity of this effect is still pending replication.

These data suggest that the effects of specific SNPs vary across different ethnic groups and that case-control studies must be careful to control for ethnicity in their analyses. The ethnic differences may be the result of environment that varies across cultures, ethnicity-specific genetic variation, or a combination of both factors. Although this chapter is focused specifically on *OPRD1* polymorphisms and human disease, it is important to remember that the variation in *OPRD1* occurs in the context of a large number of SNPs and copy number variations scattered throughout the genome. This other variation may alter the effects of the *OPRD1* polymorphisms being studied. For example, an analysis of 31 variants across the 3 opioid receptor genes found many interactions between SNPs in *OPRD1* and those in the other opioid receptor genes, *OPRM1* and *OPRK1* (Li and Zhang 2013). Subsets of these interactions were significantly associated with alcohol, cocaine, and opioid addiction (Li and Zhang 2013).

The most replicated results addressing *OPRD1* polymorphisms in addiction phenotypes deal with three SNPs in intron 1: rs2236861, rs2236857, and rs3766951. Levran et al. performed a case-control study of 412 heroin-addicted Caucasians and 184 controls, genotyping 1,350 variants across a range of addiction-related genes (Levran et al. 2008). The three *OPRD1* SNPs showed nominally significant associations with heroin addiction, although none of these associations were significant after multiple testing, likely due to the relatively small number of samples (Levran et al. 2008). The combined effect of rs510769 in *OPRM1* and rs2236861 was also nominally significant (Levran et al. 2008). A later study of polymorphisms in opioid receptor genes analyzed 1,459 heroin addicts and 1,495 controls from an Australian cohort and was able to confirm some of the findings described by Levran et al. (Nelson et al. 2012). Both rs2236857 and rs3766951 were significantly associated with case-control status, as were two additional variants in high LD with rs2236857 (Nelson et al. 2012). Although rs2236861 was not statistically significant after correction for multiple testing, the SNP was still nominally significant (Nelson et al. 2012). In addition, another analysis of rs2236861 in a study of Austrian heroin users undergoing treatment found a significant association with addiction status (Beer et al. 2013). These three studies, when taken together, indicate that two or three variants within intron 1 of *OPRD1* are affecting susceptibility to opioid addiction in people of European descent.

4 Pharmacogenetics of Addiction Treatment

Genetic differences can identify individuals with increased susceptibility to particular addictions. SNPs that are associated with this kind of susceptibility may also be involved in the efficacy of addiction treatments, especially in cases in which these treatments interact with targets similar to those bound by the drug of abuse. Two examples of such medications are methadone and buprenorphine, opioid addiction pharmacotherapeutics that target the same primary receptor as illicit opioids. Two pharmacogenetic analyses focused on *OPRD1* were performed on the START clinical trial, which compared methadone and Suboxone (buprenorphine/naloxone) in a multi-site trial across the United States. The first study analyzed six polymorphisms and identified one intronic SNP, rs678849, that predicted the efficacy of both medications in African-Americans as measured by opioid-positive urine drug screens (Crist et al. 2013b). Methadone patients carrying the T allele were more likely to test positive than C/C individuals, while the opposite effect was observed in the Suboxone arm of the trial (Crist et al. 2013b). If confirmed, this finding would indicate that rs678849 could be genotyped prospectively in African-American opioid addicts and used to personalize treatments most likely to work for specific patients. It is important to note that the effect of rs678849 was not observed in patients of European descent, further supporting data from addiction case-control studies that indicate significant genetic differences between ethnic groups.

A second study of the START cohort looked for sex-specific pharmacogenetic effects of the same six variants in European-Americans, who made up the majority of the patient population in the trial. Another polymorphism in intron 1, rs581111, predicted outcome in the female Suboxone patients but not males (Clarke et al. 2014). Women carrying the G allele had a significantly higher percentage of opioid-positive urine tests than those with the T/T genotype (Clarke et al. 2014). As with the initial analysis in African-Americans, replications in independent patient populations and additional prospective pharmacogenetic trials are imperative before the findings can be translated into FDA-approved prognostic indicators.

5 Eating Disorders and Obesity

The breadth of addiction-related findings in the *OPRD1* field is unsurprising given the intimate involvement of the opioid system with the reward pathways that underlie most drugs of abuse. However, the opioid receptors are involved in many other aspects of neurobiology and behavior. Eating disorders and obesity might have connections with the opioid system given the rewarding effects of food consumption. Supporting this hypothesis, rodent studies have demonstrated increased opioid receptor binding following consumption of fatty foods and liquids and some opioid antagonists have been shown to prevent binge eating in rats (Mathes et al. 2009).

A linkage study of anorexia nervosa (AN) patients and unaffected family members identified a significant locus on chromosome 1p33-36, a region which includes the *OPRD1* gene (Grice et al. 2002; Kaye et al. 2000). Subsequent research in AN cases and controls found associations between three of the five tested SNPs in *OPRD1* and AN status (Bergen et al. 2003). Allelic and genotypic association analyses were nominally significant for rs204081, while only the allelic association tests were nominally significant for rs536706 and rs760589 (Bergen et al. 2003). Several haplotypes containing those variants were also nominally associated with AN status.

Brown et al. attempted to confirm the association between *OPRD1* and AN by genotyping six SNPs. An upstream SNP, rs569356, had highly significant allelic ($p = 0.0008$) and genotypic ($p = 0.0011$) associations with the disorder (Brown et al. 2007). The variant was also significant when the cases were limited to either restricting anorexia nervosa or binge-purging anorexia nervosa, suggesting that *OPRD1* may be relevant across disease subtypes (Brown et al. 2007). rs569356 is ~2 kb upstream of the first exon of *OPRD1* and has been shown to affect transcription in vitro. However, the SNP is in perfect linkage disequilibrium ($r^2 = 1.0$) with a number of other variants in populations of European descent. Included in this haplotype block is the non-synonymous coding SNP rs1042114, which has also been shown to affect DOR function (1000 Genomes Project, http://www.1000genomes.org).

Two genome-wide association studies (GWAS) have compared AN cases to healthy controls. Neither GWAS found any associations that reached a genome-wide level of significance ($p = 5 \times 10^{-8}$), although this may be due primarily to the relatively small number of samples (Boraska et al. 2014; Wang et al. 2011). Recent meta-analyses of schizophrenia by the Psychiatric Genetics Consortium have suggested that psychiatric disease is likely the result of many loci with modest odds ratios. These studies have further indicated that tens of thousands of samples are required to find more than a handful of significant loci (Ripke 2014). The two AN GWAS analyzed only 1,033 and 5,551 cases and the studies are therefore likely to be statistically underpowered to detect associations (Boraska et al. 2014; Wang et al. 2011).

Although no SNPs reached genome-wide significance, Wang et al. reported that rs533123 in OPRD1 was nominally associated with AN status ($p = 0.0015$, OR = 1.2) (Wang et al. 2011). This SNP is in high LD ($r^2 = 0.95$) with rs569356 and the direction of the effect was the same as previously reported (Brown et al. 2007). A meta-analysis of these rs533123 data and the previous rs569356 data from Brown et al. was also significant ($p = 1.76 \times 10^{-5}$), again adding support for an association between this *OPRD1* haplotype block and AN (Wang et al. 2011).

An additional GWAS analyzed the TwinsUK cohort dataset for associations with six phenotypes related to eating disorders: (1) drive for thinness, (2) body dissatisfaction, (3) bulimia, (4) weight fluctuation, (5) breakfast-skipping behavior, and (6) childhood obsessive-compulsive personality disorder (Boraska et al. 2012). As with the other AN GWAS, no SNPs were significant at a genome-wide level likely due to statistical power issues. However, rs1042114 had a nominally significant association ($p = 0.0048$) with body dissatisfaction rating (Boraska et al. 2012).

This effect fits well with the previous connections to AN since sufferers of the disorder report high rates of body dissatisfaction (Sala et al. 2012). More targeted studies of select OPRD1 polymorphisms in independent AN populations are necessary to attempt replication of these findings.

On the opposite end of the spectrum from anorexia, obesity is now a significant problem in many countries throughout the world with the spread of the Western diet and sedentary lifestyles. A 2013 study of the Norwegian HUNT cohort attempted to identify associations between BMI and genotype at polymorphisms previously linked to obesity and/or eating disorders (Kvaloy et al. 2013). The list of analyzed SNPs included rs569356, and a significant interaction between sex and rs569356 genotype was identified in the study (Kvaloy et al. 2013). Males carrying the minor allele were more likely to have higher body mass indices in both adolescence and adulthood but this effect was not present in females in the cohort (Kvaloy et al. 2013). In contrast, the previous AN studies found the minor allele of rs569356 to be associated with increased risk of anorexia. It is important to note, however, that most cases of anorexia occur in women and more than 95% of the samples analyzed in the anorexia studies described above were provided by female patients. These data therefore suggest that this SNP has highly divergent effects in men and women that warrant additional studies. Confirmation of this finding would also indicate that *OPRD1* has differential regulation between the sexes and that sex-specific effects may be relevant to phenotypes beyond those related to weight.

6 Alzheimer's Disease

Alzheimer's disease (AD) is the most common cause of dementia worldwide and the disease has a significant hereditary component (Ballard et al. 2011). Unfortunately only 11 significant loci were identified in the most recent GWAS meta-analysis, even though genetics have been estimated to account for 70% of susceptibility to the disease (Ballard et al. 2011; Lambert et al. 2013). The study included more than 17,000 cases and 37,000 controls, suggesting that AD is a highly polygenic disease with many genetic variants contributing small effects (Lambert et al. 2013). More targeted studies of genes known to be involved in the pathogenesis of the disease may therefore be more effective until studies with significantly larger samples sizes can be performed.

Mouse model experiments have indicated a strong connection between DOR and AD. During the development of AD, problems arise with the cleavage of the amyloid precursor protein (APP) and subsequent clearance of the beta-amyloid product. These issues result in the formation of neurotoxic amyloid plaques in the brains of AD patients. DOR has been shown to be essential for proper internalization and localization of the enzymes responsible for cleaving APP (Teng et al. 2010). In a mouse model of AD caused by dual expression of mutant APP and mutant presenilin-1, knockdown of DOR results in reduced plaque formation and treatment with the DOR antagonist naltrindole prevents the learning and memory issues normally observed in this model system (Teng et al. 2010). In contrast, imaging data indicates that decreased levels of DOR are already present in some regions of AD postmortem brains, perhaps suggesting some kind of negative feedback mechanism or spatial regulation (Mathieu-Kia et al. 2001).

The effects of several *OPRD1* polymorphisms on fMRI data were analyzed in healthy elderly European-Americans and Australians between 20 and 30 years of age (Roussotte et al. 2014). One of the SNPs previously linked to drug-dependence phenotypes, rs678849, was found to be associated with brain volume differences in both the American and Australian cohorts (Roussotte et al. 2014). The authors also found the SNP to be associated with the tau/beta-amyloid ratio in the healthy elderly population (Roussotte et al. 2014). Tau and beta-amyloid are biomarkers that have previously been linked to AD and the ratio of the two is thought to be associated with the risk of developing the disease (Maddalena et al. 2003). Healthy individuals with the C/C genotype at rs678849 had increased tau/beta-amyloid ratios compared to individuals with the C/T or T/T genotypes ($p = 0.027$), suggesting that they may have increased susceptibility to AD (Roussotte et al. 2014). As with the addiction findings for rs678849, there is currently no understanding of the specific mechanism by which the SNP is related to these phenotypes. However, the location of rs678849 in the first intron of *OPRD1* would suggest that a change in gene expression is a potential functional consequence of carrying the minor allele. Such a change might alter the risk of developing AD through the mechanism outlined by Teng et al. (2010).

A more specific role in AD has been suggested for the non-synonymous variant rs1042114. As mentioned above, the presence of the minor allele of rs1042114 changes a phenylalanine to a cysteine. The cysteine-containing protein has impaired trafficking to the cell surface and constitutive internalization once there, resulting in reduced numbers of mature receptors and decreased amyloid beta (Leskela et al. 2009; Sarajarvi et al. 2011). The variant protein can also impair trafficking of the phenylalanine-containing receptor through dominant negative effects (Leskela et al. 2012). Given the findings of Teng et al. (2010), the rs1042114 effect is likely a loss of function: reduced receptor density at the cell surface might mimic *OPRD1* knockdown and the subsequent reductions in amyloid plaque accumulation. Sarajarvi et al. however, have demonstrated that rs1042114 heterozygotes are overrepresented in Alzheimer's patients and that heterozygous patients have increased activity of the secretase enzymes responsible for producing beta-amyloid (Sarajarvi et al. 2011, 2015). Since the effect of a loss-of-function allele would likely be most notable in individuals homozygous for the minor allele, these data suggest that all or part of the observed phenotype may be a gain-of-function effect (Leskela et al. 2009, 2012; Sarajarvi et al. 2011). Further research is necessary to settle this issue and determine if naltrindole or related compounds are potentially relevant treatments for patients depending on their rs1042114 genotype. Given the clearly defined function of DOR in the pathogenesis of AD, other SNPs found to be associated with *OPRD1* expression or function may also warrant further study.

7 Conclusion

Associations have certainly been observed between *OPRD1* polymorphisms and several human diseases. However, two key issues that plague many genetic studies have kept these findings from being translated into the clinical world: replication and mechanism. In a minority of cases, such as those linking certain *OPRD1* polymorphisms and opioid addiction, a genetic association study has been supported by subsequent studies in other cohorts. This is the exception rather than the rule. In many cases replication never occurs for logistical reasons. For example, an independent population of appropriate size to reproduce the findings may not exist and it may not be financially feasible to collect a new set of samples. Problems can also occur even if the replication study is performed: many genetic associations do not replicate in other sample sets. These failures may represent false positives in the initial analyses but may also reflect confounding variables or statistical power issues. For example, significant associations may not replicate in different ethnic groups if the SNP of interest is not in linkage disequilibrium with the causative variant. Differences between cohorts in subject age or sex may also affect analyses. Comorbidities, such as significant levels of polydrug abuse in addict populations, can present additional problems. Statistical power problems are especially prevalent when researching relatively uncommon diseases or minority populations, which limits the number of available subjects and decreases the chances of collecting samples from a sufficiently large number of patients. Small follow-up studies may fail to replicate previous findings solely because the studies are underpowered to detect associations with the expected effect sizes. This problem is particularly noticeable in GWAS since those studies analyze hundreds of thousands of variants and require significant correction for multiple testing.

The lack of mechanistic data is another problem that prevents the translational relevance of many of these *OPRD1* findings. Few of the genetic variants found to be significantly associated with human disease have clearly defined functions. Only rs1042114 and rs569356 have functional effects that are not only hypothesized, but have also been demonstrated. In in vitro systems, rs1042114 genotype affects DOR maturation, while rs569356 genotype is associated with *OPRD1* expression. However, there is currently no in vivo data confirming that either SNP affects DOR function in humans, particularly in the brain. Furthermore, the specific role of DOR in human diseases other than Alzheimer's disease has not been well established. While it is logical to hypothesize that DOR is relevant in drug addiction or eating disorders, given the links those diseases have with reward pathways, there is limited understanding of the receptor's actual involvement. Nor has there been any significant differentiation between the functions of DOR and the mu-opioid receptor, even in opioid addiction in which the drugs of abuse target the mu-opioid receptor. The role of DOR in specific diseases and the functions of the SNPs involved will have to be determined before therapeutics can be intelligently designed around these *OPRD1* genotypes.

References

Ballard C, Gauthier S, Corbett A, Brayne C, Aarsland D, Jones E (2011) Alzheimer's disease. Lancet 377(9770):1019–1031

Beer B, Erb R, Pavlic M, Ulmer H, Giacomuzzi S, Riemer Y et al (2013) Association of polymorphisms in pharmacogenetic candidate genes (OPRD1, GAL, ABCB1, OPRM1) with opioid dependence in European population: a case-control study. PLoS One 8(9):e75359

Bergen AW, van den Bree MB, Yeager M, Welch R, Ganjei JK, Haque K et al (2003) Candidate genes for anorexia nervosa in the 1p33-36 linkage region: serotonin 1D and delta opioid receptor loci exhibit significant association to anorexia nervosa. Mol Psychiatry 8(4):397–406

Boraska V, Davis OS, Cherkas LF, Helder SG, Harris J, Krug I et al (2012) Genome-wide association analysis of eating disorder-related symptoms, behaviors, and personality traits. Am J Med Genet B Neuropsychiatr Genet 159B(7):803–811

Boraska V, Franklin CS, Floyd JA, Thornton LM, Huckins LM, Southam L et al (2014) A genome-wide association study of anorexia nervosa. Mol Psychiatry 19(10):1085–1094

Brown KM, Bujac SR, Mann ET, Campbell DA, Stubbins MJ, Blundell JE (2007) Further evidence of association of OPRD1 & HTR1D polymorphisms with susceptibility to anorexia nervosa. Biol Psychiatry 61(3):367–373

Clarke TK, Crist RC, Ang A, Ambrose-Lanci LM, Lohoff FW, Saxon AJ et al (2014) Genetic variation in OPRD1 and the response to treatment for opioid dependence with buprenorphine in European-American females. Pharmacogenomics J 14(3):303–308

Crist RC, Ambrose-Lanci LM, Vaswani M, Clarke TK, Zeng A, Yuan C et al (2013a) Case-control association analysis of polymorphisms in the delta-opioid receptor, OPRD1, with cocaine and opioid addicted populations. Drug Alcohol Depend 127(1–3):122–128

Crist RC, Clarke TK, Ang A, Ambrose-Lanci LM, Lohoff FW, Saxon AJ et al (2013b) An intronic variant in OPRD1 predicts treatment outcome for opioid dependence in African-Americans. Neuropsychopharmacology 38(10):2003–2010

Edenberg HJ (2011) Common and rare variants in alcohol dependence. Biol Psychiatry 70(6):498–499

Franke P, Nothen MM, Wang T, Neidt H, Knapp M, Lichtermann D et al (1999) Human delta-opioid receptor gene and susceptibility to heroin and alcohol dependence. Am J Med Genet 88(5):462–464

Grice DE, Halmi KA, Fichter MM, Strober M, Woodside DB, Treasure JT et al (2002) Evidence for a susceptibility gene for anorexia nervosa on chromosome 1. Am J Hum Genet 70(3):787–792

Hadjiconstantinou M, Neff NH (2011) Nicotine and endogenous opioids: neurochemical and pharmacological evidence. Neuropharmacology 60(7–8):1209–1220

Herz A (1997) Endogenous opioid systems and alcohol addiction. Psychopharmacology (Berl) 129(2):99–111

Janecka A, Fichna J, Janecki T (2004) Opioid receptors and their ligands. Curr Top Med Chem 4(1):1–17

Kaye WH, Lilenfeld LR, Berrettini WH, Strober M, Devlin B, Klump KL et al (2000) A search for susceptibility loci for anorexia nervosa: methods and sample description. Biol Psychiatry 47(9):794–803

Kendler KS, Jacobson KC, Prescott CA, Neale MC (2003) Specificity of genetic and environmental risk factors for use and abuse/dependence of cannabis, cocaine, hallucinogens, sedatives, stimulants, and opiates in male twins. Am J Psychiatry 160(4):687–695

Kvaloy K, Kulle B, Romundstad P, Holmen TL (2013) Sex-specific effects of weight-affecting gene variants in a life course perspective – the HUNT study, Norway. Int J Obes (Lond) 37 (9):1221–1229

Lambert JC, Ibrahim-Verbaas CA, Harold D, Naj AC, Sims R, Bellenguez C et al (2013) Meta-analysis of 74,046 individuals identifies 11 new susceptibility loci for Alzheimer's disease. Nat Genet 45(12):1452–1458

Leskela TT, Markkanen PM, Alahuhta IA, Tuusa JT, Petaja-Repo UE (2009) Phe27Cys polymorphism alters the maturation and subcellular localization of the human delta opioid receptor. Traffic 10(1):116–129

Leskela TT, Lackman JJ, Vierimaa MM, Kobayashi H, Bouvier M, Petaja-Repo UE (2012) Cys-27 variant of human delta-opioid receptor modulates maturation and cell surface delivery of Phe-27 variant via heteromerization. J Biol Chem 287(7):5008–5020

Levran O, Londono D, O'Hara K, Nielsen DA, Peles E, Rotrosen J et al (2008) Genetic susceptibility to heroin addiction: a candidate gene association study. Genes Brain Behav 7(7):720–729

Li Z, Zhang H (2013) Analyzing interaction of mu-, delta- and kappa-opioid receptor gene variants on alcohol or drug dependence using a pattern discovery-based method. J Addict Res Ther Suppl 7:007

Loh el W, Fann CS, Chang YT, Chang CJ, Cheng AT (2004) Endogenous opioid receptor genes and alcohol dependence among Taiwanese Han. Alcohol Clin Exp Res 28(1):15–19

Maddalena A, Papassotiropoulos A, Muller-Tillmanns B, Jung HH, Hegi T, Nitsch RM et al (2003) Biochemical diagnosis of Alzheimer disease by measuring the cerebrospinal fluid ratio of phosphorylated tau protein to beta-amyloid peptide42. Arch Neurol 60(9):1202–1206

Maher BS, Vladimirov VI, Latendresse SJ, Thiselton DL, McNamee R, Kang M et al (2011) The AVPR1A gene and substance use disorders: association, replication, and functional evidence. Biol Psychiatry 70(6):519–527

Mathes WF, Brownley KA, Mo X, Bulik CM (2009) The biology of binge eating. Appetite 52 (3):545–553

Mathieu-Kia AM, Fan LQ, Kreek MJ, Simon EJ, Hiller JM (2001) Mu-, delta- and kappa-opioid receptor populations are differentially altered in distinct areas of postmortem brains of Alzheimer's disease patients. Brain Res 893(1–2):121–134

Mayer P, Rochlitz H, Rauch E, Rommelspacher H, Hasse HE, Schmidt S et al (1997) Association between a delta opioid receptor gene polymorphism and heroin dependence in man. Neuroreport 8(11):2547–2550

Merikangas KR, Stolar M, Stevens DE, Goulet J, Preisig MA, Fenton B et al (1998) Familial transmission of substance use disorders. Arch Gen Psychiatry 55(11):973–979

Nelson EC, Lynskey MT, Heath AC, Wray N, Agrawal A, Shand FL et al (2012) Association of OPRD1 polymorphisms with heroin dependence in a large case-control series. Addict Biol 19:111–121

Ripke S (2014) Biological insights from 108 schizophrenia-associated genetic loci. Nature 511 (7510):421–427

Roussotte FF, Jahanshad N, Hibar DP, Sowell ER, Kohannim O, Barysheva M et al (2014) A commonly carried genetic variant in the delta opioid receptor gene, OPRD1, is associated with smaller regional brain volumes: replication in elderly and young populations. Hum Brain Mapp 35(4):1226–1236

Sala L, Mirabel-Sarron C, Pham-Scottez A, Blanchet A, Rouillon F, Gorwood P (2012) Body dissatisfaction is improved but the ideal silhouette is unchanged during weight recovery in anorexia nervosa female inpatients. Eat Weight Disord 17(2):e109–e115

Sarajarvi T, Tuusa JT, Haapasalo A, Lackman JJ, Sormunen R, Helisalmi S et al (2011) Cysteine 27 variant of the delta-opioid receptor affects amyloid precursor protein processing through altered endocytic trafficking. Mol Cell Biol 31(11):2326–2340

Sarajarvi T, Marttinen M, Natunen T, Kauppinen T, Makinen P, Helisalmi S et al (2015) Genetic variation in delta-opioid receptor associates with increased beta- and gamma-secretase activity in the late stages of Alzheimer's disease. J Alzheimers Dis 48(2):507–516

Teng L, Zhao J, Wang F, Ma L, Pei G (2010) A GPCR/secretase complex regulates beta- and gamma-secretase specificity for Abeta production and contributes to AD pathogenesis. Cell Res 20(2):138–153

Wang K, Zhang H, Bloss CS, Duvvuri V, Kaye W, Schork NJ et al (2011) A genome-wide association study on common SNPs and rare CNVs in anorexia nervosa. Mol Psychiatry 16(9):949–959

Xu K, Liu XH, Nagarajan S, Gu XY, Goldman D (2002) Relationship of the delta-opioid receptor gene to heroin abuse in a large Chinese case/control sample. Am J Med Genet 110(1):45–50

Xuei X, Flury-Wetherill L, Bierut L, Dick D, Nurnberger J Jr, Foroud T et al (2007) The opioid system in alcohol and drug dependence: family-based association study. Am J Med Genet B Neuropsychiatr Genet 144B(7):877–884

Yoo JH, Kitchen I, Bailey A (2012) The endogenous opioid system in cocaine addiction: what lessons have opioid peptide and receptor knockout mice taught us? Br J Pharmacol 166(7):1993–2014

Zhang H, Kranzler HR, Yang BZ, Luo X, Gelernter J (2008) The OPRD1 and OPRK1 loci in alcohol or drug dependence: OPRD1 variation modulates substance dependence risk. Mol Psychiatry 13(5):531–543

Zhang H, Gelernter J, Gruen JR, Kranzler HR, Herman AI, Simen AA (2010) Functional impact of a single-nucleotide polymorphism in the OPRD1 promoter region. J Hum Genet 55(5):278–284

The Delta Opioid Receptor in Pain Control

Khaled Abdallah and Louis Gendron

Contents

K. Abdallah
Département de pharmacologie-physiologie, Université de Sherbrooke, Sherbrooke, QC, Canada

Institut de pharmacologie de Sherbrooke, Université de Sherbrooke, Sherbrooke, QC, Canada

Faculté de médecine et des sciences de la santé, Université de Sherbrooke, Sherbrooke, QC, Canada

Centre de recherche du CHUS, Sherbrooke, QC, Canada

L. Gendron (✉)
Département de pharmacologie-physiologie, Université de Sherbrooke, Sherbrooke, QC, Canada

Institut de pharmacologie de Sherbrooke, Université de Sherbrooke, Sherbrooke, QC, Canada

Faculté de médecine et des sciences de la santé, Université de Sherbrooke, Sherbrooke, QC, Canada

Centre de recherche du CHUS, Sherbrooke, QC, Canada

Département d'anesthésiologie, Université de Sherbrooke, Sherbrooke, QC, Canada

Quebec Pain Research Network, Sherbrooke, QC, Canada
e-mail: Louis.Gendron@USherbrooke.ca

E. M. Jutkiewicz (ed.), *Delta Opioid Receptor Pharmacology and Therapeutic Applications*, Handbook of Experimental Pharmacology 247, DOI 10.1007/164_2017_32

Abstract

Nowadays, the delta opioid receptor (DOPr) represents a promising target for the treatment of chronic pain and emotional disorders. Despite the fact that they produce limited antinociceptive effects in healthy animals and in most acute pain models, DOPr agonists have shown efficacy in various chronic pain models. In this chapter, we review the progresses that have been made over the last decades in understanding the role played by DOPr in the control of pain. More specifically, the distribution of DOPr within the central nervous system and along pain pathways is presented. We also summarize the literature supporting a role for DOPr in acute, tonic, and chronic pain models, as well as the mechanisms regulating its activity under specific conditions. Finally, novel compounds that have make their way to clinical trials are discussed.

Keywords

Analgesia • Antinociception • Delta opioid receptor • Pain • Pain pathways

1 The Opioid System

While opium has been used for centuries, the endogenous opioid system was only discovered in the mid-1970s. Two pentapeptides named methionine- and leucine-enkephalin were first identified and characterized by Hughes and colleagues in 1975 (Hughes et al. 1975). Rapidly thereafter, endorphin (Li and Chung 1976; Loh et al. 1976) and dynorphin (Goldstein et al. 1979) were also discovered. More recently, an opioid-like peptide named nociceptin/orphanin was also identified (Meunier et al. 1995; Reinscheid et al. 1995). The endogenous opioid/opioid-like ligands are derived from four distinct precursors, namely proenkephalin, pro-opiomelanocortin (POMC), prodynorphin, and pronociceptin/orphanin. The synthesis of the opioid peptides therefore depends on the activity of endo- or carboxypeptidases (for more details on the discovery of the opioid peptides, see Akil et al. 1998; Darland et al. 1998; Snyder and Pasternak 2003).

Opioid peptides bind to three major receptor subtypes, namely mu (MOPr), delta (DOPr), and kappa (KOPr) opioid receptors. These receptors are, respectively, encoded by the *oprm1*, *oprd1*, and *oprk1* genes. The opioid-like peptide nociceptin/orphanin rather binds to the opioid-like receptor called Orphanin FQ/nociceptin receptor (NOPr). All four receptors belong to the superfamily of G protein-coupled receptors (GPCRs) and exhibit a high sequence homology in their protein structure and genomic organization (Kieffer and Gaveriaux Ruff 2002; Stevens 2009).

Opioids (and opioid receptors) are particularly well known for their important effects in controlling pain. However, opioids are also involved in reward, addiction, neuroprotection, and many other physiological processes such as respiration, gastrointestinal motility, as well as in the endocrine and the immune systems (for reviews, see Kieffer and Gaveriaux-Ruff 2002; Kieffer and Evans 2009; Pradhan et al. 2011; Sauriyal et al. 2011; Chu Sin Chung and Kieffer 2013; Lutz and Kieffer 2013; Gendron et al. 2015). Indeed, MOPr agonists are widely prescribed for the management of pain, although their pronounced unwanted effects (constipation, respiratory depression, sedation, tolerance) often limit their usage (McQuay 1999; Al-Hasani and Bruchas 2011). By contrast, DOPr agonists also produce pain relief but they were shown to have fewer unwanted effects than MOPr agonists (Dondio et al. 2001; Petrillo et al. 2003; Gallantine and Meert 2005) and do not induce tolerance in various animal models (Dondio et al. 2001; Mika et al. 2001; Beaudry et al. 2009). DOPr agonists therefore appear as a good and promising alternative for the treatment of chronic pain (Pradhan et al. 2011). It is worth noting that DOPr agonists also have anxiolytic, anti-depressive, analgesic, and cardio- and neuroprotective effects (Pradhan et al. 2011; Headrick et al. 2015).

In this chapter, we will discuss the role and the functions of DOPr in pain control. In particular, we will describe the distribution of DOPr along the pain pathways and summarize the literature supporting a role for DOPr in the treatment of pain. An overview of novel compounds and their effects in clinical trials will also be provided.

2 The Delta Opioid Receptor: A One-of-a-Kind

2.1 Cloning of DOPr

In 1992, two distinct groups successfully identified an opioid binding site from NG108-15 cells (Evans et al. 1992; Kieffer et al. 1992). Using similar approaches, both groups concomitantly cloned the mouse DOPr by creating a random cDNA library from the RNA of these cells. The cloning of MOPr and KOPr followed soon afterward (Chen et al. 1993; Meng et al. 1993; Minami et al. 1993; Wang et al. 1993).

Oprd1, the gene encoding DOPr, was identified and its chromosomal localization determined. In humans, *oprd1* is located in the distal part of the short arm of chromosome 1. Interestingly, although they share a similar genomic structure and a high sequence homology (Zaki et al. 1996; Chaturvedi et al. 2000), genes encoding MOPr and KOPr are found on different chromosomes, namely in the long arm of chromosomes 6 and 8, respectively (Befort et al. 1994; Wang et al. 1994; Yasuda et al. 1994). The detailed structure of *oprd1* and its translational and epigenetic regulation have been recently reviewed elsewhere (Wei and Loh 2011; Gendron et al. 2016).

2.2 DOPr Structure and Signaling

As a member of the GPCR superfamily, DOPr contains seven hydrophobic transmembrane domains connected by intra- and extracellular loops. DOPr also possesses N- and C-terminal tails, respectively, at its extra- and intracellular ends.

The crystal structures of all three opioid receptors have been recently resolved (Granier et al. 2012; Manglik et al. 2012; Wu et al. 2012). The structures revealed a well-conserved amino acid backbone among the three receptors in the lower part of the binding pocket, a region important for the recognition of the morphinan group. This portion of the binding pocket interacts with the "message segment" of the ligand which is responsible for its efficacy. The upper part of the binding pocket is, however, divergent among the three receptors and its interaction with the distinct "address" segment of the ligand is responsible for receptor selectivity (Granier et al. 2012). DOPr was also found to contain a sodium allosteric binding site regulating biased signaling and constitutive activity (Fenalti et al. 2014).

As MOPr and KOPr, DOPr interacts with numerous proteins and signaling partners (Gendron et al. 2016). When activated, conformational changes within the receptor and its transmembrane domains are leading to the activation of multiple signaling pathways. In particular, the G protein subunits $G\alpha_{i/o}$ and $G\beta\gamma$ dissociate from each other and act on various intracellular effectors. The activation of the G protein modifies the activity of calcium (*P*/*Q*-, *N*- and *L*-type) and potassium channels (G protein gated inwardly rectifying potassium, Kir3) and inhibits adenylyl cyclase activity (reducing the level of intracellular cAMP). These events produce a decrease in neuronal excitability and modifications of gene expression (Kieffer and Evans 2009; Al-Hasani and Bruchas 2011; Gendron et al. 2016). Following agonist stimulation, DOPr also undergoes rapid phosphorylation by G protein-regulated kinases (GRKs). Phosphorylation of DOPr on its C-terminal tail by GRKs is followed by the recruitment of β-arrestins and internalization of the receptor via clathrin-coated vesicles (endocytosis). After internalization, GPCRs are either recycled back to the plasma membrane or undergo degradation (Bie and Pan 2007). While MOPr is mainly recycled back to the plasma membrane, DOPr was shown to be primarily degraded through the lysosomal pathway (Tsao and von Zastrow 2000; Finn and Whistler 2001; Whistler et al. 2002). Several motifs within the receptor are involved in controlling this process. In particular, specific interactions with distinct sorting proteins are routing DOPr either to the degradation or to the recycling pathways. As an example, the Na+/H+ exchanger regulatory factor (NHERF), also called ERM-binding phosphoprotein 50 (EBP50), and the *N*-ethylmaleimide sensitive factor (NSF) were found to be important for the recycling of DOPr (Heydorn et al. 2004; Bie et al. 2010). By contrast, GPCR associated sorting protein (GASP) and sorting nexin-1 (SNX-1) were shown to sort the receptor to the degradation pathway. The C-terminal tail of DOPr seems to have a high affinity for GASP and SNX-1 (Whistler et al. 2002; Heydorn et al. 2004; Simonin et al. 2004). Indeed, swapping the C-terminal tail of DOPr with that of MOPr was shown to shift the fate of the receptor from the degradation toward the recycling pathway (Whistler et al. 2002). Similarly, GASP inactivation was shown to reduce the amount of DOPr in lysosomal compartments and to inhibit its downregulation following agonist stimulation (Whistler et al. 2002).

2.3 DOPr Expression and Distribution

The expression of DOPrs in the mammalian central nervous system has been widely investigated. Precisely, the distribution of DOPr mRNA and binding sites was studied by in situ hybridization, autoradiography, and/or immunohistochemistry (Mansour et al. 1987, 1993, 1994; Sharif and Hughes 1989; George et al. 1994; Cahill et al. 2001a; Pradhan and Clarke 2005; Peng et al. 2012). In mice, the expression of the opioid receptors begins at early developmental stages in the CNS and in peripheral tissues (Zhu et al. 1998). MOPr and KOPr mRNA were, respectively, detected in basal ganglia and midbrain as early as at embryonic stage E11.5. By contrast, the expression of DOPr mRNA only begins at E13.5 in the pons and the hypothalamus. Compared to MOPr and KOPr, DOPr remains restricted within a few brain regions including the caudate–putamen, the olfactory tubercle, and the parabrachial nucleus until late in the development. In dorsal root ganglia (DRGs), DOPr mRNA was detected as early as the embryonic stage E12.5 while it only appears at E15.5 in the ventral part of the spinal cord. Surprisingly, the expression of DOPr mRNA in the mouse dorsal horn of the spinal cord only appears at E17.5 (Zhu et al. 1998).

Although opioid binding sites have been observed by autoradiography in the developing embryo (Kent et al. 1981), specific binding for DOPr has not been observed in rodents before the second week after birth (McDowell and Kitchen 1986; Negri et al. 1997). However, DPDPE-induced GTPγS binding was reported in the caudate–putamen at E12.5 and at E17.5 in the pons and the hypothalamus, suggesting the existence of functional DOPrs at these stages, at least in mice (Nitsche and Pintar 2003).

In the adults, the three opioid receptors are not evenly distributed throughout the CNS, suggesting that they have distinct physiological roles (for reviews see Mansour et al. 1995; Le Merrer et al. 2009). Whereas MOPrs are widely distributed in the brain with an enrichment in the thalamus, striatum, interpeduncular complex, habenula, cortex, superior and inferior colliculi, DOPrs are mainly expressed in distinct areas of the forebrain, predominantly in the olfactory tubercle, cerebral cortex, amygdala, nucleus accumbens, and striatum (Fig. 1a, b). KOPrs are mainly found in the cortex, olfactory tubercle, striatum, nucleus accumbens, hypothalamus, amygdala, and periaqueductal gray (PAG) (Mansour et al. 1987; Sharif and Hughes 1989; Slowe et al. 1999). Interestingly, in the areas where the three receptors were found, their cellular distribution often differs. This is exemplified in the striatum, where DOPr and KOPr are diffusely distributed while MOPrs are expressed in patch-like clusters (Mansour et al. 1987).

Of a particular interest for this chapter, DOPrs are known to be largely expressed along the pain pathways in all animal species studied to date, including humans. DOPrs are indeed present in primary afferents (i.e., DRGs), in the spinal cord, as well as in important structures along the ascending and descending pain pathways (Fig. 1). Among these structures, it is worth noting that DOPr is expressed in the PAG, the rostro-ventral medulla (RVM), the cerebral cortex, and the amygdala (Mansour et al. 1994, 1995; Cahill et al. 2001a; Mennicken et al. 2003; Poulin et al.

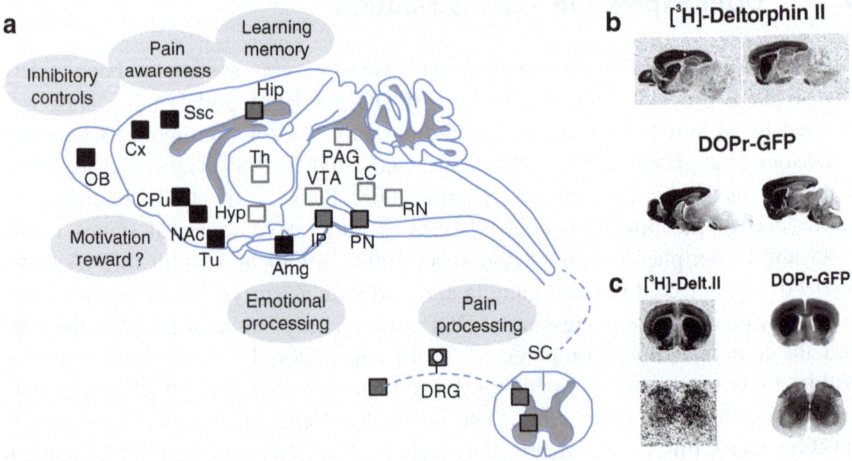

Fig. 1 DOPr distribution and functions in the central nervous system. (**a**) Schematic representation of DOPr binding sites and presumed functions in the central nervous system (CNS). DOPrs are highly expressed in the rostral part of the brain (*black squares*), especially in the olfactory bulb, cortex, striatum, and amygdala suggesting a role in pain processing and awareness, in emotional disorders (depression and anxiety), in addiction and impulsivity. Moderate (*grey squares*) to weak (*open squares*) expression of DOPr is also observed throughout the caudal part of the brain, including the spinal cord and the DRGs. (**b**, **c**) Representative photomicrographs showing a similar pattern of expression between DOPr binding sites labeled with [^{3}H]-Deltorphin II and DOPr immunofluorescence from DOPr-GFP knock-in mice (modified with permission from Pradhan et al. 2011; Bardoni et al. 2014)

2006; Peng et al. 2012). Interestingly, the distribution of DOPr in the spinal cord – and possibly in other structures as well – significantly differs among species. In the rodent spinal cord (mice and rats), DOPr is diffusely distributed in the gray matter (Sharif and Hughes 1989; Arvidsson et al. 1995; Mennicken et al. 2003) (Fig. 2). In monkeys, although DOPr binding sites are also found in all lamina of the spinal cord, a higher density of binding could be observed in the superficial lamina. Most interestingly, DOPr binding sites are restricted to the superficial lamina as well as in the Clark's column in the human spinal cord (Mennicken et al. 2003) (Fig. 2). The fact that the DOPr transcript is virtually absent in the human spinal cord (Fig. 2) suggests that DOPr binding sites are exclusively present on presynaptic primary afferent axon terminals. These observations strongly advise for a specialization of DOPr toward the pain pathways in higher species. For instance, the intrathecal injection of DOPr agonists in various acute and chronic pain models has been shown to produce antinociception, supporting a role for DOPr in pain.

The exact distribution of DOPr in primary afferents remains a matter of controversy and most certainly differs among species. While DOPr was commonly shown to be expressed in all three types of DRG neurons (Dado et al. 1993; Mansour et al. 1994; Ji et al. 1995; Minami et al. 1995; Zhang et al. 1998; Wang and Wessendorf 2001; Mennicken et al. 2003; Gendron et al. 2006), in DOPr-GFP knock-in mice

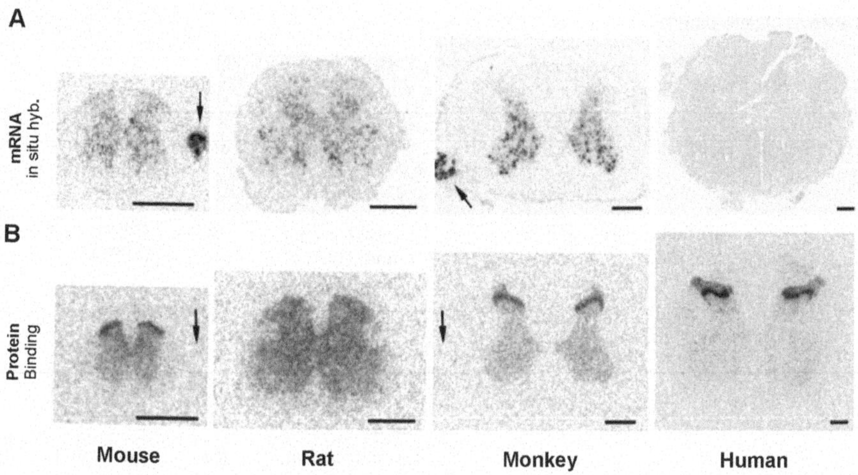

Fig. 2 Phylogenetic changes in DOPr expression in spinal cord and DRGs. Representative photomicrographs illustrating (**A**) DOPr mRNA expression (in situ hybridization) and (**B**) [^{125}I]-Deltorphin labeled DOPr binding sites in spinal cord and DRGs of mice, rats, monkeys, and humans (modified with permission from Mennicken et al. 2003)

DOPr was rather found to be primarily expressed in large myelinated non-peptidergic neurons and around hair follicles supporting a role in the perception of mechanical stimuli and light touch (Bardoni et al. 2014). These observations support the fact that MOPr and DOPr were, respectively, shown to specifically inhibit thermal and mechanical pain (Scherrer et al. 2009). This is, however, in sharp contrast with the work of others. Firstly, DOPr has often been found in substance P-containing neurons by a number of independent groups, and with different experimental approaches (Guan et al. 2005; Zhang et al. 2006, 2010; Riedl et al. 2009; Wang et al. 2010; Zhao et al. 2011). In particular, a role for an interaction between DOPr and preprotachykinin A (the precursor for substance P) in the targeting of DOPr to the cell surface through the regulated secretory pathway has been described (Guan et al. 2005). DOPr was indeed found to be present in large dense core vesicles (LDCV) containing substance P (Guan et al. 2005; Zhao et al. 2011). The presence of DOPr in substance P-containing neurons was also confirmed by single-cell RT-PCR (Wang et al. 2010). Also contrasting with the work cited above is the fact that the activation of DOPr by various agonists and in different animal models of pain was not only found to inhibit noxious mechanical stimuli but also heat-induced pain (Tables 1 and 2).

For a GPCR, DOPr was found to have an uncommon subcellular location. Indeed, under normal conditions DOPr was found to be retained in the cytoplasm, in association with intracellular compartments (Pasquini et al. 1992; Arvidsson et al. 1995; Cheng et al. 1995, 1997; Zhang et al. 1998; Cahill et al. 2001a; Commons et al. 2001; Wang and Pickel 2001) (Fig. 3a). Interestingly, it was observed that the density of cell surface DOPr can be increased under certain conditions such as in chronic pain models, or following prolonged morphine

Table 1 Acute pain modulation by DOPr agonists

Type	Test	Agonist	Effective doses (route of administration)	Animal species	References
Thermal pain	Tail flick	Deltorphin II	2.5–5 μg (i.t.)	Mouse	Dubois and Gendron (2010)
	Tail flick (warm water)	SNC80	104 nmol (i.c.v.) 69 nmol (i.t.) 57 mg/kg (i.p.)	Mouse	Bilsky et al. (1995)
	Tail flick (light beam)	SB-235863	100–300 mg/kg (p.o.) no effect	Rat	Petrillo et al. (2003)
		Deltorphin II	30–45–60 nmol (i.c.v.)	Rat	Fraser et al. (2000a)
		SNC80	200–300–400 nmol (i.c.v.)	Rat	Fraser et al. (2000a)
	Tail flick	SNC80	80 mg/kg (s.c.) no effect	Rat	Gallantine and Meert (2005)
		DPDPE	20 μg no effect Intra-RVM Intra-PAG	Rat	Rossi et al. (1994)
		Deltorphin II	20 μg Intra-RVM Intra-PAG	Rat	Rossi et al. (1994)
	Hot plate	Deltorphin II	10 μg (i.t.)	Rat	Cahill et al. (2001b)
		SNC80	100 nmol (i.c.v.)	Mouse	Bilsky et al. (1995)
		SB-235863	No effect (p.o.)	Rat	Petrillo et al. (2003)
	Hargreaves test	SNC80	No effect at 200 μg (i.t.)	Rat	Kouchek et al. (2013)
		Deltorphin II	No effect at 50 μg (i.pl.)	Rat	Kabli and Cahill (2007)
Mechanical pain	Von Frey test	SNC80	No effect (10 mg/kg, i.p.)	Mouse	Pradhan et al. (2013)
	Paw pressure	Deltorphin II	EC80 60 nmol (i.c.v.)	Rat	Fraser et al. (2000b)
		SNC80	EC80 400 nmol (i.c.v.)	Rat	Fraser et al. (2000b)
Chemical pain	Capsaicin	Deltorphin II	10 μg (i.t.)	Rat	Beaudry et al. (2011)
	Capsaicin (tail thermal hypersensitivity)	SNC80	1–10 mg/kg (s.c.)	Monkey	Brandt et al. (2001)
	Capsaicin (mechanical)	DPDPE	10–100–300 μg (i.m.)	Rat	Saloman et al. (2011)
	Prostaglandin E2	SNC80	3.2 mg/kg (s.c.)	Monkey	Brandt et al. (2001)

(continued)

Table 1 (continued)

Type	Test	Agonist	Effective doses (route of administration)	Animal species	References
	Formalin	Deltorphin II	10 μg (i.t.)	Rat	Beaudry et al. (2011)
		Deltorphin II	5 μg (i.t.)	Mouse	Morinville et al. (2003)
		Deltorphin II	ED50 7.7 μg/phase I and 32.4 μg/phase II (i.t.)	Rat	Cahill et al. (2001b)
		Deltorphin II	20 nmol (i.t.) 100 nmol (ipl)	Rat	Bilsky et al. (1996b)
		Deltorphin II	1–10 μg (i.t.)	Rat	Pradhan et al. (2006)
		SNC80	200 μg (i.t.)	Rat	Kouchek et al. (2013)
		Deltorphin II	50 μg (ipl)	Rat	Kabli and Cahill (2007)
		SNC80	11–44–111 nmol (ipl)	Rat	Obara et al. (2009)
		DSLET	14–42–70 nmol (ipl)	Rat	Obara et al. (2009)
		KNT-127	3 mg/kg (s.c.)	Mouse	Saitoh et al. (2011)
		SNC80	3 μmol/kg (i.v.)	Mouse	Barn et al. (2001)
	Acetic acid	KNT-127	3 mg/kg (s.c.)	Mouse	Saitoh et al. (2011)
		SNC80	10 mg/kg (s.c.)	Rat	Gallantine and Meert (2005)

treatment (Cahill et al. 2001a, 2003; Commons 2003; Morinville et al. 2003, 2004; Lucido et al. 2005; Gendron et al. 2006) (Fig. 3b, c). Since it is not the purpose of this chapter, the distinct mechanisms involved in the regulation of DOPr trafficking will not be discussed here. This topic has, however, recently been extensively reviewed elsewhere (Gendron et al. 2016). Simply, it should be kept in mind that the subcellular localization of DOPr and the possibility to increase its density at the cell surface could explain why DOPr agonists are more potent under certain conditions than in control/naïve animals.

Table 2 Chronic pain modulation by DOPr agonists

Type	Test	Agonist	Effective doses (route of administration)	Animal species	References
Inflammatory CFA	*Thermal pain*				
	Hargreaves Plantar test	Deltorphin II	1–3–10 μg (i.t.)	Rat	Cahill et al. (2003), Gendron et al. (2007a), and Beaudry et al. (2009, 2015b)
		Deltorphin II	1–2.5 μg (i.t.)	Mouse	Gendron et al. (2007b), Beaudry et al. (2009, 2015b), and Dubois and Gendron (2010)
		DPDPE Deltorphin II	50 nM (i.t.)	Mouse	Qiu et al. (2000)
		Deltorphin II	3–10–30–60 nmol (i.c.v.)	Rat	Fraser et al. (2000a)
		SNC80	100–300 nmol (i.c.v.)	Rat	Fraser et al. (2000a)
		SB-235863	30–70 mg/kg (s.c.)	Rat	Beaudry et al. (2009)
		SNC80	40 mg/kg (s.c.)	Rat	Gallantine and Meert (2005)
		SNC80	10 mg/kg (s.c.)	Mouse	Gaveriaux-Ruff et al. (2008)
		DPDPE	77.4–154.8 nmol (ipl)	Mouse	Hervera et al. (2009)
	Tail flick	SNC80	3.2 mg/kg (s.c.)	Monkey	Brandt et al. (2001)
	Mechanical pain				
	Von Frey filament	Deltorphin II	10–30 μg (i.t.)	Rat	Otis et al. (2011)
		SNC80	10 mg/kg (i.p.)	Mouse	Pradhan et al. (2013)
		SNC80	10 mg/kg (s.c.)	Mouse	Gaveriaux-Ruff et al. (2008)
	Paw pressure test (Randall–Stilotto)	DPDPE	10–100 μg (ipl)	Rat	Zhou et al. (1998)
Carrageenan	*Thermal pain*				
	Hargreaves Plantar test	SB-235863	10 mg/kg (p.o.)	Rat	Petrillo et al. (2003)
		Deltorphin II	10 μg (i.t.)	Rat	Stewart and Hammond (1994)

(continued)

Table 2 (continued)

Type	Test	Agonist	Effective doses (route of administration)	Animal species	References
		DPDPE	30 μg (i.t.)	Rat	Stewart and Hammond (1994)
	Mechanical pain				
	Von Frey filament	SNC80	200 μg (i.t.)	Rat	Kouchek et al. (2013)
Cancer pain	*Mechanical pain*				
	Von Frey filament	Deltorphin II	3–10–30 μg (i.t.)	Rat	Otis et al. (2011)
		DVal Ala-E	1.3 mg/kg (i.p.)	Mouse	Brainin-Mattos et al. (2006)
		SNC80	10 nmol (ipl)	Mouse	Ye et al. (2012)
	Thermal pain				
	Unilateral hotplate test	DPDPE	30 μg (peritumoral)	Mouse	Baamonde et al. (2005)
Diabetic neuropathy	Tail flick	TAN-67	ED50 ~6 μg (i.c.v.)	Mouse	Kamei et al. (1997b)
	Formalin	TAN67	30 mg/kg (s.c.)	Mouse	Kamei et al. (1997a)
Neuropathic pain	*Thermal pain*				
	Hargreaves Plantar test	DSLET	111 nmol (ipl)	Rat	Obara et al. (2009)
		SNC80	111 nmol (ipl)	Rat	Obara et al. (2009)
		SB-235863	10 mg/kg (p.o.)	Rat	Petrillo et al. (2003)
	Noxious thermal stimuli (paw)	Deltorphin II	10 μg (i.t.)	Rat	Holdridge and Cahill (2007)
	Tail flick (cold allodynia)	Deltorphin II	15–25 μg (i.t.)	Rat	Mika et al. (2001)
	Tail flick (heat and cold stimuli)	Deltorphin II	1.5–15–25 μg (i.t.)	Rat	Mika et al. (2001)
	Tail flick (cold allodynia)	DPDPE	25 μg (i.t.)	Rat	Mika et al. (2001)
	Tail flick (heat and cold stimuli)	DPDPE	5–25 μg (i.t.)	Rat	Mika et al. (2001)
	Acetone application	DPDPE	20 μg intra-PAG	Rat	Sohn et al. (2000)
	Mechanical pain				
	Von Frey filament	Deltorphin II	10–15–30 μg (i.t.)	Rat	Holdridge and Cahill (2007)

(continued)

Table 2 (continued)

Type	Test	Agonist	Effective doses (route of administration)	Animal species	References
		Deltorphin II	50 μg (ipl)	Rat	Kabli and Cahill (2007)
		SNC80	22–66–111 nmol (ipl)	Rat	Obara et al. (2009)
		DSLET	22–56–111 nmol (ipl)	Rat	Obara et al. (2009)
		DPDPE	20 μg intra-PAG	Rat	Sohn et al. (2000)
		BUBU	1.5–6 mg/kg (i.v.)	Rat	Desmeules et al. (1993)

ipl intraplantar, *s.c.* subcutaneous, *i.t.* intrathecal, *i.m.* intramuscular, *i.c.v.* intracerebroventricular, *i.p.* intraperitoneal, *p.o.* perorally, *i.v.* intravenous

3 DOPr and Pain Modulation

The analgesic efficacy of DOPr agonists was widely investigated using pharmacological and genetic approaches. Before describing the effects of DOPr agonists in acute and chronic pain models, it is worth noting that mice deficient for DOPr (DOPr knockout mice) did not show any significant change in pain perception following acute noxious stimuli (thermal, mechanical, or chemical stimuli) (Zhu et al. 1999; Filliol et al. 2000). However, sensitivity to thermal and mechanical stimuli is increased in inflammatory and neuropathic pain models in DOPr knockout mice (Nadal et al. 2006; Gaveriaux-Ruff et al. 2008). These observations therefore suggest that a constitutive tone of endogenous opioid release acting on DOPr would prevent exacerbation of chronic pain. This hypothesis is supported by the fact that the selective ablation of DOPr in NaV1.8 sensory neurons increases chronic pain (Gaveriaux-Ruff et al. 2011).

3.1 DOPr-Mediated Analgesia in Acute Pain Models

Agonists acting at DOPr are known to produce antinociception (Gaveriaux-Ruff and Kieffer 2011). Although first evidence for DOPr-mediated antinociception was provided in the early 1980s (Brantl et al. 1982), the lack of highly selective DOPr ligands prevented a clear demonstration of the physiological effects of this receptor. A pioneer study used [2-D-penicillamine, 5-D-penicillamine]enkephalin (DPDPE), a highly selective DOPr agonist (Mosberg et al. 1983), and confirmed that DOPr could mediate antinociception in the hot plate test (Porreca et al. 1984). Although their antinociceptive effects have been commonly reported thereafter, it is generally accepted that DOPr agonists, no matter the route of administration, only have weak

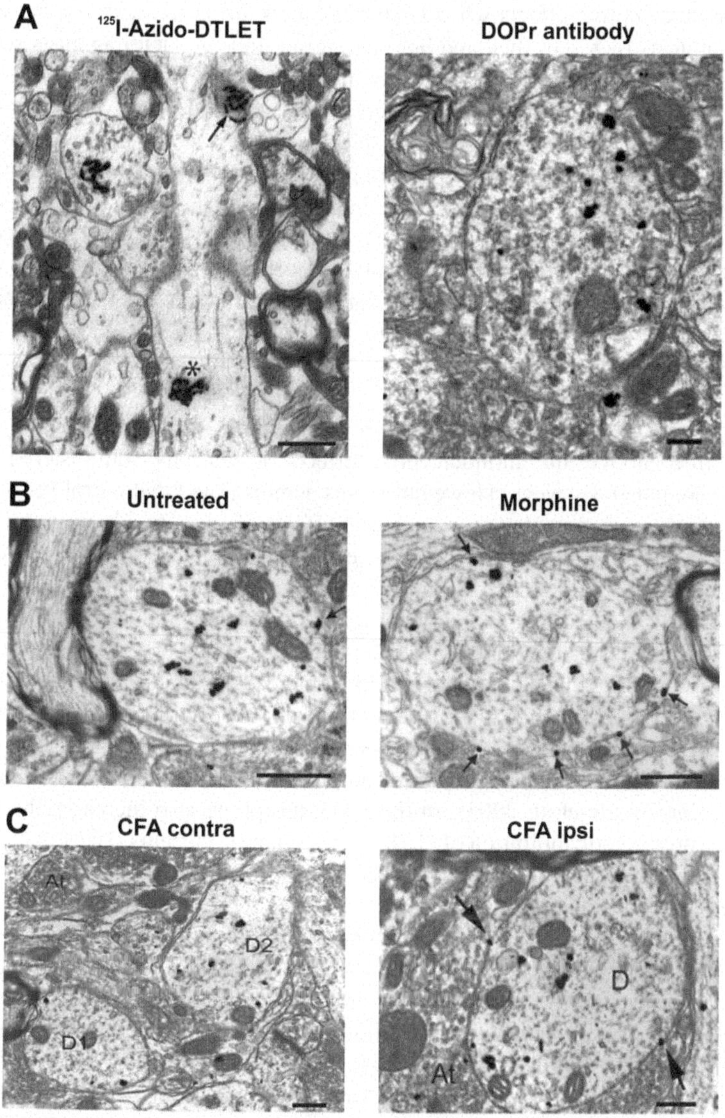

Fig. 3 Enhancement of DOPr expression at the plasma membrane of neurons under inflammatory or chronic morphine conditions. (**a**) Intracellular localization of DOPr in the neostriatum labeled with [^{125}I]-Azido-DTLET (a DOPr selective agonist, *left panel*) or by a DOPr antibody directed against the 3–17 segment (*left panel*) in untreated animals. (**b**, **c**) Electron microscopy of immunolabeled DOPr in the superficial laminae of lumbar spinal cord dorsal horn in animals treated with morphine (**b**, *right panel*) or in the CFA pain model (**c**, *right panel*) showing an increase in immunogold particles associated with the plasma membrane (adapted with permission from Pasquini et al. 1992; Cahill et al. 2001b, 2003, Lucido et al. 2005)

or no antinociceptive effects when healthy animals are tested with routinely used acute pain tests (e.g., tail flick and hot plate tests). Indeed, DOPr agonists seem to have only modest antinociceptive effects in acute pain models when compared to MOPr agonists. As an example, the i.c.v. administration of [D-Ala2, *N*-methyl-Phe4, Gly5-ol]enkephalin (DAMGO; a selective MOPr agonist) produced a more profound analgesia than DPDPE in the hot plate test (Porreca et al. 1984). Similarly, DAMGO was also shown to reduce mechanical nociception by 80% at a dose of 0.2 nmol (i.c.v.) in the paw pressure test while doses of 60 and 400 nmol of the DOPr agonists Deltorphin II and SNC80, respectively, were needed to produce similar effects (Fraser et al. 2000b). Table 1 summarizes the DOPr-mediated antinociceptive effects in acute pain tests.

3.1.1 Regulation of DOPr by MOPr in Healthy Animals

As described above, the antinociceptive effects of centrally administered DOPr agonists are generally weak. However, it was demonstrated that morphine or other MOPr agonists can potentiate the analgesic effects of spinally administered DOPr agonists (Cahill et al. 2001b; Morinville et al. 2003; Gendron et al. 2007a). The cellular mechanisms involved in the potentiation of DOPr functions are unclear. However, it was noted that the administration of morphine induces a translocation of DOPr to the plasma membrane in DRG (Gendron et al. 2006), spinal cord (Cahill et al. 2001b; Morinville et al. 2003; Gendron et al. 2007a), and central gray neurons (Lucido et al. 2005; see also Fig. 3b). As of to date, the exact mechanisms involved in this process have not been totally unveiled. We do know, however, that it involves MOPr as these effects are completely abolished in MOPr knockout animals (Morinville et al. 2003). In the PAG, morphine also increases the DOPr-mediated presynaptic inhibition of GABAergic synaptic currents (Hack et al. 2005). Both MOPr and β-arrestin 2 have been shown to be important for the upregulation of DOPr functions in the PAG (Hack et al. 2005). A more recent study also suggests that morphine induces a cdk5-mediated phosphorylation of the threonine 161 residue located in the second intracellular loop of DOPr (Xie et al. 2009). Phosphorylation of this residue by cdk5 would indeed increase the membrane expression of DOPr and, ultimately, enhance the antinociceptive effects of DOPr agonists in morphine-treated animals (Beaudry et al. 2015b). The phosphorylation of DOPr by cdk5 was further hypothesized to disrupt the formation of the MOPr-DOPr heterodimer (Xie et al. 2009). This is consistent with the observations made by others who show that chronic morphine treatment potentiates MOPr and DOPr heterodimerization throughout the CNS including areas involved in pain processing and in the DRGs (Gupta et al. 2010). It was further demonstrated that a mixture of the MOPr agonist methadone and the DOPr antagonist naltriben can stabilize DOPr at the cell surface in a heterodimer form preventing its endocytosis and therefore avoiding degradation (Milan-Lobo and Whistler 2011; Milan-Lobo et al. 2013).

3.2 DOPr-Mediated Analgesia in Acute Inflammatory Pain Models

By contrast to their minor effects in acute pain tests, DOPr agonists were found to be more efficient at alleviating acute inflammatory pain. Formalin- and capsaicin-induced pain behaviors are indeed efficiently inhibited following DOPr activation (Bilsky et al. 1996b; Cahill et al. 2001b; Morinville et al. 2003; Pradhan et al. 2006; Beaudry et al. 2011). As an example, the intrathecal administration of Deltorphin II was shown to significantly reduce the typical biphasic nociceptive response induced by formalin as well as the spontaneous pain-like behaviors – licking, biting, and flinching – induced by capsaicin. These effects were completely antagonized by the DOPr selective antagonist naltrindole, therefore supporting a role for DOPr (Bilsky et al. 1996b; Cahill et al. 2001b; Morinville et al. 2003; Pradhan et al. 2006; Beaudry et al. 2011). At the spinal level, DOPr-mediated analgesia was shown to involve an inhibition of substance P release (Beaudry et al. 2011; Kouchek et al. 2013). Substance P is produced in peptidergic C fibers and released upon activation of these neurons by peripheral noxious stimuli (Cao et al. 1998). As revealed by an inhibition of c-fos expression, the intrathecal administration of Deltorphin II or SNC80 reduces the activation of spinal neurons (Beaudry et al. 2011; Kouchek et al. 2013). Activation of DOPr in the spinal cord also prevents substance P release and blocks NK1 internalization in the superficial lamina of the lumbar dorsal horn induced by intraplantar formalin and capsaicin (Beaudry et al. 2011; Kouchek et al. 2013).

The systemic administration of DOPr agonists also produces antinociception in response to chemical stimuli (Barn et al. 2001; Brandt et al. 2001; Saitoh et al. 2011; Saloman et al. 2011). In primates, thermal hypersensitivity induced by capsaicin and prostaglandin E2 was completely reversed following subcutaneous administration of SNC80 (Brandt et al. 2001). KNT127 and SNC80 were also found to inhibit the biphasic nociceptive response induced by formalin as well as the acetic acid induced abdominal constrictions in mice (Barn et al. 2001; Saitoh et al. 2011).

Beside their central mechanism of actions, DOPr agonists also produce analgesia via receptors expressed in the periphery (Stein et al. 2001). The intraplantar administration of Deltorphin II, SNC80, and DSLET was shown to effectively suppress formalin-induced pain behaviors in rodents (Bilsky et al. 1996b; Kabli and Cahill 2007; Obara et al. 2009). This antinociceptive effect was completely reversed by intraplantar treatment with the DOPr antagonist naltrindole or by a pretreatment with DOPr antisense oligodeoxynucleotides (Bilsky et al. 1996a, b). Again, DOPr agonists had no significant effect on thermal or mechanical nociceptive thresholds under normal conditions (Bilsky et al. 1996b; Kabli and Cahill 2007; Obara et al. 2009).

3.2.1 Regulation of DOPr in Acute Inflammatory Pain

A recent study by Doyle Brackley and collaborators has provided mechanistic insights for the lack of efficacy of peripheral DOPr agonists in naïve animals.

They indeed observed that peripheral DOPr expressed on sensory nerves is constitutively desensitized by a GRK2-dependent mechanism. They found that under basal conditions GRK2 is constitutively associated with DOPr, therefore preventing its coupling to G proteins or other signaling partners. The knockdown of GRK2 was found to be sufficient to increase peripheral DOPr-mediated analgesia. Interestingly, in inflamed tissues PKC activation leads to RKIP phosphorylation which in turn sequesters GRK2. In this model, the sequestering of GRK2 "awakens" DOPr in sensory neurons and increases the analgesic effects of peripheral DOPr agonists (Brackley et al. 2016).

3.3 DOPr-Mediated Analgesia in Chronic Pain Models

The antinociceptive effects of various DOPr agonists in animal models of chronic pain are summarized in Table 2. In general, the efficacy and the potency of DOPr agonists at producing antinociception in chronic pain models are more important, when compared to acute pain models. Overall, the observations made in preclinical models of chronic pain suggest that DOPr agonists efficiently inhibit inflammatory, neuropathic, diabetic, as well as cancer pain. Interestingly, recent observations also support a role for DOPr in the treatment of migraine.

3.3.1 DOPr in Inflammatory Pain Models

Complete Freund's adjuvant (CFA) and carrageenan are commonly used to induce inflammation or as rodent models of arthritis (Klareskog 1989; Hansra et al. 2000). In these inflammatory pain models, spinal DOPr activation was shown to alleviate hyperalgesia (Hylden et al. 1991; Stewart and Hammond 1994; Qiu et al. 2000). In the CFA model of inflammation, Deltorphin II is effective at reducing both thermal hyperalgesia and mechanical allodynia in a dose-dependent manner (Cahill et al. 2003; Gendron et al. 2007a, b; Beaudry et al. 2009, 2015a; Dubois and Gendron 2010; Otis et al. 2011). The effects of Deltorphin II are DOPr-mediated since they are completely antagonized by DOPr selective antagonists. Interestingly, Deltorphin II has no analgesic effect on the uninflamed paw, supporting a lack of effects in healthy tissues (Cahill et al. 2003; Gendron et al. 2007a; Beaudry et al. 2009; Dubois and Gendron 2010; Otis et al. 2011). Indeed, in these chronic pain models, DOPr agonists commonly display a leftward shift of their dose-response effects when compared to dose-response curves in healthy animals.

The enhancement of DOPr analgesic potency in inflammatory pain models is thought to be the result of an increase in DOPr expression at the plasma membrane of spinal neurons. As stated above, the subcellular distribution of DOPr in the lumbar dorsal horn, as assessed by electron microscopy, revealed a predominant localization of the receptor within the intracellular compartments of neurons (Fig. 3b, c). However, following some inflammatory/pain state, an increase in DOPr distribution at the plasma membrane was seen in the ipsilateral lumbar spinal cord and the DRG neurons (Cahill et al. 2003; Gendron et al. 2006) (see also Fig. 3c). Studies based on fluorescent ligand internalization (used as a tool to evaluate the density of membrane receptors) further revealed an increased level of internalization in the lumbar spinal cord and in small and medium DRG neurons in inflamed animals, thus supporting an upregulation of DOPr at the neuronal plasmalemma (Gendron et al. 2006, 2007a). Again, the mechanisms involved in this process are still unclear. As for the morphine-induced regulation of DOPr, CFA-induced inflammation requires MOPr to increase the membrane density of DOPr and the antinociceptive effects of DOPr agonists (Cahill et al. 2003; Gendron et al. 2007b). Similarly, the inhibition of cdk5-induced phosphorylation of DOPr prevents the enhancement of DOPr-mediated analgesia (Beaudry et al. 2015b). However, if an interaction between DOPr and preprotachykinin A appears to be essential for the membrane expression of DOPr in non-treated mice, the upregulation of DOPr function in the CFA model was shown to be independent of substance P (Dubois and Gendron 2010).

Centrally administered (i.c.v.) DOPr agonists also produce analgesia in the CFA model of inflammatory pain. SNC-80 and Deltorphin II were shown to increase the time to paw withdrawal in response to a thermal stimulus (Hargreaves test) (Fraser et al. 2000a). The potency of centrally administered DOPr agonists was also found to be improved in animals with persistent inflammation when compared to healthy animals. Indeed, the effective i.c.v. dose of SNC80 and Deltorphin II required to produce antihyperalgesic effect in the rat CFA model of inflammation is three times lower than that needed to induce analgesia to thermal stimulus in acute pain models (Fraser et al. 2000a). Other groups have also demonstrated an involvement of DOPr in the descending pain pathways. In fact, DOPr activation in supraspinal sites such as the RVM and PAG was found to produce analgesia in various animal models of chronic pain (for review, see Bie and Pan 2007). As an example, microinjection of Deltorphin II into the RVM was shown to dose-dependently reverse thermal hyperalgesia in CFA-treated rats (Hurley and Hammond 2000).

Activation of peripheral DOPr produces antinociception under pathological pain conditions. DPDPE, when administered in the periphery, produces antinociception in CFA and neuropathic pain models (Zhou et al. 1998; Hervera et al. 2009; Obara et al. 2009). Following inflammation, opioid receptors were shown to be upregulated in primary afferents in which they are highly transported toward the free nerve endings in the periphery. As a consequence, the potency of peripheral opioid agonists in mediating analgesia is enhanced. The low pH in the inflamed tissues is also thought to facilitating ligand/receptor coupling (for reviews, see Stein et al. 2001; Stein and Lang 2009). The local administration of the DOPr agonist SNC80 was also shown to dose-dependently reduce the mechanical hyperalgesia induced by the subcutaneous injection of prostaglandin E2 in the hindpaw (Pacheco and Duarte 2005). DOPr-induced analgesia in the periphery is thought to be mediated by the nitric oxide/cGMP pathway. Indeed, in the CFA model of inflammation the nitric oxide donor NOC-18 potentiates the antihyperalgesic effect of DPDPE (Hervera et al. 2009). Nitric oxide synthase or guanylate cyclase inhibitors also prevent the SNC80-mediated analgesia in the prostaglandin E2 pain model (Pacheco and Duarte 2005). Interestingly, the intraplantar administration of glibenclamide and tolbutamide, two ATP-sensitive K+ channel blockers, was shown to reduce the analgesia produced by the local administration of SNC80, suggesting that the antinociceptive effect of this compound is specifically mediated by ATP-sensitive K+ channels (Pacheco and Duarte 2005).

3.3.2 DOPr in Neuropathic Pain Models

As reported in Table 2, DOPr agonists are also efficient at alleviating neuropathic pain in various preclinical models. The intrathecal administration of DPDPE or Deltorphin II was shown to significantly relieve allodynia and hyperalgesia in the sciatic nerve ligation model (Mika et al. 2001; Holdridge and Cahill 2007). DPDPE injected into the ventral PAG also effectively reduces mechanical and thermal allodynia in a neuropathic pain model where both the tibial and sural nerves are completely cut (Sohn et al. 2000). In other nerve injury models, peripherally administered DOPr agonists also produce anti-allodynic effects (Kabli and Cahill 2007; Obara et al. 2009). In the peripheral nerve injury model, the increased analgesic effects of DOPr agonists may be the consequence of a higher level of DOPr expression or as a relocalization of DOPr at the cell surface (Kabli and Cahill 2007; Obara et al. 2009).

3.3.3 DOPr in Diabetic Neuropathy

Diabetic neuropathy represents another disease in which DOPr agonists may be used to reduce pain. For instance, the non-peptide TAN-67 was found to produce a dose-dependent antinociception in the mouse tail flick assay when administered i.c.v. (Kamei et al. 1997b). In diabetic mice, an increase in the endogenous tone of the spinal DOPr system was demonstrated. Indeed, in these mice, the inflammatory phase of the formalin test is greatly reduced, an effect reversed by naltrindole (Kamei et al. 1997a).

3.3.4 DOPr in Bone Cancer Pain

Although it has a unique set of characteristics, bone cancer-induced pain includes an important neuropathic component (Honore et al. 2000). It is therefore not surprising to see that DOPr agonists are effective in bone cancer models. In a rat model of metastatic bone cancer-induced pain (Dore-Savard et al. 2010), the intrathecal administration of Deltorphin II was shown to dose-dependently reverse mechanical allodynia, an effect completely blocked by a pretreatment with the DOPr antagonist naltrindole (Otis et al. 2011). The intraperitoneal administration of [dVal(L)2,Ala(L)5]E, another selective DOPr agonist, also produces analgesia in a mouse model of bone cancer-induced pain (Brainin-Mattos et al. 2006). The analgesic effect of DOPr was also demonstrated in mice bearing a tibial osteosarcoma (Baamonde et al. 2005). In this model the peritumoral injection of DPDPE induced a naltrindole-sensitive increase in the paw thermal withdrawal latencies (Baamonde et al. 2005). Similarly, DOPr activation alleviates mechanical hypersensitivity in an orthotopic mouse oral cancer model (Ye et al. 2012). In humans coping with cancer pain, intrathecally administered DADLE, a DOPr-preferred agonist, has also been shown to produce analgesia, even in patients who had developed tolerance to morphine (Onofrio and Yaksh 1983; Moulin et al. 1985; Krames et al. 1986).

3.3.5 DOPr Agonists in Trigeminal and Migraine Pain Models

DOPr appears to play a crucial role in the modulation of trigeminal pain. In rodents and humans, DOPr mRNA can be found in small-, medium-, and large-sized trigeminal ganglia neurons (Mennicken et al. 2003). In the trigeminal nucleus caudalis (Sp5C), a structure involved in modulating and processing somatosensory and nociceptive inputs originating from the orofacial region, DOPr binding sites have a more widespread distribution in rodents than humans. In humans, DOPr binding sites are confined to the superficial laminae of the Sp5C (Mennicken et al. 2003; Ichikawa et al. 2005). Supporting a role for DOPr in trigeminal pain, the activation of peripheral DOPrs was found to attenuate the capsaicin-induced mechanical hypersensitivity in the masseter muscle via the activation of GIRK channels in rats (Saloman et al. 2011; Chung et al. 2014). The activation of DOPr with low doses of DPDPE was also found to reduce substance P release from Sp5C slices, a hallmark of opioid-mediated analgesia (Suarez-Roca and Maixner 1992). Yet, the systemic administration of Deltorphin II produces a pronounced inhibition of C fiber-evoked responses in wide dynamic range neurons of the Sp5C (Wang et al. 1996). Under inflammatory conditions, DOPr-mediated trigeminal analgesia is also enhanced. As measured by a reduction in CGRP release and adenylate cyclase activity, a pretreatment with bradykinin increases the potency of DPDPE to inhibit the activity of trigeminal nociceptors (Patwardhan et al. 2005). In trigeminal nociceptors, this effect is concomitant to an increase of cell surface DOPr (Patwardhan et al. 2005).

In patients unresponsive to classical treatments, opioids acting on MOPrs are often prescribed to treat severe cases of migraine headaches[1] (for review see Becker 2015). In some cases, however, extensive treatments with opioids can lead to an exacerbation of the frequency and the intensity of migraine episodes in addition to interfere with other migraine therapies (Bigal and Lipton 2008; Bigal et al. 2008; Ansari and Kouti 2016). Recent reports revealed a promising therapeutic potential for DOPr in alleviating migraine headaches (Charles and Pradhan 2016). In an animal model of migraine induced by nitroglycerine (NTG), DOPr activation efficiently reduces thermal hyperalgesia and mechanical allodynia, two symptoms often observed in humans coping with migraine (Pradhan et al. 2014). In this model, the antinociceptive effect of SNC80 is similar to that obtained with sumatriptan, a classical serotonergic receptor (5-HT_{1B} and 5-HT_{1D}) agonist used to treat migraine headaches. SNC80 was also found to be efficient in reducing the aversive state

[1]Migraine is the most common and disabling neurological disorder that occurs as recurrent, pulsatile, episodic headaches with or without aura. It is thought to be the result of trigeminal nerve activation leading to distension in cerebral and meningeal blood vessels. The cortical spreading depression (CSD) is defined as a slowly propagated wave of depolarization originating from the occipital to the frontal part of the brain which is followed by a suppression of brain activity (Goadsby et al. 2009; Olesen et al. 2009).

evoked by NTG in the conditioned place preference test, further supporting a role for DOPr in alleviating migraine headaches (Pradhan et al. 2014). These observations are supported by the fact that cortical spreading depression events (CSD; a phenomenon thought to be responsible for the occurrence of migraine with aura, Charles and Baca 2013) evoked by KCl were reduced by the systemic administration of SNC80 (Pradhan et al. 2014).

4 Novel Compounds and Clinical Trials

DOPr is commonly considered as a potential target for the development of novel therapies for the management of chronic pain and emotional disorders (Pradhan et al. 2011). One of the major challenges in the development of novel DOPr agonists for the clinic is the propensity of such ligands to induce nonlethal convulsions at analgesic doses (Comer et al. 1993; Dykstra et al. 1993; Pakarinen et al. 1995; Broom et al. 2002a, b; Chung et al. 2015). Still, a few drugs targeting DOPr have been moved to clinical trials. ADL5747 and ADL5859 are two orally bioavailable compounds (Le Bourdonnec et al. 2008, 2009) that were tested in small cohorts of patients. These compounds, which are devoid of pro-convulsive actions in preclinical models, have indeed been tested for acute (NCT00993863) and chronic (NCT00979953) pain management in Phase 2 clinical trials. Unfortunately, none of the compounds were more effective than the placebo in patients suffering from osteoarthritic pain.

More recently, it was proposed that DOPr-induced seizures are mediated by the activation of the β-arrestin 2 pathway. Exploiting the concept of biased ligands, Trevena, Inc. has developed a novel orally available DOPr-selective compound with a robust bias toward the G protein signaling pathway. This is to say that the compound has virtually no ability to recruit β-arrestin 2 but still activates G proteins with high efficiency. The preclinical evaluation of TRV250 is promising for the treatment of migraine headaches. TRV250 was found to preserve the analgesic properties of common DOPr agonists without producing seizures (http://www.trevenainc.com/TRV250.php).

Although this was not thoroughly covered in this chapter, one should note that DOPr can form dimers or interact with other GPCRs (reviewed in Gendron et al. 2016). Because of their unique pharmacology, GPCR dimers represent a novel class of targets for the development of new drugs and/or therapies (Fujita et al. 2014, 2015). One such target is the MOPr-DOPr heteromer (Fujita et al. 2015). A library screening for this target led to the identification of CYM51010, a selective MOPr-DOPr agonist (Gomes et al. 2013). In the tail flick test, CYM51010 was found to produce antinociception without inducing tolerance (Gomes et al. 2013). This target is thought to induce analgesia without causing the common unwanted effects associated with opioids (Fujita et al. 2015). Bivalent ligands designed to have a high affinity for MOPr and DOPr or KOPr and DOPr were found to exhibit good analgesic properties. Compounds targeting MOPr and DOPr such as L2, L4 (Harvey et al. 2012), MDAN (Daniels et al. 2005b), or RV-JIM-C3 (Podolsky et al. 2013) as well as compounds targeting KOPr and DOPr such as KDAN-18 (Daniels et al. 2005a) and KDN-21 (Bhushan et al. 2004) were found to produce robust analgesia with no apparent signs of tolerance, physical dependence, or sedation (Daniels et al. 2005b; Ansonoff et al. 2010; Podolsky et al. 2013). Two such bivalent ligands targeting MOPr and DOPr are currently in clinical trials. Compound 51 and MuDelta (both acting as MOPr agonist and DOPr antagonist) completed clinical trials in patients suffering from irritable bowel syndrome (Breslin et al. 2012; Wade et al. 2012). The MuDelta was approved recently by the US authorities to be commercialized under the name of Eluxadoline (Garnock-Jones 2015). This compound proved to be efficient in relieving abdominal pain symptoms and diarrhea (Garnock-Jones 2015; Lembo et al. 2016).

5 Conclusion

DOPr represents a promising therapeutic target for the treatment of chronic pain and emotional disorders. Although DOPr agonists produce only weak analgesic effects in healthy animals and in acute pain models, numerous groups have previously described an increase in their analgesic potency in chronic pain models (e.g., inflammatory, neuropathic, and bone cancer-induced pain models). Interestingly, the increased analgesic effects of DOPr agonists are paralleled by a translocation of DOPr from the intracellular compartments to the plasma membrane of spinal cord and DRG neurons.

Acknowledgments This work was supported by grants from the Natural Sciences and Engineering Research Council of Canada [RGPIN-2015-05213] and the Canadian Institutes of Health Research (CIHR) [MOP 123399] to L.G. L.G. is the recipient of a Chercheur-boursier Senior from the Fonds de la Recherche du Québec-Santé. The authors are grateful to Véronique Blais for her help in editing the manuscript.

References

Akil H, Owens C, Gutstein H, Taylor L, Curran E, Watson S (1998) Endogenous opioids: overview and current issues. Drug Alcohol Depend 51:127–140

Al-Hasani R, Bruchas MR (2011) Molecular mechanisms of opioid receptor-dependent signaling and behavior. Anesthesiology 115:1363–1381

Ansari H, Kouti L (2016) Drug interaction and serotonin toxicity with opioid use: another reason to avoid opioids in headache and migraine treatment. Curr Pain Headache Rep 20:50

Ansonoff MA, Portoghese PS, Pintar JE (2010) Consequences of opioid receptor mutation on actions of univalent and bivalent kappa and delta ligands. Psychopharmacology 210:161–168

Arvidsson U, Dado RJ, Riedl M, Lee JH, Law PY, Loh HH, Elde R, Wessendorf MW (1995) Delta-opioid receptor immunoreactivity: distribution in brainstem and spinal cord, and relationship to biogenic amines and enkephalin. J Neurosci 15:1215–1235

Baamonde A, Lastra A, Juarez L, Garcia V, Hidalgo A, Menendez L (2005) Effects of the local administration of selective mu-, delta-and kappa-opioid receptor agonists on osteosarcoma-induced hyperalgesia. Naunyn Schmiedebergs Arch Pharmacol 372:213–219

Bardoni R, Tawfik VL, Wang D, Francois A, Solorzano C, Shuster SA, Choudhury P, Betelli C, Cassidy C, Smith K, de Nooij JC, Mennicken F, O'Donnell D, Kieffer BL, Woodbury CJ, Basbaum AI, MacDermott AB, Scherrer G (2014) Delta opioid receptors presynaptically regulate cutaneous mechanosensory neuron input to the spinal cord dorsal horn. Neuron 81:1312–1327

Barn DR, Caulfield WL, Cottney J, McGurk K, Morphy JR, Rankovic Z, Roberts B (2001) Parallel synthesis and biological activity of a new class of high affinity and selective delta-opioid ligand. Bioorg Med Chem 9:2609–2624

Beaudry H, Proteau-Gagne A, Li S, Dory Y, Chavkin C, Gendron L (2009) Differential noxious and motor tolerance of chronic delta opioid receptor agonists in rodents. Neuroscience 161:381–391

Beaudry H, Dubois D, Gendron L (2011) Activation of spinal mu- and delta-opioid receptors potently inhibits substance P release induced by peripheral noxious stimuli. J Neurosci 31:13068–13077

Beaudry H, Gendron L, Moron JA (2015a) Implication of delta opioid receptor subtype 2 but not delta opioid receptor subtype 1 in the development of morphine analgesic tolerance in a rat model of chronic inflammatory pain. Eur J Neurosci 41:901–907

Beaudry H, Mercier-Blais AA, Delaygue C, Lavoie C, Parent JL, Neugebauer W, Gendron L (2015b) Regulation of mu and delta opioid receptor functions: involvement of cyclin-dependent kinase 5. Br J Pharmacol 172:2573–2587

Becker WJ (2015) Acute migraine treatment in adults. Headache 55:778–793

Befort K, Mattei MG, Roeckel N, Kieffer B (1994) Chromosomal localization of the delta opioid receptor gene to human 1p34.3-p36.1 and mouse 4D bands by in situ hybridization. Genomics 20:143–145

Bhushan RG, Sharma SK, Xie Z, Daniels DJ, Portoghese PS (2004) A bivalent ligand (KDN-21) reveals spinal delta and kappa opioid receptors are organized as heterodimers that give rise to delta(1) and kappa(2) phenotypes. Selective targeting of delta-kappa heterodimers. J Med Chem 47:2969–2972

Bie B, Pan ZZ (2007) Trafficking of central opioid receptors and descending pain inhibition. Mol Pain 3:37

Bie B, Zhang Z, Cai YQ, Zhu W, Zhang Y, Dai J, Lowenstein CJ, Weinman EJ, Pan ZZ (2010) Nerve growth factor-regulated emergence of functional delta-opioid receptors. J Neurosci 30:5617–5628

Bigal ME, Lipton RB (2008) Excessive acute migraine medication use and migraine progression. Neurology 71:1821–1828

Bigal ME, Serrano D, Buse D, Scher A, Stewart WF, Lipton RB (2008) Acute migraine medications and evolution from episodic to chronic migraine: a longitudinal population-based study. Headache 48:1157–1168

Bilsky EJ, Calderon SN, Wang T, Bernstein RN, Davis P, Hruby VJ, McNutt RW, Rothman RB, Rice KC, Porreca F (1995) SNC 80, a selective, nonpeptidic and systemically active opioid delta agonist. J Pharmacol Exp Ther 273:359–366

Bilsky EJ, Bernstein RN, Hruby VJ, Rothman RB, Lai J, Porreca F (1996a) Characterization of antinociception to opioid receptor selective agonists after antisense oligodeoxynucleotide-mediated "knock-down" of opioid receptor in vivo. J Pharmacol Exp Ther 277:491–501

Bilsky EJ, Wang T, Lai J, Porreca F (1996b) Selective blockade of peripheral delta opioid agonist induced antinociception by intrathecal administration of delta receptor antisense oligodeoxynucleotide. Neurosci Lett 220:155–158

Brackley AD, Gomez R, Akopian AN, Henry MA, Jeske NA (2016) GRK2 constitutively governs peripheral delta opioid receptor activity. Cell Rep 16:2686–2698

Brainin-Mattos J, Smith ND, Malkmus S, Rew Y, Goodman M, Taulane J, Yaksh TL (2006) Cancer-related bone pain is attenuated by a systemically available delta-opioid receptor agonist. Pain 122:174–181

Brandt MR, Furness MS, Mello NK, Rice KC, Negus SS (2001) Antinociceptive effects of delta-opioid agonists in Rhesus monkeys: effects on chemically induced thermal hypersensitivity. J Pharmacol Exp Ther 296:939–946

Brantl V, Pfeiffer A, Herz A, Henschen A, Lottspeich F (1982) Antinociceptive potencies of beta-casomorphin analogs as compared to their affinities towards mu and delta opiate receptor sites in brain and periphery. Peptides 3:793–797

Breslin HJ, Diamond CJ, Kavash RW, Cai C, Dyatkin AB, Miskowski TA, Zhang SP, Wade PR, Hornby PJ, He W (2012) Identification of a dual delta OR antagonist/mu OR agonist as a potential therapeutic for diarrhea-predominant irritable bowel syndrome (IBS-d). Bioorg Med Chem Lett 22:4869–4872

Broom DC, Jutkiewicz EM, Folk JE, Traynor JR, Rice KC, Woods JH (2002a) Convulsant activity of a non-peptidic delta-opioid receptor agonist is not required for its antidepressant-like effects in Sprague-Dawley rats. Psychopharmacology 164:42–48

Broom DC, Nitsche JF, Pintar JE, Rice KC, Woods JH, Traynor JR (2002b) Comparison of receptor mechanisms and efficacy requirements for delta-agonist-induced convulsive activity and antinociception in mice. J Pharmacol Exp Ther 303:723–729

Cahill CM, McClellan KA, Morinville A, Hoffert C, Hubatsch D, O'Donnell D, Beaudet A (2001a) Immunohistochemical distribution of delta opioid receptors in the rat central nervous system: evidence for somatodendritic labeling and antigen-specific cellular compartmentalization. J Comp Neurol 440:65–84

Cahill CM, Morinville A, Lee MC, Vincent JP, Collier B, Beaudet A (2001b) Prolonged morphine treatment targets delta opioid receptors to neuronal plasma membranes and enhances delta-mediated antinociception. J Neurosci 21:7598–7607

Cahill CM, Morinville A, Hoffert C, O'Donnell D, Beaudet A (2003) Up-regulation and trafficking of delta opioid receptor in a model of chronic inflammation: implications for pain control. Pain 101:199–208

Cao YQ, Mantyh PW, Carlson EJ, Gillespie AM, Epstein CJ, Basbaum AI (1998) Primary afferent tachykinins are required to experience moderate to intense pain. Nature 392:390–394

Charles AC, Baca SM (2013) Cortical spreading depression and migraine. Nat Rev Neurol 9:637–644

Charles A, Pradhan AA (2016) Delta-opioid receptors as targets for migraine therapy. Curr Opin Neurol 29:314–319

Chaturvedi K, Christoffers KH, Singh K, Howells RD (2000) Structure and regulation of opioid receptors. Biopolymers 55:334–346

Chen Y, Mestek A, Liu J, Hurley JA, Yu L (1993) Molecular cloning and functional expression of a mu-opioid receptor from rat brain. Mol Pharmacol 44:8–12

Cheng PY, Svingos AL, Wang H, Clarke CL, Jenab S, Beczkowska IW, Inturrisi CE, Pickel VM (1995) Ultrastructural immunolabeling shows prominent presynaptic vesicular localization of delta-opioid receptor within both enkephalin- and nonenkephalin-containing axon terminals in the superficial layers of the rat cervical spinal cord. J Neurosci 15:5976–5988

Cheng PY, Liu-Chen LY, Pickel VM (1997) Dual ultrastructural immunocytochemical labeling of mu and delta opioid receptors in the superficial layers of the rat cervical spinal cord. Brain Res 778:367–380

Chu Sin Chung P, Kieffer BL (2013) Delta opioid receptors in brain function and diseases. Pharmacol Ther 140:112–120

Chung MK, Cho YS, Bae YC, Lee J, Zhang X, Ro JY (2014) Peripheral G protein-coupled inwardly rectifying potassium channels are involved in delta-opioid receptor-mediated anti-hyperalgesia in rat masseter muscle. Eur J Pain 18:29–38

Chung PC, Boehrer A, Stephan A, Matifas A, Scherrer G, Darcq E, Befort K, Kieffer BL (2015) Delta opioid receptors expressed in forebrain GABAergic neurons are responsible for SNC80-induced seizures. Behav Brain Res 278:429–434

Comer SD, Hoenicke EM, Sable AI, McNutt RW, Chang KJ, De Costa BR, Mosberg HI, Woods JH (1993) Convulsive effects of systemic administration of the delta opioid agonist BW373U86 in mice. J Pharmacol Exp Ther 267:888–895
Commons KG (2003) Translocation of presynaptic delta opioid receptors in the ventrolateral periaqueductal gray after swim stress. J Comp Neurol 464:197–207
Commons KG, Beck SG, Rudoy C, Van Bockstaele EJ (2001) Anatomical evidence for presynaptic modulation by the delta opioid receptor in the ventrolateral periaqueductal gray of the rat. J Comp Neurol 430:200–208
Dado RJ, Law PY, Loh HH, Elde R (1993) Immunofluorescent identification of a delta (delta)-opioid receptor on primary afferent nerve terminals. Neuroreport 5:341–344
Daniels DJ, Kulkarni A, Xie Z, Bhushan RG, Portoghese PS (2005a) A bivalent ligand (KDAN-18) containing delta-antagonist and kappa-agonist pharmacophores bridges delta2 and kappa1 opioid receptor phenotypes. J Med Chem 48:1713–1716
Daniels DJ, Lenard NR, Etienne CL, Law PY, Roerig SC, Portoghese PS (2005b) Opioid-induced tolerance and dependence in mice is modulated by the distance between pharmacophores in a bivalent ligand series. Proc Natl Acad Sci U S A 102:19208–19213
Darland T, Heinricher MM, Grandy DK (1998) Orphanin FQ/nociceptin: a role in pain and analgesia, but so much more. Trends Neurosci 21:215–221
Desmeules JA, Kayser V, Gacel G, Guilbaud G, Roques BP (1993) The highly selective delta agonist BUBU induces an analgesic effect in normal and arthritic rat and this action is not affected by repeated administration of low doses of morphine. Brain Res 611:243–248
Dondio G, Ronzoni S, Farina C, Graziani D, Parini C, Petrillo P, Giardina GA (2001) Selective delta opioid receptor agonists for inflammatory and neuropathic pain. Farmaco 56:117–119
Dore-Savard L, Otis V, Belleville K, Lemire M, Archambault M, Tremblay L, Beaudoin JF, Beaudet N, Lecomte R, Lepage M, Gendron L, Sarret P (2010) Behavioral, medical imaging and histopathological features of a new rat model of bone cancer pain. PLoS One 5:e13774
Dubois D, Gendron L (2010) Delta opioid receptor-mediated analgesia is not altered in preprotachykinin A knockout mice. Eur J Neurosci 32:1921–1929
Dykstra LA, Schoenbaum GM, Yarbrough J, McNutt R, Chang KJ (1993) A novel delta opioid agonist, BW373U86, in squirrel monkeys responding under a schedule of shock titration. J Pharmacol Exp Ther 267:875–882
Evans CJ, Keith DE Jr, Morrison H, Magendzo K, Edwards RH (1992) Cloning of a delta opioid receptor by functional expression. Science 258:1952–1955
Fenalti G, Giguere PM, Katritch V, Huang XP, Thompson AA, Cherezov V, Roth BL, Stevens RC (2014) Molecular control of delta-opioid receptor signalling. Nature 506:191–196
Filliol D, Ghozland S, Chluba J, Martin M, Matthes HW, Simonin F, Befort K, Gaveriaux-Ruff C, Dierich A, LeMeur M, Valverde O, Maldonado R, Kieffer BL (2000) Mice deficient for delta- and mu-opioid receptors exhibit opposing alterations of emotional responses. Nat Genet 25:195–200
Finn AK, Whistler JL (2001) Endocytosis of the mu opioid receptor reduces tolerance and a cellular hallmark of opiate withdrawal. Neuron 32:829–839
Fraser GL, Gaudreau GA, Clarke PB, Menard DP, Perkins MN (2000a) Antihyperalgesic effects of delta opioid agonists in a rat model of chronic inflammation. Br J Pharmacol 129:1668–1672
Fraser GL, Holmgren J, Clarke PB, Wahlestedt C (2000b) Antisense inhibition of delta-opioid receptor gene function in vivo by peptide nucleic acids. Mol Pharmacol 57:725–731
Fujita W, Gomes I, Devi LA (2014) Revolution in GPCR signalling: opioid receptor heteromers as novel therapeutic targets: IUPHAR review 10. Br J Pharmacol 171:4155–4176
Fujita W, Gomes I, Devi LA (2015) Heteromers of mu-delta opioid receptors: new pharmacology and novel therapeutic possibilities. Br J Pharmacol 172:375–387
Gallantine EL, Meert TF (2005) A comparison of the antinociceptive and adverse effects of the mu-opioid agonist morphine and the delta-opioid agonist SNC80. Basic Clin Pharmacol Toxicol 97:39–51
Garnock-Jones KP (2015) Eluxadoline: first global approval. Drugs 75:1305–1310

Gaveriaux-Ruff C, Kieffer BL (2011) Delta opioid receptor analgesia: recent contributions from pharmacology and molecular approaches. Behav Pharmacol 22:405–414

Gaveriaux-Ruff C, Karchewski LA, Hever X, Matifas A, Kieffer BL (2008) Inflammatory pain is enhanced in delta opioid receptor-knockout mice. Eur J Neurosci 27:2558–2567

Gaveriaux-Ruff C, Nozaki C, Nadal X, Hever XC, Weibel R, Matifas A, Reiss D, Filliol D, Nassar MA, Wood JN, Maldonado R, Kieffer BL (2011) Genetic ablation of delta opioid receptors in nociceptive sensory neurons increases chronic pain and abolishes opioid analgesia. Pain 152:1238–1248

Gendron L, Lucido AL, Mennicken F, O'Donnell D, Vincent JP, Stroh T, Beaudet A (2006) Morphine and pain-related stimuli enhance cell surface availability of somatic delta-opioid receptors in rat dorsal root ganglia. J Neurosci 26:953–962

Gendron L, Esdaile MJ, Mennicken F, Pan H, O'Donnell D, Vincent J-P, Devi LA, Cahill CM, Stroh T, Beaudet A (2007a) Morphine priming in rats with chronic inflammation reveals a dichotomy between antihyperalgesic and antinociceptive properties of deltorphin. Neuroscience 144:263–274

Gendron L, Pintar JE, Chavkin C (2007b) Essential role of mu opioid receptor in the regulation of delta opioid receptor-mediated antihyperalgesia. Neuroscience 150:807–817

Gendron L, Mittal N, Beaudry H, Walwyn W (2015) Recent advances on the delta opioid receptor: from trafficking to function. Br J Pharmacol 172:403–419

Gendron L, Cahill CM, von Zastrow M, Schiller PW, Pineyro G (2016) Molecular pharmacology of delta-opioid receptors. Pharmacol Rev 68:631–700

George SR, Zastawny RL, Briones-Urbina R, Cheng R, Nguyen T, Heiber M, Kouvelas A, Chan AS, O'Dowd BF (1994) Distinct distributions of mu, delta and kappa opioid receptor mRNA in rat brain. Biochem Biophys Res Commun 205:1438–1444

Goadsby PJ, Charbit AR, Andreou AP, Akerman S, Holland PR (2009) Neurobiology of migraine. Neuroscience 161:327–341

Goldstein A, Tachibana S, Lowney LI, Hunkapiller M, Hood L (1979) Dynorphin-(1-13), an extraordinarily potent opioid peptide. Proc Natl Acad Sci U S A 76:6666–6670

Gomes I, Fujita W, Gupta A, Saldanha SA, Negri A, Pinello CE, Eberhart C, Roberts E, Filizola M, Hodder P, Devi LA (2013) Identification of a mu-delta opioid receptor heteromer-biased agonist with antinociceptive activity. Proc Natl Acad Sci U S A 110:12072–12077

Granier S, Manglik A, Kruse AC, Kobilka TS, Thian FS, Weis WI, Kobilka BK (2012) Structure of the delta-opioid receptor bound to naltrindole. Nature 485:400–404

Guan JS, Xu ZZ, Gao H, He SQ, Ma GQ, Sun T, Wang LH, Zhang ZN, Lena I, Kitchen I, Elde R, Zimmer A, He C, Pei G, Bao L, Zhang X (2005) Interaction with vesicle luminal protachykinin regulates surface expression of delta-opioid receptors and opioid analgesia. Cell 122:619–631

Gupta A, Mulder J, Gomes I, Rozenfeld R, Bushlin I, Ong E, Lim M, Maillet E, Junek M, Cahill CM, Harkany T, Devi LA (2010) Increased abundance of opioid receptor heteromers after chronic morphine administration. Sci Signal 3:ra54

Hack SP, Bagley EE, Chieng BC, Christie MJ (2005) Induction of delta-opioid receptor function in the midbrain after chronic morphine treatment. J Neurosci 25:3192–3198

Hansra P, Moran EL, Fornasier VL, Bogoch ER (2000) Carrageenan-induced arthritis in the rat. Inflammation 24:141–155

Harvey JH, Long DH, England PM, Whistler JL (2012) Tuned-affinity bivalent ligands for the characterization of opioid receptor heteromers. ACS Med Chem Lett 3:640–644

Headrick JP, See Hoe LE, Du Toit EF, Peart JN (2015) Opioid receptors and cardioprotection – 'opioidergic conditioning' of the heart. Br J Pharmacol 172:2026–2050

Hervera A, Leanez S, Negrete R, Pol O (2009) The peripheral administration of a nitric oxide donor potentiates the local antinociceptive effects of a DOR agonist during chronic inflammatory pain in mice. Naunyn Schmiedebergs Arch Pharmacol 380:345–352

Heydorn A, Sondergaard BP, Ersboll B, Holst B, Nielsen FC, Haft CR, Whistler J, Schwartz TW (2004) A library of 7TM receptor C-terminal tails. Interactions with the proposed post-endocytic sorting proteins ERM-binding phosphoprotein 50 (EBP50), N-ethylmaleimide-sensitive factor (NSF), sorting nexin 1 (SNX1), and G protein-coupled receptor-associated sorting protein (GASP). J Biol Chem 279:54291–54303

Holdridge SV, Cahill CM (2007) Spinal administration of a delta opioid receptor agonist attenuates hyperalgesia and allodynia in a rat model of neuropathic pain. Eur J Pain 11:685–693

Honore P, Rogers SD, Schwei MJ, Salak-Johnson JL, Luger NM, Sabino MC, Clohisy DR, Mantyh PW (2000) Murine models of inflammatory, neuropathic and cancer pain each generates a unique set of neurochemical changes in the spinal cord and sensory neurons. Neuroscience 98:585–598

Hughes J, Smith TW, Kosterlitz HW, Fothergill LA, Morgan BA, Morris HR (1975) Identification of two related pentapeptides from the brain with potent opiate agonist activity. Nature 258:577–580

Hurley RW, Hammond DL (2000) The analgesic effects of supraspinal mu and delta opioid receptor agonists are potentiated during persistent inflammation. J Neurosci 20:1249–1259

Hylden JL, Thomas DA, Iadarola MJ, Nahin RL, Dubner R (1991) Spinal opioid analgesic effects are enhanced in a model of unilateral inflammation/hyperalgesia: possible involvement of noradrenergic mechanisms. Eur J Pharmacol 194:135–143

Ichikawa H, Schulz S, Hollt V, Sugimoto T (2005) Delta-opioid receptor-immunoreactive neurons in the rat cranial sensory ganglia. Brain Res 1043:225–230

Ji RR, Zhang Q, Law PY, Low HH, Elde R, Hokfelt T (1995) Expression of mu-, delta-, and kappa-opioid receptor-like immunoreactivities in rat dorsal root ganglia after carrageenan-induced inflammation. J Neurosci 15:8156–8166

Kabli N, Cahill CM (2007) Anti-allodynic effects of peripheral delta opioid receptors in neuropathic pain. Pain 127:84–93

Kamei J, Kashiwazaki T, Hitosugi H, Nagase H (1997a) The role of spinal delta1-opioid receptors in inhibiting the formalin-induced nociceptive response in diabetic mice. Eur J Pharmacol 326:31–36

Kamei J, Kawai K, Mizusuna A, Saitoh A, Morita K, Narita M, Tseng LF, Nagase H (1997b) Supraspinal delta 1-opioid receptor-mediated antinociceptive properties of (-)-TAN-67 in diabetic mice. Eur J Pharmacol 322:27–30

Kent JL, Pert CB, Herkenham M (1981) Ontogeny of opiate receptors in rat forebrain: visualization by in vitro autoradiography. Brain Res 254:487–504

Kieffer BL, Evans CJ (2009) Opioid receptors: from binding sites to visible molecules in vivo. Neuropharmacology 56(Suppl 1):205–212

Kieffer BL, Gaveriaux-Ruff C (2002) Exploring the opioid system by gene knockout. Prog Neurobiol 66:285–306

Kieffer BL, Befort K, Gaveriaux-Ruff C, Hirth CG (1992) The delta-opioid receptor: isolation of a cDNA by expression cloning and pharmacological characterization. Proc Natl Acad Sci U S A 89:12048–12052

Klareskog L (1989) What can we learn about rheumatoid arthritis from animal models? Springer Semin Immunopathol 11:315–333

Kouchek M, Takasusuki T, Terashima T, Yaksh TL, Xu Q (2013) Effects of intrathecal SNC80, a delta receptor ligand, on nociceptive threshold and dorsal horn substance p release. J Pharmacol Exp Ther 347:258–264

Krames ES, Wilkie DJ, Gershow J (1986) Intrathecal D-Ala2-D-Leu5-enkephalin (DADL) restores analgesia in a patient analgetically tolerant to intrathecal morphine sulfate. Pain 24:205–209

Le Bourdonnec B, Windh RT, Ajello CW, Leister LK, Gu M, Chu GH, Tuthill PA, Barker WM, Koblish M, Wiant DD, Graczyk TM, Belanger S, Cassel JA, Feschenko MS, Brogdon BL, Smith SA, Christ DD, Derelanko MJ, Kutz S, Little PJ, DeHaven RN, DeHaven-Hudkins DL, Dolle RE (2008) Potent, orally bioavailable delta opioid receptor agonists for the treatment of pain: discovery of N,N-diethyl-4-(5-hydroxyspiro[chromene-2,4'-piperidine]-4-yl)benzamide (ADL5859). J Med Chem 51:5893–5896

Le Bourdonnec B, Windh RT, Leister LK, Zhou QJ, Ajello CW, Gu M, Chu GH, Tuthill PA, Barker WM, Koblish M, Wiant DD, Graczyk TM, Belanger S, Cassel JA, Feschenko MS, Brogdon BL, Smith SA, Derelanko MJ, Kutz S, Little PJ, DeHaven RN, DeHaven-Hudkins DL, Dolle RE (2009) Spirocyclic delta opioid receptor agonists for the treatment of pain: discovery of N,N-diethyl-3-hydroxy-4-(spiro[chromene-2,4′-piperidine]-4-yl) benzamide (ADL5747). J Med Chem 52:5685–5702

Le Merrer J, Becker JA, Befort K, Kieffer BL (2009) Reward processing by the opioid system in the brain. Physiol Rev 89:1379–1412

Lembo AJ, Lacy BE, Zuckerman MJ, Schey R, Dove LS, Andrae DA, Davenport JM, McIntyre G, Lopez R, Turner L, Covington PS (2016) Eluxadoline for irritable bowel syndrome with diarrhea. N Engl J Med 374:242–253

Li CH, Chung D (1976) Isolation and structure of an untriakontapeptide with opiate activity from camel pituitary glands. Proc Natl Acad Sci U S A 73:1145–1148

Loh HH, Tseng LF, Wei E, Li CH (1976) beta-endorphin is a potent analgesic agent. Proc Natl Acad Sci U S A 73:2895–2898

Lucido AL, Morinville A, Gendron L, Stroh T, Beaudet A (2005) Prolonged morphine treatment selectively increases membrane recruitment of delta-opioid receptors in mouse basal ganglia. J Mol Neurosci 25:207–214

Lutz PE, Kieffer BL (2013) Opioid receptors: distinct roles in mood disorders. Trends Neurosci 36:195–206

Manglik A, Kruse AC, Kobilka TS, Thian FS, Mathiesen JM, Sunahara RK, Pardo L, Weis WI, Kobilka BK, Granier S (2012) Crystal structure of the micro-opioid receptor bound to a morphinan antagonist. Nature 485:321–326

Mansour A, Khachaturian H, Lewis ME, Akil H, Watson SJ (1987) Autoradiographic differentiation of mu, delta, and kappa opioid receptors in the rat forebrain and midbrain. J Neurosci 7:2445–2464

Mansour A, Thompson RC, Akil H, Watson SJ (1993) Delta opioid receptor mRNA distribution in the brain: comparison to delta receptor binding and proenkephalin mRNA. J Chem Neuroanat 6:351–362

Mansour A, Fox CA, Burke S, Meng F, Thompson RC, Akil H, Watson SJ (1994) Mu, delta, and kappa opioid receptor mRNA expression in the rat CNS: an in situ hybridization study. J Comp Neurol 350:412–438

Mansour A, Fox CA, Akil H, Watson SJ (1995) Opioid-receptor mRNA expression in the rat CNS: anatomical and functional implications. Trends Neurosci 18:22–29

McDowell J, Kitchen I (1986) Ontogenesis of delta-opioid receptors in rat brain using [3H] [D-Pen2,D-Pen5]enkephalin as a binding ligand. Eur J Pharmacol 128:287–289

McQuay H (1999) Opioids in pain management. Lancet 353:2229–2232

Meng F, Xie GX, Thompson RC, Mansour A, Goldstein A, Watson SJ, Akil H (1993) Cloning and pharmacological characterization of a rat kappa opioid receptor. Proc Natl Acad Sci U S A 90:9954–9958

Mennicken F, Zhang J, Hoffert C, Ahmad S, Beaudet A, O'Donnell D (2003) Phylogenetic changes in the expression of delta opioid receptors in spinal cord and dorsal root ganglia. J Comp Neurol 465:349–360

Meunier JC, Mollereau C, Toll L, Suaudeau C, Moisand C, Alvinerie P, Butour JL, Guillemot JC, Ferrara P, Monsarrat B et al (1995) Isolation and structure of the endogenous agonist of opioid receptor-like ORL1 receptor. Nature 377:532–535

Mika J, Przewlocki R, Przewlocka B (2001) The role of delta-opioid receptor subtypes in neuropathic pain. Eur J Pharmacol 415:31–37

Milan-Lobo L, Whistler JL (2011) Heteromerization of the mu- and delta-opioid receptors produces ligand-biased antagonism and alters mu-receptor trafficking. J Pharmacol Exp Ther 337:868–875

Milan-Lobo L, Enquist J, van Rijn RM, Whistler JL (2013) Anti-analgesic effect of the mu/delta opioid receptor heteromer revealed by ligand-biased antagonism. PLoS One 8:e58362

Minami M, Toya T, Katao Y, Maekawa K, Nakamura S, Onogi T, Kaneko S, Satoh M (1993) Cloning and expression of a cDNA for the rat kappa-opioid receptor. FEBS Lett 329:291–295

Minami M, Maekawa K, Yabuuchi K, Satoh M (1995) Double in situ hybridization study on coexistence of mu-, delta- and kappa-opioid receptor mRNAs with preprotachykinin A mRNA in the rat dorsal root ganglia. Brain Res Mol Brain Res 30:203–210

Morinville A, Cahill CM, Esdaile MJ, Aibak H, Collier B, Kieffer BL, Beaudet A (2003) Regulation of delta-opioid receptor trafficking via mu-opioid receptor stimulation: evidence from mu-opioid receptor knock-out mice. J Neurosci 23:4888–4898

Morinville A, Cahill CM, Aibak H, Rymar VV, Pradhan A, Hoffert C, Mennicken F, Stroh T, Sadikot AF, O'Donnell D, Clarke PB, Collier B, Henry JL, Vincent JP, Beaudet A (2004) Morphine-induced changes in delta opioid receptor trafficking are linked to somatosensory processing in the rat spinal cord. J Neurosci 24:5549–5559

Mosberg HI, Hurst R, Hruby VJ, Gee K, Yamamura HI, Galligan JJ, Burks TF (1983) Bis-penicillamine enkephalins possess highly improved specificity toward delta opioid receptors. Proc Natl Acad Sci U S A 80:5871–5874

Moulin DE, Max MB, Kaiko RF, Inturrisi CE, Maggard J, Yaksh TL, Foley KM (1985) The analgesic efficacy of intrathecal D-Ala2-D-Leu5-enkephalin in cancer patients with chronic pain. Pain 23:213–221

Nadal X, Banos JE, Kieffer BL, Maldonado R (2006) Neuropathic pain is enhanced in delta-opioid receptor knockout mice. Eur J Neurosci 23:830–834

Negri L, Severini C, Lattanzi R, Potenza RL, Melchiorri P (1997) Postnatal development of delta-opioid receptor subtypes in mice. Br J Pharmacol 120:989–994

Nitsche JF, Pintar JE (2003) Opioid receptor-induced GTPgamma35S binding during mouse development. Dev Biol 253:99–108

Obara I, Parkitna JR, Korostynski M, Makuch W, Kaminska D, Przewlocka B, Przewlocki R (2009) Local peripheral opioid effects and expression of opioid genes in the spinal cord and dorsal root ganglia in neuropathic and inflammatory pain. Pain 141:283–291

Olesen J, Burstein R, Ashina M, Tfelt-Hansen P (2009) Origin of pain in migraine: evidence for peripheral sensitisation. Lancet Neurol 8:679–690

Onofrio BM, Yaksh TL (1983) Intrathecal delta-receptor ligand produces analgesia in man. Lancet 1:1386–1387

Otis V, Sarret P, Gendron L (2011) Spinal activation of delta opioid receptors alleviates cancer-related bone pain. Neuroscience 183:221–229

Pacheco DF, Duarte ID (2005) Delta-opioid receptor agonist SNC80 induces peripheral antinociception via activation of ATP-sensitive K+ channels. Eur J Pharmacol 512:23–28

Pakarinen ED, Woods JH, Moerschbaecher JM (1995) Repeated acquisition of behavioral chains in squirrel monkeys: comparisons of a mu, kappa and delta opioid agonist. J Pharmacol Exp Ther 272:552–559

Pasquini F, Bochet P, Garbay-Jaureguiberry C, Roques BP, Rossier J, Beaudet A (1992) Electron microscopic localization of photoaffinity-labelled delta opioid receptors in the neostriatum of the rat. J Comp Neurol 326:229–244

Patwardhan AM, Berg KA, Akopain AN, Jeske NA, Gamper N, Clarke WP, Hargreaves KM (2005) Bradykinin-induced functional competence and trafficking of the delta-opioid receptor in trigeminal nociceptors. J Neurosci 25:8825–8832

Peng J, Sarkar S, Chang SL (2012) Opioid receptor expression in human brain and peripheral tissues using absolute quantitative real-time RT-PCR. Drug Alcohol Depend 124:223–228

Petrillo P, Angelici O, Bingham S, Ficalora G, Garnier M, Zaratin PF, Petrone G, Pozzi O, Sbacchi M, Stean TO, Upton N, Dondio GM, Scheideler MA (2003) Evidence for a selective role of the delta-opioid agonist [8R-(4bS*,8aalpha,8abeta, 12bbeta)]7,10-dimethyl-1-methoxy-11-(2-methylpropyl)oxycarbonyl 5,6,7,8,12,12b-hexahydro-(9H)-4,8-methanobenzofuro[3,2-e] pyrrolo[2,3-g]isoquinoli ne hydrochloride (SB-235863) in blocking hyperalgesia associated with inflammatory and neuropathic pain responses. J Pharmacol Exp Ther 307:1079–1089

Podolsky AT, Sandweiss A, Hu J, Bilsky EJ, Cain JP, Kumirov VK, Lee YS, Hruby VJ, Vardanyan RS, Vanderah TW (2013) Novel fentanyl-based dual mu/delta-opioid agonists for the treatment of acute and chronic pain. Life Sci 93:1010–1016

Porreca F, Mosberg HI, Hurst R, Hruby VJ, Burks TF (1984) Roles of mu, delta and kappa opioid receptors in spinal and supraspinal mediation of gastrointestinal transit effects and hot-plate analgesia in the mouse. J Pharmacol Exp Ther 230:341–348

Poulin JF, Chevalier B, Laforest S, Drolet G (2006) Enkephalinergic afferents of the centromedial amygdala in the rat. J Comp Neurol 496:859–876

Pradhan AA, Clarke PB (2005) Comparison between delta-opioid receptor functional response and autoradiographic labeling in rat brain and spinal cord. J Comp Neurol 481:416–426

Pradhan AA, Siau C, Constantin A, Clarke PB (2006) Chronic morphine administration results in tolerance to delta opioid receptor-mediated antinociception. Neuroscience 141:947–954

Pradhan AA, Befort K, Nozaki C, Gaveriaux-Ruff C, Kieffer BL (2011) The delta opioid receptor: an evolving target for the treatment of brain disorders. Trends Pharmacol Sci 32:581–590

Pradhan A, Smith M, McGuire B, Evans C, Walwyn W (2013) Chronic inflammatory injury results in increased coupling of delta opioid receptors to voltage-gated Ca^{2+} channels. Mol Pain 9:8

Pradhan AA, Smith ML, McGuire B, Tarash I, Evans CJ, Charles A (2014) Characterization of a novel model of chronic migraine. Pain 155:269–274

Qiu C, Sora I, Ren K, Uhl G, Dubner R (2000) Enhanced delta-opioid receptor-mediated antinociception in mu-opioid receptor-deficient mice. Eur J Pharmacol 387:163–169

Reinscheid RK, Nothacker HP, Bourson A, Ardati A, Henningsen RA, Bunzow JR, Grandy DK, Langen H, Monsma FJ Jr, Civelli O (1995) Orphanin FQ: a neuropeptide that activates an opioidlike G protein-coupled receptor. Science 270:792–794

Riedl MS, Schnell SA, Overland AC, Chabot-Dore AJ, Taylor AM, Ribeiro-da-Silva A, Elde RP, Wilcox GL, Stone LS (2009) Coexpression of alpha 2A-adrenergic and delta-opioid receptors in substance P-containing terminals in rat dorsal horn. J Comp Neurol 513:385–398

Rossi GC, Pasternak GW, Bodnar RJ (1994) Mu and delta opioid synergy between the periaqueductal gray and the rostro-ventral medulla. Brain Res 665:85–93

Saitoh A, Sugiyama A, Nemoto T, Fujii H, Wada K, Oka J, Nagase H, Yamada M (2011) The novel delta opioid receptor agonist KNT-127 produces antidepressant-like and antinociceptive effects in mice without producing convulsions. Behav Brain Res 223:271–279

Saloman JL, Niu KY, Ro JY (2011) Activation of peripheral delta-opioid receptors leads to anti-hyperalgesic responses in the masseter muscle of male and female rats. Neuroscience 190:379–385

Sauriyal DS, Jaggi AS, Singh N (2011) Extending pharmacological spectrum of opioids beyond analgesia: multifunctional aspects in different pathophysiological states. Neuropeptides 45:175–188

Scherrer G, Imamachi N, Cao YQ, Contet C, Mennicken F, O'Donnell D, Kieffer BL, Basbaum AI (2009) Dissociation of the opioid receptor mechanisms that control mechanical and heat pain. Cell 137:1148–1159

Sharif NA, Hughes J (1989) Discrete mapping of brain Mu and delta opioid receptors using selective peptides: quantitative autoradiography, species differences and comparison with kappa receptors. Peptides 10:499–522

Simonin F, Karcher P, Boeuf JJ, Matifas A, Kieffer BL (2004) Identification of a novel family of G protein-coupled receptor associated sorting proteins. J Neurochem 89:766–775

Slowe SJ, Simonin F, Kieffer B, Kitchen I (1999) Quantitative autoradiography of mu-,delta- and kappa1 opioid receptors in kappa-opioid receptor knockout mice. Brain Res 818:335–345

Snyder SH, Pasternak GW (2003) Historical review: opioid receptors. Trends Pharmacol Sci 24:198–205

Sohn JH, Lee BH, Park SH, Ryu JW, Kim BO, Park YG (2000) Microinjection of opiates into the periaqueductal gray matter attenuates neuropathic pain symptoms in rats. Neuroreport 11:1413–1416

Stein C, Lang LJ (2009) Peripheral mechanisms of opioid analgesia. Curr Opin Pharmacol 9:3–8

Stein C, Machelska H, Binder W, Schafer M (2001) Peripheral opioid analgesia. Curr Opin Pharmacol 1:62–65
Stevens CW (2009) The evolution of vertebrate opioid receptors. Front Biosci (Landmark Ed) 14:1247–1269
Stewart PE, Hammond DL (1994) Activation of spinal delta-1 or delta-2 opioid receptors reduces carrageenan-induced hyperalgesia in the rat. J Pharmacol Exp Ther 268:701–708
Suarez-Roca H, Maixner W (1992) Delta-opioid-receptor activation by [D-Pen2,D-Pen5]enkephalin and morphine inhibits substance P release from trigeminal nucleus slices. Eur J Pharmacol 229:1–7
Tsao PI, von Zastrow M (2000) Type-specific sorting of G protein-coupled receptors after endocytosis. J Biol Chem 275:11130–11140
Wade PR, Palmer JM, McKenney S, Kenigs V, Chevalier K, Moore BA, Mabus JR, Saunders PR, Wallace NH, Schneider CR, Kimball ES, Breslin HJ, He W, Hornby PJ (2012) Modulation of gastrointestinal function by MuDelta, a mixed micro opioid receptor agonist/ micro opioid receptor antagonist. Br J Pharmacol 167:1111–1125
Wang H, Pickel VM (2001) Preferential cytoplasmic localization of delta-opioid receptors in rat striatal patches: comparison with plasmalemmal mu-opioid receptors. J Neurosci 21:3242–3250
Wang H, Wessendorf MW (2001) Equal proportions of small and large DRG neurons express opioid receptor mRNAs. J Comp Neurol 429:590–600
Wang JB, Imai Y, Eppler CM, Gregor P, Spivak CE, Uhl GR (1993) mu opiate receptor: cDNA cloning and expression. Proc Natl Acad Sci U S A 90:10230–10234
Wang JB, Johnson PS, Persico AM, Hawkins AL, Griffin CA, Uhl GR (1994) Human mu opiate receptor. cDNA and genomic clones, pharmacologic characterization and chromosomal assignment. FEBS Lett 338:217–222
Wang XM, Yan JQ, Zhang KM, Mokha SS (1996) Role of opioid receptors (mu, delta 1, delta 2) in modulating responses of nociceptive neurons in the superficial and deeper dorsal horn of the medulla (trigeminal nucleus caudalis) in the rat. Brain Res 739:235–243
Wang HB, Zhao B, Zhong YQ, Li KC, Li ZY, Wang Q, Lu YJ, Zhang ZN, He SQ, Zheng HC, Wu SX, Hokfelt TG, Bao L, Zhang X (2010) Coexpression of delta- and mu-opioid receptors in nociceptive sensory neurons. Proc Natl Acad Sci U S A 107:13117–13122
Wei LN, Loh HH (2011) Transcriptional and epigenetic regulation of opioid receptor genes: present and future. Annu Rev Pharmacol Toxicol 51:75–97
Whistler JL, Enquist J, Marley A, Fong J, Gladher F, Tsuruda P, Murray SR, Von Zastrow M (2002) Modulation of postendocytic sorting of G protein-coupled receptors. Science 297:615–620
Wu H, Wacker D, Mileni M, Katritch V, Han GW, Vardy E, Liu W, Thompson AA, Huang XP, Carroll FI, Mascarella SW, Westkaemper RB, Mosier PD, Roth BL, Cherezov V, Stevens RC (2012) Structure of the human kappa-opioid receptor in complex with JDTic. Nature 485:327–332
Xie WY, He Y, Yang YR, Li YF, Kang K, Xing BM, Wang Y (2009) Disruption of Cdk5-associated phosphorylation of residue threonine-161 of the delta-opioid receptor: impaired receptor function and attenuated morphine antinociceptive tolerance. J Neurosci 29:3551–3564
Yasuda K, Espinosa R 3rd, Takeda J, Le Beau MM, Bell GI (1994) Localization of the kappa opioid receptor gene to human chromosome band 8q11.2. Genomics 19:596–597
Ye Y, Dang D, Viet CT, Dolan JC, Schmidt BL (2012) Analgesia targeting IB4-positive neurons in cancer-induced mechanical hypersensitivity. J Pain 13:524–531
Zaki PA, Bilsky EJ, Vanderah TW, Lai J, Evans CJ, Porreca F (1996) Opioid receptor types and subtypes: the delta receptor as a model. Annu Rev Pharmacol Toxicol 36:379–401
Zhang X, Bao L, Arvidsson U, Elde R, Hokfelt T (1998) Localization and regulation of the delta-opioid receptor in dorsal root ganglia and spinal cord of the rat and monkey: evidence for association with the membrane of large dense-core vesicles. Neuroscience 82:1225–1242
Zhang X, Bao L, Guan JS (2006) Role of delivery and trafficking of delta-opioid peptide receptors in opioid analgesia and tolerance. Trends Pharmacol Sci 27:324–329
Zhang X, Bao L, Ma GQ (2010) Sorting of neuropeptides and neuropeptide receptors into secretory pathways. Prog Neurobiol 90:276–283

Zhao B, Wang HB, Lu YJ, Hu JW, Bao L, Zhang X (2011) Transport of receptors, receptor signaling complexes and ion channels via neuropeptide-secretory vesicles. Cell Res 21:741–753

Zhou L, Zhang Q, Stein C, Schafer M (1998) Contribution of opioid receptors on primary afferent versus sympathetic neurons to peripheral opioid analgesia. J Pharmacol Exp Ther 286:1000–1006

Zhu Y, Hsu MS, Pintar JE (1998) Developmental expression of the mu, kappa, and delta opioid receptor mRNAs in mouse. J Neurosci 18:2538–2549

Zhu Y, King MA, Schuller AG, Nitsche JF, Reidl M, Elde RP, Unterwald E, Pasternak GW, Pintar JE (1999) Retention of supraspinal delta-like analgesia and loss of morphine tolerance in delta opioid receptor knockout mice. Neuron 24:243–252

Delta Opioid Receptors and Modulation of Mood and Emotion

Isaac J. Dripps and Emily M. Jutkiewicz

Contents

Abstract

Depression is a pervasive and debilitating mental disorder that is inadequately treated by current pharmacotherapies in a majority of patients. Although opioids have long been known to regulate mood states, the use of opioids to treat depression is rarely discussed. This chapter explores the preclinical and clinical evidence supporting the antidepressant-like effects of opioid ligands, and in particular, delta opioid receptor (DOR) agonists. DOR agonists have been shown to produce antidepressant-like effects in a number of animal models. Some DOR agonists also produce convulsions which has limited their clinical utility. However, DOR agonists that generate antidepressant-like effects without convulsions have recently been developed and these drugs are beginning to be evaluated in humans. Work investigating potential mechanisms of action for the antidepressant-like effects of DOR agonists is also explored. Understanding mechanisms that give rise to DOR-mediated behaviors is critical for the development of DOR drugs with improved safety and clinical utility, and future work should be devoted to elucidating these pathways.

I.J. Dripps • E.M. Jutkiewicz (✉)
Department of Pharmacology, University of Michigan Medical School, Ann Arbor, MI 48109, USA
e-mail: ejutkiew@umich.edu

E. M. Jutkiewicz (ed.), *Delta Opioid Receptor Pharmacology and Therapeutic Applications*, Handbook of Experimental Pharmacology 247, DOI 10.1007/164_2017_42

Keywords
Animal models • Antidepressant-like effects • Convulsions • Delta opioid receptor • Delta opioid receptor agonists • Enkephalin • Intracellular molecules • Mood • Neurotransmitters

1 Introduction

Major depressive disorder (MDD) is a psychiatric disease in which those affected experience a depressed mood (i.e., feelings of sadness, emptiness, or hopelessness) and/or a loss of interest or pleasure in everyday activities. The recently released fifth edition of the Diagnostic and Statistical Manual of Mental Disorders defines a person as having a major depressive episode when they exhibit one of the former symptoms and at least four of the following within a 2-week period: (1) significant changes in weight or (2) sleeping pattern, (3) agitation, (4) fatigue, (5) feelings of worthlessness or excessive guilt, (6) diminished concentration, and/or (7) recurring thoughts of death or suicide (American Psychiatric Association 2013). In addition, MDD often presents with comorbidities such as chronic pain and anxiety disorders.

Major depression is a pervasive disease, affecting around 350 million (1 in 20) people worldwide (WHO 2012). It is the top cause of disability in terms of total years lost to disability and is the leading cause of disease burden in women regardless of income (WHO 2008). Furthermore, evidence points to a mother with depression being a risk factor for poor growth and development in children, suggesting this disease can have lasting effects in subsequent generations (Rahman et al. 2008). However, the gravest concern with a depressed patient is an increased risk of suicide. MDD is responsible for approximately half of all suicides in the United States with 15% of people with depression eventually committing suicide (Loosen and Shelton 2008).

Although there are currently several methods for treating major depression, each has proved problematic. The selective serotonin reuptake inhibitors (SSRIs) are the first line therapy and most commonly used drug class for the treatment of major depression. SSRIs act by blocking the reuptake of serotonin into presynaptic nerve terminals thereby increasing neurotransmitter signaling. Unfortunately, it can take up to 6 weeks for an SSRI to reach a full effect, and 70% of patients do not achieve remission with an SSRI alone (Trivedi et al. 2006a). Augmenting SSRI treatment with a second antidepressant can be helpful although only one-third of patients who do not respond to SSRI monotherapy achieve remission with combination therapy (Trivedi et al. 2006b). Furthermore, SSRI treatment can lead to adverse events and complications including tinnitus, insomnia, akathisia, and sexual dysfunction.

The tricyclic antidepressants (TCAs) are an alternative treatment for major depression and were the primary therapy prior to the development of SSRIs. TCAs act through a variety of mechanisms; however, the majority block serotonin and/or norepinephrine reuptake. Like SSRIs, TCAs can also take several weeks to reach their full effect and effectively treat a small percentage of patients. In the

Sequenced Treatment Alternatives to Relieve Depression (STAR*D) clinical trial, about 20% of patients not adequately treated with SSRI mono or combination therapy achieved remission with a TCA (Fava et al. 2006). TCAs can also have effects on the cardiovascular system including changes in heart rate or rhythm and orthostatic hypotension. Furthermore, the therapeutic index of TCAs is small and overdoses can be lethal making administration of these drugs to suicidal patients concerning. Other traditional medications used for the treatment of depression include serotonin receptor agonists, serotonin and norepinephrine reuptake inhibitors (SNRIs), and monoamine oxidase inhibitors (MAOIs).

An alternative to pharmacotherapy, and arguably the most effective treatment for major depression currently available, is electroconvulsive therapy (ECT). ECT works by using an electric current to induce a generalized seizure in the central nervous system of the patient. The mechanism by which this seizure alleviates depressive symptoms is currently unknown, although there is evidence that it involves modulation of the opioid system (Emrich et al. 1979; Inturrisi et al. 1982). The primary side effects associated with ECT are confusion and memory loss. Although rare, this memory loss is potentially permanent. Other problems surrounding ECT include insufficient patient understanding and public disapproval of its use (Eisendrath and Lichtmacher 2014). Given the inadequacy of available treatments, there is a demonstrable need for alternative therapies for major depression.

Although the majority of antidepressant therapies function via augmentation of aminergic neurotransmission, the monoamine deficiency hypothesis of depression is likely overly simplistic as several alternative targets including GABA, glutamate, adenosine, stress hormones, and opioids have been proposed to be involved in mediating depressive symptoms. Changes in opioid signaling have already been observed with currently used treatments for depression, such as electroconvulsive therapy (ECT), and the atypical antidepressant tianeptine, which was recently shown to be an agonist at both the mu opioid receptors (MOR) and delta opioid receptors (DOR), albeit at large concentrations (Gassaway et al. 2014). Despite these findings, the use of opioids for the treatment of depression is rarely discussed.

Opioid receptors are a family of G protein coupled receptors that signal through inhibitory Gi/o proteins. The three classical subtypes of opioid receptors (mu, delta, and kappa) are activated by several endogenous opioid peptides. β-endorphin is derived from prepro-opiomelanocortin (POMC) and activates MOR and DOR with relatively equal potency. Leu- and met-enkephalin are cleaved from preproenkephalin and bind and activate DOR with approximately tenfold selectivity over MOR. Dynorphin A and B are derived from preprodynorphin and potently activate kappa opioid receptors (KOR) with minor activity at MOR.

2 Opioids in Depression

It is well established that opioids alter mood states. MOR agonists such as morphine are euphorigenic and analgesic. Conversely, chronic use of and subsequent withdrawal from MOR agonists produce depressive-like symptoms. Agonists at the kappa opioid receptor have been shown to induce feelings of dysphoria, while KOR antagonists appear to have antidepressant-like effects (Lutz and Kieffer 2013). Furthermore, endogenous opioids are thought to be involved in neuronal circuits mediating reward, pleasure, and dopaminergic signaling (Jutkiewicz and Roques 2012). Endogenous opioids and opioid receptors are also highly expressed in brain regions implicated in mood disorders, specifically the prefrontal cortex, ventral striatum, and amygdala (Le Merrer et al. 2009; Lutz and Kieffer 2013). Based on these findings, it is thought that endogenous opioids play a role in regulating mood states and that dysregulation of the opioid system may be responsible for the anhedonia observed in depressed patients.

There are conflicting reports regarding the levels of endogenous β-endorphin levels in depressed patients. In fact, different studies have found β-endorphin levels in depressed patients to be higher (Goodwin et al. 1993), lower (Djurović et al. 1999), and no different (Naber et al. 1981) than healthy controls. Those studies sampled β-endorphin from the plasma, serum, and CSF, respectively, which could account for some of these discrepancies. Elevated β-endorphin levels may correlate with other mood disorders including primary affective disorders (Genazzani et al. 1984), the manic phase of manic-depressive disorder (Lindström et al. 1978), as well as severe anxiety, phobias, and obsessive-compulsive behaviors in depressed patients (Darko et al. 1992). Despite the inconsistent findings regarding endogenous β-endorphin levels in depressed patients, treatment with known antidepressants has been shown to increase endogenous opioid levels. Administration of the selective serotonin reuptake inhibitor (SSRI) fluvoxamine or the 5-HT_{1A} partial agonist gepirone (Anderson et al. 1990) increased β-endorphin levels in depressed patients and normal volunteers, respectively, suggesting an interaction between the opioid and serotonergic systems. Furthermore, the tricyclic antidepressant doxepin improved pain and depression scores in patients diagnosed with a chronic pain condition and clinical depression in a randomized double-blinded study (Hameroff et al. 1982). These patients also had increased nonspecific enkephalin-like activity, but not β-endorphin plasma levels, in response to doxepin. Taken together, these data suggest that excessive as well as insufficient levels of β-endorphin can lead to disease states and therefore tight control of β-endorphin levels, or other endogenous opioids, may be necessary for proper mood regulation.

Opioid ligands have also been administered to depressed patients to investigate their antidepressant-like effects. The mixed opioid ligand cyclazocine improved depressive symptoms in clinical trials with severely depressed, chronically ill mental patients and patients who did not respond to treatment with the tricyclic antidepressant imipramine (Fink and Shapiro 1969). Kline et al. (1977) and Gerner et al. (1980) both found that intravenous infusions of β-endorphin rapidly (2–4 h) improved symptoms in depressed patients. Intravenous infusions of the met-enkephalin analogue FK 33–824 acutely improved depressive symptoms on the first day of treatment but not on the second (Jungkunz et al. 1983), but it is unclear which opioid receptor is responsible for these effects. However, other studies found small, but non-significant, increases in depression scores in patients after acute infusions of the MOR selective agonists morphine or methadone (Extein et al.

1981), suggesting that the MOR alone cannot account for the antidepressant-like effects of these opioids. The mixed opioid ligand buprenorphine was found to significantly improve depression scores in patients with major depressive disorder (Emrich et al. 1981; Bodkin et al. 1995; Nyhuis et al. 2008) and in patients over 50 years old with treatment resistant depression (Karp et al. 2014). Consistent with the effects in humans, the opioid ligand buprenorphine decreased immobility scores in the mouse forced swim test. While the antinociceptive effects of buprenorphine are mediated by the MOR (Hayes et al. 1986; Kamei et al. 1995), the anti-depressant-like effects of buprenorphine were absent in KOR knockout mice, but not in MOR and DOR knockout mice, suggesting that these actions are due to its antagonist activity at kappa receptors (Falcon et al. 2016).

There is also evidence that changes at the receptor level of the opioid system play a role in depression. In a study utilizing positron emission tomography (PET), the binding potential of the MOR selective radiotracer [11C]-carfentanil was significantly increased – indicating a decrease in endogenous opioid neurotransmission – in healthy patients during a sustained sadness state in which participants were asked to recall an autobiographical event that would induce a negative emotional state (Zubieta et al. 2003). A follow-up study found that the binding potential of [11C]-carfentanil was decreased in depressed patients during a sustained sadness challenge and that MOR receptor availability in depressed patients was significantly decreased relative to healthy controls during a neutral emotional state (Kennedy et al. 2006). Single nucleotide polymorphisms of the MOR were also associated with increased efficacy of the SSRI citalopram in patients with major depressive disorder (Garriock et al. 2010). Expression of *OPRK1*, the gene encoding the kappa opioid receptor, was found to be increased in the dorsolateral prefrontal cortex of depressed patients (Deo et al. 2013). Changes in DOR expression and *OPRD1* and their effects on depression have not been explored.

Alterations in response to ECT have been observed in several systems, including serotonin, norepinephrine, and dopamine. Furthermore, no changes in one individual system can fully explain the therapeutic benefits. Repeated ECT treatments also produce compensatory anticonvulsant effects such as increased seizure threshold, decreased seizure duration, and enhanced inhibitory neurotransmission (Sackeim 1999; Baghai 2008). It has been proposed that these effects may bring about the antidepressant-like effects of ECT. The magnitude of change in seizure threshold and seizure duration likely impact therapeutic outcome and endogenous opioids have been proposed to play a role in the changing of seizure threshold. Individual and repeated ECT treatments resulted in elevated plasma immunoreactive β-endorphin in depressed patients (Emrich et al. 1979; Inturrisi et al. 1982). Electroconvulsive shock (ECS) has been shown to elevate proenkephalin-derived peptide and DOR levels (Tortella et al. 1989). In addition, these changes were blocked by the DOR selective antagonist ICI-174,864, suggesting a DOR-mediated effect. Based on these findings, the opioid system could be a useful target for the development of novel therapeutics for depression.

3 Antidepressant-Like Effects of the DOR

In 1976, Plotnikoff et al. showed that met-enkephalin potentiated DOPA-induced increases in motor activity. Because the tricyclic antidepressants were also effective in this assay, it was used as an early screening technique for antidepressant drugs (Everett 1966). The antidepressant-like effects of opioids were further supported when enkephalins and endorphins were shown to decrease immobility in the forced swim test and learned helplessness paradigm, again demonstrating effects similar to clinically used antidepressants (Kastin et al. 1978; Tejedor-Real et al. 1995). Later, numerous experiments showed that preventing the breakdown of endogenous opioid peptides using enkephalinase inhibitors produced antidepressant-like effects. Tejedor-Real et al. (1993) demonstrated that RB38A, a mixed enkephalinase inhibitor, and RB38B, a selective endopeptidase EC 3.4.24.11 inhibitor, reduced escape failures in the learned helplessness paradigm, and that these effects were blocked by the nonselective opioid receptor antagonist naloxone, suggesting an opioid receptor-mediated effect. In the mouse forced swim test, the enkephalinase inhibitor BL-2401 produced naloxone-reversible antidepressant-like effects, again indicating an opioid receptor-mediated effect (Kita et al. 1997).

However, these experiments did not necessarily demonstrate a role for the DOR in mediating these behaviors. The antidepressant-like effects of RB101, a mixed enkephalinase inhibitor, and the DOR selective peptide agonist BUBU (Tyr-D.Ser-(*O-tert*-butyl)-Gly-Phe-Leu-Thr(O-Tet-butyl-OH)), in the learned helplessness paradigm were attenuated by the DOR selective antagonist naltrindole, suggested that the antidepressant-like effects of these drugs were DOR-mediated in mice (Baamonde et al. 1992) and rats (Tejedor-Real et al. 1998). RB101 was later shown to consistently produce DOR-mediated antidepressant-like effects in the rat forced swim test (Jutkiewicz et al. 2006a). Recently, opiorphin, an endogenously expressed inhibitor of human neutral endopeptidase and aminopeptidase-N, was found to induce antidepressant-like effects in the rat forced swim test (Javelot et al. 2010). These effects were blocked by naltrindole as well as the mu opioid selective antagonist β-funaltrexamine (β-FNA), indicating a role for both delta and MORs (Javelot et al. 2010; Yang et al. 2011). Although these studies demonstrate that stimulation of the DOR produces antidepressant-like effects in animal models, the effect of opioid tone on mood states had not been examined. Filliol et al. (2000) demonstrated a role for endogenous tone at DORs in depression and depressive-like symptoms by showing that DOR knockout mice (*OPRD1*-deficient) exhibited anxiogenic and prodepressive behaviors.

The development of nonpeptidic DOR selective agonists greatly aided the investigation of DOR-mediated antidepressant-like effects by allowing for the study of centrally mediated behaviors using peripherally administered compounds that directly stimulate the receptor. The nonpeptidic DOR selective agonists (+) BW373U86 ((±)-4-((α-R*)-α-(2S*,5R*)-4-allyl-2,5-dimethyl-1-piperazinyl)-3-hydroxybenzyl)-*N*,*N*-diethylbenzamide), SNC80 ((+)-4-[(alpha R)-alpha-((2S,5R)-4-allyl-2,5-dimethyl-1-piperazinyl)-3-methoxybenzyl]-*N*,*N*-diethylbenzamide), and TAN-67 ((−)-2-methyl-4aα-(3-hydroxyphenyl)-1,2,3,4,4a,5,12,12aα-octahydorquinolino[2,3,3g]isoquiniline dihydrobromide) all demonstrated antidepressant-like effects in the rat forced swim test (Broom et al. 2002a; Nagase et al. 2002). The effects of (+)BW373U86 and SNC80 were shown to be naltrindole-reversible indicating a DOR-mediated effect (Broom et al. 2002a). Additionally, SNC80 has been found to elicit other potential antidepressant-like effects, including improving the emotionality score of olfactory bulbectomized rats (Saitoh et al. 2008) and reversing pain depressed responding of intracranial self-stimulation in rats (Negus et al. 2012). Furthermore, SNC80 is not self-administered by monkeys (Negus et al. 1998), does not facilitate intracranial self-stimulation (Do Carmo et al. 2009), and does not promote dopamine efflux in the nucleus accumbens (Longoni et al. 1998), suggesting a low abuse potential. Unlike typical antidepressants, which require multiple administrations to generate an effect in many animal models of depression, these DOR agonists were effective after a single, acute dose, suggesting a faster onset of action. Although tolerance develops to some of the effects of delta agonists after a single dose, delta agonists continue to produce antidepressant-like effects after repeated administration (Jutkiewicz et al. 2005a; Saitoh et al. 2008; Nozaki et al. 2014).

Due to the efficacy of electroconvulsive therapy (ECT) in treating depression in humans, it was hypothesized that DOR agonist-induced convulsions were required for their antidepressant-like effects, similar to that produced by other convulsive agents such as metrazol or insulin-induced seizure. DOR agonist-induced convulsions consist of brief, nonlethal, generalized seizure activity, are naltrindole-sensitive, and absent in DOR knockout mice (Comer et al. 1993; Broom et al. 2002b; Jutkiewicz et al. 2006b). Broom et al. (2002b) showed that pretreatment with the short acting benzodiazepine midazolam blocked (+)BW373U86-induced convulsions without affecting (+)BW373U86-induced antidepressant-like effects. By slowing the rate at which SNC80 was administered, Jutkiewicz et al. (2005b) eliminated the convulsive effects of SNC80 while maintaining its antidepressant-like effects. Additionally, they were able to elicit convulsions without observable antidepressant-like effects in the forced swim test in rats at a dose of 1 mg/kg SNC80 via rapid (20 s) intravenous infusions. Moreover, tolerance to the convulsive effects of delta opioid agonists develops after a single administration, whereas the antidepressant-like effects remain after chronic administration. Taken together, these data suggest that delta opioid agonist-induced convulsions do not drive or induce their antidepressant-like effects.

In recent years, nonpeptidic DOR agonists that do not produce convulsions have been developed. Promisingly, several of these drugs have demonstrable antidepressant-like effects in animal models. (−)-NIH 11082 ((−)-(1R,5R,9R)-5,9-dimethyl-2′-hydroxy-2-(6-hydroxyhexyl)-6,7-benzomorphan hydrochloride) decreased immobility in the mouse tail suspension test, although this effect was not as robust as that observed with the tricyclic antidepressant desipramine (Naidu et al. 2007). In addition, (−)-NIH 11082 produced antidepressant-like effects without stimulating locomotor activity or eliciting convulsions. The reduction in immobility was reversed by naltrindole in a dose-dependent manner, but not by receptor subtype selective doses of naltrexone (mu) or nor-BNI (kappa), suggesting that the antidepressant-like effects of (−)-NIH 11082 are DOR-mediated. (−)-NIH

11082 did not substitute for morphine or exacerbate withdrawal symptoms in morphine-dependent monkeys, further indicating lack of activity at the MOR (Aceto et al. 2007). Surprisingly, Traynor et al. (2005) found that (−)-NIH 11082 bound the DOR with low affinity and had little to no efficacy in a $[^{35}S]GTP\gamma S$ assay. These findings suggest that (−)-NIH 11082 may exert its antidepressant-like effects via an allosteric or indirect means, though the exact mechanism has yet to be determined.

UFP-512 (H-Dmt-Tic-NH-CH(CH2-COOH)-Bid) is a systemically active peptidomimetic and is effective in multiple animal models of depression. In the mouse forced swim test, UFP-512 produced significant decreases in immobility time when administered intraperitoneally or intracerebroventricularly (Vergura et al. 2008). Tolerance to these effects is likely to be minimal as this reduction in immobility was maintained in mice after 7 days of UFP-512 treatment i.p. (Aguila et al. 2007). In an open field test, UFP-512 did not significantly alter the locomotor activity in mice at doses capable of reducing immobility in the forced swim test, suggesting that the antidepressant-like effects of UFP-512 are not due to any stimulant properties of the compound (Vergura et al. 2008). In the rat forced swim test, i.p. administration of UFP-512 significantly reduced immobility without significantly altering climbing and swimming behaviors (Vergura et al. 2008; Kabli et al. 2014). Kabli et al. also demonstrated that injection of UFP-512 bilaterally into the rat nucleus accumbens was sufficient to reduce immobility in the forced swim test and reduces the latency to drink milk in a novelty-induced hypophagia assay. Interestingly, these effects could be blocked with either the delta opioid antagonist naltrindole or the mu opioid antagonist CTOP, suggesting that both delta and MORs are required for the antidepressant-like effects of UFP-512 or that the doses used for intra-accumbens injections were nonselective. There are no reports of UFP-512 eliciting convulsions.

ADL5859 (*N,N*-diethyl-4-(5-hydroxyspiro[chromene-2,4′-piperidine]-4-yl) benzamide) significantly reduced immobility and increased swimming when administered at 3 mg/kg orally in a rat forced swim test (Le Bourdonnec et al. 2008). These antidepressant-like effects were not accompanied by any convulsions, hyperlocomotion, or stereotypy in rats or mice at doses up to 1 g/kg. In addition, no EEG disturbances were observed in rats at doses up to 30 mg/kg i.v. When given concurrently, ADL5859 and the kappa opioid antagonist LY2444296 produced a synergistic antidepressant-like effect in the mouse forced swim test (Huang et al. 2016). ADL5859 passed phase I clinical trials and was evaluated in human clinical trials for the treatment of rheumatoid arthritis and neuropathic pain but was not found effective (Spahn and Stein 2016). To this point, studies evaluating ADL5859 as a treatment for depression in humans have not been conducted.

AZD2327 (4-{(R)-(3-Aminophenyl)[4-(4-fluorobenzyl)-piperazin-1-yl] methyl}-*N,N*-diethylbenzamide) has been shown to be effective in a number of models of anxiety and depression. During the avoidance phase of a learned helplessness test, a rodent model of depression, AZD2327 significantly reduced the number of escape failures by rats previously exposed to inescapable shock (Hudzik et al. 2011). Pairs of rats treated with AZD2327 spent significantly more time engaging in social interaction compared to vehicle treated pairs. In a model of acute anxiety, AZD2327 blocked the increase in norepinephrine release in the medial prefrontal cortex typically observed after delivery of a footshock. No tolerance to this effect developed even after 21 days of AZD2327 administration. In a modified Geller-Seifter conflict test in rats, AZD2327 increased response rates during the suppressed component of the test without altering responses in the unsuppressed component. Pretreatment with the delta opioid antagonist naltrindole blocked the AZD2327-induced reversal of suppressed response rate, suggesting that the anxiolytic effects of AZD2327 are mediated by the DOR (Hudzik et al. 2011).

AZD2327 also possesses stimulant activity as it dose-dependently increased locomotor activity in rats in an open field test. However, the dose necessary to stimulate locomotor activity was threefold to tenfold higher than that required to generate antidepressant-like and anxiolytic effects, so it is unlikely that the results in those experiments were confounded by the stimulant properties of AZD2327. Like SNC80 and (+)BW373U86, AZD2327 readily produces convulsions in mice, dogs, and primates (Hudzik et al. 2014). Preclinical and phase I studies suggest that maintaining plasma levels of AZD2327 below 15 ng/mL minimizes the incidence of these convulsive events (Richards et al. 2016). Interestingly, there was no evidence of convulsions in rats treated with AZD2327 alone; however, it did lower the threshold for pentylenetetrazole-induced clonic seizures in rats (Hudzik et al. 2011). In a small pilot study of patients with anxious major depressive disorder, 7 of 12 patients responded to AZD2327 while only 3 of 9 responded to placebo (Richards et al. 2016). This study was terminated early by the company purportedly due to business strategy reasons, and was therefore underpowered and did not find a significant drug effect. Nevertheless, AZD2327 showed some therapeutic promise and was well tolerated with no seizures or changes in epileptiform activity.

KNT-127 (1,2,3,4,4a,5,12,12a-octahydro-2-methyl-4aβ,1β-([1,2]benzenomethano)-2,6-diazanaphthacene-12aβ,17-diol) is structurally similar to TAN67 (see Table 1) and has been extensively investigated in animal models of depression and anxiety. In the mouse forced swim test, KNT-127 significantly decreased immobility and increased swimming behavior without affecting overall locomotor activity or eliciting convulsions (Saitoh et al. 2011). These antidepressant like effects of KNT 127 were reversed by the DOR selective antagonist naltrindole as well as the putative delta 2 opioid receptor antagonist naltriben. Daily injections of 5 mg/kg KNT-127 did not affect the ability of acute administration of 3 mg/kg KNT-127 to reduce immobility in the mouse forced swim test suggesting that chronic administration of KNT-127 does not induce tolerance to its antidepressant-like effects (Nozaki et al. 2014). Furthermore, daily administration of KNT-127 significantly decreased hyperemotionality scores in olfactory-bulbectomized rats throughout the 14-day test period (Gotoh et al. 2016).

KNT-127 has also been shown to have strong anxiolytic properties. In an elevated plus maze test, KNT-127 dose-dependently increased the amount of time rats spent in the open arms of the maze. Rats treated with KNT-127 also spent significantly more time in the light in a light/dark test and more time in the center space in an open field test (Saitoh et al. 2013). In each of these models, the effects of

Table 1 Chemical structures of representative delta opioid receptor agonists

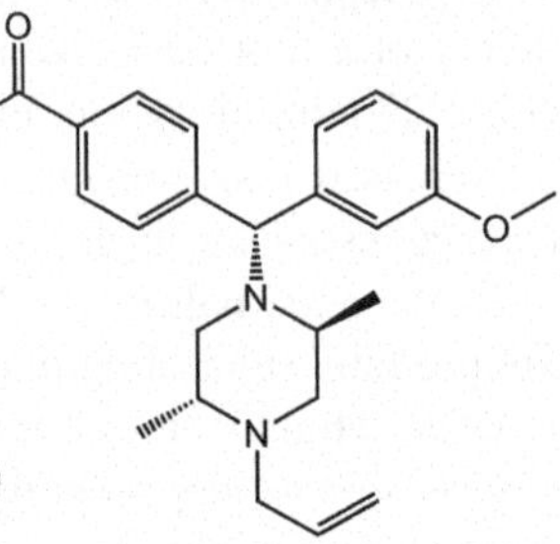 SNC80	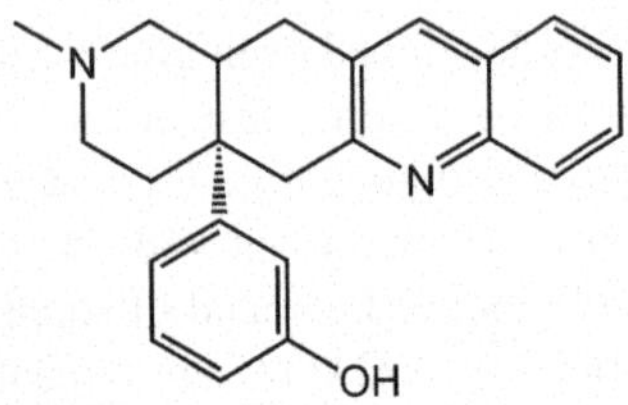 TAN67
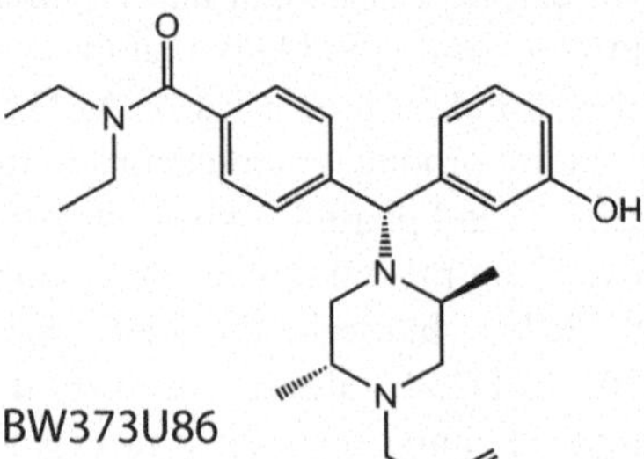 BW373U86	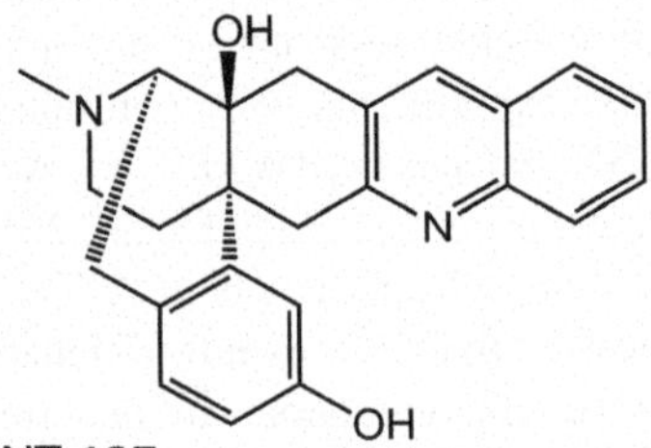 KNT-127
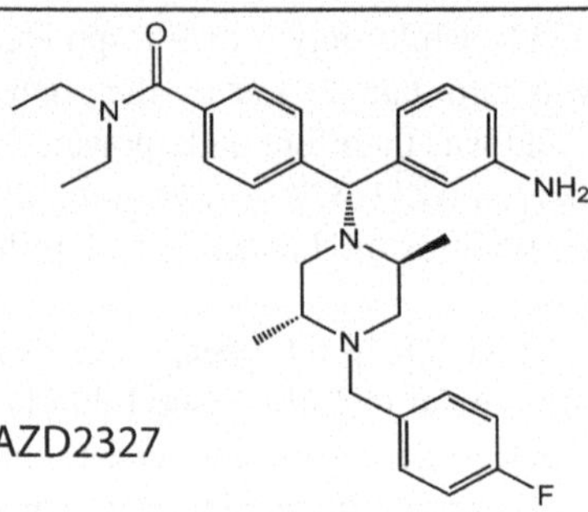 AZD2327	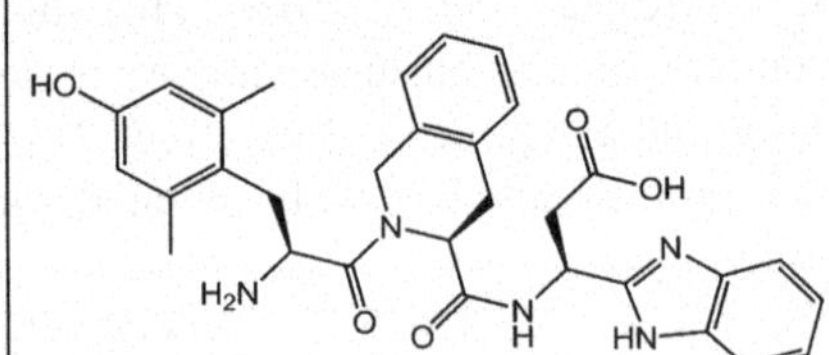 UFP-512
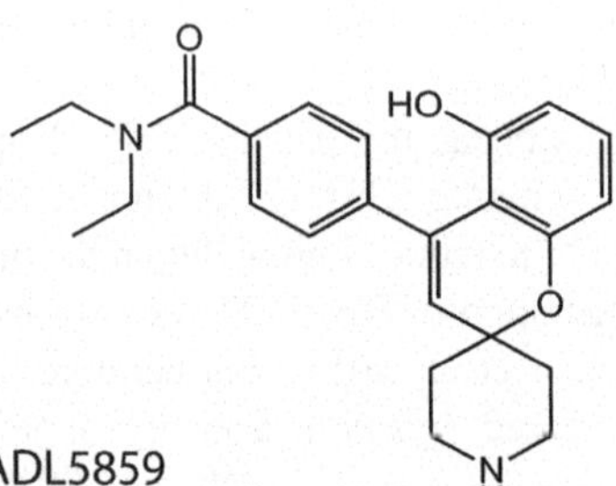 ADL5859	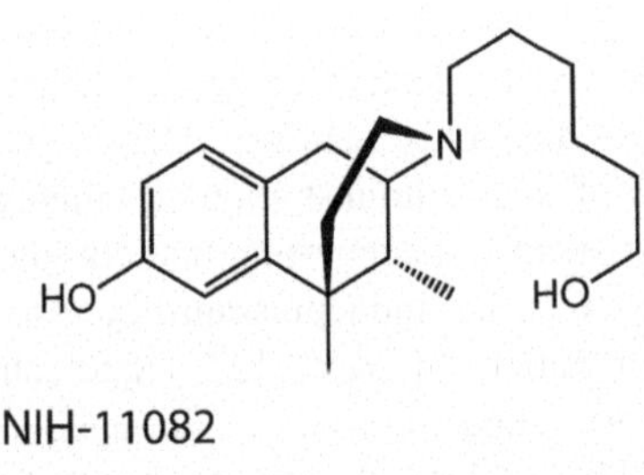 NIH-11082

KNT-127 were blocked by both naltrindole and the putative delta 2 opioid receptor antagonist naltriben indicating that the anxiolytic effects of KNT-127 are mediated by DORs (Saitoh et al. 2013; Sugiyama et al. 2014).

Microdialysis analysis after i.p. administration of KNT-127 in male Sprague-Dawley rats revealed increases in dopamine and L-glutamate release within the striatum, nucleus accumbens, and medial prefrontal cortex, all of which are brain regions where the DOR is highly expressed (Tanahashi et al. 2012). Decreases in GABA release were also observed in the nucleus accumbens and medial prefrontal cortex. Interestingly, these KNT-127-induced changes in neurotransmission were blocked by the putative delta 1 opioid receptor antagonist 7-benzylidenenaltrexone (BNTX) but not by the putative delta 2 opioid receptor antagonist naltriben. Therefore it is unclear as to whether these changes in neurotransmitter release are responsible for the mood enhancing effects of KNT-127 because naltriben was capable of blocking the antidepressant-like and anxiolytic effects of KNT-127 in behaving rodents whereas BNTX was not.

Overall, DOR agonists produce antidepressant-like effects in a number of animal models. These effects are DOR-mediated and are not dependent on convulsive or stimulant activity. However, the mechanism responsible for the antidepressant-like effects of DOR agonists is under investigation.

4 Possible Mechanisms of Action

Although monoamines, namely dopamine, norepinephrine, and serotonin, play a well-established role in regulating emotion and cognition (Robbins and Arnsten 2009), the etiology of depression likely goes beyond deficiencies in the levels of these neurotransmitters in the brain. There are two primary problems with the monoamine deficiency hypothesis of depression. First, although clinically used antidepressants typically take weeks to achieve a therapeutic effect, they block reuptake and/or metabolism of monoamines within hours or days of first use. Second, loss of serotonin or norepinephrine does not readily cause depression in healthy controls suggesting that monoamine deficiency is not sufficient to produce depression. Therefore, alternatives to the monoamine hypothesis of depression have been put forward.

One such hypothesis proposes that depression is caused by dysfunction of glutamatergic neurotransmission. Excess glutamate leads to neurotoxicity and this loss of neurons is thought to promote a depressive phenotype. Many clinical studies have shown elevated levels of glutamate in depressed patients that are reduced after antidepressant treatment (for review see Sanacora et al. 2012). In addition, low doses of the noncompetitive NMDA receptor antagonist ketamine have been shown to elicit rapid antidepressant-like effects in human patients (Monteggia and Zarate 2015) and animal models (Browne and Lucki 2013). These effects last for up to 2 weeks after a single dose of ketamine suggesting a synaptic plasticity-mediated mechanism.

The interactions between the delta opioid and glutamatergic systems are not well characterized and differ across brain regions. The peptidic delta agonist DPDPE enhanced the glutamate content of intrastriatal dialysate (Billet et al. 2004), but also inhibited glutamate release in the rat anterior cingulate cortex (Tanaka and North 1994) and the amygdala of morphine treated rats (Bie et al. 2009). SNC80 has been found to increase glutamate release in rat striatum (Bosse et al. 2014). UFP-512 decreased glutamate release in the rat substantia nigra (Mabrouk et al. 2009). KNT-127 increased glutamate release within the striatum, nucleus accumbens, and medial prefrontal cortex of male Sprague-Dawley rats (Tanahashi et al. 2012). Further research is needed to characterize the role of glutamate in eliciting DOR-mediated antidepressant-like effects.

Another putative mechanism for the actions of antidepressant drugs is through upregulation of brain derived neurotrophic factor (BDNF). BDNF is a member of the neurotrophin family of growth factors and promotes the growth, survival, and differentiation of neurons. Many studies have shown that stress decreases BDNF expression and promotes cell death in brain regions that regulate mood (Duman 2003; Lee and Kim 2010). Serum levels of BDNF in depressed patients are significantly lower compared to healthy controls (Bocchio-Chiavetto et al. 2010). In postmortem studies, BDNF expression was decreased in the hippocampus and prefrontal cortex of depressed patients and suicide victims (Dwivedi et al. 2003; Karege et al. 2005). Furthermore, antidepressant treatment has been shown to increase BDNF expression in preclinical and clinical studies (Duman 2003; Lee and Kim 2010).

There are few reports examining the effects of DOR agonists on BDNF. DPDPE increased BDNF mRNA expression in the rat frontal cortex (Torregrossa et al. 2006) while (+)BW373U86 increased BDNF mRNA expression in the hippocampus, amygdala, and frontal cortex (Torregrossa et al. 2004). For both drugs, these changes were naltrindole-sensitive and occurred at doses that also produced antidepressant-like effects in the forced swim test. Interestingly, upregulation of BDNF in response to delta opioid agonists was observed before increases could be observed with traditional antidepressants, suggesting a faster onset of action. Elevated levels of BDNF mRNA in the frontal cortex persisted after 8 days of daily (+)BW373U86 injections but returned to basal levels after 21 days of treatment indicating tolerance to this effect (Torregrossa et al. 2005). AZD2327 significantly increased BDNF expression in the rat hippocampus but not in the frontal cortex and plasma BDNF levels remained unchanged (Richards et al. 2016). AZD2327 also failed to alter plasma BDNF levels in human patients, albeit in a small, underpowered cohort (Richards et al. 2016). Taken together, these data suggest that BDNF expression may correlate with the antidepressant actions of delta opioid agonists, although future studies should examine whether BDNF plays a causal role in mediating these effects.

Little is known about the intracellular signaling molecules and pathways involved in DOR-mediated antidepressant-like effects. Regulator of G protein signaling 4 (RGS4), a negative modulator of G protein signaling that accelerates Gα-mediated GTP hydrolysis, was implicated in the regulation of DOR-mediated antidepressant-like effects when loss of RGS4 potentiated the decreases in immobility produced by a single dose of SNC80 in the mouse forced swim test (Stratinaki et al. 2013). Recently, loss of RGS4 was found to functionally increase the therapeutic index of SNC80 by improving the potency of SNC80 to induce antidepressant-like effects in the mouse forced swim test without altering SNC80-induced convulsions (Dripps et al. 2017). These data suggest that DOR-mediated antidepressant-like effects in the forced swim test are generated via a G protein-dependent signaling mechanism. It is possible that RGS proteins other than RGS4 modulate DOR-induced convulsive effects. Alternatively, this specific DOR-mediated behavior may be generated by a G protein-independent, arrestin-mediated signaling mechanism (Violin 2014). Previous studies have demonstrated that DOR activation leads to signaling through G protein-dependent and -independent pathways (Bradbury et al. 2009; Charfi et al. 2014, 2015). However, there are few reports connecting these distinct signaling mechanisms to specific behavioral outputs (Chiang et al. 2016; Pradhan et al. 2016).

In summary, depression is a serious and intractable psychiatric disease that is not adequately treated with current therapies. DOR agonists have been shown to produce antidepressant-like effects in a variety of animal models for depression. The convulsive activity of several DOR agonists has heretofore limited their clinical utility; however, the development of DOR agonists that do not produce convulsions should allow for more thorough testing in humans. Although some DOR agonists alter glutamatergic neurotransmission and others upregulate BDNF expression, consistent with modern hypotheses on the etiology of depression, the mechanism by which DOR-mediated antidepressant-like effects are generated is not known and should be investigated further. At the intracellular level, DOR-mediated behaviors are likely generated via distinct intracellular signaling pathways. Determining the signal transduction mechanisms that give rise to DOR-mediated behaviors is critical for the development of DOR drugs with improved safety and clinical utility and future work should be devoted to elucidating these pathways.

References

Aceto MD, May EL, Harris LS, Bowman ER, Cook CD (2007) Pharmacological studies with a nonpeptidic, delta-opioid (-)-(1R,5R,9R)-5,9-dimethyl-2'-hydroxy-2-(6-hydroxyhexyl)-6,7-benzomorphan hydrochloride ((-)-NIH 11082). Eur J Pharmacol 566:88–93

Aguila B, Coulbault L, Boulouard M, Léveillé F, Davis A, Tóth G, Borsodi A, Balboni G, Salvadori S, Jauzac P, Allouche S (2007) In vitro and in vivo pharmacological profile of UFP-512, a novel selective delta-opioid receptor agonist; correlations between desensitization and tolerance. Br J Pharmacol 152:1312–1324. Erratum in: Br J Pharmacol. 152: 1325

American Psychiatric Association (2013) Major depressive disorder. In: Desk reference to the diagnostic criteria from DSM-5, Washington

Anderson IM, Cowen PJ, Grahame-Smith DG (1990) The effects of gepirone on neuroendocrine function and temperature in humans. Psychopharmacology 100:498–503

Baamonde A, Daugé V, Ruiz-Mayo M, Fulga IG, Turcaud S, Fournié-Zaluski MC, Roques BP (1992) Antidepressant-type effects of endogenous enkephalins protected by systemic RB 101 are mediated by opioid delta and dopamine D1 stimulation. Eur J Pharmacol 216:157–166
Baghai TC (2008) Electroconvulsive therapy and its different indications. Dialogues Clin Neurosci 10:105–117
Bie B, Zhu W, Zhizhong ZP (2009) Rewarding morphine-induced synaptic function of δ-opioid receptors on central glutamate synapses. J Pharmacol Exp Ther 329:290–296
Billet F, Dourmap N, Costentin J (2004) Involvement of corticostriatal glutamatergic terminals in striatal dopamine release elicited by stimulation of δ-opioid receptors. Eur J Neurosci 20:2629–2638
Bocchio-Chiavetto L, Bagnardi V, Zanardini R, Molteni R, Nielsen MG, Placentino A, Giovannini C, Rillsosi L, Ventriglia M, Riva MA, Gennarelli M (2010) Serum and plasma BDNF levels in major depression: a replication study and meta-anaylses. World J Biol Psychiatry 11:763–773
Bodkin JA, Zornberg GL, Lukas SE, Cole JO (1995) Buprenorphine treatment of refractory depression. J Clin Psychopharmacol 15:49–57
Bosse KE, Jutkiewicz EM, Schultz-Kuszak KN, Mabrouk OS, Kennedy RT, Gnegy ME, Traynor JR (2014) Synergistic activity between the delta-opioid agonist SNC80 and amphetamine occurs via a glutamatergic NMDA-receptor dependent mechanism. Neuropharmacology 77:19–27
Bradbury FA, Zelnik JC, Traynor JR (2009) G protein independent phosphorylation and internalization of the delta-opioid receptor. J Neurochem 109:1526–1535
Broom DC, Jutkiewicz EM, Folk JE, Traynor JR, Rice KC, Woods JH (2002a) Nonpeptidic delta-opioid agonists reduce immobility in forced swim assay in rats. Neuropsychopharmacology 26:744–755
Broom DC, Jutkiewicz EM, Folk JE, Traynor JR, Rice KC, Woods JH (2002b) Convulsant activity of a non-peptidic delta-opioid receptor agonist is not required for its antidepressant-like effects in Sprague-Dawley rats. Psychopharmacology 164:42–48
Browne CA, Lucki I (2013) Antidepressant effects of ketamine: mechanisms underlying fast-acting novel antidepressants. Front Pharmacol 4:161
Charfi I, Nagi K, Mnie-Filali O, Thibault D, Balboni G, Schiller PW, Trudeau LE, Pineyro G (2014) Ligand- and cell-dependent determinants of internalization and cAMP modulation by DOR (DOR) agonists. Cell Mol Life Sci 71:1529–1546
Charfi L, Audet N, Bagheri Tudashki H, Pineyro G (2015) Identifying ligand-specific signaling within biased responses: focus on d opioid receptor ligands. Br J Pharmacol 172:435–448
Chiang T, Sansuk K, van Rijn RM (2016) β-Arrestin 2 dependence of δ opioid receptor agonists is correlated with alcohol intake. Br J Pharmacol 173:332–343
Comer SD, Hoenicke EM, Sable AI, McNutt RW, Chang KJ, De Costa BR, Mosberg HI, Woods JH (1993) Convulsive effects of systemic administration of the delta opioid agonist BW373U86 in mice. J Pharmacol Exp Ther 267:888–895
Darko DF, Risch SC, Gillan JC, Golshan S (1992) Association of beta-endorphin with specific clinical symptoms of depression. Am J Psychiatry 149:1162–1167
Deo AJ, Huang YY, Hodgkinson CA, Xin Y, Oguendo MA, Dwork AJ, Arango V, Brent DA, Goldman D, Mann JJ, Haghighi F (2013) A large-scale candidate gene analysis of mood disorders: evidence of neurotrophic tyrosine kinase receptor and opioid receptor signaling dysregulation. Psychiatr Genet 23:47–55
Djurović D, Milić-Askrabić J, Majkić-Singh N (1999) Serum beta-endorphin level in patients with depression on fluvoxamine. Farmaco 54:130–133
Do Carmo GP, Folk JE, Rice KC, Chartoff E, Carlezon WA Jr, Negus SS (2009) The selective non-peptidic delta opioid agonist SNC80 does not facilitate intracranial self-stimulation in rats. Eur J Pharmacol 604:58–65

Dripps IJ, Wang Q, Neubig RR, Rice KC, Traynor JR, Jutkiewicz EM (2017) The role of regulator of G protein signaling 4 in delta-opioid receptor-mediated behaviors. Psychopharmacology 234(1):29–39. https://doi.org/10.1007/s00213-016-4432-5. [Epub ahead of print]

Duman RS (2003) Role of neurotrophic factors in the etiology and treatment of mood disorders. NeuroMolecular Med 5:11–25

Dwivedi Y, Rizavi HS, Conley RR, Roberts RC, Tamminga CA, Pandey GN (2003) Altered gene expression of brain-derived neurotrophic factor and receptor tyrosine kinase B in postmortem brain of suicide subjects. Arch Gen Psychiatry 60:804–815

Eisendrath SJ, Lichtmacher JE (2014) Psychiatric disorders. In: Papadakis MA, SJ MP, Rabow MW (eds) Current medical diagnosis & treatment 2014. McGraw-Hill, New York. Chapter 25

Emrich HM, Höllt V, Kissling W, Fischler M, Laspe H, Heinemann H, von Zerssen D, Herz A (1979) Beta-endorphin-like immunoreactivity in cerebrospinal fluid and plasma of patients with schizophrenia and other neuropsychiatric disorders. Pharmakopsychiatr Neuropsychopharmakol 12:269–276

Emrich HM, Vogt P, Herz A (1981) A possible role of opioids in depression: significant improvement after buprenorphine. Biol Psychiatry 16:380–385

Everett GM (1966) The Dopa response potentiation test and its use in screening for anti-depressant drugs. In: Garattini S, Dukes MNG (eds) Anti-depressant drugs. Excerpta Medica Foundation, Amsterdam

Extein I, Pickar D, Gold MS, Gold PW, Pottash AL, Sweeney DR, Ross RJ, Rebard R, Martin D, Goodwin FK (1981) Methadone and morphine in depression. Psychopharmacol Bull 17:29–33

Falcon E, Browne CA, Leon RM, Fleites VC, Sweeney R, Kirby LG, Lucki I (2016) Antidepressant-like effects of buprenorphine are mediated by kappa opioid receptors. Neuropsychopharmacology 41:2344–2351

Fava M, Rush AJ, Wisniewski SR, Nierenberg AA, Alpert JE, McGrath PJ, Thase ME, Wardern D, Biggs M, Luther JF, Niederehe G, Ritz L, Trivedi MH (2006) A comparison of mirtazapine and nortriptyline following two consecutive failed medication treatments for depressed outpatients: a STAR*D report. Am J Psychiatry 163:1161–1172

Filliol D, Ghozland S, Chluba J, Martin M, Matthes HW, Simonin F, Befort K, Gavériaux-Ruff C, Dierich A, LeMeur M, Valverde O, Maldonado R, Kieffer BL (2000) Mice deficient for delta- and mu-opioid receptors exhibit opposing alterations of emotional responses. Nat Genet 25:195–200

Fink M, Shapiro DM (1969) EEG patterns as an index of clinical activity of psychoactive drugs. Electroencephalogr Clin Neurophysiol 27(7):710

Garriock HA, Tanowitz M, Kraft JB, Dang VC, Peters EJ, Jenkins GD, Reinalda MS, McGrath PJ, van Zastrow M, Slager SL, Hamilton SP (2010) Association of mu-opioid receptor variants and response to citalopram in major depressive disorder. Am J Psychiatry 167:565–573

Gassaway MM, Rives ML, Kruegel AC, Javitch JA, Sames D (2014) The atypical antidepressant and neurorestorative agent tianeptine is a μ-opioid receptor agonist. Transl Psychiatry 4:e411. https://doi.org/10.1038/tp.2014.30

Genazzani AR, Petraglia F, Facchinetti F, Monittola C, Scarone S, Brambilia F (1984) Opioid plasma levels in primary affective disorders. Effect of desimipramine therapy. Neuropsychobiology 12:78–85

Gerner RH, Catlin DH, Gorelick DA, Hui KK, Li CH (1980) Beta-endorphin infusion causes behavioral change in psychiatric inpatients. Arch Gen Psychiatry 37:642–647

Goodwin GM, Austin MP, Curran SM, Ross M, Murray C, Prentice N, Ebmeier KP, Bennie J, Carrol S, Dick H, Fink G (1993) The elevation of plasma β-endorphin levels in major depression. J Affect Disord 29:281–289

Gotoh L, Saitoh A, Yamada M, Fujii H, Nagase H, Yamada M (2016) Effects of repeated treatment with a DOR agonist KNT-127 on hyperemotionality in olfactory-bulbectomized rats. Behav Brain Res 32:11–14. https://doi.org/10.1016/j.bbr.2016.11.008

Hameroff SR, Cork RC, Scherer K, Crago BR, Neuman C, Womble JR, Davis TP (1982) Doxepin effects on chronic pain, depression and plasma opioids. J Clin Psychiatry 43:22–27

Hayes AG, Skingle M, Tyers MB (1986) Reversal by beta-funaltrexamine of the antinociceptive effect of opioid agonists in the rat. Br J Pharmacol 88:867–872

Huang P, Tunis J, Parry C, Tallarida R, Liu-Chen LY (2016) Synergistic antidepressant-like effects between a kappa opioid antagonist (LY2444296) and a delta opioid agonist (ADL5859) in the mouse forced swim test. Eur J Pharmacol 781:53–59

Hudzik TJ, Maciag C, Smith MA, Caccese R, Pietras MR, Bui KH, Coupal M, Adam L, Payza K, Griffin A, Smagin G, Song D, Swedberg MD, Brown W (2011) Preclinical pharmacology of AZD2327: a highly selective agonist of the δ-opioid receptor. J Pharmacol Exp Ther 338:195–204

Hudzik TJ, Pietras MR, Caccese R, Bui KH, Yocca F, Paronis CA, Swedberg MD (2014) Effects of the δ opioid agonist AZD2327 upon operant behaviors and assessment of its potential for abuse. Pharmacol Biochem Behav 124:48–57

Inturrisi CE, Alexopoulos G, Lipman R, Foley K, Roosier J (1982) Beta-endorphin immunoreactivity in the plasma of psychiatric patients receiving electroconvulsive treatment. Ann N Y Acad Sci 398:413–423

Javelot H, Messaoudi M, Garnier S, Rougeot C (2010) Human opiorphin is a naturally occurring antidepressant acting selectively on enkephalin-dependent delta-opioid pathways. J Physiol Pharmacol 61:355–362

Jungkunz G, Nedopil N, Rüther E (1983) Acute effects of the synthetic analogue of methionine enkephalin analogue FK 33-824 on depressive symptoms. Pharmacopsychiatria 16:90–92

Jutkiewicz EM, Roques BO (2012) Endogenous opioids as physiological antidepressants: complementary role of δ receptors and dopamine. Neuropsychopharmacology 37:303–304

Jutkiewicz EM, Kaminsky ST, Rice KC, Traynor JR, Woohs JH (2005a) Differential behavioral tolerance to the delta-opioid agonist SNC80 ([(+)-4-[(alphaR)-alpha-[(2S,5R)-2,5-dimethyl-4-(2-propenyl)-1-piperazinyl]-(3-methoxyphenyl)methyl]-N,N-diethylbenzamide) in Sprague-Dawley rats. J Pharmacol Exp Ther 315:414–422

Jutkiewicz EM, Rice KC, Traynor JR, Woods JH (2005b) Separation of the convulsions and antidepressant-like effects produced by the delta-opioid agonist SNC80 in rats. Psychopharmacology 182:588–596

Jutkiewicz EM, Baladi MG, Folk JE, Rice KC, Woods JH (2006a) The convulsive and electroencephalographic changes produced by nonpeptidic d-opioid agonists in rats: comparison with pentylenetetrazol. J Pharmacol Exp Ther 317:1337–1348

Jutkiewicz EM, Torregrossa MM, Sobczyk-Kojiro K, Mosberg HI, Folk JE, Rice KC, Watson SJ, Woods JH (2006b) Behavioral and neurobiological effects of the enkephalinase inhibitor RB101 relative to its antidepressant-like effects. Eur J Pharmacol 531:151–159

Kabli N, Nguyen T, Balboni G, O'Dowd BF, George SR (2014) Antidepressant-like and anxiolytic-like effects following activation of the μ-δ opioid receptor heteromer in the nucleus accumbens. Mol Psychiatry 19:986–994

Kamei J, Saitoh A, Suzuki T, Misawa M, Nagase H, Kasuya Y (1995) Buprenorphine exerts its antinociceptive activity via mu 1-opioid receptors. Life Sci 56:PL285–PL290

Karege F, Vaudan G, Schwald M, Perroud N, La Harpe R (2005) Neurotrophin levels in postmortem brains of suicide victims and the effects of antemortem diagnosis and psychotropic drugs. Brain Res Mol Brain Res 136:29–37

Karp JF, Butters MA, Begley AE, Miller MD, Lenze EJ, Blumberger DM, Mulsant BH, Reynolds CF III (2014) Safety, tolerability, and clinical effect of low-dose buprenorphine for treatment-resistant depression in midlife and older adults. J Clin Psychiatry 75:785–793

Kastin AJ, Scollan EL, Ehrensing RH, Schally AV, Coy DH (1978) Enkephalin and other peptides reduce passiveness. Pharmacol Biochem Behav 9:515–519

Kennedy SE, Koepper RA, Young EA, Zubiet JK (2006) Dysregulation of endogenous opioid emotion regulatory circuitry in major depression in women. Arch Gen Psychiatry 63:1199–1208

Kita A, Imano K, Seto Y, Yakuo I, Deguchi T, Nakamura H (1997) Antinociceptive and antidepressant-like profiles of BL-2401, a novel enkephalinase inhibitor, in mice and rats. Jpn J Pharmacol 75:337–346

Kline NS, Li CH, Lehmann HE, Lajtha A, Laski E, Cooper T (1977) Beta-endorphin-induced changes in schizophrenic and depressed patients. Arch Gen Psychiatry 34:1111–1113

Le Bourdonnec B, Windh RT, Ajello CW, Leister LK, Gu M, Chu GH, Tuthill PA, Barker WM, Koblish M, Wiant DD, Graczyk TM, Belanger S, Cassel JA, Feschenko MS, Brogdon BL, Smith SA, Christ DD, Derelanko MJ, Kutz S, Little PJ, DeHaven RN, DeHaven-Hudkins DL, Dolle RE (2008) Potent, orally bioavailable DOR agonists for the treatment of pain: discovery of N,N-diethyl-4-(5-hydroxyspiro[chromene-2,4'-piperidine]-4-yl)benzamide (ADL5859). J Med Chem 51:5893–5896

Le Merrer J, Becker JAJ, Befort K, Kieffer BL (2009) Reward processing by the opioid system in the brain. Physiol Rev 89:1379–1412

Lee BH, Kim YK (2010) The roles of BDNF in the pathophysiology of major depression and in antidepressant treatment. Psychiatry Investig 7:231–235

Lindström LH, Widerlöv E, Gunne LM, Wahlström A, Terenius L (1978) Endorphins in human cerebrospinal fluid: clinical correlations to some psychotic states. Acta Psychiatr Scand 57:153–164

Longoni R, Cadoni C, Mulas A, Di Chiara G, Spina L (1998) Dopamine-dependent behavioural stimulation by non-peptidic delta opioids BW373U86 and SNC80: 2. Place-preference and brain microdialysis studies in rats. Behav Pharmacol 9:9–14

Loosen PT, Shelton RC (2008) Mood disorders. In: Ebert MH, Loosen PT, Nurcombe B, Leckman JF (eds) CURRENT diagnosis & treatment: psychiatry, 2e. McGraw-Hill, New York. Chapter 18

Lutz PE, Kieffer BL (2013) Opioid receptors: distinct roles in mood disorders. Trends Neurosci 36:195–206

Mabrouk OS, Marti M, Salvadori S, Morari M (2009) The novel DOR agonist UFP-512 dually modulates motor activity in hemiparkinsonian rats via control of the nigro-thalamic pathway. Neuroscience 164:360–369

Monteggia LM, Zarate C Jr (2015) Antidepressant actions of ketamine: from molecular mechanisms to clinical practice. Curr Opin Neurobiol 30:139–143

Naber D, Pickar D, Post RM, Van Kammen DP, Waters RN, Ballenger JC, Goodwin FK, Bunney WE Jr (1981) Endogenous opioid activity and beta-endorphin immunoreactivity in CSF of psychiatric patients and normal volunteers. Am J Psychiatry 138:1457–1462

Nagase H, Tanaka T, Saitoh A (2002) Therapeutics of preventive medicines for mood disorders or anxiert disorders. U.S. Patent WO/2002/002117

Naidu PS, Lichtman AH, Archer CC, May EL, Harris LS, Aceto MD (2007) NIH 11082 produces anti-depressant-like activity in the mouse tail-suspension test through a delta-opioid receptor mechanism of action. Eur J Pharmacol 566:132–136

Negus SS, Gatch MB, Mello NK, Zhang X, Rice K (1998) Behavioral effects of the delta-selective opioid agonist SNC80 and related compounds in monkeys. J Pharmacol Exp Ther 286:362–375

Negus SS, Rosenberg MB, Altarifi AA, O'Connell RH, Folk JE, Rice KC (2012) Effects of the δ opioid receptor agonist SNC80 on pain-related depression of intracranial self-stimulation (ICSS) in rats. J Pain 13:317–327

Nozaki C, Nagase H, Nemoto T, Matifas A, Kieffer BL, Gaveriaux-Ruff C (2014) In vivo properties of KNT-127, a novel δ opioid receptor agonist: receptor internalization, antihyperalgesia, and antidepressant effects in mice. Br J Pharmacol 171:5376–5386

Nyhuis PW, Gastpar M, Scherbaum N (2008) Opiate treatment in depression refractory to antidepressants and electroconvulsive therapy. J Clin Psychopharmacol 28:593–595

Plotnikoff NP, Kastin AJ, Coy DH, Christensen CW, Schally AV, Sprites MA (1976) Neuropharmacological actions of enkephalin after systemic administration. Life Sci 19:1283–1288

Pradhan AA, Perro J, Walwyn WM, Smith ML, Vicente-Sanchez A, Segura L, Bana A, Keiffer BL, Evans CJ (2016) Agonist-specific recruitment of arrestin isforms differentially modify DOR function. J Neurosci 36:3541–3551

Rahman A, Patel V, Maselko J, Kirkwood B (2008) The neglected 'm' in MCH programmes–why mental health of mothers is important for child nutrition. Trop Med Int Health 13:579–583

Richards EM, Mathews DC, Luckenbaugh DA, Ionescu DF, Machado-Vieira R, Niciu MJ, Duncan WC, Nolan NM, Franco-Chaves JA, Hudzik T, Maciag C, Li S, Cross A, Smith MA, Zarate CA Jr (2016) A randomized, placebo-controlled pilot trial of the DOR agonist AZD2327 in anxious depression. Psychopharmacology 233:1119–1130

Robbins TW, Arnsten AF (2009) The neuropsychopharmacology of executive function: monoaminergic modulation. Annu Rev Neurosci 32:267–287

Sackeim HA (1999) The anticonvulsant hypothesis of the mechanisms of ECT: current status. J ECT 15:5–26

Saitoh A, Yamada M, Yamada M, Takahashi K, Yamaguchi K, Murasawa H, Nakatani A, Tatsumi Y, Hirose N, Kamei J (2008) Antidepressant-like effects of the delta-opioid receptor agonist SNC80 ([(+)-4-[(alphaR)-alpha-[(2S,5R)-2,5-dimethyl-4-(2-propenyl)-1-piperazinyl]-(3-methoxyphenyl)methyl]-N,N-diethylbenzamide) in an olfactory bulbectomized rat model. Brain Res 1208:160–169

Saitoh A, Sugiyama A, Nemoto T, Fujii H, Wada K, Oka J, Nagase H, Yamada M (2011) The novel δ opioid receptor agonist KNT-127 produces antidepressant-like and antinociceptive effects in mice without producing convulsions. Behav Brain Res 223:271–279

Saitoh A, Sugiyama A, Yamada M, Inagaki M, Oka J, Nagase H, Yamada M (2013) The novel δ opioid receptor agonist KNT-127 produces distinct anxiolytic-like effects in rats without producing the adverse effects associated with benzodiazepines. Neuropharmacology 67:485–493

Sanacora G, Treccani G, Popoli M (2012) Towards a glutamate hypothesis of depression: an emerging frontier of neuropsychopharmacology for mood disorders. Neuropharmacology 62:63–77

Spahn V, Stein C (2016) Targeting DORs for pain treatment: drugs in phase I and II clinical development. Expert Opin Investig Drugs. https://doi.org/10.1080/13543784.2017.1275562. [Epub ahead of print]

Stratinaki M, Varidaki A, Mitsi V, Ghose S, Magida J, Dias C, Russo SJ, Vialou V, Caldarone BJ, Tamminga CA, Nestler EJ, Zachariou V (2013) Regulator of G protein signaling 4 is a crucial modulator of antidepressant drug action in depression and neuropathic pain models. Proc Natl Acad Sci U S A 110:8254–8259

Sugiyama A, Nagase H, Oka J, Yamada M, Saitoh A (2014) DOR(2)-selective but not DOR(1)-selective antagonist abolishes anxiolytic-like effects of the δ opioid receptor agonist KNT-127. Neuropharmacology 79:314–320

Tanahashi S, Ueda Y, Nakajima A, Yamamura S, Nagase H, Okada M (2012) Novel δ1-receptor agonist KNT-127 increases the release of dopamine and L-glutamate in the striatum, nucleus accumbens and median pre-frontal cortex. Neuropharmacology 62:2057–2067

Tanaka E, North RA (1994) Opioid actions on rat anterior cingulate cortex neurons in vitro. J Neruroscì 14:1106–1113

Tejedor-Real P, Mico JA, Maldonado R, Roques BP, Gilbert-Rahola J (1993) Effect of mixed (RB 38A) and selective (RB 38B) inhibitors of enkephalin degrading enzymes on a model of depression in the rat. Biol Psychiatry 34:100–107

Tejedor-Real P, Mico JA, Maldonado R, Roques BP, Gilbert-Rahola J (1995) Implication of endogenous opioid system in the learned helplessness model of depression. Pharmacol Biochem Behav 52:145–152

Tejedor-Real P, Micó JA, Smadja C, Maldonado R, Roques BP, Gilbert-Rahola J (1998) Involvement of delta-opioid receptors in the effects induced by endogenous enkephalins on learned helplessness model. Eur J Pharmacol 354:1–7

Torregrossa MM, Isgor C, Folk JE, Rice KC, Watson SJ, Woods JH (2004) The delta-opioid receptor agonist (+)BW373U86 regulates BDNF mRNA expression in rats. Neuropsychopharmacology 29:649–659

Torregrossa MM, Folk JE, Rice KC, Watson SJ, Woods JH (2005) Chronic administration of the DOR agonist (+)BW373U86 and antidepressants on behavior in the forced swim test and BDNF mRNA expression in rats. Psychopharmacology 183:31–40

Torregrossa MM, Jutkiewicz EM, Mosberg HI, Balboni G, Watson SJ, Woods JH (2006) Peptidic DOR agonists produce antidepressant-like effects in the forced swim test and regulate BDNF mRNA expression in rats. Brain Res 1069:172–181

Tortella FC, Long JB, Hong JS, Holaday JW (1989) Modulation of endogenous opioid systems by electroconvulsive shock. Convuls Ther 5:261–273

Traynor JR, Fantegrossi W, Woods JH (2005) Evaluation of compounds for opioid activity. In: Dewey WL (ed) Problems of drug dependence, 2004: proceedings of the 66th annual scientific meeting; the Committee on Problems of Drug Dependence, Inc., NIDA Research Monograph. U.S. Department of Health and Human Services, Rockville, pp 131–159

Trivedi MH, Fava M, Wisniewski SR, Thase ME, Quitkin F, Warden D, Ritz L, Nierenberg AA, Lebowitz B, Biggs MM, Luther JF, Shores-Wilson K, Rush AJ, STAR*D Study Team (2006a) Medication augmentation after the failure of SSRIs for depression. N Engl J Med 354:1243–1252

Trivedi MH, Rush AJ, Wisniewski SR, Nierenberg AA, Warden D, Ritz L, Norquist G, Howland RH, Lebowitz B, McGrath PJ, Shores-Wilson K, Biggs MM, Balasubramani GK, Fava M, STAR*D Study Team (2006b) Evaluation of outcomes with citalopram for depression using measurement-based care in STAR*D: implications for clinical practice. Am J Psychiatry 163:28–40

Vergura R, Balboni G, Spagnolo B, Gavioli E, Lambert DG, McDonald J, Trapella C, Lazarus LH, Regoli D, Guerrini R, Salvadori S, Caló G (2008) Anxiolytic- and antidepressant-like activities of H-Dmt-Tic-NH-CH(CH2-COOH)-Bid (UFP-512), a novel selective DOR agonist. Peptides 29:93–103

Violin JD (2014) Biased ligands at mu and DORs: targeting selective signalling to develop improved therapeutics. In: International narcotic research conference 2014, p 22

World Health Organization (2008) The global burden of disease 2004 update. http://www.who.int/healthinfo/global_burden_disease/GBD_report_2004update_full.pdf

World Health Organization (2012) Depression: a global public health concern. http://www.who.int/mental_health/management/depression/who_paper_depression_wfmh_2012.pdf

Yang QZ, Lu SS, Tian XZ, Yang AM, Ge WW, Chen Q (2011) The antidepressant-like effect of human opiorphin via opioid-dependent pathways in mice. Neurosci Lett 489:131–135

Zubieta JK, Ketter TA, Bueller JA, Xu Y, Kilbourn MR, Young EA, Koeppe RA (2003) Regulation of human affective responses by anterior cingulate and limbic mu-opioid neurotransmission. Arch Gen Psychiatry 60:1145–1153

Delta Opioid Pharmacology in Relation to Alcohol Behaviors

Doungkamol Alongkronrusmee, Terrance Chiang, and Richard M. van Rijn

Contents

D. Alongkronrusmee • T. Chiang • R.M. van Rijn (✉)
Department of Medicinal Chemistry and Molecular Pharmacology, Purdue University, West Lafayette, IN, USA
e-mail: rvanrijn@purdue.edu

E. M. Jutkiewicz (ed.), *Delta Opioid Receptor Pharmacology and Therapeutic Applications*, Handbook of Experimental Pharmacology 247, DOI 10.1007/164_2016_30

Abstract

Delta opioid receptors (DORs) are heavily involved in alcohol-mediated processes in the brain. In this chapter we provide an overview of studies investigating how alcohol directly impacts DOR pharmacology and of early studies indicating DOR modulation of alcohol behavior. We will offer a brief summary of the different animal species used in alcohol studies investigating DORs followed by a broader overview of the types of alcohol behaviors modulated by DORs. We will highlight a small set of studies investigating the relationship between alcohol and DORs in analgesia. We will then provide an anatomical overview linking DOR expression in specific brain regions to different alcohol behaviors. In this section, we will provide two models that try to explain how endogenous opioids acting at DORs may influence alcohol behaviors. Next, we will provide an overview of studies investigating certain new aspects of DOR pharmacology, including the formation of heteromers and biased signaling. Finally, we provide a short overview of the genetics of the DORs in relation to alcohol use disorders (AUDs) and a short statement on the potential of using DOR-based therapeutics for treatment of AUDs.

Keywords

Alcohol use disorder • Delta opioid receptor • Pharmacology • Behavioral models • Genetics • Medication development • Enkephalin

1 Direct Impact of Alcohol on Delta Opioid Receptor Pharmacology

Acute and chronic alcohol exposure can impact DOR pharmacology on several different levels: (1) by changes to endogenous opioid levels, (2) changes in DOR expression level, or (3) modifying affinity or potency of endogenous and exogenous ligands for DORs in particular brain regions. Acute alcohol exposure has been shown to reduce the affinity of DOR by various mechanisms. Hiller and Hoffman have demonstrated that alcohol and other aliphatic alcohols selectively inhibit the binding of enkephalins, the DOR preferring endogenous opioids, to DOR binding sites as observed by a reduction in the affinity or by an increase in the ligand's dissociation rate (Hiller et al. 1984; Hoffman et al. 1984). This reversible inhibition may result from the cell membrane perturbation by alcohols since the potency of inhibitory effect and the degree of membrane disorganization are correlated with the alcohol chain length (Hiller et al. 1984). Acute alcohol treatment also decreases the binding of the DOR agonist [^{3}H]DPDPE, agonist-stimulated [^{35}S]GTPγS binding, and rate of receptor internalization in brain tissue or in N18TG2 cells expressing mouse DOR (Khatami et al. 1987; Gomes et al. 2000; Tabakoff and Hoffman 1983). However, the effects of acute alcohol on DOR pharmacology may be more pronounced when directly administered to the tissue as brains from rats with prior exposure to a single dose of alcohol did not exhibit changes in DOR affinity (Jorgensen and Hole 1986).

While the effects of acute alcohol exposure may be detectable only in vitro, chronic alcohol exposure has been reported to modify DOR affinity and expression both in cell lines and in animal tissue. Chronic alcohol exposure in neuronal cell lines endogenously expressing DORs has been reported to increase DOR levels (Charness et al. 1983, 1986, 1993) in part due to increasing DOR mRNA levels (Charness et al. 1993; Jenab and Inturrisi 1994). Ex vivo, functional studies on brain tissue from chronic alcohol-exposed rats found decreased DOR agonist-stimulated [^{35}S]GTPγS binding in the alcohol-exposed rats compared to controls, possibly as a consequence of receptor internalization and phosphorylation (Saland et al. 2004). A small number of studies have reported minor changes in DOR affinity, expression, and/or functionality with chronic alcohol exposure (Lucchi et al. 1984, 1985; Sim-Selley et al. 2002). On the other hand, several studies found that chronic alcohol exposure increased DOR expression levels, either without affecting affinity (Rossi et al. 1988) or while slightly decreasing affinity (Hynes et al. 1983). In vivo experiments have also found evidence for increased expression of DORs after chronic alcohol exposure (van Rijn et al. 2012a; Bie et al. 2009). Some of the differential DOR expression in response to chronic alcohol exposure may be ascribed to different species, strains, or age of animals, specific brain regions studied, and alcohol intake paradigms (e.g., duration, dose/concentration used). Previous studies have shown that young rats express higher levels of DORs than older rats (Rossi et al. 1988; Nielsen et al. 2012a). The influence of stress, for example, stress induced by alcohol withdrawal (Becker 2012), may be another factor that can impact DOR expression in some of the animal models. It has been shown that stressful events can increase DOR expression (Margolis et al. 2011; Commons 2003; Nielsen et al. 2012b), whereas stress prior to alcohol consumption may prevent an increase in DOR affinity for DPDPE (Przewlocka and Lason 1990).

2 Initial Behavioral Evidence for a Role of Delta Opioid Receptors in Alcohol Use

Initial behavioral evidence for an interventional role of opioid receptors in alcohol use came from studies using the nonselective opioid receptor antagonists naloxone (Froehlich et al. 1991) and naltrexone (Le et al. 1993). In order to identify which of the opioid receptor subtypes is involved in alcohol consumption, various groups have tested antagonists selective for different subtypes of opioid receptor. Beta-funaltrexamine, an irreversible mu-opioid receptor (MOR) antagonist, did not decrease alcohol consumption at a dose previously shown to antagonize various effects of morphine (Le et al. 1993). On the other hand, selective DOR antagonists,

ICI 174,864 and naltrindole, did reduce voluntary alcohol consumption (Le et al. 1993). ICI 174,864 is a peptide and thus susceptible to endogenous proteases (Cotton et al. 1984), whereas naltrindole is a nonpeptide small molecule, which has a longer period of effectiveness in vivo (Portoghese et al. 1988, 1990). Naltrindole also has greater potency, selectivity, and binding affinity for DOR than ICI 174,864 (Portoghese et al. 1990) and is therefore more commonly used as a tool to study DOR pharmacology. However, naltrindole has been reported to decrease saccharin intake as well as alcohol consumption, suggesting that intake of sweet solutions may also release endogenous opioids (Krishnan-Sarin et al. 1995a). Studies using the enkephalinase inhibitor thiorphan provided another line of support for a role of DORs in alcohol intake. Thiorphan, by enhancing endogenous activation of DORS through their ability to increase enkephalin tone, can elevate alcohol intake in rats (Froehlich et al. 1991). To better understand the mechanism of action behind the opioid modulation of alcohol intake, Widdowson and Holman looked at alcohol's effect on dopamine release in the brain. Basal dopamine release from striatal slices is dose-dependently increased in the presence of DOR agonist. Bath application of alcohol onto the striatal slices raised the dopamine level in a dose-dependent manner. This effect is reversible by the use of the DOR antagonist ICI 174,864, linking this response to DORs (Widdowson and Holman 1992).

Additional data supporting a role for DORs in alcohol consumption came from studies performed by de Waele et al. (1996, 1997) using a portacaval anastomosis (PCA) model in rats. In this model, a cirrhosis-like state was induced using portal-systemic shunts causing portal-systemic encephalopathy (de Waele et al. 1996). PCA rats exhibited enhanced voluntary alcohol consumption compared to sham controls, and PCA rats also had a reported increased density of DORs in the nucleus accumbens (de Waele et al. 1996). These increases in alcohol consumption due to PCA are also reversible by the use of naloxone (de Waele et al. 1997).

3 Use of Preclinical Animal Models to Study the Role of Delta Opioid Receptors in Alcohol Use Disorders

Delta opioid receptors form a novel target to treat AUDs. In order to develop novel DOR therapeutics, DOR-selective ligands need to be tested in preclinical animal models. To properly interpret results of DOR modulation of alcohol behaviors in animal models, it is important to understand pharmacological differences and similarities between the animal species used. The species most commonly used in preclinical studies are mice, rats, and primates. Reports have identified modulation of alcohol behaviors in all three species by DOR-selective drugs. Additionally, certain strains of mice and rats are known to prefer alcohol more than other substrains (Belknap et al. 1993; Yoneyama et al. 2008; Simms et al. 2008). Several research groups have selectively bred mice and rats to either prefer or not prefer alcohol including high- and low-alcohol-preferring (HAP, LAP) mice (Grahame et al. 1999), Finnish Alko alcohol and non-alcohol (AA, ANA) rats (Eriksson 1968), high- and low-alcohol-drinking (HAD, LAD) rats (Li et al. 1993), and

Sardinian alcohol-preferring and alcohol-non-preferring (sP, sNP) rats (Fadda et al. 1989). Opioid and enkephalin expressions can differ within these species (Nylander et al. 1994). Alcohol-preferring C57BL/6 mice show higher DOR expression in brain regions mediating drug reward compared to low-alcohol-preferring DBA/2 mice (de Waele and Gianoulakis 1997). The more alcohol-preferring C57BL/6J mice expressed met-enkephalin at lower levels than the less alcohol-preferring C57BL/N mice (Blum et al. 1982). This may in part be due to increased enkephalinase activity in C57BL/6 mice (George et al. 1991). It is important to note that several studies using selectively bred alcohol-preferring AA rats did not find DOR antagonists (naltrindole, Me2-Dmt-Tic-OH, and ICI 174,864) to decrease alcohol intake (Hyytia 1993; Honkanen et al. 1996; Ingman et al. 2003). Only two studies were identified that have investigated DOR modulation of alcohol behaviors in primates. A study in rhesus monkeys (Williams and Woods 1998) did not find that naltrindole modulated alcohol intake. In squirrel monkeys MOR agonists were found to be more effective than DOR agonists in attenuating the discriminative stimulus of alcohol, although the DOR agonist SNC80 was able to increase the discriminative stimulus effects of low-to-intermediate doses of alcohol (Platt and Bano 2011).

4 Delta Opioid Receptor Modulation of Alcohol Behaviors

There are many facets to what constitutes as well as what causes AUDs. In order to understand the therapeutic potential of a drug target in the treatment of AUDs, it is important to model these different aspects of alcohol use. The therapeutic potential of DORs has been studied in a diverse set of alcohol paradigms, which we will try to summarize here.

4.1 Volitional Alcohol Intake

Alcohol use in rodents is frequently studied using a two-bottle choice paradigm, in which mice or rats are given a choice between water and alcohol with a certain percentage, usually in the range of 1–20%, for a specified period of time per day. Frequently these studies are performed under a reversed light cycle to enable the animals to drink the alcohol during their active cycle (drink in the dark, DiD paradigm). It has been shown that DOR antagonists like ICI 174,864, naltrindole and naltriben, and SoRI-9409 can reduce voluntary alcohol intake in these models (Krishnan-Sarin et al. 1995a, b; Franck et al. 1998; Nielsen et al. 2008; Henderson-Redmond and Czachowski 2014). In a similar mouse model, the DOR agonist TAN-67 has also been reported to decrease alcohol intake (van Rijn and Whistler 2009), whereas the DOR agonist SNC80 increased alcohol consumption (van Rijn et al. 2010). Some studies however report not finding that naltrindole significantly decreases volitional alcohol intake in rodents (van Rijn and Whistler 2009; Stromberg et al. 1998). This may be due to differences in endogenous opioid tone based on the age, genetic background, and alcohol history of the animals tested. For example, naltrindole was found to decrease alcohol intake when, during alcohol exposure, no water is available (Kim et al. 2000).

4.2 Taste Aversion

The bitter taste of alcohol can serve as a deterrent for alcohol use. Behavioral models exist that measure palatability of alcohol using a taste-reactivity test in which animals are monitored for their physical response when they come in contact with alcohol. Relatively high dose of naltrindole can cause alcohol taste aversion and reduce alcohol intake in rats (Higley and Kiefer 2006; Froehlich et al. 1998). However, it remains to be seen whether naltrindole-induced taste aversion would be a viable strategy to promote alcohol cessation considering the high dose necessary to accomplish this effect.

4.3 Alcohol Uptake and Metabolism

Delta opioid receptor agonists were not shown to interfere with alcohol uptake from the stomach or modify alcohol metabolism (van Rijn and Whistler 2009; Chiang et al. 2016).

4.4 Loss of Righting Reflex

High doses of alcohol will cause sedation. In rats microinjections of the DOR antagonist ICI 174,864 in the periaqueductal gray (PAG), nucleus accumbens (NAcc), and septum decreased alcohol-induced loss of righting reflex, suggesting a role for endogenous opioids in alcohol-induced sedation (Widdowson 1987).

4.5 Behavioral Sensitization

Repeated use of drugs of abuse and alcohol can cause behavioral sensitization, i.e., a stronger behavioral response compared with that observed during the first exposure. Repeated alcohol exposure can lead to sensitized alcohol-induced locomotion. Several studies have shown that this type of behavioral sensitization can be blocked by a MOR but not a DOR antagonist (Arias et al. 2010; Pastor and Aragon 2006; Pastor et al. 2005).

4.6 Motivated Responding for Alcohol

In animals trained to lever press for alcohol, naltrindole did not reduce responding on the alcohol-paired lever in mice (Middaugh et al. 2000), rats (Spanagel 1996), or monkeys (Williams and Woods 1998). However, in infant rats or alcohol-preferring rats, the DOR antagonists (naltrindole or naltriben) can decrease alcohol responding (Henderson-Redmond and Czachowski 2014; Hyytia and Kiianmaa 2001; June et al. 1999; Miranda-Morales et al. 2012). These results again highlight how experimental factors can affect the ability to observe DOR modulation of alcohol behaviors.

4.7 Conditioned Place Preference

Alcohol, at nonsedating doses, can cause release of dopamine from dopaminergic ventral tegmental area (VTA) neurons. This dopamine is thought to valuate alcohol as a pleasurable and rewarding substance worth seeking (Mirenowicz and Schultz 1996; Brodie et al. 1999; Bromberg-Martin et al. 2010). A commonly used paradigm to study the rewarding properties of drugs is the conditioned place preference (CPP) test (Tzschentke 1998). In rats the expression of alcohol-induced CPP could also be blocked by naltrindole (Gibula-Bruzda et al. 2015). In C57BL/6 mice the expression of alcohol CPP can be blocked by the DOR agonist SNC80 but can be slightly enhanced by another DOR agonist TAN-67 (van Rijn et al. 2012b). This corresponds with earlier findings showing that SNC80 increases but TAN-67 decreases alcohol consumption (van Rijn and Whistler 2009; van Rijn et al. 2010). Van Rijn et al. (2010) proposed that, by reducing the rewarding effects of alcohol, mice treated with SNC80 will need to drink more alcohol to obtain the same rewarding effects they would normally obtain when drinking alcohol. Not only can DOR agonists modulate alcohol CPP, alcohol exposure can also impact CPP of DOR agonists. Mitchell et al. found that rats exposed to alcohol, but not alcohol-naïve rats, displayed conditioned place preference to the DOR2 agonist deltorphin II (2014). It has been shown that stress (induced by foot shock) increases the rewarding properties of alcohol, which can be blocked by naltrindole, suggesting the involvement of endogenous enkephalins. In this paradigm TAN-67 enhanced stress-induced alcohol place preference (Matsuzawa et al. 1998, 1999), consistent with the finding that TAN-67 enhances alcohol CPP (van Rijn et al. 2012b).

4.8 Reinstatement of Alcohol-Seeking Behavior

Like other types of drug addiction, alcoholism is characterized as a chronic relapsing condition. One method of studying alcohol relapse is to train mice to self-administer alcohol followed by a period of abstinence and then use either drugs, stress, or contextual cues to reinstate alcohol-seeking behavior. In this paradigm the DOR antagonists naltrindole and SoRI-9409 were able to inhibit drug seeking in rats (Nielsen et al. 2012b; Ciccocioppo et al. 2002; Marinelli et al. 2009). Reinstatement can also be measured in a CPP paradigm. A study by Gibula-Bruzda et al. found that the enkephalin derivative cUENK6 (cyclo[Nε,Nbeta-carbonyl-D-Lys2,Dap5] enkephalinamide) could reinstate alcohol CPP in a naltrindole reversible manner (2015).

4.9 Alcohol Withdrawal-Induced Seizures

One of the well-known consequences of alcohol withdrawal is the occurrence of potentially life-threatening seizures (Becker et al. 1997; McKeon et al. 2008). Even though both DOR agonists (van Rijn and Whistler 2009; van Rijn et al. 2010, 2012b, 2013) and DOR antagonists (Nielsen et al. 2008, 2012b; Krishnan-Sarin et al. 1995a, b; June et al. 1999) are considered possible options for reducing alcohol use, a potential limiting side effect of some DOR agonists is that they can produce seizures in naïve animals (Broom et al. 2002a, b; Dykstra et al. 1993; Jutkiewicz et al. 2006; Negus et al. 1998; Yajima et al. 2000). However, this is not the case for all DOR agonists (Naidu et al. 2007; Saitoh et al. 2011). It is thus more encouraging than surprising that certain DOR agonists are able to reduce alcohol withdrawal-induced audiogenic seizures (Kotlinska and Langwinski 1986). Given that enkephalin levels are decreased during alcohol withdrawal (Borg et al. 1982), the use of DOR agonists to prevent alcohol withdrawal-induced seizures can be considered a form of replacement therapy.

5 Impact of Alcohol Use and Withdrawal on Delta Opioid Receptor-Mediated Analgesia

Certain doses of alcohol produce analgesia (Woodrow and Eltherington 1988), potentially through the release of endogenous opioids. Indeed alcohol-induced thermal analgesia can be blocked by opioid receptor subtype-selective antagonists (Campbell et al. 2007). Additionally, alcohol exposure can modulate opioid-induced antinociception. For example, chronic but not acute alcohol consumption decreases thermal analgesic potency of MOR (morphine) and DOR (DSLET) agonists without changes in opioid receptor expression or affinity in both brain and spinal cord (Shah et al. 1997).

Recent studies have revealed that, under naïve conditions, DORs are selectively expressed in nonpeptidergic pain circuits that regulate mechanical sensitivity, whereas MORs are localized in peptidergic pain circuits that process thermal nociception (Scherrer et al. 2009). Van Rijn and coworkers have illustrated that chronic alcohol may increase DOR cell surface expression in the spinal cord neurons modulating thermal pain (van Rijn et al. 2012a). The newly translocated DORs could potentially modulate MOR analgesia by forming a DOR-MOR heteromeric complex (van Rijn et al. 2012a; Milan-Lobo et al. 2013; He et al. 2011) or potentially by competing for downstream mediators or by synergistic cross talk (Overland et al. 2009; Rowan et al. 2014). Given that alcohol dependence and alcohol withdrawal can induce a state of hyperalgesia (Egli et al. 2012; Dina et al. 2008; Gatch 2009; Jochum et al. 2010) and upregulate DOR cell surface expressions, DORs may serve as a promising target to treat hyperalgesia caused by alcoholic neuropathy or alcohol withdrawal. Yet, surprisingly few studies have investigated the role of opioid receptors in the mechanism by which alcohol dependence modulates analgesia. Further investigations are thus needed to understand how DORs and other opioid receptor subtypes impact the dynamic process of alcohol modulation of pain states.

6 Neuroanatomical Analysis of Delta Opioid Receptor-Induced Modulation of Alcohol Behaviors

The dopaminergic neurons in the VTA play a central role in the mechanism of action of drugs of abuse and alcohol (Pierce and Kumaresan 2006; Koob and Volkow 2010). VTA dopaminergic neurons project to the NAcc, striatum, prefrontal cortex, hypothalamus, amygdala, and hippocampus (Fields et al. 2007). Studies using autoradiography and in situ hybridization or using a transgenic DOR receptor linked to a green fluorescent protein (DOR-eGFP) have provided evidence that DORs are expressed in these important brain areas (de Waele and Gianoulakis 1997; Dilts and Kalivas 1990; Goodman et al. 1980; Codd et al. 2010; Kitchen et al. 1995; Mansour et al. 1987; Blackburn et al. 1988; Shivers et al. 1986; Harlan et al. 1987; Le Moine et al. 1994; Mansour et al. 1994; Erbs et al. 2015). Different studies have investigated the role of DORs on alcohol behaviors in each of these brain regions. Here we summarize the unique functions of DORs in these brain regions in relation to alcohol behaviors (Fig. 1).

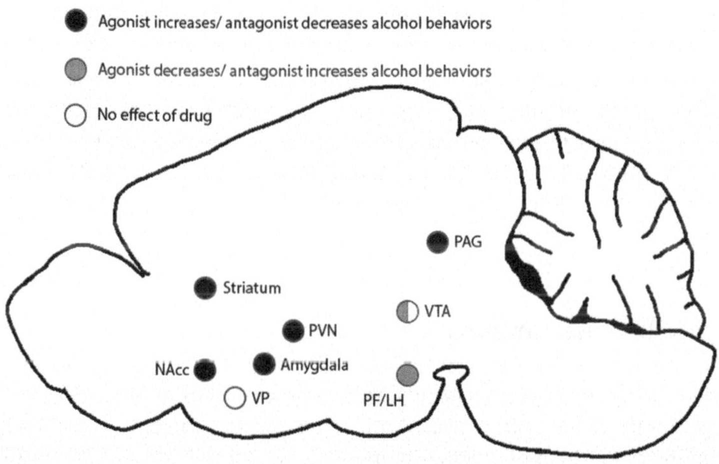

Fig. 1 Impact of local modulation of DORs on alcohol behaviors. In the ventral tegmental area (VTA), the DOR agonist DPDPE decreases and the DOR antagonist TIPPψ increases alcohol intake (Margolis et al. 2008), although naltrindole reportedly had no effect (Hyytia and Kiianmaa 2001). The DOR agonist SNC80 increases and the DOR antagonist naltrindole decreases alcohol intake in the dorsal striatum (Nielsen et al. 2012a). In the nucleus accumbens (NAcc), the DOR agonist DALA increases alcohol intake (Barson et al. 2009) and naltrindole decreases alcohol intake (Hyytia and Kiianmaa 2001) and alcohol-induced dopamine release (Acquas et al. 1993), whereas the DOR antagonist ICI 174,864 attenuates alcohol-induced loss of righting reflex (Widdowson 1987). Naltrindole decreases alcohol intake (Hyytia and Kiianmaa 2001) and alcohol place preference (Bie et al. 2009) in the amygdala. The DOR agonist DALA increases alcohol intake (Barson et al. 2010) in the paraventricular nucleus (PVN) but decreases alcohol intake (Chen et al. 2013) in the perifornical lateral hypothalamus (PF/LH). Naltrindole injection in the ventral pallidum (VP) does not modulate alcohol intake (Kemppainen et al. 2012)

6.1 Ventral Tegmental Area

Opioid receptors located on presynaptic GABA terminals in the VTA are well known for their ability to disinhibit dopaminergic neurons upon activation (Fields et al. 2007; Fields and Margolis 2015). It is known that rats will lever press to receive intra-VTA infusions of not only morphine and other MOR agonists but also DOR agonists (Devine and Wise 1994). This indicates that DORs, like MORs, play a role in drug reinforcement of opioid self-administration. Interesting evidence of DOR function modulating dopamine release from VTA neurons came from a study investigating the use of acupuncture in alcohol withdrawal. Alcohol withdrawal increases the excitability of the VTA GABA neurons (Zhao 2008), which can be reduced by acupuncture in rats in a naltrindole reversible manner (Yang et al. 2010), suggesting that acupuncture releases DOR-selective endogenous opioids in the VTA. Studies by the Fields group have provided important insight into the role of DORs in alcohol consumption. Margolis et al. found that low-alcohol-drinking rats exhibited stronger DOR inhibition of $GABA_A$ signaling in VTA slices. In these low-alcohol-drinking rats, intra-VTA injection of a DOR agonist decreased alcohol intake, whereas injection of the DOR antagonist TIPP-ψ significantly increased alcohol intake in low but not in high-alcohol-drinking rats (Margolis et al. 2008). This result could explain how microinjection of naltrindole in the VTA of alcohol-preferring rats and high-alcohol-drinking Wistar rats did not affect alcohol intake (Hyytia and Kiianmaa 2001). It appears that rats that are more anxious, more stressed, and/or more intoxicated by alcohol express higher levels of DORs in the VTA (Margolis et al. 2011; Mitchell et al. 2012). It remains uncertain how translatable the findings of an association between DOR expression and stress levels in alcohol-consuming rodents are to humans as results from PET study using the DOR radiotracer in alcohol-dependent patients found no correlation between $[^{11}C]$ methyl-naltrindole-binding potential and cortisol or adrenocorticotropin (Wand et al. 2013).

6.2 Nucleus Accumbens

The nucleus accumbens plays a critical role in processes of drug reinforcement and stress (Marinelli et al. 2005). Mesolimbic dopamine projections connecting the VTA with the NAcc also play a facilitatory role in alcohol self-administration. The DOR agonist deltorphin II directly infused into the NAcc has been shown to mimic the effect of alcohol-induced dopamine release (Acquas et al. 1993). This effect can be inhibited by using naltrindole, suggesting that endogenous opioids are involved and that DORs are locally expressed on the terminals of dopaminergic neurons, as previously suggested by Borg and Taylor (Borg and Taylor 1997). Microinjection of naltrindole into the NAcc also decreases alcohol responding in alcohol-preferring Wistar rats (Hyytia and Kiianmaa 2001). Several studies have reported that DOR and enkephalin expressions are lower in the NAcc of alcohol-preferring animals compared to alcohol-non-preferring subjects (Soini et al. 1998; Fadda et al. 1999; Strother et al. 2001). However, alcohol exposure can increase pro-enkephalin (Mendez and Morales-Mulia 2006) and enkephalin mRNA levels in the NAcc of the alcohol-preferring but not of the alcohol-non-preferring animals (Nylander et al. 1994; Marinelli et al. 2005; Mendez et al. 2010; Li et al. 1998). Despite release of endogenous opioids upon alcohol use, Turchan found no changes in DOR expression in the NAcc and striatum after access to 1–6% alcohol over a 1-month period (Turchan et al. 1999). It seems that genotype has a substantial influence on whether or not alcohol exposure increases endogenous opioid levels in the NAcc and whether those endogenous opioids will impact DOR expression levels or can be effectively blocked by a DOR antagonist to modulate alcohol responding.

6.3 Striatum

Already early on, DORs in the striatum were identified to modulate dopamine release (Dourmap et al. 1990; Petit et al. 1986). These striatal DORS are involved in the dynamic interplay between alcohol and the delta opioidergic system. For example, acute alcohol can increase met-enkephalin levels in the striatum (Seizinger et al. 1983; Schulz et al. 1980). Prolonged alcohol exposure on the other hand decreases met-enkephalin levels in the striatum of rats (Seizinger et al. 1983; Schulz et al. 1980), but will return to baseline after withdrawal (Schulz et al. 1980). However, in mice, relatively short access to 7% alcohol did not have a large effect on DOR expression or activity in the striatum (Shen et al. 1997). Still, in rats, endogenous opioids acting on DORs in the striatum are suggested to increase alcohol intake. For example, microinfusions of naltrindole into the dorsal striatum inhibit alcohol intake in Long-Evans rats, whereas microinjection of the DOR agonist SNC80 increases alcohol intake (Nielsen et al. 2012a). While striatopallidal neurons are known to contain enkephalins, naltrindole injection into the ventral pallidum did not affect alcohol intake in AA rats (Kemppainen et al. 2012).

6.4 Hypothalamus

The paraventricular nucleus (PVN) of the hypothalamus plays a coordinating role with regard to neuroendocrine responses. The PVN neurons are involved in stress management and appetitive behavior and are innervated by enkephalinergic fibers (Beaulieu et al. 1996). It has been shown that dietary fat releases endogenous opioids in the PVN (Chang et al. 2007a). Alcohol similarly increases enkephalin levels in the PVN (Chang et al. 2007b). In an apparent positive feedback loop, direct microinjection of the DOR agonist DALA in the PVN increases alcohol intake (Barson et al. 2010). The actions of this feedback loop were also apparent by findings showing that ingestion of a fatty meal can increase alcohol intake (Carrillo et al. 2004). Interestingly, microinjection of DALA in the perifornical lateral hypothalamus decreases alcohol intake. This may be caused by local inhibition of orexin function in the perifornical lateral hypothalamus that indirectly controls opioid action in other brain areas (Chen et al. 2013).

6.5 Amygdala

The amygdala is important for processing fearful as well as rewarding stimuli (Koob and Volkow 2010; Janak and Tye 2015). Studies have revealed that in the absence of DORs, alcohol increases $GABA_A$ inhibitory postsynaptic currents (IPSCs) in the central amygdala. This effect could also be mimicked using a DOR antagonist (Kang-Park et al. 2007). These results suggest that in wild-type mice, alcohol causes release of endogenous enkephalins in the central amygdala that block GABA release by acting on inhibitory coupled DORs. On the other hand, the DOR agonist DPDPE could decrease GABA IPSCs (Kang-Park et al. 2007). Interestingly, functional DORs in the central amygdala are only detectable in alcohol-exposed, but not naïve, rats (Bie et al. 2009). This may be an underlying reason why microinjection of naltrindole in the amygdala of alcohol-preferring Wistar rats can decrease alcohol responding (Hyytia and Kiianmaa 2001). Microinjection of naltrindole into the central amygdala could also decrease alcohol-conditioned place preference in rats (Bie et al. 2009), highlighting the important role of DORs in reward processing in the amygdala of alcohol-dependent subjects.

7 Role of Enkephalins and Endogenous Opioids on Delta Opioid Receptor Modulation of Alcohol Use

Endogenous opioids are thought to play an important role in the development of AUDs. Two hypotheses exist describe the association between endogenous opioids and alcohol use. The "opioid compensation hypothesis of alcoholism" presumes that subsequent alcohol intake can compensate a lack of endorphinergic activity during alcohol withdrawal (Ulm et al. 1995). The second hypothesis which we will call the "opioid reward hypothesis of alcoholism" proposes that alcohol-increased release of endogenous opioids in response to alcohol can enhance dopamine release in the NAcc by disinhibiting dopaminergic neurons (Cowen and Lawrence 1999).

7.1 The "Opioid Compensation Hypothesis of Alcoholism"

This hypothesis is supported by evidence showing that intracerebroventricular injection of met-enkephalin into rat brains decreases alcohol consumption (Ho and Rossi 1982). Additionally, increased enkephalinase activity is associated with increased alcohol consumption (George et al. 1991). Chronic alcohol consumption seems to decrease levels of met-enkephalin in the striatum of Sprague Dawley rats (Lucchi et al. 1984; Nylander et al. 1994; Cowen and Lawrence 1999). Lower levels of met-enkephalin in the nucleus accumbens (NAcc) and lower levels of leu-enkephalin in the VTA have been observed in alcohol-preferring AA rats and C57BL/6J mice relative to alcohol-avoiding ANA rats and C57BL/6N mice (Nylander et al. 1994; Blum et al. 1982). Alcohol has been shown to elevate levels of met-enkephalin in these rats, which may, in turn, regulate its reinforcement (Nylander et al. 1994). The mRNA expression of preproenkephalin in the striatum and NAcc of alcohol-preferring FH rats has also been demonstrated to be lower than that of alcohol-non-preferring WKY rats (Cowen et al. 1998).

7.2 The "Opioid Reward Hypothesis of Alcoholism"

This hypothesis is supported by the fact that medications used to manage alcohol dependence such as the nonselective opioid antagonist Revia® (naltrexone) block the effects of endogenous opioids released by alcohol intake (Benjamin et al. 1993; Gonzales and Weiss 1998; Zalewska-Kaszubska et al. 2006; Zalewska-Kaszubska et al. 2008). The selective DOR antagonists naltrindole and naltriben, which block endogenous opioids binding to DORs, have been reported to decrease alcohol consumption (Krishnan-Sarin et al. 1995a, b; van Rijn and Whistler 2009). Concordantly, elevated endogenous enkephalin tone using the enkephalinase inhibitor thiorphan (Froehlich et al. 1991), microinjection of enkephalin analogues into mesolimbic and hypothalamic regions (Barson et al. 2009, 2010), or microinjection of SNC80 in rat striatum (Nielsen et al. 2012a) increases alcohol consumption. It has been suggested that alcohol-induced endogenous opioid peptide release may counteract the aversive effects of alcohol and ultimately lead to high alcohol drinking (Froehlich et al. 1991).

7.3 Potential Reasons for the Bidirectional Effect of Enkephalins on Alcohol Behaviors

It appears that the role of the endogenous opioid system in alcohol reward is still not unequivocally understood. Interpretation of studies investigating the role of the endogenous opioid system in alcohol behaviors may be complicated by anxiety (Mitchell et al. 2012; Roberts et al. 2001) or stress (Pohorecky et al. 1999) mechanisms, dissimilar distribution patterns of enkephalins in the brain (Lugo et al. 2006), or a joint action of enkephalins and beta-endorphin (Tseng et al. 2013). The DORs have been implicated in modulating anxiety-like behavior; for example, DOR KO mice are reported to have enhanced anxiety-like behavior relative to wild-type mice (Filliol et al. 2000), and DOR agonists can reduce anxiety-like behavior (van Rijn et al. 2010; Saitoh et al. 2013). The anxiety-like response in DOR KO mice can be reversed by the self-administered alcohol (Roberts et al. 2001). The increased anxiety-like state of the DOR KO mice could be a major reason why DOR KO mice show increased alcohol intake and alcohol self-administration (van Rijn and Whistler 2009; Roberts et al. 2001), a result which would not have been predicted based on the observed decrease in alcohol intake by several DOR antagonists. In addition to anxiety, stress may also modulate DOR responses by potentiating the effect of DOR agonists (Pohorecky et al. 1999). Despite no changes in alcohol consumption or alcohol place preference between preproenkephalin knockout and wild-type mice (Koenig and Olive 2002), stress-induced alcohol intake was decreased in preproenkephalin knockout mice, suggesting the importance of stress in the interpretation of these results (Racz et al. 2008). Thus it may be difficult to dissociate the individual contribution of opioids, anxiety, stress, and dopamine from each other for their effect on the reinforcing effects of alcohol (Cowen and Lawrence 1999; Herz 1997). Furthermore, alcohol exposure does not unidirectionally alter enkephalin levels throughout the brain. Lugo et al. found that alcohol exposure increased met-enkephalin levels in the VTA, but decreased them in the central nucleus of the amygdala (2006). Moreover, even though the single enkephalin or endorphin knockout mice did not show altered alcohol CPP, mice lacking both enkephalins and beta-endorphin exhibited a decrease in alcohol-induced CPP when compared to wild-type controls, suggesting that alcohol may at least exert its rewarding action through a joint action of enkephalins and beta-endorphin (Tseng et al. 2013). These results are supported by data showing that only a high dose of naloxone that blocks both DORs and MORs attenuates alcohol CPP (Tseng et al. 2013).

8 Impact of Heteromerization and Biased Signaling of the Delta Opioid Receptor on Alcohol Use

Before the DOR was cloned, suggestions of the existence of two DOR subtypes had appeared based on differential behavioral responses of a set of DOR agonists and DOR antagonists. Responses that were induced by DPDPE and blocked by 7-benzylidenenaltrexone were labeled DOR1, whereas effects stemming from deltorphin II activation that could be blocked by naltriben were labeled DOR2. However, the pharmacology of these DOR subtypes has been difficult to reproduce in vitro and remains unclear (van Rijn et al. 2013). Still that hasn't prevented researchers from investigating how these DOR subtype-selective compounds modulate alcohol behaviors. Initial studies showed that the DOR2-selective antagonist naltriben could reduce alcohol intake in rats (Krishnan-Sarin et al. 1995b) and in mice (van Rijn and Whistler 2009). Interestingly, there appears to be a dichotomy between DOR subtype-selective ligands and their ability to modulate alcohol intake as the DOR1 agonist reduces alcohol intake in mice (van Rijn and Whistler 2009). DOR1 and DOR2 modulate alcohol intake through different mechanisms as coadministration of naltriben and TAN-67 synergistically decreased alcohol intake (van Rijn and Whistler 2009). Margolis and coworkers found that the DOR1 agonist DPDPE when injected into the VTA decreased alcohol intake in rats (Margolis et al. 2008). Later, van Rijn et al. reported that the DOR-selective agonist SNC80 increased alcohol intake in mice and labeled this as a DOR2 response (van Rijn et al. 2010). This was confirmed by data showing that intra-striatal microinjection of SNC80 in rats led to an increase in alcohol consumption (Nielsen et al. 2012a). The opposing alcohol modulatory responses of TAN-67 and SNC80 were confirmed in alcohol CPP studies (van Rijn et al. 2012b). A similar distinction in DOR1 and DOR2 modulation of alcohol behavior came from a CPP study by Mitchell et al. showing that alcohol-exposed rats displayed CPP for the DOR2 agonist deltorphin II CPP, but not the DOR1 agonist DPDPE when the drugs were injected directly into the VTA (Mitchell et al. 2014).

Three hypotheses, which are not necessarily mutually exclusive, have been proposed to explain the DOR subtypes: the "location hypothesis," the "receptor hypothesis," and the "drug hypothesis." The "location hypothesis" originates from findings that in the VTA DOR1 actions are primarily presynaptic, whereas DOR2 effects are most likely both pre- and postsynaptic (Margolis et al. 2011). Moreover, Mitchell et al. have proposed that alcohol exposure can change expression and function of the presynaptic DORs with time (Mitchell et al. 2012).

The "receptor hypothesis" suggests that the existence of the DOR subtypes arises through receptor dimerization. More precisely DOR1 is a DOR-MOR heteromer, whereas DOR2 are DOR homomers. Transgenic studies using knockout animals revealed that the effects of TAN-67 on alcohol intake, but not those of naltriben, were abolished in MOR KO mice (van Rijn and Whistler 2009). Both compounds were ineffectual in DOR KO mice (van Rijn and Whistler 2009).

The "drug hypothesis" proposes that differences in ligand-induced signal transduction pathways underlie the differential responses of DOR agonists and antagonists. This hypothesis is based on the notion of biased G protein-coupled receptor signaling, where certain drugs may only activate G-proteins, while other drugs solely signal by recruiting beta-arrestins (McDonald et al. 2000; Miller and Lefkowitz 2001). A very recent publication has shown that there is a very strong correlation between the degree of beta-arrestin 2 recruitment and the ability of a DOR-selective drug to decrease or increase alcohol intake (Chiang et al. 2016). Chiang et al. proposed that SNC80 and drugs with similar chemical structures increase alcohol intake in mice because of their strong degree of beta-arrestin 2 recruitment. On the other hand, TAN-67 is a very weak recruiter of beta-arrestin 2 and decreases alcohol intake in a mechanism that is not beta-arrestin 2 dependent (Chiang et al. 2016), most likely G-protein mediated.

9 Genetic Variations in the Delta Opioid Receptor and Its Association with Alcohol Dependence

It has been well documented that certain mice strains consume more alcohol than others (Belknap et al. 1993; Yoneyama et al. 2008). Scientists have performed genetic studies aimed at finding genes that are associated with increased alcohol intake in hopes of translating these findings to human studies. For example, C57BL/6 mice more readily consume alcohol than DBA/2J mice (Belknap et al. 1993; Yoneyama et al. 2008). A study performing quantitative trait locus (QTL) analysis on these two strains of mice found that a QTL for taste aversion was, among several other genes on multiple chromosomes, in close proximity of the DOR gene (Risinger and Cunningham 1998). However, to our knowledge no studies have investigated if the DOR gene of C57BL/6 mice contains different single nucleotide polymorphisms (SNPs) than those found in DBA/2J mice. The rise of CRISPR technology would make it easier to change a C57BL/6 DOR SNP, if it were to exist, into the DBA/2J counterpart and vice versa and observe alcohol behaviors. Another study in mice found that the C320T polymorphism (Ala107Val) in exon 2 of the DOR gene conferred higher alcohol preference to CT heterozygous mice compared to homozygous CC mice (Sacharczuk et al. 2014). The C320T SNP was more prevalent in mice selectively bred to experience limited swim stress-induced analgesia that tend to have lower opioid receptor system activity, but enhanced basal and stress-induced alcohol drinking (Sacharczuk et al. 2008), suggesting that the C320T SNP may be partially responsible for these phenotypical changes in behavior. Interestingly the Ala107Val mutation in the DOR changes the analgesic potency of the DOR agonist SNC80 (Sacharczuk et al. 2010). In humans a relatively common point mutation in the MOR gene (A118G/Asn40Asp) has been linked with affecting the effectiveness of Revia® (naltrexone) in treatment-seeking patients suffering from AUDs, although clinical evidence is not uniform (Gelernter et al. 2007; Oslin et al. 2003). Therefore, a number of genetic studies have investigated SNPs in the DOR and their relation to naltrexone efficacy, as well as their correlation with alcohol and substance use disorders. One study found a significant interaction between the DOR SNP (rs4654327) and naltrexone-induced reduction in alcohol craving and stimulatory effects (Ashenhurst et al. 2012). However other studies reported that alcohol-dependent males carrying one of three DOR SNPs (rs1042114, rs2234918, rs678849) did not significantly impact naltrexone therapeutic efficacy (Gelernter et al. 2007). Two of those SNPs (rs2234918, rs678849) also did not alter the therapeutic efficacy of the opioid antagonist nalmefene in heavy drinkers (Arias et al. 2008). A linkage study identified the DOR gene as a candidate gene for heavy alcohol use (Hansell et al. 2009). Indeed a couple of studies identified a specific haplotype (G/C/A/A/C/T), consisting of six alleles (rs1042114, rs678849, rs2298896, rs12749204, rs2234918, rs204076) to be a potential risk factor for AUDs (Zhang et al. 2008) as well as haplotypes of DOR SNPs (rs2298896, rs12749204, rs2236857, rs421300) with MOR SNPs to associate with AUDs in European Americans (Li and Zhang 2013). However, most studies investigating SNPs in the DOR gene have not found significant associations with AUD. For example, the rs2234918 SNP which is a silent mutation (C921T/Gly307Gly) in the DOR gene was shown by two research groups not to be associated with increased alcohol dependence (Franke et al. 1999; el Loh et al. 2004). Another study investigating a population of European Americans found no positive association between 11 individual DOR SNPs and AUD (Zhang et al. 2008). Further studies that included several more DOR SNPs, as well as single-point mutations in the preproenkephalin gene, did not find genetic association with AUDs in large cohort of alcohol-dependent European Americans (Xuei et al. 2007). Despite the limited association between human DOR SNPs and AUDs, DOR SNPs (rs1042114, rs678849, rs2236857, and rs581111) and preproenkephalin SNPs (rs1437277, rs1975285, and rs2609997) have been associated with opioid/heroin and cocaine use disorders (Zhang et al. 2008; Xuei et al. 2007; Crist et al. 2013; Nelson et al. 2014).

10 Therapeutic Potential of DOR Agonists for Treatment of Alcohol Use Disorders

Alcohol, enkephalins, and DORs are dynamically linked making drugs that target DORs interesting therapeutics for treatment of AUDs. Interestingly both DOR antagonists and agonists are capable of decreasing alcohol intake. Yet, we hypothesize that "DOR1"-selective agonists like TAN-67 are the most suitable option for developing efficacious therapy for alcohol dependence (van Rijn et al. 2013). In particular DOR agonists have anxiolytic-like, antidepressive-like, and antinociceptive properties (van Rijn et al. 2013; Pradhan et al. 2011), which would help reduce alcohol relapse (George et al. 2008; Heilig et al. 2010; Sinha and Li 2007). Whether TAN-67 is particularly effective because it has limited beta-arrestin2 efficacy (Chiang et al. 2016) or it can preferentially engage DOR-MOR heteromers (van Rijn and Whistler 2009), or both, remains under investigation. While certain DOR agonists may decrease seizure threshold (Broom et al. 2002a, b; Dykstra et al. 1993; Jutkiewicz et al. 2006; Negus et al. 1998; Yajima et al. 2000), several DOR agonists are available that have limited to no observable seizure activity at therapeutic dose (Naidu et al. 2007; Saitoh et al. 2011). Moreover, DOR agonists may have less abuse potential, based on several reports showing absence or limited place preference (van Rijn et al. 2012b; Mitchell et al. 2014; Suzuki et al. 1996).

The variability in alcohol behaviors observed with DOR agonists and antagonists is most likely due to the dynamic nature of the relationship between alcohol and the delta opioid system. In mouse and rat models, there appears to be important differences in DOR functionality depending on age, genotype, environmental factors (e.g., stress), and history of alcohol exposure. We believe there is a great potential for DOR-selective drugs to be beneficial in the treatment of AUDs. However, at this point in time, only a handful of studies have modulated DORs in primates to investigate the impact on alcohol behavior. Given the weak association of DOR SNPs with alcohol-dependent patients, it is crucial to perform more primate studies to obtain a better sense of the translatability of the rodent findings. In particular studies using primates with a history of alcohol use and/or in a subset of high-alcohol-drinking primates will be valuable additions to the current knowledge on DOR modulation of alcohol behavior.

Acknowledgments RMvR is supported by a K99/R00 pathway to independence award from the National Institute on Alcohol Abuse and Alcoholism and a research grant from the Ralph W. and Grace M. Showalter Research Trust.

References

Acquas E, Meloni M, Di Chiara G (1993) Blockade of delta-opioid receptors in the nucleus accumbens prevents ethanol-induced stimulation of dopamine release. Eur J Pharmacol 230:239–241

Arias AJ, Armeli S, Gelernter J, Covault J, Kallio A, Karhuvaara S, Koivisto T, Makela R, Kranzler HR (2008) Effects of opioid receptor gene variation on targeted nalmefene treatment in heavy drinkers. Alcohol Clin Exp Res 32:1159–1166

Arias C, Molina JC, Spear NE (2010) Differential role of mu, delta and kappa opioid receptors in ethanol-mediated locomotor activation and ethanol intake in preweanling rats. Physiol Behav 99:348–354

Ashenhurst JR, Bujarski S, Ray LA (2012) Delta and kappa opioid receptor polymorphisms influence the effects of naltrexone on subjective responses to alcohol. Pharmacol Biochem Behav 103:253–259

Barson JR, Carr AJ, Soun JE, Sobhani NC, Leibowitz SF, Hoebel BG (2009) Opioids in the nucleus accumbens stimulate ethanol intake. Physiol Behav 98:453–459

Barson JR, Carr AJ, Soun JE, Sobhani NC, Rada P, Leibowitz SF, Hoebel BG (2010) Opioids in the hypothalamic paraventricular nucleus stimulate ethanol intake. Alcohol Clin Exp Res 34:214–222

Beaulieu J, Champagne D, Drolet G (1996) Enkephalin innervation of the paraventricular nucleus of the hypothalamus: distribution of fibers and origins of input. J Chem Neuroanat 10:79–92

Becker HC (2012) Effects of alcohol dependence and withdrawal on stress responsiveness and alcohol consumption. Alcohol Res 34:448–458

Becker HC, Diaz-Granados JL, Weathersby RT (1997) Repeated ethanol withdrawal experience increases the severity and duration of subsequent withdrawal seizures in mice. Alcohol 14:319–326

Belknap JK, Crabbe JC, Young ER (1993) Voluntary consumption of ethanol in 15 inbred mouse strains. Psychopharmacology (Berl) 112:503–510

Benjamin D, Grant ER, Pohorecky LA (1993) Naltrexone reverses ethanol-induced dopamine release in the nucleus accumbens in awake, freely moving rats. Brain Res 621:137–140

Bie B, Zhu W, Pan ZZ (2009) Ethanol-induced delta-opioid receptor modulation of glutamate synaptic transmission and conditioned place preference in central amygdala. Neuroscience 160:348–358

Blackburn TP, Cross AJ, Hille C, Slater P (1988) Autoradiographic localization of delta opiate receptors in rat and human brain. Neuroscience 27:497–506

Blum K, Briggs AH, DeLallo L, Elston SF, Ochoa R (1982) Whole brain methionine-enkephalin of ethanol-avoiding and ethanol-preferring c57BL mice. Experientia 38:1469–1470

Borg PJ, Taylor DA (1997) Involvement of mu- and delta-opioid receptors in the effects of systemic and locally perfused morphine on extracellular levels of dopamine, DOPAC and HVA in the nucleus accumbens of the halothane-anaesthetized rat. Naunyn Schmiedebergs Arch Pharmacol 355:582–588

Borg S, Kvande H, Rydberg U, Terenius L, Wahlstrom A (1982) Endorphin levels in human cerebrospinal fluid during alcohol intoxication and withdrawal. Psychopharmacology (Berl) 78:101–103

Brodie MS, Pesold C, Appel SB (1999) Ethanol directly excites dopaminergic ventral tegmental area reward neurons. Alcohol Clin Exp Res 23:1848–1852

Bromberg-Martin ES, Matsumoto M, Hikosaka O (2010) Dopamine in motivational control: rewarding, aversive, and alerting. Neuron 68:815–834

Broom DC, Nitsche JF, Pintar JE, Rice KC, Woods JH, Traynor JR (2002a) Comparison of receptor mechanisms and efficacy requirements for delta-agonist-induced convulsive activity and antinociception in mice. J Pharmacol Exp Ther 303:723–729

Broom DC, Jutkiewicz EM, Folk JE, Traynor JR, Rice KC, Woods JH (2002b) Convulsant activity of a non-peptidic delta-opioid receptor agonist is not required for its antidepressant-like effects in Sprague-Dawley rats. Psychopharmacology (Berl) 164:42–48

Campbell VC, Taylor RE, Tizabi Y (2007) Effects of selective opioid receptor antagonists on alcohol-induced and nicotine-induced antinociception. Alcohol Clin Exp Res 31:1435–1440

Carrillo CA, Leibowitz SF, Karatayev O, Hoebel BG (2004) A high-fat meal or injection of lipids stimulates ethanol intake. Alcohol 34:197–202

Chang GQ, Karatayev O, Ahsan R, Gaysinskaya V, Marwil Z, Leibowitz SF (2007a) Dietary fat stimulates endogenous enkephalin and dynorphin in the paraventricular nucleus: role of circulating triglycerides. Am J Physiol Endocrinol Metab 292:E561–E570

Chang GQ, Karatayev O, Ahsan R, Avena NM, Lee C, Lewis MJ, Hoebel BG, Leibowitz SF (2007b) Effect of ethanol on hypothalamic opioid peptides, enkephalin, and dynorphin: relationship with circulating triglycerides. Alcohol Clin Exp Res 31:249–259

Charness ME, Gordon AS, Diamond I (1983) Ethanol modulation of opiate receptors in cultured neural cells. Science 222:1246–1248

Charness ME, Querimit LA, Diamond I (1986) Ethanol increases the expression of functional delta-opioid receptors in neuroblastoma x glioma NG108-15 hybrid cells. J Biol Chem 261:3164–3169

Charness ME, Hu G, Edwards RH, Querimit LA (1993) Ethanol increases delta-opioid receptor gene expression in neuronal cell lines. Mol Pharmacol 44:1119–1127

Chen YW, Barson JR, Chen A, Hoebel BG, Leibowitz SF (2013) Opioids in the perifornical lateral hypothalamus suppress ethanol drinking. Alcohol 47:31–38

Chiang T, Sansuk K, Van Rijn RM (2016) Beta-arrestin 2 dependence of delta opioid receptor agonists is correlated with alcohol intake. Br J Pharmacol 173:332–343

Ciccocioppo R, Martin-Fardon R, Weiss F (2002) Effect of selective blockade of mu(1) or delta opioid receptors on reinstatement of alcohol-seeking behavior by drug-associated stimuli in rats. Neuropsychopharmacology 27:391–399

Codd EE, Ma J, Zhang SP, Stone DJ Jr, Colburn RW, Brandt MR, Chevalier KM, Zhu Y, Peng S, Cai C, Acton PD, Flores CM (2010) Ex vivo delta opioid receptor autoradiography: CNS receptor occupancy of two novel compounds over their antihyperalgesic dose range. Pharmacol Biochem Behav 96:130–135

Commons KG (2003) Translocation of presynaptic delta opioid receptors in the ventrolateral periaqueductal gray after swim stress. J Comp Neurol 464:197–207

Cotton R, Giles MG, Miller L, Shaw JS, Timms D (1984) ICI 174864: a highly selective antagonist for the opioid delta-receptor. Eur J Pharmacol 97:331–332

Cowen MS, Lawrence AJ (1999) The role of opioid-dopamine interactions in the induction and maintenance of ethanol consumption. Prog Neuropsychopharmacol Biol Psychiatry 23:1171–1212

Cowen MS, Rezvani A, Jarrott B, Lawrence AJ (1998) Distribution of opioid peptide gene expression in the limbic system of Fawn-Hooded (alcohol-preferring) and Wistar-Kyoto (alcohol-non-preferring) rats. Brain Res 796:323–326

Crist RC, Ambrose-Lanci LM, Vaswani M, Clarke TK, Zeng A, Yuan C, Ferraro TN, Hakonarson H, Kampman KM, Dackis CA, Pettinati HM, O'Brien CP, Oslin DW, Doyle GA, Lohoff FW, Berrettini WH (2013) Case-control association analysis of polymorphisms in the delta-opioid receptor, OPRD1, with cocaine and opioid addicted populations. Drug Alcohol Depend 127:122–128

de Waele JP, Gianoulakis C (1997) Characterization of the mu and delta opioid receptors in the brain of the C57BL/6 and DBA/2 mice, selected for their differences in voluntary ethanol consumption. Alcohol Clin Exp Res 21:754–762

de Waele JP, Audet RM, Leong DK, Butterworth RF (1996) Portacaval anastomosis induces region-selective alterations of the endogenous opioid system in the rat brain. Hepatology 24:895–901

de Waele JP, Audet RM, Rose C, Butterworth RF (1997) The portacaval-shunted rat: a new model for the study of the mechanisms controlling voluntary ethanol consumption and ethanol preference? Alcohol Clin Exp Res 21:305–310

Devine DP, Wise RA (1994) Self-administration of morphine, DAMGO, and DPDPE into the ventral tegmental area of rats. J Neurosci 14:1978–1984

Dilts RP, Kalivas PW (1990) Autoradiographic localization of delta opioid receptors within the mesocorticolimbic dopamine system using radioiodinated [2-D-penicillamine, 5-D-penicillamine]enkephalin (125I-DPDPE). Synapse 6:121–132

Dina OA, Khasar SG, Alessandri-Haber N, Green PG, Messing RO, Levine JD (2008) Alcohol-induced stress in painful alcoholic neuropathy. Eur J Neurosci 27:83–92

Dourmap N, Michael-Titus A, Costentin J (1990) Local enkephalins tonically modulate dopamine release in the striatum: a microdialysis study. Brain Res 524:153–155

Dykstra LA, Schoenbaum GM, Yarbrough J, McNutt R, Chang KJ (1993) A novel delta opioid agonist, BW373U86, in squirrel monkeys responding under a schedule of shock titration. J Pharmacol Exp Ther 267:875–882

Egli M, Koob GF, Edwards S (2012) Alcohol dependence as a chronic pain disorder. Neurosci Biobehav Rev 36:2179–2192

el Loh W, Fann CS, Chang YT, Chang CJ, Cheng AT (2004) Endogenous opioid receptor genes and alcohol dependence among Taiwanese Han. Alcohol Clin Exp Res 28:15–19

Erbs E, Faget L, Scherrer G, Matifas A, Filliol D, Vonesch JL, Koch M, Kessler P, Hentsch D, Birling MC, Koutsourakis M, Vasseur L, Veinante P, Kieffer BL, Massotte D (2015) A mu-delta opioid receptor brain atlas reveals neuronal co-occurrence in subcortical networks. Brain Struct Funct 220:677–702

Eriksson K (1968) Genetic selection for voluntary alcohol consumption in the albino rat. Science 159:739–741

Fadda F, Mosca E, Colombo G, Gessa GL (1989) Effect of spontaneous ingestion of ethanol on brain dopamine metabolism. Life Sci 44:281–287

Fadda P, Tronci S, Colombo G, Fratta W (1999) Differences in the opioid system in selected brain regions of alcohol-preferring and alcohol-nonpreferring rats. Alcohol Clin Exp Res 23:1296–1305

Fields HL, Margolis EB (2015) Understanding opioid reward. Trends Neurosci 38:217–225

Fields HL, Hjelmstad GO, Margolis EB, Nicola SM (2007) Ventral tegmental area neurons in learned appetitive behavior and positive reinforcement. Annu Rev Neurosci 30:289–316

Filliol D, Ghozland S, Chluba J, Martin M, Matthes HW, Simonin F, Befort K, Gaveriaux-Ruff C, Dierich A, LeMeur M, Valverde O, Maldonado R, Kieffer BL (2000) Mice deficient for delta- and mu-opioid receptors exhibit opposing alterations of emotional responses. Nat Genet 25:195–200

Franck J, Lindholm S, Raaschou P (1998) Modulation of volitional ethanol intake in the rat by central delta-opioid receptors. Alcohol Clin Exp Res 22:1185–1189

Franke P, Nothen MM, Wang T, Neidt H, Knapp M, Lichtermann D, Weiffenbach O, Mayer P, Hollt V, Propping P, Maier W (1999) Human delta-opioid receptor gene and susceptibility to heroin and alcohol dependence. Am J Med Genet 88:462–464

Froehlich JC, Zweifel M, Harts J, Lumeng L, Li TK (1991) Importance of delta opioid receptors in maintaining high alcohol drinking. Psychopharmacology (Berl) 103:467–472

Froehlich JC, Badia-Elder NE, Zink RW, McCullough DE, Portoghese PS (1998) Contribution of the opioid system to alcohol aversion and alcohol drinking behavior. J Pharmacol Exp Ther 287:284–292

Gatch MB (2009) Ethanol withdrawal and hyperalgesia. Curr Drug Abuse Rev 2:41–50

Gelernter J, Gueorguieva R, Kranzler HR, Zhang H, Cramer J, Rosenheck R, Krystal JH, VA Cooperative Study #425 Study Group (2007) Opioid receptor gene (OPRM1, OPRK1, and OPRD1) variants and response to naltrexone treatment for alcohol dependence: results from the VA Cooperative Study. Alcohol Clin Exp Res 31:555–563

George SR, Roldan L, Lui A, Naranjo CA (1991) Endogenous opioids are involved in the genetically determined high preference for ethanol consumption. Alcohol Clin Exp Res 15:668–672

George DT, Gilman J, Hersh J, Thorsell A, Herion D, Geyer C, Peng X, Kielbasa W, Rawlings R, Brandt JE, Gehlert DR, Tauscher JT, Hunt SP, Hommer D, Heilig M (2008) Neurokinin 1 receptor antagonism as a possible therapy for alcoholism. Science 319:1536–1539

Gibula-Bruzda E, Marszalek-Grabska M, Gawel K, Witkowska E, Izdebski J, Kotlinska JH (2015) The influence of the new enkephalin derivative, cyclo[N(epsilon), N(beta)-carbonyl-d-Lys(2),

Dap(5)] enkephalinamide (cUENK6), on reinstatement of ethanol-induced conditioned place preference in rats. Physiol Behav 145:50–56

Gomes I, Trapaidze N, Turndorf H, Devi LA, Bansinath M (2000) Acute ethanol treatment modulates delta opioid receptors in N18TG2 cells. Anesthesiology 92:1789–1798

Gonzales RA, Weiss F (1998) Suppression of ethanol-reinforced behavior by naltrexone is associated with attenuation of the ethanol-induced increase in dialysate dopamine levels in the nucleus accumbens. J Neurosci 18:10663–10671

Goodman RR, Snyder SH, Kuhar MJ, Young WS 3rd (1980) Differentiation of delta and mu opiate receptor localizations by light microscopic autoradiography. Proc Natl Acad Sci U S A 77:6239–6243

Grahame NJ, Li TK, Lumeng L (1999) Selective breeding for high and low alcohol preference in mice. Behav Genet 29:47–57

Hansell NK, Agrawal A, Whitfield JB, Morley KI, Gordon SD, Lind PA, Pergadia ML, Montgomery GW, Madden PA, Todd RD, Heath AC, Martin NG (2009) Can we identify genes for alcohol consumption in samples ascertained for heterogeneous purposes? Alcohol Clin Exp Res 33:729–739

Harlan RE, Shivers BD, Romano GJ, Howells RD, Pfaff DW (1987) Localization of preproenkephalin mRNA in the rat brain and spinal cord by in situ hybridization. J Comp Neurol 258:159–184

He SQ, Zhang ZN, Guan JS, Liu HR, Zhao B, Wang HB, Li Q, Yang H, Luo J, Li ZY, Wang Q, Lu YJ, Bao L, Zhang X (2011) Facilitation of mu-opioid receptor activity by preventing delta-opioid receptor-mediated codegradation. Neuron 69:120–131

Heilig M, Thorsell A, Sommer WH, Hansson AC, Ramchandani VA, George DT, Hommer D, Barr CS (2010) Translating the neuroscience of alcoholism into clinical treatments: from blocking the buzz to curing the blues. Neurosci Biobehav Rev 35:334–344

Henderson-Redmond A, Czachowski C (2014) Effects of systemic opioid receptor ligands on ethanol- and sucrose seeking and drinking in alcohol-preferring (P) and Long Evans rats. Psychopharmacology (Berl) 231:4309–4321

Herz A (1997) Endogenous opioid systems and alcohol addiction. Psychopharmacology (Berl) 129:99–111

Higley AE, Kiefer SW (2006) Delta receptor antagonism, ethanol taste reactivity, and ethanol consumption in outbred male rats. Alcohol 40:143–150

Hiller JM, Angel LM, Simon EJ (1984) Characterization of the selective inhibition of the delta subclass of opioid binding sites by alcohols. Mol Pharmacol 25:249–255

Ho AK, Rossi N (1982) Suppression of ethanol consumption by MET-enkephalin in rats. J Pharm Pharmacol 34:118–119

Hoffman PL, Chung CT, Tabakoff B (1984) Effects of ethanol, temperature, and endogenous regulatory factors on the characteristics of striatal opiate receptors. J Neurochem 43:1003–1010

Honkanen A, Vilamo L, Wegelius K, Sarviharju M, Hyytia P, Korpi ER (1996) Alcohol drinking is reduced by a mu 1- but not by a delta-opioid receptor antagonist in alcohol-preferring rats. Eur J Pharmacol 304:7–13

Hynes MD, Lochner MA, Bemis KG, Hymson DL (1983) Chronic ethanol alters the receptor binding characteristics of the delta opioid receptor ligand, D-Ala2-D-Leu5 enkephalin in mouse brain. Life Sci 33:2331–2337

Hyytia P (1993) Involvement of mu-opioid receptors in alcohol drinking by alcohol-preferring AA rats. Pharmacol Biochem Behav 45:697–701

Hyytia P, Kiianmaa K (2001) Suppression of ethanol responding by centrally administered CTOP and naltrindole in AA and Wistar rats. Alcohol Clin Exp Res 25:25–33

Ingman K, Salvadori S, Lazarus L, Korpi ER, Honkanen A (2003) Selective delta-opioid receptor antagonist N, N(CH3)2-Dmt-Tic-OH does not reduce ethanol intake in alcohol-preferring AA rats. Addict Biol 8:173–179

Janak PH, Tye KM (2015) From circuits to behaviour in the amygdala. Nature 517:284–292

Jenab S, Inturrisi CE (1994) Ethanol and naloxone differentially upregulate delta opioid receptor gene expression in neuroblastoma hybrid (NG108-15) cells. Brain Res Mol Brain Res 27:95–102

Jochum T, Boettger MK, Burkhardt C, Juckel G, Bar KJ (2010) Increased pain sensitivity in alcohol withdrawal syndrome. Eur J Pain 14:713–718

Jorgensen HA, Hole K (1986) Evidence from behavioural and in vitro receptor binding studies that the enkephalinergic system does not mediate acute ethanol effects. Eur J Pharmacol 125:249–256

June HL, McCane SR, Zink RW, Portoghese PS, Li TK, Froehlich JC (1999) The delta 2-opioid receptor antagonist naltriben reduces motivated responding for ethanol. Psychopharmacology (Berl) 147:81–89

Jutkiewicz EM, Baladi MG, Folk JE, Rice KC, Woods JH (2006) The convulsive and electroencephalographic changes produced by nonpeptidic delta-opioid agonists in rats: comparison with pentylenetetrazol. J Pharmacol Exp Ther 317:1337–1348

Kang-Park MH, Kieffer BL, Roberts AJ, Siggins GR, Moore SD (2007) Presynaptic delta opioid receptors regulate ethanol actions in central amygdala. J Pharmacol Exp Ther 320:917–925

Kemppainen H, Raivio N, Suo-Yrjo V, Kiianmaa K (2012) Opioidergic modulation of ethanol self-administration in the ventral pallidum. Alcohol Clin Exp Res 36:286–293

Khatami S, Hoffman PL, Shibuya T, Salafsky B (1987) Selective effects of ethanol on opiate receptor subtypes in brain. Neuropharmacology 26:1503–1507

Kim SG, Stromberg MF, Kim MJ, Volpicelli JR, Park JM (2000) The effect of antagonists selective for mu- and delta-opioid receptor subtypes on alcohol consumption in C57BL/6 mice. Alcohol 22:85–90

Kitchen I, Leslie FM, Kelly M, Barnes R, Crook TJ, Hill RG, Borsodi A, Toth G, Melchiorri P, Negri L (1995) Development of delta-opioid receptor subtypes and the regulatory role of weaning: radioligand binding, autoradiography and in situ hybridization studies. J Pharmacol Exp Ther 275:1597–1607

Koenig HN, Olive MF (2002) Ethanol consumption patterns and conditioned place preference in mice lacking preproenkephalin. Neurosci Lett 325:75–78

Koob GF, Volkow ND (2010) Neurocircuitry of addiction. Neuropsychopharmacology 35:217–238

Kotlinska J, Langwinski R (1986) Audiogenic seizures during ethanol withdrawal can be blocked by a delta opioid agonist. Drug Alcohol Depend 18:361–367

Krishnan-Sarin S, Jing SL, Kurtz DL, Zweifel M, Portoghese PS, Li TK, Froehlich JC (1995a) The delta opioid receptor antagonist naltrindole attenuates both alcohol and saccharin intake in rats selectively bred for alcohol preference. Psychopharmacology (Berl) 120:177–185

Krishnan-Sarin S, Portoghese PS, Li TK, Froehlich JC (1995b) The delta 2-opioid receptor antagonist naltriben selectively attenuates alcohol intake in rats bred for alcohol preference. Pharmacol Biochem Behav 52:153–159

Le Moine C, Kieffer B, Gaveriaux-Ruff C, Befort K, Bloch B (1994) Delta-opioid receptor gene expression in the mouse forebrain: localization in cholinergic neurons of the striatum. Neuroscience 62:635–640

Le AD, Poulos CX, Quan B, Chow S (1993) The effects of selective blockade of delta and mu opiate receptors on ethanol consumption by C57BL/6 mice in a restricted access paradigm. Brain Res 630:330–332

Li Z, Zhang H (2013) Analyzing interaction of mu-, delta- and kappa-opioid receptor gene variants on alcohol or drug dependence using a pattern discovery-based method. J Addict Res Ther (Suppl 7):007

Li TK, Lumeng L, Doolittle DP (1993) Selective breeding for alcohol preference and associated responses. Behav Genet 23:163–170

Li XW, Li TK, Froehlich JC (1998) Enhanced sensitivity of the nucleus accumbens proenkephalin system to alcohol in rats selectively bred for alcohol preference. Brain Res 794:35–47

Lucchi L, Rius RA, Uzumaki H, Govoni S, Trabucchi M (1984) Chronic ethanol changes opiate receptor function in rat striatum. Brain Res 293:368–371

Lucchi L, Rius RA, Govoni S, Trabucchi M (1985) Chronic ethanol induces changes in opiate receptor function and in met-enkephalin release. Alcohol 2:193–195

Lugo JN Jr, Wilson MA, Kelly SJ (2006) Perinatal ethanol exposure alters met-enkephalin levels of male and female rats. Neurotoxicol Teratol 28:238–244

Mansour A, Khachaturian H, Lewis ME, Akil H, Watson SJ (1987) Autoradiographic differentiation of mu, delta, and kappa opioid receptors in the rat forebrain and midbrain. J Neurosci 7:2445–2464

Mansour A, Fox CA, Burke S, Meng F, Thompson RC, Akil H, Watson SJ (1994) Mu, delta, and kappa opioid receptor mRNA expression in the rat CNS: an in situ hybridization study. J Comp Neurol 350:412–438

Margolis EB, Fields HL, Hjelmstad GO, Mitchell JM (2008) Delta-opioid receptor expression in the ventral tegmental area protects against elevated alcohol consumption. J Neurosci 28:12672–12681

Margolis EB, Mitchell JM, Hjelmstad GO, Fields HL (2011) A novel opioid receptor-mediated enhancement of GABAA receptor function induced by stress in ventral tegmental area neurons. J Physiol 589:4229–4242

Marinelli PW, Bai L, Quirion R m, Gianoulakis C (2005) A microdialysis profile of Met-enkephalin release in the rat nucleus accumbens following alcohol administration. Alcohol Clin Exp Res 29:1821–1828

Marinelli PW, Funk D, Harding S, Li Z, Juzytsch W, Le AD (2009) Roles of opioid receptor subtypes in mediating alcohol-seeking induced by discrete cues and context. Eur J Neurosci 30:671–678

Matsuzawa S, Suzuki T, Misawa M, Nagase H (1998) Involvement of mu- and delta-opioid receptors in the ethanol-associated place preference in rats exposed to foot shock stress. Brain Res 803:169–177

Matsuzawa S, Suzuki T, Misawa M, Nagase H (1999) Different roles of mu-, delta- and kappa-opioid receptors in ethanol-associated place preference in rats exposed to conditioned fear stress. Eur J Pharmacol 368:9–16

McDonald PH, Chow CW, Miller WE, Laporte SA, Field ME, Lin FT, Davis RJ, Lefkowitz RJ (2000) Beta-arrestin 2: a receptor-regulated MAPK scaffold for the activation of JNK3. Science 290:1574–1577

McKeon A, Frye MA, Delanty N (2008) The alcohol withdrawal syndrome. J Neurol Neurosurg Psychiatry 79:854–862

Mendez M, Morales-Mulia M (2006) Ethanol exposure differentially alters pro-enkephalin mRNA expression in regions of the mesocorticolimbic system. Psychopharmacology (Berl) 189:117–124

Mendez M, Barbosa-Luna IG, Perez-Luna JM, Cupo A, Oikawa J (2010) Effects of acute ethanol administration on methionine-enkephalin expression and release in regions of the rat brain. Neuropeptides 44:413–420

Middaugh LD, Kelley BM, Groseclose CH, Cuison ER Jr (2000) Delta-opioid and 5-HT3 receptor antagonist effects on ethanol reward and discrimination in C57BL/6 mice. Pharmacol Biochem Behav 65:145–154

Milan-Lobo L, Enquist J, van Rijn RM, Whistler JL (2013) Anti-analgesic effect of the mu/delta opioid receptor heteromer revealed by ligand-biased antagonism. PLoS One 8:e58362

Miller WE, Lefkowitz RJ (2001) Expanding roles for beta-arrestins as scaffolds and adapters in GPCR signaling and trafficking. Curr Opin Cell Biol 13:139–145

Miranda-Morales RS, Spear NE, Nizhnikov ME, Molina JC, Abate P (2012) Role of mu, delta and kappa opioid receptors in ethanol-reinforced operant responding in infant rats. Behav Brain Res 234:267–277

Mirenowicz J, Schultz W (1996) Preferential activation of midbrain dopamine neurons by appetitive rather than aversive stimuli. Nature 379:449–451

Mitchell JM, Margolis EB, Coker AR, Fields HL (2012) Alcohol self-administration, anxiety, and cortisol levels predict changes in delta opioid receptor function in the ventral tegmental area. Behav Neurosci 126:515–522

Mitchell JM, Margolis EB, Coker AR, Allen DC, Fields HL (2014) Intra-VTA deltorphin, but not DPDPE, induces place preference in ethanol-drinking rats: distinct DOR-1 and DOR-2 mechanisms control ethanol consumption and reward. Alcohol Clin Exp Res 38:195–203

Naidu PS, Lichtman AH, Archer CC, May EL, Harris LS, Aceto MD (2007) NIH 11082 produces anti-depressant-like activity in the mouse tail-suspension test through a delta-opioid receptor mechanism of action. Eur J Pharmacol 566:132–136

Negus SS, Gatch MB, Mello NK, Zhang X, Rice K (1998) Behavioral effects of the delta-selective opioid agonist SNC80 and related compounds in rhesus monkeys. J Pharmacol Exp Ther 286:362–375

Nelson EC, Lynskey MT, Heath AC, Wray N, Agrawal A, Shand FL, Henders AK, Wallace L, Todorov AA, Schrage AJ, Madden PA, Degenhardt L, Martin NG, Montgomery GW (2014) Association of OPRD1 polymorphisms with heroin dependence in a large case-control series. Addict Biol 19:111–121

Nielsen CK, Simms JA, Pierson HB, Li R, Saini SK, Ananthan S, Bartlett SE (2008) A novel delta opioid receptor antagonist, SoRI-9409, produces a selective and long-lasting decrease in ethanol consumption in heavy-drinking rats. Biol Psychiatry 64:974–981

Nielsen CK, Simms JA, Li R, Mill D, Yi H, Feduccia AA, Santos N, Bartlett SE (2012a) Delta-opioid receptor function in the dorsal striatum plays a role in high levels of ethanol consumption in rats. J Neurosci 32:4540–4552

Nielsen CK, Simms JA, Bito-Onon JJ, Li R, Ananthan S, Bartlett SE (2012b) The delta opioid receptor antagonist, SoRI-9409, decreases yohimbine stress-induced reinstatement of ethanol-seeking. Addict Biol 17:224–234

Nylander I, Hyytia P, Forsander O, Terenius L (1994) Differences between alcohol-preferring (AA) and alcohol-avoiding (ANA) rats in the prodynorphin and proenkephalin systems. Alcohol Clin Exp Res 18:1272–1279

Oslin DW, Berrettini W, Kranzler HR, Pettinati H, Gelernter J, Volpicelli JR, O'Brien CP (2003) A functional polymorphism of the mu-opioid receptor gene is associated with naltrexone response in alcohol-dependent patients. Neuropsychopharmacology 28:1546–1552

Overland AC, Kitto KF, Chabot-Dore AJ, Rothwell PE, Fairbanks CA, Stone LS, Wilcox GL (2009) Protein kinase C mediates the synergistic interaction between agonists acting at alpha2-adrenergic and delta-opioid receptors in spinal cord. J Neurosci 29:13264–13273

Pastor R, Aragon CM (2006) The role of opioid receptor subtypes in the development of behavioral sensitization to ethanol. Neuropsychopharmacology 31:1489–1499

Pastor R, Sanchis-Segura C, Aragon CM (2005) Effect of selective antagonism of mu(1)-, mu (1/2)-, mu(3)-, and delta-opioid receptors on the locomotor-stimulating actions of ethanol. Drug Alcohol Depend 78:289–295

Petit F, Hamon M, Fournie-Zaluski MC, Roques BP, Glowinski J (1986) Further evidence for a role of delta-opiate receptors in the presynaptic regulation of newly synthesized dopamine release. Eur J Pharmacol 126:1–9

Pierce RC, Kumaresan V (2006) The mesolimbic dopamine system: the final common pathway for the reinforcing effect of drugs of abuse? Neurosci Biobehav Rev 30:215–238

Platt DM, Bano KM (2011) Opioid receptors and the discriminative stimulus effects of ethanol in squirrel monkeys: mu and delta opioid receptor mechanisms. Eur J Pharmacol 650:233–239

Pohorecky LA, Skiandos A, Zhang X, Rice KC, Benjamin D (1999) Effect of chronic social stress on delta-opioid receptor function in the rat. J Pharmacol Exp Ther 290:196–206

Portoghese PS, Sultana M, Takemori AE (1988) Naltrindole, a highly selective and potent non-peptide delta opioid receptor antagonist. Eur J Pharmacol 146:185–186

Portoghese PS, Sultana M, Takemori AE (1990) Design of peptidomimetic delta opioid receptor antagonists using the message-address concept. J Med Chem 33:1714–1720

Pradhan AA, Befort K, Nozaki C, Gaveriaux-Ruff C, Kieffer BL (2011) The delta opioid receptor: an evolving target for the treatment of brain disorders. Trends Pharmacol Sci 32:581–590
Przewlocka B, Lason W (1990) Stress prevents the chronic ethanol-induced delta opioid receptor supersensitivity in the rat brain. Pol J Pharmacol Pharm 42:137–142
Racz I, Schurmann B, Karpushova A, Reuter M, Cichon S, Montag C, Furst R, Schutz C, Franke PE, Strohmaier J, Wienker TF, Terenius L, Osby U, Gunnar A, Maier W, Bilkei-Gorzo A, Nothen M, Zimmer A (2008) The opioid peptides enkephalin and beta-endorphin in alcohol dependence. Biol Psychiatry 64:989–997
Risinger FO, Cunningham CL (1998) Ethanol-induced conditioned taste aversion in BXD recombinant inbred mice. Alcohol Clin Exp Res 22:1234–1244
Roberts AJ, Gold LH, Polis I, McDonald JS, Filliol D, Kieffer BL, Koob GF (2001) Increased ethanol self-administration in delta-opioid receptor knockout mice. Alcohol Clin Exp Res 25:1249–1256
Rossi F, Quirico N, Carillo C (1988) Ethanol retains its ability to modify endogenous opiate system in aged rats. Riv Neurol 58:127–130
Rowan MP, Szteyn K, Doyle AP, Gomez R, Henry MA, Jeske NA (2014) Beta-arrestin-2-biased agonism of delta opioid receptors sensitizes transient receptor potential vanilloid type 1 (TRPV1) in primary sensory neurons. Mol Pain 10:50
Sacharczuk M, Juszczak G, Sliwa AT, Tymosiak-Zielinska A, Lisowski P, Jaszczak K, Pluta R, Lipkowski A, Sadowski B, Swiergiel AH (2008) Differences in ethanol drinking between mice selected for high and low swim stress-induced analgesia. Alcohol 42:487–492
Sacharczuk M, Lesniak A, Korostynski M, Przewlocki R, Lipkowski A, Jaszczak K, Sadowski B (2010) A polymorphism in exon 2 of the delta-opioid receptor affects nociception in response to specific agonists and antagonists in mice selectively bred for high and low analgesia. Pain 149:506–513
Sacharczuk M, Lesniak A, Lipkowski AW, Korostynski M, Przewlocki R, Sadowski B (2014) Association between the A107V substitution in the delta-opioid receptors and ethanol drinking in mice selected for high and low analgesia. Addict Biol 19:643–651
Saitoh A, Sugiyama A, Nemoto T, Fujii H, Wada K, Oka J, Nagase H, Yamada M (2011) The novel delta opioid receptor agonist KNT-127 produces antidepressant-like and antinociceptive effects in mice without producing convulsions. Behav Brain Res 223:271–279
Saitoh A, Sugiyama A, Yamada M, Inagaki M, Oka J, Nagase H, Yamada M (2013) The novel delta opioid receptor agonist KNT-127 produces distinct anxiolytic-like effects in rats without producing the adverse effects associated with benzodiazepines. Neuropharmacology 67:485–493
Saland LC, Abeyta A, Frausto S, Raymond-Stintz M, Hastings CM, Carta M, Valenzuela CF, Savage DD (2004) Chronic ethanol consumption reduces delta-and mu-opioid receptor-stimulated G-protein coupling in rat brain. Alcohol Clin Exp Res 28:98–104
Scherrer G, Imamachi N, Cao YQ, Contet C, Mennicken F, O'Donnell D, Kieffer BL, Basbaum AI (2009) Dissociation of the opioid receptor mechanisms that control mechanical and heat pain. Cell 137:1148–1159
Schulz R, Wuster M, Duka T, Herz A (1980) Acute and chronic ethanol treatment changes endorphin levels in brain and pituitary. Psychopharmacology (Berl) 68:221–227
Seizinger BR, Bovermann K, Maysinger D, Hollt V, Herz A (1983) Differential effects of acute and chronic ethanol treatment on particular opioid peptide systems in discrete regions of rat brain and pituitary. Pharmacol Biochem Behav 18(Suppl 1):361–369
Shah S, Duttaroy A, Sehba F, Chen B, Philippe J, Monderson T, Lau-Cam C, Carroll J, Yoburn BC (1997) The effect of ethanol drinking on opioid analgesia and receptors in mice. Alcohol 14:361–366
Shen J, Chan KW, Chen BT, Philippe J, Sehba F, Duttaroy A, Carroll J, Yoburn BC (1997) The effect of in vivo ethanol consumption on cyclic AMP and delta-opioid receptors in mouse striatum. Brain Res 770:65–71

Shivers BD, Harlan RE, Romano GJ, Howells RD, Pfaff DW (1986) Cellular localization of proenkephalin mRNA in rat brain: gene expression in the caudate-putamen and cerebellar cortex. Proc Natl Acad Sci U S A 83:6221–6225

Simms JA, Steensland P, Medina B, Abernathy KE, Chandler LJ, Wise R, Bartlett SE (2008) Intermittent access to 20% ethanol induces high ethanol consumption in Long-Evans and Wistar rats. Alcohol Clin Exp Res 32:1816–1823

Sim-Selley LJ, Sharpe AL, Vogt LJ, Brunk LK, Selley DE, Samson HH (2002) Effect of ethanol self-administration on mu- and delta-opioid receptor-mediated G-protein activity. Alcohol Clin Exp Res 26:688–694

Sinha R, Li CS (2007) Imaging stress- and cue-induced drug and alcohol craving: association with relapse and clinical implications. Drug Alcohol Rev 26:25–31

Soini SL, Ovaska T, Honkanen A, Hyytia P, Korpi ER (1998) Brain opioid receptor binding of [3H]CTOP and [3H]naltrindole in alcohol-preferring AA and alcohol-avoiding ANA rats. Alcohol 15:227–232

Spanagel R (1996) The influence of opioid antagonists on the discriminative stimulus effects of ethanol. Pharmacol Biochem Behav 54:645–649

Stromberg MF, Casale M, Volpicelli L, Volpicelli JR, O'Brien CP (1998) A comparison of the effects of the opioid antagonists naltrexone, naltrindole, and beta-funaltrexamine on ethanol consumption in the rat. Alcohol 15:281–289

Strother WN, Chernet EJ, Lumeng L, Li TK, McBride WJ (2001) Regional central nervous system densities of delta-opioid receptors in alcohol-preferring P, alcohol-nonpreferring NP, and unselected Wistar rats. Alcohol 25:31–38

Suzuki T, Tsuji M, Mori T, Misawa M, Endoh T, Nagase H (1996) Effect of the highly selective and nonpeptide delta opioid receptor agonist TAN-67 on the morphine-induced place preference in mice. J Pharmacol Exp Ther 279:177–185

Tabakoff B, Hoffman PL (1983) Alcohol interactions with brain opiate receptors. Life Sci 32:197–204

Tseng A, Nguyen K, Hamid A, Garg M, Marquez P, Lutfy K (2013) The role of endogenous beta-endorphin and enkephalins in ethanol reward. Neuropharmacology 73:290–300

Turchan J, Przewlocka B, Toth G, Lason W, Borsodi A, Przewlocki R (1999) The effect of repeated administration of morphine, cocaine and ethanol on mu and delta opioid receptor density in the nucleus accumbens and striatum of the rat. Neuroscience 91:971–977

Tzschentke TM (1998) Measuring reward with the conditioned place preference paradigm: a comprehensive review of drug effects, recent progress and new issues. Prog Neurobiol 56:613–672

Ulm RR, Volpicelli JR, Volpicelli LA (1995) Opiates and alcohol self-administration in animals. J Clin Psychiatry 56(Suppl 7):5–14

van Rijn RM, Whistler JL (2009) The delta(1) opioid receptor is a heterodimer that opposes the actions of the delta(2) receptor on alcohol intake. Biol Psychiatry 66:777–784

van Rijn RM, Brissett DI, Whistler JL (2010) Dual efficacy of delta opioid receptor-selective ligands for ethanol drinking and anxiety. J Pharmacol Exp Ther 335:133–139

van Rijn RM, Brissett DI, Whistler JL (2012a) Emergence of functional spinal delta opioid receptors after chronic ethanol exposure. Biol Psychiatry 71:232–238

van Rijn RM, Brissett DI, Whistler JL (2012b) Distinctive modulation of ethanol place preference by delta opioid receptor-selective agonists. Drug Alcohol Depend 122:156–159

van Rijn RM, Defriel JN, Whistler JL (2013) Pharmacological traits of delta opioid receptors: pitfalls or opportunities? Psychopharmacology (Berl) 228:1–18

Wand GS, Weerts EM, Kuwabara H, Wong DF, Xu X, McCaul ME (2013) The relationship between naloxone-induced cortisol and delta opioid receptor availability in mesolimbic structures is disrupted in alcohol-dependent subjects. Addict Biol 18:181–192

Widdowson PS (1987) The effect of neurotensin, TRH and the delta-opioid receptor antagonist ICI 174864 on alcohol-induced narcosis in rats. Brain Res 424:281–289

Widdowson PS, Holman RB (1992) Ethanol-induced increase in endogenous dopamine release may involve endogenous opiates. J Neurochem 59:157–163
Williams KL, Woods JH (1998) Oral ethanol-reinforced responding in rhesus monkeys: effects of opioid antagonists selective for the mu-, kappa-, or delta-receptor. Alcohol Clin Exp Res 22:1634–1639
Woodrow KM, Eltherington LG (1988) Feeling no pain: alcohol as an analgesic. Pain 32:159–163
Xuei X, Flury-Wetherill L, Bierut L, Dick D, Nurnberger J Jr, Foroud T, Edenberg HJ (2007) The opioid system in alcohol and drug dependence: family-based association study. Am J Med Genet B Neuropsychiatr Genet 144B:877–884
Yajima Y, Narita M, Takahashi-Nakano Y, Misawa M, Nagase H, Mizoguchi H, Tseng LF, Suzuki T (2000) Effects of differential modulation of mu-, delta- and kappa-opioid systems on bicuculline-induced convulsions in the mouse. Brain Res 862:120–126
Yang CH, Yoon SS, Hansen DM, Wilcox JD, Blumell BR, Park JJ, Steffensen SC (2010) Acupuncture inhibits GABA neuron activity in the ventral tegmental area and reduces ethanol self-administration. Alcohol Clin Exp Res 34:2137–2146
Yoneyama N, Crabbe JC, Ford MM, Murillo A, Finn DA (2008) Voluntary ethanol consumption in 22 inbred mouse strains. Alcohol 42:149–160
Zalewska-Kaszubska J, Gorska D, Dyr W, Czarnecka E (2006) Effect of acute administration of ethanol on beta-endorphin plasma level in ethanol preferring and non-preferring rats chronically treated with naltrexone. Pharmacol Biochem Behav 85:155–159
Zalewska-Kaszubska J, Gorska D, Dyr W, Czarnecka E (2008) Voluntary alcohol consumption and plasma beta-endorphin levels in alcohol-preferring rats chronically treated with naltrexone. Physiol Behav 93:1005–1010
Zhang H, Kranzler HR, Yang BZ, Luo X, Gelernter J (2008) The OPRD1 and OPRK1 loci in alcohol or drug dependence: OPRD1 variation modulates substance dependence risk. Mol Psychiatry 13:531–543
Zhao ZQ (2008) Neural mechanism underlying acupuncture analgesia. Prog Neurobiol 85:355–375

Delta Opioid Receptors: Learning and Motivation

L. P. Pellissier, C. N. Pujol, J. A. J. Becker, and J. Le Merrer

Contents

Abstract

Delta opioid receptor (DOR) displays a unique, highly conserved, structure and an original pattern of distribution in the central nervous system, pointing to a distinct and specific functional role among opioid peptide receptors. Over the last 15 years, in vivo pharmacology and genetic models have allowed significant advances in the understanding of this role. In this review, we will focus on the

L.P. Pellissier • J.A.J. Becker • J. Le Merrer (✉)
Physiologie de la Reproduction et des Comportements, INRA UMR-0085, CNRS UMR-7247, INSERM, Université François Rabelais, IFCE, 37380 Nouzilly, France
e-mail: Julie.lemerrer@tours.inra.fr

C.N. Pujol
Département de Neurosciences, Institut de Génomique fonctionnelle, INSERM U-661, CNRS UMR-5203, 34094 Montpellier, France

E. M. Jutkiewicz (ed.), *Delta Opioid Receptor Pharmacology and Therapeutic Applications*, Handbook of Experimental Pharmacology 247, DOI 10.1007/164_2016_89

involvement of DOR in modulating different types of hippocampal- and striatal-dependent learning processes as well as motor function, motivation, and reward. Remarkably, DOR seems to play a key role in balancing hippocampal and striatal functions, with major implications for the control of cognitive performance and motor function under healthy and pathological conditions.

Keywords
Associative learning • Drug–context associations • GPR88 • Hippocampus • Procedural learning • Striatum

1 Introduction

The opioid receptors belong to the large family of G-protein-coupled receptors (GPCRs) and include four members: mu (MOR), delta (DOR), and kappa (KOR) opioid receptors as well as the opioid receptor-like nociceptin/orphanin FQ receptor (NOP/ORL1). Four genes encode these receptors: *Oprm1*, *Oprd1*, *Oprk1*, and *Oprl1*. Endogenous opioid ligands, namely enkephalins, dynorphins, and endorphins, derive from large precursor proteins encoded by three genes, *Penk*, *Pdyn*, and *Pomc*, respectively. The *Pnoc* gene encodes nociceptin/orphanin FQ, the endogenous ligand of NOP/ORL1.

Among opioid receptors, DORs display a highly conserved sequence of amino acids across vertebrate species, especially within transmembrane and intracellular domains. This is particularly true among mammals, for which differences in the 372-amino acid sequence occur essentially at the C-terminus level. Resulting variability in the number of phosphorylation sites suggests quantitative differences in the recruitment of intracellular signaling pathways. Such remarkable conservation of DOR sequence in mammalian species points to a strong selection pressure for this receptor, and highly preserved functions between species. When comparing DOR sequences between non-mammals and mammals, a major divergence can be found in the extracellular loops and the N-terminus domain. These differences indicate that a shift in DOR function has probably occurred during the evolution from non-mammals to mammals. Furthermore, differences at extracellular domains suggest that DOR interacts with different extracellular partners (ligands) in these animals.

Regarding intracellular signaling, DOR activation stimulates the Gi/Go-associated pathway (Childers 1991), which inhibits cAMP production (Pei et al. 1995), recruits β-arrestins (Cen et al. 2001), activates signaling kinases such as ERK and src (Shahabi et al. 1999; Zhang et al. 1999), inhibits voltage-gated calcium channels (Buzas et al. 1998), and opens inward rectifying K^+ channels (Kovoor et al. 1997). DORs are then internalized (Ko et al. 1999) and recycled or degraded in lysosomes (Tsao and von Zastrow 2000). The recently resolved crystal structure of the human DOR reveals the presence of a sodium ion pocket, with sodium ions acting as positive allosteric modulators of the receptor (Fenalti et al. 2014). Interestingly, this sodium binding pocket would play a role in β-arrestin signaling, which suggests that fine tune signaling of DORs could be more complex than initially stated.

2 Brain Expression of the Delta Opioid Receptors: Hints for Multiple Roles in Brain Function

DORs are broadly expressed in the brain. Their distribution was assessed using either in situ hybridization (ISH) to localize neuronal cell bodies expressing *Oprdl* transcripts or ligand autoradiography and fluorescent DORs (DOR-eGFP) to detect the receptors themselves. In this review, we will focus on ISH data to provide a summarized view of cerebral DOR distribution, based on the literature (Mansour et al. 1993, 1994, 1995) and open resources (Allen Brain Atlas: http://www.brain-map.org/), having in mind that experimental data show only few mismatches between DOR mRNA and protein distribution (Erbs et al. 2015; Kitchen et al. 1997; Pradhan and Clarke 2005; Scherrer et al. 2006; Slowe et al. 1999). Interestingly, the general pattern of DOR distribution appears highly informative regarding the multiple roles proposed for these receptors in brain function (Fig. 1).

Oprdl transcripts are prominently expressed in cortical regions, including whole neocortex (frontal, parietal, temporal, and occipital), where expression is preferentially detected in median layers, cortical regions of the amygdala (basolateral, cortical, and medial nuclei – BLA, CoA, and MeA, respectively), claustrum (Cl), endopiriform (En), and entorhinal (Ent) cortices, subiculum (S), presubiculum (PrS), and parasubiculum (PaS) as well as dorsal and ventral hippocampus (HPC) and dentate gyrus (DG). This is consistent with previously demonstrated implication of DORs in high order cognitive functions, such as decision-making and associative learning (Laurent et al. 2015; Le Merrer et al. 2013) and emotional processes such as anxiety (Filliol et al. 2000; Perrine et al. 2006).

In the subcortical forebrain, hotspots of *Oprdl* expression are found in the medial septum (MS) and diagonal band of Broca (ventral: VDB and horizontal: HDB, see Fig. 1) and striatum (caudate putamen and nucleus accumbens – CPu and NAc, respectively), with a remarkable punctiform distribution strikingly matching this of mRNAs coding for choline acetyltransferase (*Chat*), an enzyme necessary for the synthesis of acetylcholine (Ach). Puncta of high *Oprdl* mRNA levels in these regions indeed correspond to cholinergic interneurons (CINs) (striatum) or projecting cholinergic neurons (MS, VDB, and HDB) (Bertran-Gonzalez et al. 2013; Gazyakan et al. 2000; Le Moine et al. 1994; Scherrer et al. 2006). In the striatum, however, *Oprdl* expression is not restricted to these puncta, in agreement with expression in other cell types (Jiang and North 1992; Scherrer et al. 2006). Expression in the MS and diagonal band of Broca suggests that DOR activity can fine-tune cholinergic projections to the HPC, further supporting a role in associative (spatial) learning. High levels of *Oprdl* transcripts in striatal regions point to a role of DORs in locomotion, motor coordination, motor skill learning and impulsivity (CPu), motivation and reward (NAc). Interestingly, a third region enriched in *Oprdl* transcripts is the pons, with high levels of expression detected in the pontine nucleus (PN) and adjacent reticulotegmental nucleus (RtTg). These two nuclei receive information from the motor regions of the cortex and project to the deep cerebellar nuclei (CrbN, where *Oprdl* mRNA expression is also high), which in turn project to the cerebellar cortex (CrbCx, where *Oprdl* transcripts are detected in

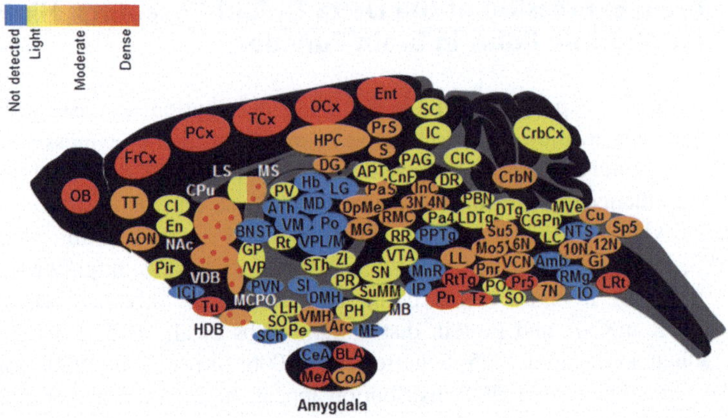

Fig. 1 Distribution of *Oprd1* transcripts in the brain. Intensity of expression varies from low (*blue*) to high (*red*) (see scale for intermediate colors). In striatal regions, septum and diagonal band, *red dots* depict particularly high level of expression in cholinergic neurons. *Abbreviations*: *3N* oculomotor nucleus, *4N* trochlear nucleus, *6N* abducens nucleus, *7N* facial nucleus, *10N* dorsal motor nucleus of vague, *11N* accessory nucleus, *12N* hypoglossal nucleus, *Amb* ambiguus nucleus, *AON* anterior olfactory nucleus, *APT* anterior pretectal nucleus, *Arc* arcuate hypothalamic nucleus, *Ath* anterior thalamus, *BLA* basolateral nucleus of the amygdala, *BNST* bed nucleus of the stria terminalis, *CeA* central nucleus of the amygdala, *CGPN* central gray of the pons, *CI* claustrum, *CIC* central nucleus of the inferior colliculus, *CnF* cuneiform nucleus, *CoA* cortical nucleus of the amygdala, *CPu* caudate putamen, *CrbCx* cerebellar cortex, *CrbN* cerebellar deep nuclei, *Cu* cuneate nucleus, *DG* dentate gyrus, *DMH* dorsomedial nucleus of the hypothalamus, *DpMe* deep mesencephalic nucleus, *DR* dorsal raphe nucleus, *DTg* dorsal tegmental nucleus, *En* endopiriform cortex, *Ent* entorhinal cortex, *FrCx* frontal cortex, *Gi* gigantocellular reticular nucleus, *GP/VP* globus pallidus/ventral pallidum, *Hb* habebula, *HDB* nucleus of the horizontal limb of the diagonal band, *HPC* hippocampus, *IC* inferior colliculus, *Icj* islands of Calleja, *InC* interstitial nucleus of Cajal, *IO* inferior olive, *IP* interpeduncular nucleus, *LC* locus coeruleus, *LDTg* laterodorsal tegmental nucleus, *LG* lateral geniculate nucleus, *LH* lateral hypothalamus, *LL* lateral lemniscus, *LRt* lateral reticular nucleus, *LS* lateral septal nucleus, *MB* mammillary bodies, *MCPO* magnocellular preoptic nucleus, *MD* mediodorsal nucleus of the thalamus, *ME* medial eminence, *MeA* medial nucleus of the amygdala, *MG* medial geniculate nucleus, *MnR* median raphe nucleus, *Mo5* motor trigeminal nucleus, *MS* medial septal nucleus, *Mve* medial vestibular nucleus, *NAc* nucleus accumbens, *NTS* nucleus tractus solitarius, *OB* olfactory bulbs, *Ocx* occipital cortex, *Pa4* paratrochlear nucleus, *PAG* periaqueductal gray, *PaS* parasubiculum, *PBN* parabrachial nucleus, *PCx* parietal cortex, *Pe* periventricular nucleus of the hypothalamus, *PH* posterior nucleus of the hypothalamus, *Pir* piriform cortex, *Pn* pontine nuclei, *Pnr* pontine reticular nucleus, *PO* paraolivary nucleus, *Po* posterior thalamic nuclear group, *PPTg* pedunculopontine tegmental nucleus, *PR* prerubral field, *Pr5* principal sensory trigeminal nucleus, *PrS* presubiculum, *PV* paraventricular nucleus of the thalamus, *PVN* paraventricular nucleus of the hypothalamus, *RMC* red nucleus - magnocellular part, *RMg* raphe magnus nucleus, *RR* retrorubral nucleus, *Rt* reticular nucleus of the thalamus, *RtTg* reticulotegmental nucleus of the pons, *S* subiculum, *SC* superior colliculus, *SCh* suprachiasmatic nucleus, *SI* substantia innominate, *SN* substantia nigra, *SO* supraoptic nucleus, *Sol* nucleus of the solitary tract, *Sp5* spinal trigeminal nucleus, *STh* subthalamic nucleus, *Su5* supratrigeminal nucleus, *SuMM* supramammillary nucleus, *TCx* temporal cortex, *TT* tenia tecta, *Tu* olfactory tubercule, *Tz* nucleus of the trapezoid body, *VCN* ventral cochlear nucleus, *VDB* nucleus of the vertical limb of the diagonal band, *VM* ventromedial nucleus of the thalamus, *VMH* ventromedial nucleus of the hypothalamus, *VPL/M* ventral postero-lateral/posteromedial nuclei of the thalamus, *VTA* ventral tegmental area, *ZI* zona incerta

the external layer of gray matter, in the surroundings of Purkinje cells), and thus contribute to motor control and motor skill learning. Finally, mRNAs coding for DORs is found particularly abundant in the lateral reticular nucleus (LRt), a brain stem nucleus receiving inputs from dorsal spinal cord and projecting to the cerebellum, also critically involved in motor function (Alstermark and Ekerot 2013). Together, these anatomical data thus suggest that DORs play a major role in motor control.

High levels of *Oprd1* mRNAs are detected all along the olfactory tract, including olfactory bulbs (OB), anterior olfactory nucleus (AON), taenia tecta (TT), piriform cortex (Pir), olfactory tubercle (Tu), MeA, ventromedial nucleus of the hypothalamus (VMH) up to the Ent cortex. Such location suggests a role for DORs in odor perception and processing. Similarly, *Oprd1* transcripts are also present along the auditory pathway, in the cochlear (CN) nucleus, in the nucleus of the trapezoid body (Tz), the superior olivary nucleus (SO), the lateral lemniscus (LL), the inferior colliculus (IC), and finally the medial geniculate nucleus (MG). Thus DORs are very likely involved in the control of auditory perception and processing. In line with a role of DORs in sensory processing, *Oprd1* mRNAs can be detected throughout the midbrain and brain stem in the origin nuclei of cranial nerves (cranial nerve III: oculomotor nucleus – 3N; IV: trochlear nucleus – 4N; V: motor trigeminal nucleus – Mo5, principal sensory nucleus – Pr5, spinal trigeminal nucleus – Sp5; VI: abducens nucleus – 6N; VII: facial nucleus – 7N; X: dorsal motor nucleus of vagus – 10N; XII: hypoglossal nucleus – 12N, but not ambiguus nor solitarius nucleus – Amb and NTS, respectively), all parts of the parasympathic system and thus sharing a cholinergic nature. Of note, DOR thus appears frequently and highly expressed in cholinergic neurons. Lastly, *Oprd1* mRNA can be found in the cuneate nucleus which carries proprioceptive information from the upper body as part of the posterior column-medial lemniscus pathway, further pointing to a participation in sensory perception.

In the midbrain, *Oprd1* transcripts are present in the substantia nigra (SN) and ventral tegmental area (VTA), red nucleus - magnocellular part (RMC), deep mesencephalic nucleus (DpMe), periaqueductal gray (PAG), and dorsal raphe nucleus (DR). Among these, SN, RN, and DpMe are involved in motor control (Rodriguez et al. 2001), VTA and DR in motivation and reward, and PAG in anxiety and pain processing. The lowest levels of *Oprd1* expression are detected in the diencephalon. *Oprd1* is expressed in the paraventricular and reticular nuclei of the thalamus (PV), as well as zona incerta and subthalamic nucleus, another key structure for motor control. *Oprd1* transcripts are more abundant in the hypothalamus, with hot spots detected in the VMH and arcuate nucleus (Arc). Regions of low but detectable expression include the magnocellular preoptic nucleus (MCPO), the lateral, periventricular, and posterior hypothalamic nuclei (LH, Pe, and PH), the supraoptic (SO) and supramammillary (SuMM) nuclei, and mammillary bodies (MB). Interestingly, several of these regions are involved in the control of sexual behavior (MCPO, LH, Pe, and VMH), endocrine function (Arc, SO, and Pe), reward (LH and SuMM), and memory (MB), arguing for a role of DORs in these different functions.

Altogether, anatomical data draw a remarkable picture of DOR location in the brain, suggestive of major roles in controlling cognitive, learning, and memory processes, motor function, motivation, and reward, anxiety and sensory/pain processing. Interestingly, this distribution pattern is original among opioid receptors, although overlaps exist with mu and kappa opioid distributions (Erbs et al. 2015; Le Merrer et al. 2009; Mansour et al. 1995), pointing to a unique, distinct role for DORs in brain function.

3 Delta Opioid Receptors and Place/Associative Learning

Abundant expression of DOR in the HPC and tightly connected structures such as subiculum, Ent cortex, or septal area (Fig. 1) points towards a crucial role of these receptors in HPC-dependent place/associative learning. Pharmacological data have strongly supported the notion that stimulating or inactivating DORs impacts memory performance. DOR agonists administered peripherally either facilitate (Martinez et al. 1984; Pavone et al. 1990; Yang et al. 2003) or impair (Jutkiewicz et al. 2003; Martinez et al. 1984; Schulteis and Martinez 1990; Ukai et al. 1997) avoidance or operant learning, while the preferential delta antagonist ICI 174,864 improves retrieval of avoidance conditioning in mice (Ilyutchenok and Dubrovina 1995; Schulteis and Martinez 1990). These studies, however, have not addressed the role of DORs in different learning paradigms, such as place/associative learning versus conditioning or motor skill learning, known to rely on distinct neurobiological mechanisms and brain substrates. Moreover, a major concern when using pharmacology is the possible cross-reactivity of delta agonists and antagonists with other opioid receptors, especially mu opioid receptors (Hutcheson et al. 2001; Scherrer et al. 2004). In this context, gene knockout, either total or partial, represents a unique tool to address the physiologic role of DORs. We will summarize here what the study of mice lacking the *Oprd1* gene and other genetically modified animals has taught us about the role of DOR in modulating spatial/associative learning and memory processes and discuss about the neurobiological substrates underlying this role. We will focus on HPC-dependent behavioral responses, having in mind that brain regions directly or indirectly connected to the HPC are likely to also contribute to these behaviors.

3.1 Spatial Navigation and Place Learning

We explored learning and memory abilities in mice lacking the DOR gene, *Oprd1*$^{-/-}$ animals (Filliol et al. 2000). We used behavioral tasks known to challenge hippocampal function to assess associative learning performance (Le Merrer et al. 2013). In a three-phase novel object recognition paradigm, we evidenced that *Oprd1* gene deletion impairs selective recognition of object location and spares novel object recognition. Such selective impairment suggests hippocampal dysfunction (Ennaceur et al. 1997; Mumby et al. 2002; Oliveira et al. 2010). Importantly, acute peripheral administration of the DOR antagonist naltrindole (0.3 mg/kg) similarly impaired recognition of object location in WT mice, indicating that deficient spatial abilities in *Oprd1*$^{-/-}$ animals result primarily from absent DOR signaling and not from developmental adaptations. In a dual solution cross-maze task, mutant mice performed at similar levels as their WT counterparts but took longer to adopt an allocentric strategy, another behavioral landmark of hippocampal dysfunction in rats and mice (Deipolyi et al. 2008; Packard 2009; Packard and McGaugh 1996). Together, these data clearly point for a crucial role of DOR activity in mediating spatial learning and memory processes, which are known to depend on hippocampal functional integrity.

Remarkably, behavioral data collected from a completely distinct mouse line, animals lacking the orphan receptor GPR88 (*Gpr88*$^{-/-}$), further support the hypothesis of such a role for DORs. GPR88 is a striatal-enriched gene critically involved in modulating dopamine neurotransmission and striatal physiology (Logue et al. 2009; Quintana et al. 2012). We created a Gpr88$^{-/-}$ mouse line and investigated the impact of Gpr88 gene deletion at multiple levels. We examined several molecular and cellular end points and revealed increased [^{35}S]-GTPγS binding mediated by the selective delta agonist SNC-80 in the striatum of *Gpr88*$^{-/-}$ mice, suggestive of facilitated DOR function, at least in this region. We also explored a vast repertoire of behavioral responses in *Gpr88*$^{-/-}$ mice using extensive phenotyping (Meirsman et al. 2016). We noticed that behavioral features of these mutants remarkably oppose several aspects of *Oprd1*$^{-/-}$ mice phenotype, notably regarding spatial navigation/learning. Indeed, when freely exploring a Y-maze, GPR88-lacking animals showed a trend towards higher spontaneous alternation and returned significantly less into the same arm, indicative of less perseverative errors. In a three-phase novel object recognition paradigm, these mutants performed better in recognizing object location. In a dual solution cross-maze task, *Gpr88*$^{-/-}$ mice not only shifted sooner from an allocentric to an egocentric strategy but also reached higher levels of performance than WT controls. Moreover, when *Gpr88*$^{-/-}$ mice were prompted to reverse their choice in the cross-maze, and they learnt this novel rule more rapidly than control animals. Together, the data suggest facilitated HPC-dependent place learning in mutant animals. To test the involvement of DOR (hyper)activity in such phenotype, we evaluated whether chronic inhibition of DORs using the antagonist naltrindole (0.3 mg/kg subcutaneous) would normalize behavior in GPR88 null mice. During Y-maze exploration, chronic naltrindole normalized spontaneous alternation in *Gpr88*$^{-/-}$ mice, by increasing significantly the number of same arm returns. This result suggests that excessive DOR signaling participates in facilitating spatial learning in these mutants. Altogether, behavioral and pharmacological data collected from *Oprd1*$^{-/-}$ and *Gpr88*$^{-/-}$ animals thus indicate that, under physiological conditions, DOR activation eases spatial navigation and place learning.

3.2 Drug–Context Associations

Interestingly, deficient spatial learning in mice lacking DORs could account for their impairment in drug-induced conditioned place preference (CPP) when spatial cues are prominent (Chefer and Shippenberg 2009; Le Merrer et al. 2011, 2012), by precluding drug/context associations (Luo et al. 2011). Involvement of DOR in drug reward and seeking will be discussed in a later Sect. 4.2.2. Place conditioning is a form of stimulus–outcome learning commonly used to assess the motivational effects of psychoactive drugs. It is based on the observation that animals will learn to approach or avoid distinct spatial environments that have previously been associated with rewarding (place preference – CPP) or aversive (place aversion) drug effects, respectively (Cunningham et al. 2011). Others and we showed that CPP induced by morphine administration is reduced in *Oprd1*$^{-/-}$ animals (Chefer and Shippenberg 2009; Le Merrer et al. 2011, 2012). Accordingly, peripheral pretreatment with the DOR antagonist naltrindole (0.3 mg/kg) before conditioning sessions was shown to abolish morphine-induced CPP (Chefer and Shippenberg 2009). These results indicate disruptive effects of DOR blockade on morphine-induced place conditioning. Several pieces of evidence argued at this stage for deficient drug/context associations rather than reduced morphine reward in DOR null mice. First, deficient place conditioning was also observed in these mutants when using an aversive stimulus, lithium chloride injections (3 mEq/kg), indicating that such deficit was not reward specific. Second, *Oprd1*$^{-/-}$ animals were able to display place preference or aversion when tested under the effects of the drug used for conditioning (morphine or lithium, respectively), signifying state dependency (Le Merrer et al. 2011, 2012). State dependency qualifies a behavioral response that can only be retrieved when the animal experiences the same (drug) state as during the acquisition of this response (Overton 1978). Interoceptive drug cues (internal state induced by drug exposure) can then function as conditioned stimuli and contribute to contextual information together with external cues. State-dependent morphine- or lithium-induced place conditioning in *Oprd1*$^{-/-}$ mice thus indicates that these animals, and not their WT counterparts, need both internal and external cues to express place preference or aversion. As long as such cues are available, however, DOR null mice can express preference for morphine-paired environment, indicating that they indeed experience morphine reward. Consistent with this, they were able to acquire intravenous (Le Merrer et al. 2011) as well as intra-VTA (David et al. 2008) morphine self-administration.

Interoceptive drug cues, interestingly, are not the only cues that DOR knockout animals can use to overcome their deficit in place conditioning. We showed that circadian time or auditory cues, whether predicting morphine injection or feeding (in food-deprived animals), could also serve *Oprd1*$^{-/-}$ animals as contextual triggers to express place conditioning (Le Merrer et al. 2012). Of note, circadian, drug, and auditory cues share a nonspatial nature, and as such should not require hippocampal functional integrity. These data further support our demonstration of deficient hippocampal-dependent learning in *Oprd1*$^{-/-}$ mice by evidencing their blunted ability to form drug–context associations and/or retrieve such associations, and the alternative strategies they can use to express their preference. Accordingly, place conditioning to nicotine is also impaired in DOR null animals (Berrendero et al. 2012). No modification of cannabinoid-induced CPP was detected in these animals, though, possibly due to the higher number of conditioning sessions (Ghozland et al. 2002) that may have facilitated drug/context associations.

Consistent with place conditioning data, studies investigating other types of context-induced conditioned responses to drugs further support the hypothesis of deficient drug–context associations in mice lacking DORs. Indeed, when *Oprd1*$^{-/-}$ animals were reexposed to an experimental context previously paired with morphine injections [see protocol in Faget et al. (2012)], they failed to demonstrate context-induced somatic signs of withdrawal (Fig. 2). High global score in vehicle-treated mutants likely reflected elevated basal levels of anxiety in these animals (Filliol et al. 2000). Regarding context-induced locomotion, however, DOR null mice, as well as mice treated with a DOR antagonist, show increased sensitization to the locomotor effects of morphine (Chefer and Shippenberg 2009), demonstrating preserved drug–context associations. In these experiments, however, the animals were tested under the effects of the drug, and thus likely displayed a state-dependent locomotor response to morphine. Conditioned activity (locomotor activity induced by exposure to the drug-paired context, in the absence of the drug) would need to be assessed in these animals to verify this point. Of note, increased locomotor sensitization to morphine in mutants may reflect enhanced motivation for the drug, as suggested by increased breaking points when tested (drug-free) for extinction of morphine self-administration (Le Merrer et al. 2011). Regarding context-induced drug seeking, we review and discuss relative experimental data in a later Sect. 4.2.2.

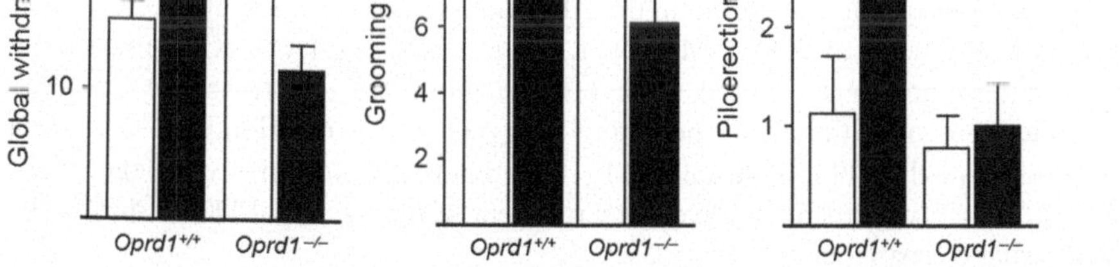

Fig. 2 Conditioned signs of withdrawal from morphine in *Oprd1*$^{-/-}$ and their WT controls. The animals (*Oprd1*$^{+/+}$: $n = 8$–10 per treatment, *Oprd1*$^{-/-}$: $n = 8$ per treatment) received daily injections of morphine (30 mg/kg) for 6 days and were immediately placed in Plexiglas transparent boxes, as previously described (Faget et al. 2012). Mutant mice fail to display somatic signs of withdrawal when exposed to the morphine-paired context under drug-free conditions. Global withdrawal score – genotype × treatment: $F_{1,32} = 758.7$, $p < 0.01$. Grooming count – genotype × treatment: $F_{1,32} = 69.5$, $p < 0.05$. Piloerection – Genotype: $F_{1,32} = 8.6$, $p < 0.01$, Treatment: $F_{1,32} = 7.9$, $p < 0.05$, genotype × treatment: $F_{1,32} = 6.2$, $p < 0.05$. *Black stars* treatment effect, *open stars* treatment × genotype interaction

Together, previous data concur to demonstrate that drug/context associations are impaired in *Oprd1*$^{-/-}$ mice, likely due to hippocampal dysfunction.

3.3 Hippocampal Delta Opioid Receptors: Implications for Associative Learning

3.3.1 Functional Role of Delta Opioid Receptors in the Hippocampus

Anatomical and pharmacological data concur to demonstrate that DORs can locally modulate hippocampal function. These receptors are indeed abundantly expressed in the HPC (Crain et al. 1986; Le Merrer et al. 2009; Mansour et al. 1995), in GABAergic interneurons (Rezai et al. 2012; Scherrer et al. 2006; Svoboda et al. 1999) where they act presynaptically to inhibit GABA release (Rezai et al. 2012; Piskorowski and Chevaleyre 2013) and consequently favor disinhibition of principal glutamatergic cells (Lupica 1995). Further electrophysiological studies have shown that pharmacological activation of DORs induces long-term depression of parvalbumin-expressing GABA interneurons within CA2 (Piskorowski and Chevaleyre 2013) and inhibits the excitatory temporoammonic pathway from the Ent cortex to CA1 (Rezai et al. 2012). Accordingly, enkephalins, among endogenous ligands of DORs, are released in the lateral perforant path (Chavkin et al. 1985), where DOR activation contributes to high frequency-induced long-term potentiation (LTP), possibly by transiently reducing GABA transmission (Bramham et al. 1991), and thus to hippocampal-dependent learning. Furthermore, DORs also play a role in the induction of LTP in dentate granule cells (Xie and Lewis 1995). Thus pharmacological or genetic inactivation of DORs in the HPC seems to prevent their endogenous ligands from inhibiting GABAergic interneurons, which makes inhibition of pyramidal cells more likely and thus reduces probability for LTP, a plausible mechanism for impaired associative learning. Under physiological conditions, activation of DORs in the HPC, conversely, would ease hippocampal function and facilitate spatial/associative learning.

Interestingly, this proposition is in agreement with gene expression data showing increased transcription of *Oprd1* in the HPC of rats trained for a spatial discrimination task (Robles et al. 2003). Moreover, when reexposing DOR eGFP mice to an environment previously paired with repeated morphine injections, we observed somatic signs of withdrawal, indicating drug–context association, and activation of hippocampal DORs as visualized by their internalization in vivo (Faget et al. 2012). These experiments unravel a recruitment of hippocampal DORs during the processing of spatial cues.

3.3.2 Delta Opioid Receptors and Hippocampal Gene Expression

To identify potential molecular mechanisms underlying impaired associative learning in *Oprd1*$^{-/-}$ mice, we quantified the expression of 67 genes of interest in the dorsal HPC of mutants as compared to wild-type controls using quantitative real-time Polymerase Chain Reaction (qRT-PCR). Interestingly, transcript levels of *Grin1* and *Grin2a*, coding for GluN1 (NR1) and GluN2A (NR2A) subunits of NMDA glutamate receptors, respectively, were low in *Oprd1*$^{-/-}$ mice (Le Merrer et al. 2013). These two subunits are crucial for spatial learning in mice (Bannerman et al. 2008; Korotkova et al. 2010; Place et al. 2012). We also detected low hippocampal mRNA levels of several genes known for their enriched expression in medium spiny neurons (MSNs – *Bcl11b/Ctip2*, *Arpp21*, *Foxp1*, *Gpr6*, *Hpca*, *Pde10a*, *Penk*, *Pdyn*, and *Tac1*). Among them, *Pdyn, Penk*, and *Tac1* code for neuropeptides (dynorphin, enkephalin, and substance P, respectively) participating in the control of hippocampal activity (McDermott and Schrader 2011; McQuiston 2011; Ogier et al. 2008). *Bcl11b/Ctip2* is involved in postnatal neurogenesis and granule cell differentiation, and its deletion in the forebrain impairs spatial learning (Simon et al. 2012). *Hpca* encodes a calcium binding protein, hippocalcin, that contributes to neuronal plasticity (Jo et al. 2010). The functional roles of *Arpp21*, *Foxp1*, *Gpr6*, and *Pde10a* in the HPC have not been explored yet or remain poorly understood (Giralt et al. 2013) despite demonstrated expression in this structure (low for *Gpr6*, see Allen Brain Atlas). In contrast, increased expression of several genes such as *Grm1* (coding metabotropic glutamate receptors, mGluR1) and *Chat* (coding for the Ach synthesizing enzyme choline acetyltransferase) may reflect compensative processes aiming at restoring hippocampal function (Aiba et al. 1994). Together, these data indicate that DOR deletion significantly impacts gene expression in the dorsal HPC, and these transcriptional modifications likely contribute to impair hippocampal function in mutant mice.

We similarly explored gene transcription in the dorsal CA1 of *Gpr88*$^{-/-}$ mice. *Gpr88* expression is too low to be detected in the HPC (Allen Brain Atlas). Nevertheless, deletion of this gene resulted in modified levels of transcripts for a few genes in the CA1: expression of *Ache* (coding the Ach degrading enzyme acetylcholinesterase) was downregulated, whereas expression of *Gabra4* (alpha4 subunit of the GABA receptor), *Foxp1* (forkhead box P1), *Wfs1* (wolframin), and *Oprd1* was upregulated (Meirsman et al. 2016). Interestingly, decreased levels of acetylcholinesterase (Hasselmo and Sarter 2011) and increased expression of wolframin (Kitamura et al. 2014; Sutt et al. 2010) may contribute to facilitate HPC-dependent associative learning in *Gpr88*$^{-/-}$ mice. Most importantly, increased *Oprd1* expression could also contribute to this facilitation, in agreement with augmented transcription in the HPC of rats trained for a spatial task (Robles et al. 2003). Moreover, pharmacological blockade of DORs normalized spatial alternation rates in *Gpr88*$^{-/-}$ mice, pointing to excessive DOR activity in these animals as an underlying mechanism of their increased spatial memory performance, although involvement of striatal DORs should not be excluded. In conclusion, these data provide further evidence for a crucial role of hippocampal DORs in underlying spatial/associative learning processes.

3.3.3 Extra-Hippocampal Delta Opioid Receptors and Associative Learning

Not only DORs in the HPC can play a role in modulating HPC-dependent place and associative learning but also DORs in other brain regions. Indeed, these receptors are highly expressed in several direct or indirect hippocampal input or output regions, such as the entorhinal , perirhinal, and prefrontal cortices, subiculum and septal area, all key brain sites for learning and memory (Dickerson and Eichenbaum 2010; White and McDonald 2002). DOR expression is also particularly high in the striatum, which functionally competes with the hippocampal formation to drive behavior (Ghiglieri et al. 2011; Packard 2009). However, although hippocampal lesion/inactivation facilitates dorsal striatal function (Middei et al. 2004; Schroeder et al. 2002), striatal lesion/inactivation fails to conversely facilitate HPC-dependent spatial learning (Castane et al. 2010; De Leonibus et al. 2007; Jacobson et al. 2012). This lack of reciprocity may result from differential implication of subpopulations of striatal MSNs. Indeed, striatal deletion of the adenosine A2a receptor (*Adora2a*) gene, whose expression is significantly enriched in D2R-MSNs (Heiman et al. 2008), decreases D2R-MSN excitability and facilitates spatial learning (Wang et al. 2006; Wei et al. 2011; Zhou et al. 2009). However stimulation of DORs in the dorsal striatum seems to repress D1R-MSN activity instead (see Sect. 4.3.1), making unlikely their involvement in facilitating HPC-dependent processes.

4 Delta Opioid Receptors, Motor Function, Response Learning, Motivation, and Reward

Oprd1 gene is highly expressed in multiple brain regions involved in motor control, including the striatum (CPu and NAc), motor cortical areas, STh, GP, SN, PN, RtTg, RN, LRt, and cerebellum (Fig. 1). Such distribution clearly designates DORs as key actors of motor function. Their role, however, appears complex, as suggested by the diverging effects of DOR ligands on motor responses. Indeed, DOR agonist SNC80 stimulates locomotor activity while other agonists fail showing such effect (Jutkiewicz et al. 2005; Le Bourdonnec et al. 2008, 2009; Nozaki et al. 2012; Saitoh et al. 2011). DOR antagonists were shown to relieve dyskinesias induced by chronic L-DOPA administration or neuroleptics (Henry et al. 2001; McCormick and Stoessl 2002). Conversely, DOR agonists were shown to improve dyskinesia in 6-hydroxydopamine (6-OHDA) hemilesioned rats, although depending on the dose (Mabrouk et al. 2009; Mabrouk et al. 2014). Discrepancies would notably lie in the brain structures primarily targeted by pharmacological compounds, such as GP versus SN (Mabrouk et al. 2009), and in differential affinity of the compounds for presynaptic versus postsynaptic receptors.

Besides their presence in motor circuits, DORs are also highly expressed in multiple brain regions modulating motivation and reward, such as the medial prefrontal cortex, NAc, VP, VTA, SuMM, and LDTg (Fig. 1), suggesting that DORs contribute to these processes. Consistent with this, pharmacological data have long suggested that MORs and DORs would play overlapping roles in mediating reward processes (Le Merrer et al. 2009). Major evidences were that DOR agonists can elicit CPP (Longoni et al. 1998; Morales et al. 2001; Shippenberg et al. 1987; Suzuki et al. 1997) and increase consumption of palatable substances (Baldo and Kelley 2007), whereas antagonists alter cocaine and nicotine self-administration (Ward and Roberts 2007), attenuate CPP to cocaine, methamphetamine, or morphine (Chefer and Shippenberg 2009; Menkens et al. 1992; Suzuki et al. 1994), and reduce heroin and cocaine self-administration (Martin et al. 2000). In these studies, however, DOR ligands may have produced part of their effects via activation of MORs (Hutcheson et al. 2001; Scherrer et al. 2004). In this context, genetically modified mice proved to be useful by allowing researchers to assess the consequences of DOR inactivation independently from that of MOR.

In this section, we will focus primarily on data obtained from mice lacking DOR or other genetically modified animals and compare with pharmacological data whenever pertinent. We will discuss the consequences of invalidating DOR on various behavioral responses for which functional integrity of the striatal regions, CPu and/or NAc, is necessary, although not sufficient, as these responses likely involve other brain sites within motor or reward circuits where DORs are also abundant.

4.1 Locomotion and Motor Function

4.1.1 Basal Locomotion and Interest for Novelty

Striatal regions are key brain sites involved in controlling locomotion and exploration (Do et al. 2012; Palmiter 2008). Deletion of the *Oprd1* gene or chronic pharmacological blockade of DOR leads to hyperlocomotion in mice (Filliol et al. 2000; Le Merrer et al. 2013). This hyperactivity fails to habituate over repeated testing (Filliol et al. 2000), although mutant mice perform as well as wild-type animals at recognizing novelty. Indeed, *Oprd1*$^{-/-}$ mice displayed similar preference for, and hyperactivity in, the novel versus familiar compartment of a place conditioning apparatus (Le Merrer et al. 2011). Furthermore, mutants visited the novel object more often during the object phase in a three-phase paradigm of object recognition (Le Merrer et al. 2013). Such facilitated novel object recognition suggests that novelty is more attractive to DOR null mice (Ennaceur 2010). Consistent with this idea, under low light conditions (15 lux), when levels of anxiety in the elevated plus-maze were similar between WT and mutant animals, *Oprd1*$^{-/-}$ mice made more head dips (Le Merrer et al. 2013), suggestive of increased novelty seeking and risk-taking behavior in these mutants. Therefore, increased locomotor activity in mice lacking DORs could result from impaired habituation, a landmark of hippocampal deficit, together with increased interest for novelty, pointing towards facilitated striatal activity.

4.1.2 Response and Motor Skill Learning

Striatal regions are critically involved in mediating procedural/response and motor skill learning (Graybiel 2008; Packard 2009; Packard and McGaugh 1996). Strikingly, such learning processes had been poorly explored in *Oprd1*$^{-/-}$ mice, despite high levels of expression in striatum. We thus examined whether DOR deletion would affect striatum-dependent learning by performing two behavioral assays. We first used a single-solution response task in the cross-maze, which solely requires a striatal-dependent egocentric strategy (response learning) (Packard 2009). Second, we tested WT and mutant mice in an accelerating rotarod task to assess skill motor learning, a form of procedural learning that was shown to tightly depend on dorsal striatum functional integrity (Dang et al. 2006; Durieux et al. 2009).

In the cross-maze, *Oprd1*$^{-/-}$ mice developed a response strategy more rapidly than WT animals under a single-solution response paradigm. This result suggests that the control of response learning is facilitated in mutants (Packard 2009; Packard and McGaugh 1996), most likely via the lateral dorsal striatum (Lovinger 2010). However, increased motivation to gain a food reward might have contributed to improve performance of mutants in this task (see Sect. 4.2.1). We thus further assessed striatal-dependent behavior using a motor skill learning task, which does not engage food seeking. *Oprd1*$^{-/-}$ mice indeed performed better than controls on the accelerating rotarod. Lateral dorsal striatal circuits are critically involved during motor skill learning (Lovinger 2010; Yin et al. 2009). Therefore our data concur to indicate that dorsal striatal function is facilitated in mice lacking DORs.

We also assessed striatum-dependent behaviors in *Gpr88*$^{-/-}$ mice, which display elevated striatal DOR activity. Interestingly, they failed to acquire a motor skill learning task on the accelerating rotarod, demonstrating a major impairment in striatal function (Meirsman et al. 2016; Quintana et al. 2012). Chronic administration of the DOR antagonist naltrindole alleviated this motor skill learning deficit, but at an early stage only, when motor learning depends on the D2 dopamine receptor-bearing medium spiny neurons (D2R-MSNs) of the dorsal striatum (Durieux et al. 2012). This particular time course of naltrindole effects suggests that excessive DOR activity in *Gpr88*$^{-/-}$ mice could compromise motor skill learning by affecting the activity of striatal D2R-MSNs. Together, behavioral data from *Oprd1*$^{-/-}$ and *Gpr88*$^{-/-}$ mice consistently point to an inhibitory influence of DORs on striatum-dependent response and skill learning processes.

4.2 Motivation, Decision-Making, and Reward

4.2.1 Motivation for Food, Food Reward, and Decision-Making

Dorsal and ventral striatum play a crucial role in regulating reward and motivation for food (Richard et al. 2013). Whereas the involvement of MORs in these processes has been extensively explored, little is known about a potential role of DORs, although the prevailing idea seems to be that MORs and DORs may play overlapping roles (Bodnar 2004; Nogueiras et al. 2012). Consistent with this, pharmacological studies have shown that DOR agonists increase the intake of palatable substances (Baldo and Kelley 2007). However, DOR antagonists, injected systematically or locally in brain areas, fail to consistently decrease palatable food intake (Bodnar et al. 2005; Katsuura and Taha 2014; Khaimova et al. 2004; Levine et al. 1994; Miner et al. 2012). In this context, the study of mice bearing genetic invalidation of *Oprd1* was useful to disentangle DOR from MOR function in motivation for food and food reward.

We assessed motivation for food in *Oprd1*$^{-/-}$ mice using two behavioral tests (Le Merrer et al. 2013). In a runway task, latency to reach sucrose reward tablets was not modified in mutant animals as compared to WT controls when the mice were confined in the end box for 20 s. When this confinement was omitted, however, knockout mice obtained more sugar pellets than WT mice, by eating and coming back faster to the start box. This result suggests increased food seeking in mutants. Of note, we omitted confinement in this experiment to reduce anxiety levels, elevated in *Oprd1*$^{-/-}$ mice (Filliol et al. 2000). We propose that, in these animals, high levels of anxiety (avoidance) compete with high motivation for food (approach) to drive behavior (Aupperle and Paulus 2010; Montgomery 1955; Powell et al. 2004). By reducing anxiety in the straight alley test, we may have unmasked the latter. Consistent with this, in a novelty-suppressed feeding (NSF) experiment, *Oprd1*$^{-/-}$ animals took as long as WT controls to start eating in the arena, but approached food pellets (lab chow) more often, and retreated, revealing conflicting avoidance and approach behaviors (Powell et al. 2004). Moreover, in the same task, *Dlx5/6*-Cre X *Oprd1*$^{fl/fl}$ (Dlx-DOR) mice, which do not express DOR in GABAergic neurons of the forebrain, and do not display high levels of anxiety (maybe due to preserved DOR expression in the amygdala), started quicker than controls to eat food in the arena (Chu Sin Chung et al. 2015). We have previously shown that latency to eat in the NSF test tightly correlates with Fos expression in the CeA, whereas the amount of food consumed when back in the home cage clusters with Fos immunostaining in the VTA (Becker et al. 2014), which illustrates, at the neurobiological level, conflictual avoidance/approach behaviors in this task. Interestingly, after the NSF assay, Fos expression was reduced in the CeA (as well as BLA) and tended to be increased in the VTA (significantly increased in the NAc) of Dlx-DOR mice, suggesting that removing DORs in GABAergic neurons of the forebrain both reduced avoidance (anxiety levels) and increased approach (motivation/reward). Together, these data suggest that DORs' activity, notably in the forebrain, represses motivation for food (palatable or not).

The fact that DORs likely decrease motivation for food does not necessarily imply that they influence food reward in the same direction. Indeed, preference for sucrose reaches similar levels in *Oprd1*$^{-/-}$ mice and WT controls (Olmstead et al. 2009). Ceiling effect (about 95% preference), however, likely made it difficult to detect increased preference in mutants. In contrast with genetic deletion, pharmacological blockade of DORs in the ventral pallidum was shown to increase saccharine palatability and consumption (Inui and Shimura 2014), while DOR inhibition in the NAc increased consumption of a sucrose solution under an anticipatory contrast paradigm (Katsuura and Taha 2014). DORs may thus exert an inhibitory control on food reward as well, especially those expressed in the ventral pallidum and NAc.

Interestingly, *Oprd1* deletion fails to significantly impact operant learning for food. *Oprd1*$^{-/-}$ animals are able to acquire an instrumental task to earn food or sucrose reward with similar levels of performance as WT controls (Gutierrez-Cuesta et al. 2014; Laurent et al. 2012; Olmstead et al. 2009). Also, Dlx-DOR mutants performed similarly to controls in acquiring chocolate-flavored pellet self-administration (Chu Sin Chung et al. 2015). Surprisingly, however, they displayed lower breaking points under a progressive-ratio schedule of reinforcement, suggestive of decreased motivation to work for this palatable food. Interestingly, *Oprd1*$^{-/-}$ mice similarly displayed a marked tendency for decreased breaking points when tested for their motivation to earn sucrose pellets under a progressive-ratio paradigm (Gutierrez-Cuesta et al. 2014). Of note, this was unlikely to result from deficient hippocampal function, as lesioning the HPC instead produces an increase in breakpoints for food (Schmelzeis and Mittleman 1996). How to reconcile increased approach of food in NSF and straight alley with decreased breakpoints for a sweet reward under a progressive-ratio schedule of reinforcement in total or conditional DOR null mice will require further investigation.

Once instrumental learning for food reward was acquired, DOR null mice were tested for impulsivity. Remarkably, these animals showed difficulties in withholding their motor response to obtain sucrose reward (Olmstead et al. 2009). This result could have reflected increased motivation for food, as evidenced previously (Le Merrer et al. 2013). Strikingly, however, comparable difficulties in waiting for a defined temporal interval to elapse were observed in rats with hippocampal lesions (Bannerman et al. 1999). Indeed, HPC is involved in controlling temporal memory (in the sense of temporal processing – succession of events – not circadian cues), likely through an inhibitory influence on the dorsal striatum (Yin and Meck 2014; Yin and Troger 2011). Remarkably, *Oprd1*$^{-/-}$ mice underestimated 15 and 45 s target durations in a bi-peak procedure, as evidenced by proportional leftward shifts of the peak functions, and similarly to mice with cytotoxic lesions of the dorsal HPC (Yin and Meck 2014). These results support the hypothesis of altered hippocampal function in DOR null mice and indicate that these animals may have difficulties in performing operant tasks when accurate timing is required, by triggering premature responses. In conclusion, impaired timing performance in DOR null mice is a more convincing candidate explanation for their increased motor impulsivity than increased motivation for food, although the latter cannot be ruled out.

Finally, DOR activity seems necessary for a previous reward experience to influence decision-making. Pavlovian incentive learning, which mediates the excitatory and inhibitory effects of conditioned stimuli (CS) based on learned associations, can influence instrumental performance. The behavioral test called Pavlovian to instrumental transfer (PIT) allows assessing the impact of such influence (Corbit and Balleine 2015). When tested in this paradigm, *Oprd1*$^{-/-}$ mice failed to increase their instrumental responding during presentation of the specific outcome-predicting stimulus (CS), proving a significant deficit in PIT (Laurent et al. 2012). Consistent with this, the DOR antagonist naltrindole abolished outcome-specific PIT in rats when injected systematically or into the shell, but not the core, of the NAc (Laurent et al. 2012, 2014). Remarkably, DOR-eGFP knockin mice trained for predictive Pavlovian responding displayed more DOR at the somatic membrane of CINs of the NAc shell. This increase correlated positively with conditioned response and later PIT performance, as well as with increased variance in action potential firing of CINs in the NAc shell (Bertran-Gonzalez et al. 2013). Connections between BLA and NAc shell are likely to be involved in this process, as their interruption causes severe impairment in outcome-specific PIT (Shiflett and Balleine 2010). Of note, BLA is one of the brain regions where DORs are most intensively expressed (Allen Brain Atlas: http://

www.brain-map.org/; Mansour et al. 1995; Scherrer et al. 2004). Together, these results thus point to a key role of DORs in modulating ongoing goal-directed behavior based on previous reward exposure.

As a conclusion, data collected from DOR null mice suggest that DOR exerts an inhibitory influence on motivation to obtain a food reward and possibly on food reward per se, but facilitates the influence of previous Pavlovian reward learning on instrumental choice performance. Invalidation of the *Oprd1* gene in restricted brain regions or neuronal populations would be useful to further explore the role of DORs in these processes. Interestingly, the notion that DOR activity may, under certain conditions, antagonize MOR-mediated effects on reward has emerged in the literature and questions the role of the former in drug addiction.

4.2.2 Drug Reward and Seeking

Animal studies using multiple models of drug exposure have drawn a complex, sometimes inconsistent, picture of DORs' role in drug reward and seeking. The dominant view appears to be that DORs would play a similar, although less critical, role than MORs in mediating these processes (Klenowski et al. 2015; Le Merrer et al. 2009). However two major concerns should be acknowledged that may have rendered functional dissociation between MORs and DORs in these processes particularly challenging. First, pharmacological tools available to target opioid receptors often lack specificity (Hutcheson et al. 2001; Scherrer et al. 2004). Second, a major difficulty when assessing drug reinforcement in animal models lies in the tight intertwining of reward and learning processes. Indeed, most animal models used to evaluate the rewarding properties of drugs also assess conditioned learning abilities (Stephens et al. 2010), and as such may notably involve HPC-dependent processes (Luo et al. 2011). Thus discrepancies between reports regarding the involvement of DORs in drug reward may reflect differential recruitment of learning processes depending on the experimental paradigm. We previously discussed the case of place preference studies, relying on drug/context associations, impaired in *Oprd1*$^{-/-}$ animals. Such impairment, however, does not exclude an effect on drug reward per se. We will discuss in this section the case of self-administration and drug seeking experiments. The former rely on operant learning, preserved in mice lacking DORs, and may thus provide useful information regarding the effects of DOR inactivation on drug reward. The latter involves both motivation for the drug and conditioning to various cues, and thus may illustrate the integrative role of DORs in these processes.

As regards self-administration studies, pharmacological investigations have evidenced a role for DORs in drug reinforcement/reward that depends on the drug tested (cocaine, nicotine, opiates, or alcohol), the route of administration (systemic, intracerebroventricular, and intracerebral), and, when relevant, the targeted brain region (Klenowski et al. 2015). Studies using DOR null mice confirmed discrepancies depending on the drug. Indeed, *Oprd1*$^{-/-}$ mice showed difficulties in acquiring cocaine or nicotine self-administration under a fixed ratio schedule of reinforcement (FR3 and FR1, respectively), reaching lower final rates of administration, and consistently achieved lower breakpoint under a progressive-ratio schedule (Berrendero et al. 2012; Gutierrez-Cuesta et al. 2014). However, *Oprd1* deletion did not prevent animals from self-administering morphine either systematically (Le Merrer et al. 2011) or into the VTA (David et al. 2008). Instead, increased breakpoints for intravenous morphine self-administration under a progressive-ratio schedule of reinforcement suggest a higher motivation for the drug in these animals (Le Merrer et al. 2011). Finally, *Oprd1*$^{-/-}$ mice self-administered more alcohol in a two-bottle choice paradigm (Roberts et al. 2001). Together, these studies suggest that cocaine and nicotine, and not morphine or alcohol, have diminished reinforcing properties in DOR null mice as compared to WT controls. Differences in drug-induced anxiety may account for these discrepancies. Cocaine and nicotine share psychostimulant properties, and as such can increase anxiety levels. These effects may detour *Oprd1*$^{-/-}$ animals, which are highly anxious under basal conditions (Filliol et al. 2000), from consuming these drugs but not narcotics, such as morphine or alcohol. Consistent with this hypothesis, a positive correlation was found between voluntary alcohol consumption in mutants and their levels of anxiety (Roberts et al. 2001). This result suggests that DOR null mice would self-administer alcohol at least in part to relieve their excessive anxiety. In this context, the study of Dlx-DOR mice (conditional DOR deletion in forebrain GABAergic neurons) represents a promising tool to disentangle increased motivation for drugs from relief of anxiety after *Oprd1* deletion, as these animals display low levels of anxiety as compared to controls but high motivation to reach food in the NSF test (Chu Sin Chung et al. 2015). Further studies using local/population-specific invalidation of *Oprd1* would be needed to better understand the role of DOR in drug self-administration.

As regards drug seeking, systemic pharmacological blockade and complete *Oprd1* knockout both result in decreased drug reinstatement. Systemic DOR antagonist administration reduced alcohol-seeking behavior elicited by drug-associated environmental stimuli in rats (Ciccocioppo et al. 2002; Marinelli et al. 2009), discrete cues (Marinelli et al. 2009), or yohimbine injections (Nielsen et al. 2012). Accordingly, *Oprd1*$^{-/-}$ mice displayed diminished cue-induced reinstatement of cocaine seeking following extinction. Furthermore, the enhancement of Fos expression triggered by cocaine reinstatement was attenuated in the dorsal striatum (CPu) and CA1 of these animals (Gutierrez-Cuesta et al. 2014). These data further document hippocampal dysfunction in *Oprd1*$^{-/-}$ mice and suggest that DOR activity in the HPC facilitates the influence of drug-paired cues to induce reinstatement of drug taking. DOR in the NAc, however, may play a different role. Indeed, intra-NAc administration of naltrindole failed to inhibit cocaine-primed reinstatement of cocaine seeking after extinction (Simmons and Self 2009) and significantly increased cue-induced cocaine-seeking behavior in rats following 24 h abstinence (Dikshtein et al. 2013). After 30 days of abstinence, DOR blockage had no longer effects on cocaine seeking by itself, but was able to prevent β-endorphin from repressing such seeking (Dikshtein et al. 2013). These last results unravel a braking activity of NAc DOR on motivation to obtain a drug of abuse, in agreement with data suggesting a similar effect on motivation for food (Le Merrer et al. 2013).

Additional studies will however be required to further explore the role of different brain populations of DOR in modulating motivation for natural or drug reinforcers.

4.3 Delta Opioid Receptor in the Striatum: Implications for Motor Function, Response Learning, Motivation, and Reward

4.3.1 Functional Role of Delta Opioid Receptors in the Striatum

Striatal regions, dorsal (CPu) and ventral (NAc), display high levels of DOR expression (see Fig. 1). In the mouse striatum, DOR transcripts were predominantly found in CINs, where receptor expression appears confined to the soma and proximal dendrites (Bertran-Gonzalez et al. 2013; Le Moine et al. 1994; Scherrer et al. 2006). Activation of DOR on these neurons produces a decrease in Ach release (Gazyakan et al. 2000). A small proportion of striatal DOR can also be detected in GABAergic (inter)neurons (Scherrer et al. 2006) and in presynaptic glutamatergic terminals (Jiang and North 1992). Finally, molecular phenotyping of MSNs expressing either D1 or D2 dopamine receptors (D1R and D2R, respectively) revealed a significant enrichment in *Oprd1* transcripts relative to the rest of the brain, but no difference between these two populations of MSNs (Heiman et al. 2008). DOR in the striatum is thus also present on MSNs, where they can fine-tune dopamine transmission (Fig. 3). Such distribution makes difficult to understand the functional consequences of DOR activation in the striatum, namely facilitation or inhibition of striatal outputs, and, eventually, which of these outputs, D1R-bearing striatonigral or D2R-bearing striatopallidal pathway, is affected. Genetically modified animals provided some cues to answer these questions.

Mice lacking DOR acquired faster a response strategy in a cross-maze and a motor skill on the accelerating rotarod (Le Merrer et al. 2013), suggesting that dorsal striatum activity would be eased in these animals (Durieux et al. 2012; Lovinger 2010). DOR activity in this region would therefore exert a braking influence on striatal function. Dorsal striatum, however, exerts a population-selective control over locomotion and motor control, D1R- and D2R-bearing MSNs being involved in distinct aspects of these functions. In order to test the reactivity of the striatonigral and striatopallidal pathways in *Oprd1*$^{-/-}$ mice, we assessed the effects of D1/D5 or D2/D3 dopamine receptors agonist administration on locomotor activity (that recruits preponderantly the dorsal part of the striatum and the NAc core) in these animals and their WT controls. We observed higher sensitivity to the locomotor stimulating effects of the D1/D5 agonist SKF-81297 in mutants. Together, our results suggest that dorsal striatal function in DOR null mice is biased towards facilitated D1R-expressing striatonigral output. Importantly, chronic naltrindole administration similarly facilitated the locomotor stimulant effects of SKF-81297 in WT animals, indicating that blocking DOR signaling is sufficient to facilitate striatonigral activity, independently from neurodevelopmental adaptations (Le Merrer et al. 2013). Finally, the locomotor effects of SKF-81297 were also found increased in Dlx-DOR mice, confirming the involvement of forebrain DOR in these processes (Chu Sin Chung et al. 2015). Together, these data point to an inhibitory role of DOR in the dorsal striatum on D1R-bearing MSNs, likely through

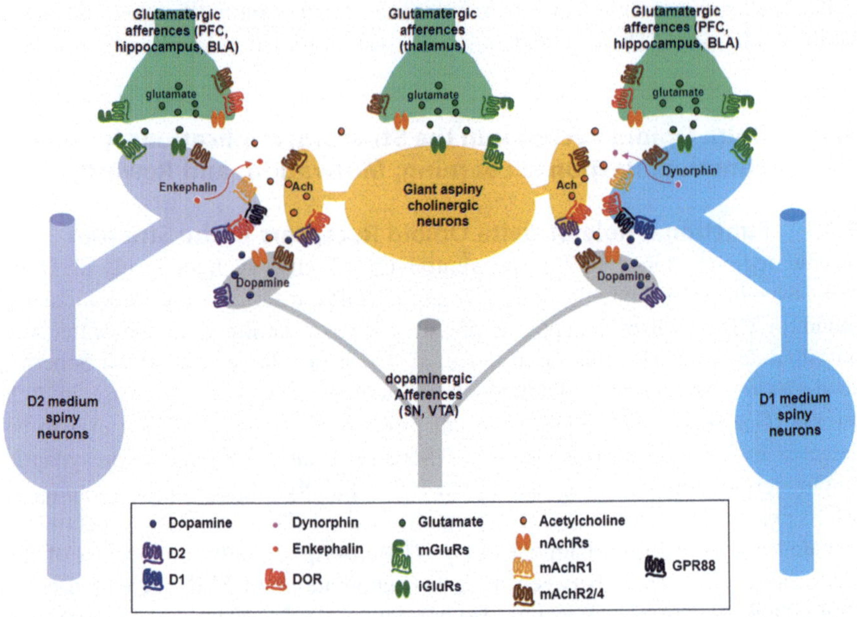

Fig. 3 Schematic representation of the localization of delta opioid receptors (DORs) and some potential G-protein-coupled receptor (GPCR) partners within local striatal microcircuitry. DORs are expressed at pre- or postsynaptic levels in most cellular types in the striatum, where they can interact functionally and/or physically with multiple other GPCRs, such as dopamine (D1R or D2R), muscarinic cholinergic (mAchRs), or GPR88 receptors. *BLA* basolateral amygdala, *iGluRs* ionotropic glutamate receptors, *mGluRs* metabotropic glutamate receptors, *nAchRs* nicotinic cholinergic receptors, *PFC* prefrontal cortex, *SN* substantia nigra, *VTA* ventral tegmental area

DOR activation at postsynaptic level (resulting in D1R-MSN hyperpolarization) although inhibition of excitatory glutamatergic afferences should also be considered (see Fig. 3). Of note, an effect at D2R-expressing MSNs cannot be excluded (Le Merrer et al. 2013).

The picture is different as regards DORs in the NAc, and notably in the shell subregion. Remarkably, Pavlovian conditioning increased DOR expression within the somatic membrane of CINs in the NAc shell of DOR-eGFP mice. This effect correlated with the level of conditioned responding and was accompanied by higher irregular/burst firing in CINs but no change in their action potential frequency (Bertran-Gonzalez et al. 2013). Increased burst firing variability would result in decreased Ach release at MSNs and thus reduce the Ach-induced bias towards cortical activation of D2R-bearing MSNs (Ding et al. 2010). Activation of DOR in the NAc shell following Pavlovian training could then indirectly facilitate the activity of D1R-bearing MSNs. Moreover, genetic deletion of *Oprd1* and systemic or intra-NAc shell injection of naltrindole abolished PIT in mice and rats, respectively (Laurent et al. 2012, 2014). PIT was similarly suppressed by intra-NAc shell pharmacological blockade of D1Rs, showing its dependence on D1R-bearing

MSNs (Laurent et al. 2014). These results indicate that, under conditions where DOR is highly expressed on CINs, their activation biases NAc shell function towards facilitated D1R-bearing MSN output. Which DOR tone, at CINs or at postsynaptic MSNs, prevails under basal conditions or following other forms of learning, however, remains to be explored.

In *Gpr88*$^{-/-}$ mice, DOR signaling is facilitated in the striatum (CPu and NAc), together with cholinergic and MOR activities (Meirsman et al. 2016). These animals thus represent a unique tool to assess the consequences of excessive DOR activity in this region. In the cross-maze, GPR88 null mice acquired earlier and better an allocentric strategy in a dual solution task and shifted sooner to a response strategy. After this shift had occurred, however, their performance started to decrease, suggesting that response learning was impaired in these animals. On the accelerating rotarod, *Gpr88*$^{-/-}$ mice completely failed to acquire a motor skill, consistent with blunted activity of D1R-expressing MSNs (Durieux et al. 2012). Chronic treatment with naltrindole restored acquisition of a motor skill in mutant mice, but surprisingly only at early stages, pointing to a restoration of D2R-bearing MSN activity under DOR blockade. Failure to maintain a high level of performance at later stages suggests that DOR blockade in this experiment was not sufficient to completely rescue D1R-bearing MSN function. Finally, *Gpr88*$^{-/-}$ mice were less sensitive to the locomotor stimulating effects of a D1R agonist, consistent with a repressive effect of DOR on D1R-bearing MSNs (Quintana et al. 2012). Together, data from *Gpr88* mutant mice suggest that excessive DOR signaling in the striatum inhibits the activity of D1R-expressing MSNs, and probably affects the D2R-expressing population of MSNs as well. Further investigation will be needed to assess NAc-dependent behavioral responses in these animals, such as motivation for food or drug reward.

4.3.2 Interactions with Other Striatal G-Protein-Coupled Receptors

The study of *Oprd1*$^{-/-}$, *Dlx5*/6-Cre x *Oprd1*$^{fl/fl}$, and *Gpr88*$^{-/-}$ mice suggests that dorsal striatal DOR inhibits the activity of D1R-expressing MSNs and may also affect the activity of D2R-bearing striatal outputs. These effects could be mediated through interactions at the level of striatal microcircuitry (Fig. 3), but may also involve direct interactions between DOR and D1R or D2R in neurons where they are co-expressed. We challenged the existence of such interactions using Bioluminescence resonance energy transfer (BRET) in heterologous cells. Remarkably, DOR appears to interact closely with D2R, suggesting the existence of potential heterodimers between these two GPCRs, but not with D1R (Fig. 4). DOR and D2R co-localize in CINs and in D2R-bearing MSNs (Fig. 3) (Ambrose et al. 2006; Calabresi et al. 2014; Heiman et al. 2008; Le Moine et al. 1994). Additional experiments will be needed to assess the pharmacological consequences of DOR and D2R co-expression. However, the rescue of D2R-MSN dependent early motor skill learning that we observed after naltrindole administration in *Gpr88*$^{-/-}$ mice points to a direct inhibitory influence of DOR on D2R signaling. This inhibition could occur at CINs by preventing D2R activation from repressing ACh release and/or directly at postsynaptic MSNs by counteracting the hyperpolarizing effect of

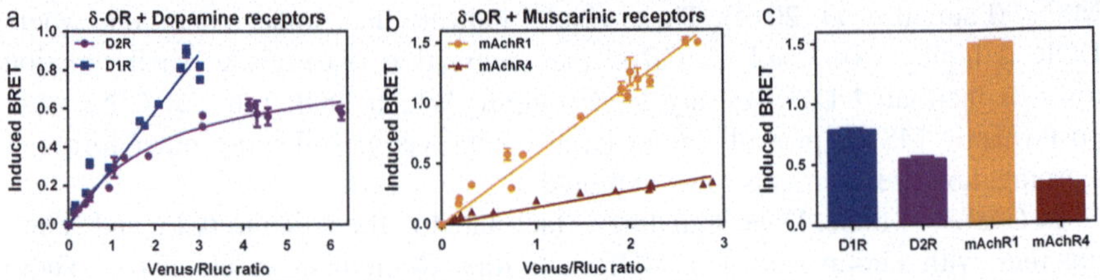

Fig. 4 Interaction between DOR and dopaminergic and cholinergic GPCRs. Twenty nanograms of DOR-Rluc8-pcDNA3 plasmids were co-expressed with increasing amount (10–120 ng) of GPCRs-Venus-pcDNA3 plasmids ($n = 3$ per condition) in HEK293FT cells to study the physical interaction of DOR with GPCR partners by Bioluminescence Resonance Energy Transfer (BRET). (**a**) BRET signals displayed specific and saturated curves with DOR-D2R whereas signals remained unsaturated with DOR-D1R co-expression. (**b**) BRET signals were not saturated with cholinergic mAchR1 or mAchR4. (**c**) DOR and mAchR1 co-expression results in remarkably high levels of energy transfer, suggesting that these receptors randomly (not as heterodimers) co-localize in the same confined cellular compartment

D2R stimulation. Interestingly, a similar mechanism could account for a trend in reduced inhibitory effects of a D2R agonist on locomotion in *Oprd1*$^{-/-}$ mice (Le Merrer et al. 2013). Further work will be required, though, to understand the molecular substrate of DOR/D1R-D2R interactions and their role in striatal function.

Dopamine receptors are obviously not the only GPCRs likely to interact directly with DORs at the striatal level. Importantly, DOR signaling was shown to modulate cholinergic tone in this region. Indeed, presynaptic DOR can inhibit Ach release in the rat striatum (Mulder et al. 1984). DORs are also abundant on CINs, where their activation should similarly reduce Ach release by hyperpolarizing these neurons. The pharmacological blockade of DOR in the shell of the NAc was shown to suppress D1R-dependent PIT in mice (Laurent et al. 2014), likely by facilitating Ach release and, consequently, D2R-MSN activity (Ding et al. 2010). Interestingly, this inhibitory effect of intra-NAc shell naltrindole on PIT was prevented by systemic administration of the M4 muscarinic cholinergic receptor (mAchR4) antagonist MT3. Although this effect of MT3 could involve postsynaptic competition for adenylate cyclase recruitment (Laurent et al. 2014), one should not exclude possible direct interactions between DOR and cholinergic GPCRs (mAchRs). We thus assessed the existence of such interactions with mAchR1 or mAchr4, both highly expressed in the striatum, in heterologous cells. We were not able to detect direct interactions (Fig. 4). However, the remarkably high levels of BRET measured between DORs and mAchr1s suggest that these receptors may randomly co-localize (not as heterodimers) in the same confined cellular compartment. Such close cellular proximity suggests that these two receptors may likely interact at functional level to modulate striatal activity.

Finally, GPR88 may also represent a direct molecular partner of DOR. These orphan receptors are among the most densely expressed GPCRs in the striatum (Ghate et al. 2007; Logue et al. 2009; Massart et al. 2009). *Gpr88* transcripts are detected in MSNs (Massart et al. 2009), with significant enrichment in D2R-bearing projections (Heiman et al. 2008). In *Gpr88*$^{-/-}$ mice, we evidenced increased DOR signaling in membrane preparations from striatal samples and a remarkable normalization of most of their behavioral features by systemic blockade of DOR (Meirsman et al. 2016). These results suggest that, under physiological conditions, GPR88 acts as a brake on DOR activity to regulate behavior. GPR88 influence at DOR activity could operate either at circuit level, or through functional competition at the level of downstream effectors within neurons, or via direct, possibly physical, interactions between these receptors. Preliminary data in our lab suggest the existence of such direct interactions. Future work will aim at assessing the pharmacological consequences of DOR/GPR88 co-expression in cells and try to understand how their interaction could contribute to the influence of DORs on striatal output balance.

4.3.3 Delta Opioid Receptors and Striatal Gene Expression

Using qRT-PCR, we assessed the levels of expression of 67 genes in the CPu and NAc of *Oprd1*$^{-/-}$ mice to identify potential molecular partners of DORs in these regions (Le Merrer et al. 2013). This analysis revealed that *Oprd1* deletion had different transcriptional consequences in these two regions, with only two genes showing commonly (up)-regulated expression (*Slc6a11* and *Grm4* – coding the GABA transporter mGAT4 and metabotropic glutamate receptor mGluR4, respectively). Interestingly, transcriptional regulations of several genes in the CPu were coherent with behavioral data pointing to facilitated D1R- and blunted D2R-bearing MSN activity in DOR null mice. Indeed, low mRNA levels of *Camk2* and *chrm4* (coding for the alpha isoform of the calcium/calmodulin-dependent protein kinase II and mAchR4, respectively) and high mRNA levels of *grin2b* (NR2B subunit of NMDA glutamate receptors) could facilitate striatonigral outputs in mutants (Gomeza et al. 1999; Guo et al. 2010; Jocoy et al. 2011; Tzavara et al. 2004). In contrast, increased expression of *Grm4* (metabotropic glutamate receptor mGluR4) and *Pdyn* (prodynorphin) and decreased expression of *Tac1* (substance P) and *Gpr6* (GPR6) would rather inhibit striatopallidal activity (Govindaiah et al. 2010; Hopkins et al. 2009; Lobo et al. 2007; Perreault et al. 2007). Of note, downregulated expression of *Bcl11b* (*Ctip2*) may represent the triggering factor for decreased expression of other MSN marker genes, such as *Foxp1* and *Chrm4* (Arlotta et al. 2008). In the NAc, and not the CPu, several genes coding major actors of Ach and monoamine degradation (*Ache*, coding acetylcholinesterase and *Maoa* – monoamine oxidase a) or their extraction from the synaptic cleft (*Slc6a4*, serotonin transporter SERT) displayed upregulated expression. These results suggest that a brake on Ach/monoamine neurotransmission is lost in *Oprd1*$^{-/-}$ mice, requiring compensatory mechanisms. Therefore, DOR in the NAc appears to exert a tonic inhibition on these systems. Altogether, these data point to a crucial role of DOR in regulating striatal functions that differ between dorsal and ventral regions.

We similarly explored the transcriptional consequences of *Gpr88* genetic invalidation for 92 genes by qRT-PCR (Meirsman et al. 2016). Remarkably, the expression of *Oprd1* was downregulated in the NAc, and not regulated in the CPu, of

Gpr88$^{-/-}$ mice, whereas DOR activity, assessed by [^{35}S]-GTPγS binding, was increased in the whole striatum. These results suggest that either increased DOR activity is restricted to the CPu in these animals, and does not involve increased gene expression, or excessive DOR activation triggers a negative feedback mechanism in both the CPu and NAc, the latter being more sensitive than the former. Of note, downregulated expression of *Rgs4*, coding Regulator of G protein signaling 4 – RGS4, in *Gpr88*$^{-/-}$ mutants suggests a close interaction between this protein and GPR88 in the striatum. Interestingly, RGS4 was shown to inhibit opioid signaling (Georgoussi et al. 2006) and may thus participate in mediating the inhibitory effects of GPR88 activation on DOR signaling. Additional investigations will be needed to better assess DOR protein levels in the absence of GPR88, such as radioactive binding using DOR selective compounds, for example.

4.3.4 Influence of Hippocampal Delta Opioid Receptors on Striatal Function

Not only striatal DOR may be involved in the control of striatal-dependent behaviors but extrastriatal DORs as well. Indeed, previous studies have evidenced a functional antagonism between the hippocampal formation and the striatum, with the dorsal HPC exerting an inhibitory influence on the dorsal striatum, whereas the ventral HPC would facilitate the activity of the ventral striatum (Yin and Meck 2014). Consequently, impaired dorsal hippocampal function in *Oprd1*$^{-/-}$ mice may ease the acquisition of response and motor skill learning tasks by biasing hippocampo-striatal balance in favor of the dorsal striatum (Ciamei and Morton 2009; Packard and McGaugh 1996; Schroeder et al. 2002). Interestingly, behavioral, pharmacological, and transcriptional data collected from *Oprd1*$^{-/-}$ and *Gpr88*$^{-/-}$ mice point a critical role for DOR in controlling the hippocampo-striatal balance, with major consequences on HPC- versus striatum-dependent learning processes (Fig. 5). Whether such role would also apply to a ventral hippocampo-accumbal balance (Hart et al. 2014) will require further investigation.

5 Conclusions and Clinical Perspectives

Over the last 15 years, in vivo pharmacology and genetically modified animals have allowed to identify a unique, original implication of DOR in high order cognitive processes, motor function, mood and emotional responses. We focused here on the involvement of this receptor in modulating learning and memory processes, motor function, and reward/motivation, notably by regulating the balance between hippocampal and striatal functions. At dorsal level, such balance ensures optimal shift between associative HPC-dependent and procedural striatum-dependent learning processes, with crucial implications for cognitive performance and motor function. In this context, pharmacological ligands selective for DOR may represent precious therapeutic tools to relieve pathologies where the hippocampo-striatal balance is compromised, such as neurodegenerative diseases affecting either the hippocampal formation or the striatum (Alzheimer disease, Parkinson disease, or Huntington disease, for example). At ventral level, DORs may contribute to a ventral HPC to

Fig. 5 DORs modulate the dorsal hippocampo-striatal balance. (**a**) Under physiological conditions, hippocampal formation and striatum compete to ensure optimal control over learning processes. (**b**) In mice lacking delta opioid receptors, hippocampus–striatum balance is tilted towards facilitated striatal function, as revealed by impaired performance in dorsal hippocampus-dependent tasks (associative learning) but facilitated acquisition of dorsal striatum-dependent tasks (procedural learning). (**c**) Conversely, the hippocampo-striatal balance is biased towards eased dorsal hippocampal-dependent processes (associative learning) and deficient dorsal striatal function (procedural learning) in mice lacking GPR88 receptors, which display increased DOR activity in the striatum. Whether DORs can similarly modulate a more ventral hippocampo-accumbal balance will deserve further investigation

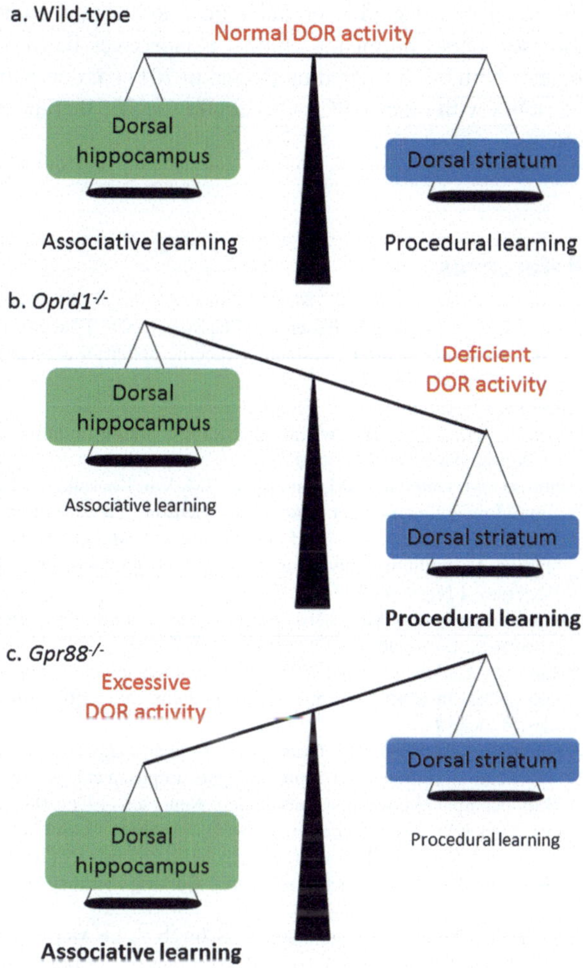

NAc crosstalk. Remarkably, within this circuit, DORs appear less involved in mediating reward processes per se than in controlling the consequences of previous reward experience on ongoing behavior. Therapeutic applications in the field of addiction thus involve the development of DOR antagonists to suppress conditioned responses to drug cues, with obvious benefit for the relief of withdrawal symptoms, reduction of drug seeking, and prevention of relapse. A caveat should be quoted here, however, as biasing the hippocampo-striatal balance towards one functional system may be detrimental to the other, as seen for spatial versus motor learning in *Oprd1*$^{-/-}$ and *Gpr88*$^{-/-}$ animals. Moreover, DORs are also involved in controlling anxiety levels and epileptogenic thresholds (Chu Sin Chung and Kieffer 2013), making them a delicate target to manipulate for therapeutic purpose. These limitations highlight the need for developing innovative pharmacological strategies to allow the targeting of specific populations of receptors, either in restricted areas of the brain or selected neuronal types, and obtain optimized treatments for CNS diseases. Other promising clinical perspectives lie in the selection of either DOR ligands with biased signaling (Kenakin 2011) or compounds targeting heterodimers of DORs with other GPCRs, to obtain specific therapeutic action with limited side effects.

References

Aiba A, Chen C, Herrup K, Rosenmund C, Stevens CF, Tonegawa S (1994) Reduced hippocampal long-term potentiation and context-specific deficit in associative learning in mGluR1 mutant mice. Cell 79:365–375

Alstermark B, Ekerot CF (2013) The lateral reticular nucleus: a precerebellar centre providing the cerebellum with overview and integration of motor functions at systems level. A new hypothesis. J Physiol 591:5453–5458

Ambrose LM, Gallagher SM, Unterwald EM, Van Bockstaele EJ (2006) Dopamine-D1 and delta-opioid receptors co-exist in rat striatal neurons. Neurosci Lett 399:191–196

Arlotta P, Molyneaux BJ, Jabaudon D, Yoshida Y, Macklis JD (2008) Ctip2 controls the differentiation of medium spiny neurons and the establishment of the cellular architecture of the striatum. J Neurosci 28:622–632

Aupperle RL, Paulus MP (2010) Neural systems underlying approach and avoidance in anxiety disorders. Dialogues Clin Neurosci 12:517–531

Baldo BA, Kelley AE (2007) Discrete neurochemical coding of distinguishable motivational processes: insights from nucleus accumbens control of feeding. Psychopharmacology (Berl) 191:439–459

Bannerman DM, Yee BK, Good MA, Heupel MJ, Iversen SD, Rawlins JN (1999) Double dissociation of function within the hippocampus: a comparison of dorsal, ventral, and complete hippocampal cytotoxic lesions. Behav Neurosci 113:1170–1188

Bannerman DM, Niewoehner B, Lyon L, Romberg C, Schmitt WB, Taylor A, Sanderson DJ, Cottam J, Sprengel R, Seeburg PH, Kohr G, Rawlins JN (2008) NMDA receptor subunit NR2A is required for rapidly acquired spatial working memory but not incremental spatial reference memory. J Neurosci 28:3623–3630

Becker JA, Clesse D, Spiegelhalter C, Schwab Y, Le Merrer J, Kieffer BL (2014) Autistic-like syndrome in mu opioid receptor null mice is relieved by facilitated mGluR4 activity. Neuropsychopharmacology 39:2049–2060

Berrendero F, Plaza-Zabala A, Galeote L, Flores A, Bura SA, Kieffer BL, Maldonado R (2012) Influence of delta-opioid receptors in the behavioral effects of nicotine. Neuropsychopharmacology 37:2332–2344

Bertran-Gonzalez J, Laurent V, Chieng BC, Christie MJ, Balleine BW (2013) Learning-related translocation of delta-opioid receptors on ventral striatal cholinergic interneurons mediates choice between goal-directed actions. J Neurosci 33:16060–16071

Bodnar RJ (2004) Endogenous opioids and feeding behavior: a 30-year historical perspective. Peptides 25:697–725

Bodnar RJ, Lamonte N, Israel Y, Kandov Y, Ackerman TF, Khalmova E (2005) Reciprocal opioid-opioid interactions between the ventral tegmental area and nucleus accumbens regions in mediating mu agonist-induced feeding in rats. Peptides 26:621–629

Bramham CR, Milgram NW, Srebro B (1991) Delta opioid receptor activation is required to induce LTP of synaptic transmission in the lateral perforant path in vivo. Brain Res 567:42–50

Buzas B, Rosenberger J, Cox BM (1998) Ca2+/calmodulin-dependent transcriptional activation of delta-opioid receptor gene expression induced by membrane depolarization in NG108-15 cells. J Neurochem 70:105–112

Calabresi P, Picconi B, Tozzi A, Ghiglieri V, Di Filippo M (2014) Direct and indirect pathways of basal ganglia: a critical reappraisal. Nat Neurosci 17:1022–1030
Castane A, Theobald DE, Robbins TW (2010) Selective lesions of the dorsomedial striatum impair serial spatial reversal learning in rats. Behav Brain Res 210:74–83
Cen B, Yu Q, Guo J, Wu Y, Ling K, Cheng Z, Ma L, Pei G (2001) Direct binding of beta-arrestins to two distinct intracellular domains of the delta opioid receptor. J Neurochem 76:1887–1894
Chavkin C, Shoemaker WJ, McGinty JF, Bayon A, Bloom FE (1985) Characterization of the prodynorphin and proenkephalin neuropeptide systems in rat hippocampus. J Neurosci 5:808–816
Chefer VI, Shippenberg TS (2009) Augmentation of morphine-induced sensitization but reduction in morphine tolerance and reward in delta-opioid receptor knockout mice. Neuropsychopharmacology 34:887–898
Childers SR (1991) Opioid receptor-coupled second messenger systems. Life Sci 48:1991–2003
Chu Sin Chung P, Kieffer BL (2013) Delta opioid receptors in brain function and diseases. Pharmacol Ther 140:112–120
Chu Sin Chung P, Keyworth HL, Martin-Garcia E, Charbogne P, Darcq E, Bailey A, Filliol D, Matifas A, Scherrer G, Ouagazzal AM, Gaveriaux-Ruff C, Befort K, Maldonado R, Kitchen I, Kieffer BL (2015) A novel anxiogenic role for the delta opioid receptor expressed in GABAergic forebrain neurons. Biol Psychiatry 77:404–415
Ciamei A, Morton AJ (2009) Progressive imbalance in the interaction between spatial and procedural memory systems in the R6/2 mouse model of Huntington's disease. Neurobiol Learn Mem 92:417–428
Ciccocioppo R, Martin-Fardon R, Weiss F (2002) Effect of selective blockade of mu(1) or delta opioid receptors on reinstatement of alcohol-seeking behavior by drug-associated stimuli in rats. Neuropsychopharmacology 27:391–399
Corbit LH, Balleine BW (2015) Learning and motivational processes contributing to Pavlovian-instrumental transfer and their neural bases: dopamine and beyond. Curr Top Behav Neurosci 27:259–289
Crain BJ, Chang KJ, McNamara JO (1986) Quantitative autoradiographic analysis of mu and delta opioid binding sites in the rat hippocampal formation. J Comp Neurol 246:170–180
Cunningham CL, Groblewski PA, Voorhees CM (2011) Place conditioning. In: Olmstead MC (ed) Animal models of addiction (neuromethods). Humana Press, Totowa, pp 167–190
Dang MT, Yokoi F, Yin HH, Lovinger DM, Wang Y, Li Y (2006) Disrupted motor learning and long-term synaptic plasticity in mice lacking NMDAR1 in the striatum. Proc Natl Acad Sci U S A 103:15254–15259
David V, Matifas A, Gavello-Baudy S, Decorte L, Kieffer BL, Cazala P (2008) Brain regional Fos expression elicited by the activation of mu- but not delta-opioid receptors of the ventral tegmental area: evidence for an implication of the ventral thalamus in opiate reward. Neuropsychopharmacology 33:1746–1759
De Leonibus E, Pascucci T, Lopez S, Oliverio A, Amalric M, Mele A (2007) Spatial deficits in a mouse model of Parkinson disease. Psychopharmacology (Berl) 194:517–525
Deipolyi AR, Fang S, Palop JJ, Yu GQ, Wang X, Mucke L (2008) Altered navigational strategy use and visuospatial deficits in hAPP transgenic mice. Neurobiol Aging 29:253–266
Dickerson BC, Eichenbaum H (2010) The episodic memory system: neurocircuitry and disorders. Neuropsychopharmacology 35:86–104
Dikshtein Y, Barnea R, Kronfeld N, Lax E, Roth-Deri I, Friedman A, Gispan I, Elharrar E, Levy S, Ben-Tzion M, Yadid G (2013) Beta-endorphin via the delta opioid receptor is a major factor in the incubation of cocaine craving. Neuropsychopharmacology 38:2508–2514
Ding JB, Guzman JN, Peterson JD, Goldberg JA, Surmeier DJ (2010) Thalamic gating of corticostriatal signaling by cholinergic interneurons. Neuron 67:294–307
Do J, Kim JI, Bakes J, Lee K, Kaang BK (2012) Functional roles of neurotransmitters and neuromodulators in the dorsal striatum. Learn Mem 20:21–28

Durieux PF, Bearzatto B, Guiducci S, Buch T, Waisman A, Zoli M, Schiffmann SN, de Kerchove d'Exaerde A (2009) D2R striatopallidal neurons inhibit both locomotor and drug reward processes. Nat Neurosci 12:393–395

Durieux PF, Schiffmann SN, de Kerchove d'Exaerde A (2012) Differential regulation of motor control and response to dopaminergic drugs by D1R and D2R neurons in distinct dorsal striatum subregions. EMBO J 31:640–653

Ennaceur A (2010) One-trial object recognition in rats and mice: methodological and theoretical issues. Behav Brain Res 215:244–254

Ennaceur A, Neave N, Aggleton JP (1997) Spontaneous object recognition and object location memory in rats: the effects of lesions in the cingulate cortices, the medial prefrontal cortex, the cingulum bundle and the fornix. Exp Brain Res 113:509–519

Erbs E, Faget L, Scherrer G, Matifas A, Filliol D, Vonesch JL, Koch M, Kessler P, Hentsch D, Birling MC, Koutsourakis M, Vasseur L, Veinante P, Kieffer BL, Massotte D (2015) A mu-delta opioid receptor brain atlas reveals neuronal co-occurrence in subcortical networks. Brain Struct Funct 220:677–702

Faget L, Erbs E, Le Merrer J, Scherrer G, Matifas A, Benturquia N, Noble F, Decossas M, Koch M, Kessler P, Vonesch JL, Schwab Y, Kieffer BL, Massotte D (2012) In vivo visualization of delta opioid receptors upon physiological activation uncovers a distinct internalization profile. J Neurosci 32:7301–7310

Fenalti G, Giguere PM, Katritch V, Huang XP, Thompson AA, Cherezov V, Roth BL, Stevens RC (2014) Molecular control of delta-opioid receptor signalling. Nature 506:191–196

Filliol D, Ghozland S, Chluba J, Martin M, Matthes HW, Simonin F, Befort K, Gaveriaux-Ruff C, Dierich A, LeMeur M, Valverde O, Maldonado R, Kieffer BL (2000) Mice deficient for delta- and mu-opioid receptors exhibit opposing alterations of emotional responses. Nat Genet 25:195–200

Gazyakan E, Hennegriff M, Haaf A, Landwehrmeyer GB, Feuerstein TJ, Jackisch R (2000) Characterization of opioid receptor types modulating acetylcholine release in septal regions of the rat brain. Naunyn Schmiedebergs Arch Pharmacol 362:32–40

Georgoussi Z, Leontiadis L, Mazarakou G, Merkouris M, Hyde K, Hamm H (2006) Selective interactions between G protein subunits and RGS4 with the C-terminal domains of the mu- and delta-opioid receptors regulate opioid receptor signaling. Cell Signal 18:771–782

Ghate A, Befort K, Becker JA, Filliol D, Bole-Feysot C, Demebele D, Jost B, Koch M, Kieffer BL (2007) Identification of novel striatal genes by expression profiling in adult mouse brain. Neuroscience 146:1182–1192

Ghiglieri V, Sgobio C, Costa C, Picconi B, Calabresi P (2011) Striatum-hippocampus balance: from physiological behavior to interneuronal pathology. Prog Neurobiol 94:102–114

Ghozland S, Matthes HW, Simonin F, Filliol D, Kieffer BL, Maldonado R (2002) Motivational effects of cannabinoids are mediated by mu-opioid and kappa-opioid receptors. J Neurosci 22:1146–1154

Giralt A, Saavedra A, Carreton O, Arumi H, Tyebji S, Alberch J, Perez-Navarro E (2013) PDE10 inhibition increases GluA1 and CREB phosphorylation and improves spatial and recognition memories in a Huntington's disease mouse model. Hippocampus 23:684–695

Gomeza J, Zhang L, Kostenis E, Felder C, Bymaster F, Brodkin J, Shannon H, Xia B, Deng C, Wess J (1999) Enhancement of D1 dopamine receptor-mediated locomotor stimulation in M (4) muscarinic acetylcholine receptor knockout mice. Proc Natl Acad Sci U S A 96:10483–10488

Govindaiah G, Wang Y, Cox CL (2010) Substance P selectively modulates GABA(A) receptor-mediated synaptic transmission in striatal cholinergic interneurons. Neuropharmacology 58:413–422

Graybiel AM (2008) Habits, rituals, and the evaluative brain. Annu Rev Neurosci 31:359–387

Guo ML, Mao LM, Wang JQ (2010) Modulation of M4 muscarinic acetylcholine receptors by interacting proteins. Neurosci Bull 26:469–473

Gutierrez-Cuesta J, Burokas A, Mancino S, Kummer S, Martin-Garcia E, Maldonado R (2014) Effects of genetic deletion of endogenous opioid system components on the reinstatement of cocaine-seeking behavior in mice. Neuropsychopharmacology 39:2974–2988

Hart G, Leung BK, Balleine BW (2014) Dorsal and ventral streams: the distinct role of striatal subregions in the acquisition and performance of goal-directed actions. Neurobiol Learn Mem 108:104–118

Hasselmo ME, Sarter M (2011) Modes and models of forebrain cholinergic neuromodulation of cognition. Neuropsychopharmacology 36:52–73

Heiman M, Schaefer A, Gong S, Peterson JD, Day M, Ramsey KE, Suarez-Farinas M, Schwarz C, Stephan DA, Surmeier DJ, Greengard P, Heintz N (2008) A translational profiling approach for the molecular characterization of CNS cell types. Cell 135:738–748

Henry B, Fox SH, Crossman AR, Brotchie JM (2001) Mu- and delta-opioid receptor antagonists reduce levodopa-induced dyskinesia in the MPTP-lesioned primate model of Parkinson's disease. Exp Neurol 171:139–146

Hopkins CR, Lindsley CW, Niswender CM (2009) mGluR4-positive allosteric modulation as potential treatment for Parkinson's disease. Future Med Chem 1:501–513

Hutcheson DM, Matthes HW, Valjent E, Sanchez-Blazquez P, Rodriguez-Diaz M, Garzon J, Kieffer BL, Maldonado R (2001) Lack of dependence and rewarding effects of deltorphin II in mu-opioid receptor-deficient mice. Eur J Neurosci 13:153–161

Ilyutchenok RY, Dubrovina NI (1995) Memory retrieval enhancement by kappa opioid agonist and mu, delta antagonists. Pharmacol Biochem Behav 52:683–687

Inui T, Shimura T (2014) Delta-opioid receptor blockade in the ventral pallidum increases perceived palatability and consumption of saccharin solution in rats. Behav Brain Res 269:20–27

Jacobson TK, Gruenbaum BF, Markus EJ (2012) Extensive training and hippocampus or striatum lesions: effect on place and response strategies. Physiol Behav 105:645–652

Jiang ZG, North RA (1992) Pre- and postsynaptic inhibition by opioids in rat striatum. J Neurosci 12:356–361

Jo J, Son GH, Winters BL, Kim MJ, Whitcomb DJ, Dickinson BA, Lee YB, Futai K, Amici M, Sheng M, Collingridge GL, Cho K (2010) Muscarinic receptors induce LTD of NMDAR EPSCs via a mechanism involving hippocalcin, AP2 and PSD-95. Nat Neurosci 13:1216–1224

Jocoy EL, Andre VM, Cummings DM, Rao SP, Wu N, Ramsey AJ, Caron MG, Cepeda C, Levine MS (2011) Dissecting the contribution of individual receptor subunits to the enhancement of N-methyl-d-aspartate currents by dopamine D1 receptor activation in striatum. Front Syst Neurosci 5:28

Jutkiewicz EM, Rice KC, Woods JH, Winsauer PJ (2003) Effects of the delta-opioid receptor agonist SNC80 on learning relative to its antidepressant-like effects in rats. Behav Pharmacol 14:509–516

Jutkiewicz EM, Kaminsky ST, Rice KC, Traynor JR, Woods JH (2005) Differential behavioral tolerance to the delta-opioid agonist SNC80 ([(+)-4-[(alphaR)-alpha-[(2S,5R)-2,5-dimethyl-4-(2-propenyl)-1-piperazinyl]-(3-me thoxyphenyl)methyl]-N,N-diethylbenzamide) in Sprague-Dawley rats. J Pharmacol Exp Ther 315:414–422

Katsuura Y, Taha SA (2014) Mu opioid receptor antagonism in the nucleus accumbens shell blocks consumption of a preferred sucrose solution in an anticipatory contrast paradigm. Neuroscience 261:144–152

Kenakin T (2011) Functional selectivity and biased receptor signaling. J Pharmacol Exp Ther 336:296–302

Khaimova E, Kandov Y, Israel Y, Cataldo G, Hadjimarkou MM, Bodnar RJ (2004) Opioid receptor subtype antagonists differentially alter GABA agonist-induced feeding elicited from either the nucleus accumbens shell or ventral tegmental area regions in rats. Brain Res 1026:284–294

Kitamura T, Pignatelli M, Suh J, Kohara K, Yoshiki A, Abe K, Tonegawa S (2014) Island cells control temporal association memory. Science 343:896–901

Kitchen I, Slowe SJ, Matthes HW, Kieffer B (1997) Quantitative autoradiographic mapping of mu-, delta- and kappa-opioid receptors in knockout mice lacking the mu-opioid receptor gene. Brain Res 778:73–88

Klenowski P, Morgan M, Bartlett SE (2015) The role of delta-opioid receptors in learning and memory underlying the development of addiction. Br J Pharmacol 172:297–310

Ko JL, Arvidsson U, Williams FG, Law PY, Elde R, Loh HH (1999) Visualization of time-dependent redistribution of delta-opioid receptors in neuronal cells during prolonged agonist exposure. Brain Res Mol Brain Res 69:171–185

Korotkova T, Fuchs EC, Ponomarenko A, von Engelhardt J, Monyer H (2010) NMDA receptor ablation on parvalbumin-positive interneurons impairs hippocampal synchrony, spatial representations, and working memory. Neuron 68:557–569

Kovoor A, Nappey V, Kieffer BL, Chavkin C (1997) Mu and delta opioid receptors are differentially desensitized by the coexpression of beta-adrenergic receptor kinase 2 and beta-arrestin 2 in xenopus oocytes. J Biol Chem 272:27605–27611

Laurent V, Leung B, Maidment N, Balleine BW (2012) Mu- and delta-opioid-related processes in the accumbens core and shell differentially mediate the influence of reward-guided and stimulus-guided decisions on choice. J Neurosci 32:1875–1883

Laurent V, Bertran-Gonzalez J, Chieng BC, Balleine BW (2014) Delta-opioid and dopaminergic processes in accumbens shell modulate the cholinergic control of predictive learning and choice. J Neurosci 34:1358–1369

Laurent V, Morse AK, Balleine BW (2015) The role of opioid processes in reward and decision-making. Br J Pharmacol 172:449–459

Le Bourdonnec B, Windh RT, Ajello CW, Leister LK, Gu M, Chu GH, Tuthill PA, Barker WM, Koblish M, Wiant DD, Graczyk TM, Belanger S, Cassel JA, Feschenko MS, Brogdon BL, Smith SA, Christ DD, Derelanko MJ, Kutz S, Little PJ, DeHaven RN, DeHaven-Hudkins DL, Dolle RE (2008) Potent, orally bioavailable delta opioid receptor agonists for the treatment of pain: discovery of N,N-diethyl-4-(5-hydroxyspiro[chromene-2,4′-piperidine]-4-yl)benzamide (ADL5859). J Med Chem 51:5893–5896

Le Bourdonnec B, Windh RT, Leister LK, Zhou QJ, Ajello CW, Gu M, Chu GH, Tuthill PA, Barker WM, Koblish M, Wiant DD, Graczyk TM, Belanger S, Cassel JA, Feschenko MS, Brogdon BL, Smith SA, Derelanko MJ, Kutz S, Little PJ, DeHaven RN, DeHaven-Hudkins DL, Dolle RE (2009) Spirocyclic delta opioid receptor agonists for the treatment of pain: discovery of N,N-diethyl-3-hydroxy-4-(spiro[chromene-2,4′-piperidine]-4-yl) benzamide (ADL5747). J Med Chem 52:5685–5702

Le Merrer J, Becker JA, Befort K, Kieffer BL (2009) Reward processing by the opioid system in the brain. Physiol Rev 89:1379–1412

Le Merrer J, Plaza-Zabala A, Del Boca C, Matifas A, Maldonado R, Kieffer BL (2011) Deletion of the delta opioid receptor gene impairs place conditioning but preserves morphine reinforcement. Biol Psychiatry 69:700–703

Le Merrer J, Faget L, Matifas A, Kieffer BL (2012) Cues predicting drug or food reward restore morphine-induced place conditioning in mice lacking delta opioid receptors. Psychopharmacology (Berl) 223:99–106

Le Merrer J, Rezai X, Scherrer G, Becker JA, Kieffer BL (2013) Impaired hippocampus-dependent and facilitated striatum-dependent behaviors in mice lacking the delta opioid receptor. Neuropsychopharmacology 38:1050–1059

Le Moine C, Kieffer B, Gaveriaux-Ruff C, Befort K, Bloch B (1994) Delta-opioid receptor gene expression in the mouse forebrain: localization in cholinergic neurons of the striatum. Neuroscience 62:635–640

Levine AS, Grace M, Portoghese PS, Billington CJ (1994) The effect of selective opioid antagonists on butorphanol-induced feeding. Brain Res 637:242–248

Lobo MK, Cui Y, Ostlund SB, Balleine BW, Yang XW (2007) Genetic control of instrumental conditioning by striatopallidal neuron-specific S1P receptor Gpr6. Nat Neurosci 10:1395–1397

Logue SF, Grauer SM, Paulsen J, Graf R, Taylor N, Sung MA, Zhang L, Hughes Z, Pulito VL, Liu F, Rosenzweig-Lipson S, Brandon NJ, Marquis KL, Bates B, Pausch M (2009) The orphan GPCR, GPR88, modulates function of the striatal dopamine system: a possible therapeutic target for psychiatric disorders? Mol Cell Neurosci 42:438–447

Longoni R, Cadoni C, Mulas A, Di Chiara G, Spina L (1998) Dopamine-dependent behavioural stimulation by non-peptide delta opioids BW373U86 and SNC 80: 2. Place-preference and brain microdialysis studies in rats. Behav Pharmacol 9:9–14

Lovinger DM (2010) Neurotransmitter roles in synaptic modulation, plasticity and learning in the dorsal striatum. Neuropharmacology 58:951–961

Luo AH, Tahsili-Fahadan P, Wise RA, Lupica CR, Aston-Jones G (2011) Linking context with reward: a functional circuit from hippocampal CA3 to ventral tegmental area. Science 333:353–357

Lupica CR (1995) Delta and mu enkephalins inhibit spontaneous GABA-mediated IPSCs via a cyclic AMP-independent mechanism in the rat hippocampus. J Neurosci 15:737–749

Mabrouk OS, Marti M, Salvadori S, Morari M (2009) The novel delta opioid receptor agonist UFP-512 dually modulates motor activity in hemiparkinsonian rats via control of the nigro-thalamic pathway. Neuroscience 164:360–369

Mabrouk OS, Viaro R, Volta M, Ledonne A, Mercuri N, Morari M (2014) Stimulation of delta opioid receptor and blockade of nociceptin/orphanin FQ receptor synergistically attenuate parkinsonism. J Neurosci 34:12953–12962

Mansour A, Thompson RC, Akil H, Watson SJ (1993) Delta opioid receptor mRNA distribution in the brain: comparison to delta receptor binding and proenkephalin mRNA. J Chem Neuroanat 6:351–362

Mansour A, Fox CA, Burke S, Meng F, Thompson RC, Akil H, Watson SJ (1994) Mu, delta, and kappa opioid receptor mRNA expression in the rat CNS: an in situ hybridization study. J Comp Neurol 350:412–438

Mansour A, Fox CA, Akil H, Watson SJ (1995) Opioid-receptor mRNA expression in the rat CNS: anatomical and functional implications. Trends Neurosci 18:22–29

Marinelli PW, Funk D, Harding S, Li Z, Juzytsch W, Le AD (2009) Roles of opioid receptor subtypes in mediating alcohol-seeking induced by discrete cues and context. Eur J Neurosci 30:671–678

Martin TJ, Kim SA, Cannon DG, Sizemore GM, Bian D, Porreca F, Smith JE (2000) Antagonism of delta(2)-opioid receptors by naltrindole-5′-isothiocyanate attenuates heroin self-administration but not antinociception in rats. J Pharmacol Exp Ther 294:975–982

Martinez JL Jr, Olson K, Hilston C (1984) Opposite effects of Met-enkephalin and Leu-enkephalin on a discriminated shock-escape task. Behav Neurosci 98:487–495

Massart R, Guilloux JP, Mignon V, Sokoloff P, Diaz J (2009) Striatal GPR88 expression is confined to the whole projection neuron population and is regulated by dopaminergic and glutamatergic afferents. Eur J Neurosci 30:397–414

McCormick SE, Stoessl AJ (2002) Blockade of nigral and pallidal opioid receptors suppresses vacuous chewing movements in a rodent model of tardive dyskinesia. Neuroscience 112:851–859

McDermott CM, Schrader LA (2011) Activation of kappa opioid receptors increases intrinsic excitability of dentate gyrus granule cells. J Physiol 589:3517–3532

McQuiston AR (2011) Mu opioid receptor activation normalizes temporo-ammonic pathway driven inhibition in hippocampal CA1. Neuropharmacology 60:472–479

Meirsman AC, Le Merrer J, Pellissier LP, Diaz J, Clesse D, Kieffer BL, Becker JA (2016) Mice lacking GPR88 show motor deficit, improved spatial learning, and low anxiety reversed by delta opioid antagonist. Biol Psychiatry 79:917–927

Menkens K, Bilsky EJ, Wild KD, Portoghese PS, Reid LD, Porreca F (1992) Cocaine place preference is blocked by the delta-opioid receptor antagonist, naltrindole. Eur J Pharmacol 219:345–346

Middei S, Geracitano R, Caprioli A, Mercuri N, Ammassari-Teule M (2004) Preserved fronto-striatal plasticity and enhanced procedural learning in a transgenic mouse model of Alzheimer's disease overexpressing mutant hAPPswe. Learn Mem 11:447–452

Miner P, Shimonova L, Khaimov A, Borukhova Y, Ilyayeva E, Ranaldi R, Bodnar RJ (2012) General, kappa, delta and mu opioid receptor antagonists mediate feeding elicited by the GABA-B agonist baclofen in the ventral tegmental area and nucleus accumbens shell in rats: reciprocal and regional interactions. Brain Res 1443:34–51

Montgomery KC (1955) The relation between fear induced by novel stimulation and exploratory behavior. J Comp Physiol Psychol 48:254–260

Morales L, Perez-Garcia C, Alguacil LF (2001) Effects of yohimbine on the antinociceptive and place conditioning effects of opioid agonists in rodents. Br J Pharmacol 133:172–178

Mulder AH, Wardeh G, Hogenboom F, Frankhuyzen AL (1984) Kappa- and delta-opioid receptor agonists differentially inhibit striatal dopamine and acetylcholine release. Nature 308:278–280

Mumby DG, Gaskin S, Glenn MJ, Schramek TE, Lehmann H (2002) Hippocampal damage and exploratory preferences in rats: memory for objects, places, and contexts. Learn Mem 9:49–57

Nielsen CK, Simms JA, Bito-Onon JJ, Li R, Ananthan S, Bartlett SE (2012) The delta opioid receptor antagonist, SoRI-9409, decreases yohimbine stress-induced reinstatement of ethanol-seeking. Addict Biol 17:224–234

Nogueiras R, Romero-Pico A, Vazquez MJ, Novelle MG, Lopez M, Dieguez C (2012) The opioid system and food intake: homeostatic and hedonic mechanisms. Obes Facts 5:196–207

Nozaki C, Le Bourdonnec B, Reiss D, Windh RT, Little PJ, Dolle RE, Kieffer BL, Gaveriaux-Ruff C (2012) Delta-opioid mechanisms for ADL5747 and ADL5859 effects in mice: analgesia, locomotion, and receptor internalization. J Pharmacol Exp Ther 342:799–807

Ogier R, Wrobel LJ, Raggenbass M (2008) Action of tachykinins in the hippocampus: facilitation of inhibitory drive to GABAergic interneurons. Neuroscience 156:527–536

Oliveira AM, Hawk JD, Abel T, Havekes R (2010) Post-training reversible inactivation of the hippocampus enhances novel object recognition memory. Learn Mem 17:155–160

Olmstead MC, Ouagazzal AM, Kieffer BL (2009) Mu and delta opioid receptors oppositely regulate motor impulsivity in the signaled nose poke task. PLoS One 4:e4410

Overton DA (1978) Basic mechanisms of state-dependent learning. Psychopharmacol Bull 14:67–68

Packard MG (2009) Exhumed from thought: basal ganglia and response learning in the plus-maze. Behav Brain Res 199:24–31

Packard MG, McGaugh JL (1996) Inactivation of hippocampus or caudate nucleus with lidocaine differentially affects expression of place and response learning. Neurobiol Learn Mem 65:65–72

Palmiter RD (2008) Dopamine signaling in the dorsal striatum is essential for motivated behaviors: lessons from dopamine-deficient mice. Ann N Y Acad Sci 1129:35–46

Pavone F, Populin R, Castellano C, Kreil G, Melchiorri P (1990) Deltorphin, a naturally occurring peptide with high selectivity for delta opioid receptors, improves memory consolidation in two inbred strains of mice. Peptides 11:591–594

Pei G, Kieffer BL, Lefkowitz RJ, Freedman NJ (1995) Agonist-dependent phosphorylation of the mouse delta-opioid receptor: involvement of G protein-coupled receptor kinases but not protein kinase C. Mol Pharmacol 48:173–177

Perreault ML, Graham D, Scattolon S, Wang Y, Szechtman H, Foster JA (2007) Cotreatment with the kappa opioid agonist U69593 enhances locomotor sensitization to the D2/D3 dopamine agonist quinpirole and alters dopamine D2 receptor and prodynorphin mRNA expression in rats. Psychopharmacology (Berl) 194:485–496

Perrine SA, Hoshaw BA, Unterwald EM (2006) Delta opioid receptor ligands modulate anxiety-like behaviors in the rat. Br J Pharmacol 147:864–872

Piskorowski RA, Chevaleyre V (2013) Delta-opioid receptors mediate unique plasticity onto parvalbumin-expressing interneurons in area CA2 of the hippocampus. J Neurosci 33:14567–14578

Place R, Lykken C, Beer Z, Suh J, McHugh TJ, Tonegawa S, Eichenbaum H, Sauvage MM (2012) NMDA signaling in CA1 mediates selectively the spatial component of episodic memory. Learn Mem 19:164–169

Powell SB, Geyer MA, Gallagher D, Paulus MP (2004) The balance between approach and avoidance behaviors in a novel object exploration paradigm in mice. Behav Brain Res 152:341–349

Pradhan AA, Clarke PB (2005) Comparison between delta-opioid receptor functional response and autoradiographic labeling in rat brain and spinal cord. J Comp Neurol 481:416–426

Quintana A, Sanz E, Wang W, Storey GP, Guler AD, Wanat MJ, Roller BA, La Torre A, Amieux PS, McKnight GS, Bamford NS, Palmiter RD (2012) Lack of GPR88 enhances medium spiny neuron activity and alters motor- and cue-dependent behaviors. Nat Neurosci 15:1547–1555

Rezai X, Faget L, Bednarek E, Schwab Y, Kieffer BL, Massotte D (2012) Mouse delta opioid receptors are located on presynaptic afferents to hippocampal pyramidal cells. Cell Mol Neurobiol 32:509–516

Richard JM, Castro DC, Difeliceantonio AG, Robinson MJ, Berridge KC (2013) Mapping brain circuits of reward and motivation: in the footsteps of Ann Kelley. Neurosci Biobehav Rev 37:1919–1931

Roberts AJ, Gold LH, Polis I, McDonald JS, Filliol D, Kieffer BL, Koob GF (2001) Increased ethanol self-administration in delta-opioid receptor knockout mice. Alcohol Clin Exp Res 25:1249–1256

Robles Y, Vivas-Mejia PE, Ortiz-Zuazaga HG, Felix J, Ramos X, Pena de Ortiz S (2003) Hippocampal gene expression profiling in spatial discrimination learning. Neurobiol Learn Mem 80:80–95

Rodriguez M, Abdala P, Barroso-Chinea P, Gonzalez-Hernandez T (2001) The deep mesencephalic nucleus as an output center of basal ganglia: morphological and electrophysiological similarities with the substantia nigra. J Comp Neurol 438:12–31

Saitoh A, Sugiyama A, Nemoto T, Fujii H, Wada K, Oka J, Nagase H, Yamada M (2011) The novel delta opioid receptor agonist KNT-127 produces antidepressant-like and antinociceptive effects in mice without producing convulsions. Behav Brain Res 223:271–279

Scherrer G, Befort K, Contet C, Becker J, Matifas A, Kieffer BL (2004) The delta agonists DPDPE and deltorphin II recruit predominantly mu receptors to produce thermal analgesia: a parallel study of mu, delta and combinatorial opioid receptor knockout mice. Eur J Neurosci 19:2239–2248

Scherrer G, Tryoen-Toth P, Filliol D, Matifas A, Laustriat D, Cao YQ, Basbaum AI, Dierich A, Vonesh JL, Gaveriaux-Ruff C, Kieffer BL (2006) Knockin mice expressing fluorescent delta-opioid receptors uncover G protein-coupled receptor dynamics in vivo. Proc Natl Acad Sci U S A 103:9691–9696

Schmelzeis MC, Mittleman G (1996) The hippocampus and reward: effects of hippocampal lesions on progressive-ratio responding. Behav Neurosci 110:1049–1066

Schroeder JP, Wingard JC, Packard MG (2002) Post-training reversible inactivation of hippocampus reveals interference between memory systems. Hippocampus 12:280–284

Schulteis G, Martinez JL Jr (1990) ICI 174,864, a selective delta opioid antagonist, reverses the learning impairment produced by [leu]enkephalin. Psychopharmacology (Berl) 100:102–109

Shahabi NA, Daaka Y, McAllen K, Sharp BM (1999) Delta opioid receptors expressed by stably transfected jurkat cells signal through the map kinase pathway in a ras-independent manner. J Neuroimmunol 94:48–57

Shiflett MW, Balleine BW (2010) At the limbic-motor interface: disconnection of basolateral amygdala from nucleus accumbens core and shell reveals dissociable components of incentive motivation. Eur J Neurosci 32:1735–1743

Shippenberg TS, Bals-Kubik R, Herz A (1987) Motivational properties of opioids: evidence that an activation of delta-receptors mediates reinforcement processes. Brain Res 436:234–239

Simmons D, Self DW (2009) Role of mu- and delta-opioid receptors in the nucleus accumbens in cocaine-seeking behavior. Neuropsychopharmacology 34:1946–1957

Simon R, Brylka H, Schwegler H, Venkataramanappa S, Andratschke J, Wiegreffe C, Liu P, Fuchs E, Jenkins NA, Copeland NG, Birchmeier C, Britsch S (2012) A dual function of Bcl11b/Ctip2 in hippocampal neurogenesis. EMBO J 31:2922–2936

Slowe SJ, Simonin F, Kieffer B, Kitchen I (1999) Quantitative autoradiography of mu-,delta- and kappa1 opioid receptors in kappa-opioid receptor knockout mice. Brain Res 818:335–345
Stephens DN, Duka T, Crombag HS, Cunningham CL, Heilig M, Crabbe JC (2010) Reward sensitivity: issues of measurement, and achieving consilience between human and animal phenotypes. Addict Biol 15:145–168
Sutt S, Raud S, Abramov U, Innos J, Luuk H, Plaas M, Koks S, Zilmer K, Mahlapuu R, Zilmer M, Vasar E (2010) Relation of exploratory behaviour to plasma corticosterone and Wfs1 gene expression in Wistar rats. J Psychopharmacol 24:905–913
Suzuki T, Yoshiike M, Mizoguchi H, Kamei J, Misawa M, Nagase H (1994) Blockade of delta-opioid receptors prevents morphine-induced place preference in mice. Jpn J Pharmacol 66:131–137
Suzuki T, Tsuji M, Mori T, Ikeda H, Misawa M, Nagase H (1997) Involvement of dopamine-dependent and -independent mechanisms in the rewarding effects mediated by delta opioid receptor subtypes in mice. Brain Res 744:327–334
Svoboda KR, Adams CE, Lupica CR (1999) Opioid receptor subtype expression defines morphologically distinct classes of hippocampal interneurons. J Neurosci 19:85–95
Tsao PI, von Zastrow M (2000) Type-specific sorting of G protein-coupled receptors after endocytosis. J Biol Chem 275:11130–11140
Tzavara ET, Bymaster FP, Davis RJ, Wade MR, Perry KW, Wess J, McKinzie DL, Felder C, Nomikos GG (2004) M4 muscarinic receptors regulate the dynamics of cholinergic and dopaminergic neurotransmission: relevance to the pathophysiology and treatment of related CNS pathologies. FASEB J 18:1410–1412
Ukai M, Takada A, Sasaki Y, Kameyama T (1997) Stimulation of delta1- and delta2-opioid receptors produces amnesia in mice. Eur J Pharmacol 338:1–6
Wang JH, Ma YY, van den Buuse M (2006) Improved spatial recognition memory in mice lacking adenosine A2A receptors. Exp Neurol 199:438–445
Ward SJ, Roberts DC (2007) Microinjection of the delta-opioid receptor selective antagonist naltrindole 5′-isothiocyanate site specifically affects cocaine self-administration in rats responding under a progressive ratio schedule of reinforcement. Behav Brain Res 182:140–144
Wei CJ, Singer P, Coelho J, Boison D, Feldon J, Yee BK, Chen JF (2011) Selective inactivation of adenosine A(2A) receptors in striatal neurons enhances working memory and reversal learning. Learn Mem 18:459–474
White NM, McDonald RJ (2002) Multiple parallel memory systems in the brain of the rat. Neurobiol Learn Mem 77:125–184
Xie CW, Lewis DV (1995) Endogenous opioids regulate long-term potentiation of synaptic inhibition in the dentate gyrus of rat hippocampus. J Neurosci 15:3788–3795
Yang S, Kawamura Y, Yoshikawa M (2003) Effect of rubiscolin, a delta opioid peptide derived from Rubisco, on memory consolidation. Peptides 24:325–328
Yin B, Meck WH (2014) Comparison of interval timing behaviour in mice following dorsal or ventral hippocampal lesions with mice having delta-opioid receptor gene deletion. Philos Trans R Soc Lond B Biol Sci 369:20120466
Yin B, Troger AB (2011) Exploring the 4th dimension: hippocampus, time, and memory revisited. Front Integr Neurosci 5:36
Yin HH, Mulcare SP, Hilario MR, Clouse E, Holloway T, Davis MI, Hansson AC, Lovinger DM, Costa RM (2009) Dynamic reorganization of striatal circuits during the acquisition and consolidation of a skill. Nat Neurosci 12:333–341
Zhang Z, Xin SM, Wu GX, Zhang WB, Ma L, Pei G (1999) Endogenous delta-opioid and ORL1 receptors couple to phosphorylation and activation of p38 MAPK in NG108-15 cells and this is regulated by protein kinase A and protein kinase C. J Neurochem 73:1502–1509
Zhou SJ, Zhu ME, Shu D, Du XP, Song XH, Wang XT, Zheng RY, Cai XH, Chen JF, He JC (2009) Preferential enhancement of working memory in mice lacking adenosine A(2A) receptors. Brain Res 1303:74–83

Delta Opioid Pharmacology in Parkinson's Disease

Omar S. Mabrouk

Contents

Abstract

Parkinson's disease (PD) is a progressive neurodegenerative disorder that compromises multiple neurochemical substrates including dopamine, norepinephrine, serotonin, acetylcholine, and glutamate systems. Loss of these transmitter systems initiates a cascade of neurological deficits beginning with motor function and ending with dementia. Current therapies primarily address the motor symptoms of the disease via dopamine replacement therapy. Exogenous dopamine replacement brings about additional challenges since after years of treatment it almost invariably gives rise to dyskinesia as a side effect. Therefore there is a clear unmet clinical need for improved PD therapeutics. Opioid receptors and their respective peptides are expressed throughout the basal ganglia and cortex where monoaminergic denervation strongly contributes to

O.S. Mabrouk, Ph.D. (✉)
Department of Chemistry, University of Michigan, 930 North University, Ann Arbor, MI 48109, USA

Department of Pharmacology, University of Michigan, 930 North University, Ann Arbor, MI 48109, USA
e-mail: omabrouk@umich.edu

E. M. Jutkiewicz (ed.), *Delta Opioid Receptor Pharmacology and Therapeutic Applications*, Handbook of Experimental Pharmacology 247, DOI 10.1007/164_2016_16

PD pathology. Delta opioid receptors are of particular interest because of their dense localization in basal ganglia and because activating this system is known to enhance locomotor activity under a variety of conditions. This chapter will outline much of the work that has demonstrated the effectiveness of delta opioid receptor activation in models of PD and its neuroprotective properties. It also discusses some of the challenges that must be addressed before moving delta opioid receptor agonists into a clinical setting.

Keywords
Basal ganglia • Delta opioid • Dyskinesias • Enkephalin • Indirect pathway • Levodopa • Movement disorders • Parkinson's disease

1 Parkinson's Disease Background

Parkinson's disease (PD) is a movement disorder that spurs a progressive degeneration of catecholamine (dopamine, norepinephrine) and other (serotonin, acetylcholine) neurochemical systems responsible for normal motor and nonmotor function. As the second most common neurodegenerative disorder, PD affects approximately 0.1% of the population totaling ~10 million people worldwide (Parkinson's Disease Foundation 2015). Nigrostriatal dopamine loss specifically impacts quality of voluntary movements and is a hallmark of PD. PD patients experience freezing (akinesia), abnormal gait, slowness of movement (bradykinesia), and tremor, all of which severely impact quality of life measures. While the loss of dopamine appears at the beginning of this neurodegenerative cascade, later loss of serotonin, norepinephrine, and acetylcholine likely plays a role in the cognitive decline seen in later stages of PD (Alexander 2004).

PD motor symptoms are stably controlled with dopamine replacement therapy (levodopa) and dopamine agonists; however, these drugs bring about additional challenges related to their side effect profile (more on this later). Additionally, this line of therapy does not improve nonmotor symptoms of PD. Therefore, there is a clear unmet clinical need for improved pharmacological tools to enhance not only motor function but also nonmotor attributes. This chapter outlines delta opioid receptor (DOR) pharmacology studies that have been performed in PD models and how this system may represent a valid drug target on several different levels.

2 Basal Ganglia and Parkinson's Disease

PD is considered a basal ganglia disorder because neurodegeneration occurs disproportionately at the nigrostriatal dopamine system. In fact, only after a majority of the cells are lost (70–80%) do the motor symptoms begin to manifest (Bernheimer et al. 1973). The neurons of the substantia nigra pars compacta (SNc) lie dorsal and lateral to the ventral tegmental area (A10) which is also a dopamine-rich brain area more closely associated with reward processing (Schultz et al. 1997). Topographically, the SNc neurons project to more dorsal lateral areas of the striatum (caudate and putamen), while the VTA projects to the ventral striatal regions, including the nucleus accumbens (Beckstead et al. 1979). Both the SNc and VTA are impacted during the course of PD, but SNc degeneration is the primary contributor to movement pathophysiology (German et al. 1989).

When released into the striatum, dopamine has several roles. Firstly, dopamine can act at D2 autoreceptors located at dopamine terminals to inhibit release of dopamine via inhibition of adenylyl cyclase and calcium channels through coupling to $G_{i/o}$ (Missale et al. 1998). Dopamine can also transmit downstream signals through dopamine D1-like (D1 and D5) and D2-like (D2, D3, D4) receptors expressed on GABAergic medium spiny neurons (MSNs). These MSNs make up the vast majority of striatal neurons – upwards of 95% (Tepper and Bolam 2004). Dopamine powerfully regulates the activity of these neurons with D1-like receptors (D1Rs) activating and D2-like receptors (D2Rs) inhibiting MSN firing (Missale et al. 1998).

The conventional view of the basal ganglia suggests a clear segregation of D1R- and D2R-expressing neurons (Fig. 1). Specifically, D1R-expressing MSNs are thought to primarily project to the substantia nigra pars reticulata (SNr) and internal segment of the globus pallidus (GPi) – the direct pathway of the basal ganglia. Meanwhile D2R-expressing MSNs are thought to only project to the external segment of the globus pallidus (GPe) – the first portion of the indirect pathway. Since the first reports of direct and indirect pathway segregation (Alexander et al. 1986; Albin et al. 1989), more recent works have made the case for less segregation of these MSNs and a more intertwined connectivity of these pathways. In fact, Lester et al. (1993) has shown that approximately 25% of MSNs express both D1Rs and D2Rs which could belong to either the direct or indirect pathway.

Dopamine release into the striatum encodes signals which activate movement. Drugs which enhance dopamine release like cocaine and amphetamine stimulate locomotor activity, while drugs which block dopamine signaling such as neuroleptics reduce movement. In PD, the near total loss of striatal dopamine causes an opposite dysregulation of the direct and indirect pathways of the basal ganglia which disrupts the normal signal processing and movement (Fig. 1b). In the direct pathway, loss of dopamine and D1R stimulation reduces striatonigral GABA release. On the other hand, in the indirect pathway, loss of dopamine and D2R stimulation enhances striatopallidal signaling. Since these neurons are GABAergic, pallidal neurons projecting from the GPe to the subthalamic nucleus (STN) are overinhibited resulting in disinhibition of STN glutamate output. Therefore the net result of dopamine loss is lower GABA (direct pathway) and enhanced glutamate (indirect pathway) feeding into the SNr/GPi (Fig. 1b). Both of these facilitatory processes impact the activity of basal ganglia output and control over thalamic activity (Deniau and Chevalier 1985). The thalamus processes inhibitory signals (GABA) from the basal ganglia on to the motor cortex via a glutamatergic relay thus controlling corticospinal output and voluntary movement.

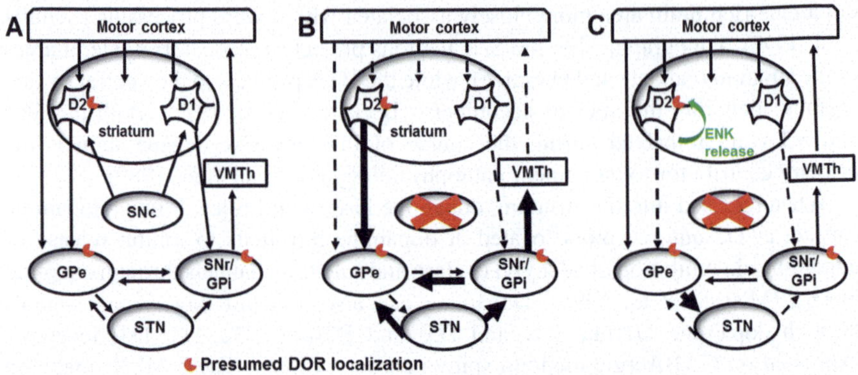

Fig. 1 Schematic representation of the direct (D1 receptor expressing) and indirect (D2 receptor expressing) pathways in the normal basal ganglia (**a**). Dopamine depletion causes an opposite dysregulation of these neurons resulting in aberrant signaling at the output level of the basal ganglia (SNr/GPi) and the entire basal ganglia-thalamocortical loop (**b**). Enkephalin increases in the dopamine-depleted state may be a compensatory mechanism through actions on delta opioid receptors (in *red*) expressed on the cell bodies of indirect pathway neurons (**c**). Inhibition of this overactive pathway would result disinhibit pallido-subthalamic projections thereby normalizing SNr/GPi output. *D1* dopamine D1-like receptors, *D2* dopamine D2-like receptors, *GPe* globus pallidus external segment, *GPi* globus pallidus internal segment, *SNc* substantia nigra pars compacta, *SNr* substantia nigra pars reticulata, *VMTh* ventromedial thalamus, *STN* subthalamic nucleus, *DOR* delta opioid receptor

For basal ganglia-thalamocortical signaling to be executed properly, striatal dopamine tone and striatal output activity must be adequately balanced. Importantly, there are other factors local to the striatum which have the ability to modulate MSN output, including endogenous opioid peptide signaling.

3 Opioid Peptides and Basal Ganglia

Like many other neuron types, MSNs not only release small molecule neurotransmitters to communicate with neighboring cells (GABA in this case) but they also express and release neuropeptides (Engber et al. 1992). Striatonigral (primarily D1R expressing) neurons express prodynorphin mRNA which generates a number of active dynorphin fragments including dynorphin1-17, dynorphin1-13, dynorphin1-8, among others (Reed et al. 2003). Dynorphin and its fragments bind to kappa opioid receptors (KORs) to inhibit firing on receptive cells. On the other hand, striatopallidal neurons (primarily expressing D2Rs) express preproenkephalin-A (PPE-A) and substance P. The products of PPE-A are the pentapeptide enkephalins (ENKs), primarily methionine-enkephalin (m-ENK) and leucine-enkephalin (l-ENK). In humans, one PPE-A mRNA will generate approximately six copies of m-ENK to one copy of l-ENK (Comb et al. 1982). Following synthesis and subsequent release, enkephalins (ENKs) bind to DORs, and to a lesser extent mu-opioid receptors (MORs), where they may produce autoinhibition or they or may inhibit adjacent neurons (again through $G_{i/o}$ signaling).

Because of the localization of its receptors on MSNs and dense nigrostriatal arborization, dopamine is well positioned to control opioid peptide activity throughout the striatum. Many studies have demonstrated how dopamine modulates the expression of striatal opioid peptides at different levels such as mRNA expression (Gerfen et al. 1991) and actual release (Mabrouk et al. 2011). In a seminal work by Charles Gerfen and colleagues, dopamine loss increased PPE-A expression, and this effect was reversed by dopamine agonist treatment (Gerfen et al. 1990). This inverse relationship between dopamine and the generation of ENK precursor mRNA has drawn considerable attention over the past two decades. One dominant theory is that the loss of dopamine drives a compensatory mechanism via ENK signaling. The theory suggests that in PD, dopamine loss causes overactivity of striatopallidal neurons which enhances inhibitory tone within the GPe (Sandyk 1988). Overinhibition of the GPe causes reduced GPe-STN GABAergic signaling, resulting in overactive STN-SNr glutamatergic signaling – which is linked to akinesia (Yoon et al. 2014). Thus in the absence of dopamine, increases in ENK may then attenuate striatopallidal activity via DOR autoreceptors which would prevent the cascade of aberrant signaling throughout the basal ganglia. Evidence for this compensatory mechanism is not clear-cut, however. Most of these studies rely on the premise that mRNA expression changes would be reflected in release, and this may not necessarily be true. Unfortunately it has proven relatively difficult to capture ENK dynamics in vivo. However, groups applying intracerebral microdialysis sampling and modern mass spectrometry measurements have been able to achieve the required sensitivity and selectivity for accurate detection (Emmett and Caprioli 1994; Li et al. 2009; Mabrouk et al. 2011).

DOR mRNA expression and autoradiographic binding studies have clearly demonstrated high DOR expression in the dorsal lateral caudate putamen, while it is relatively low in the GPe (Mansour et al. 1987, 1994). In rat striatal slices, DOR activation selectively inhibits DARPP-32 phosphorylation in adenosine A2A receptor expressing, but not dopamine D1-expressing neurons in the striatum (Lindskog et al. 1999), consistent with the notion that DORs are localized to cell bodies of striatopallidal output neurons. Taken together, it is plausible that dopamine depletion in the striatum activates ENK output by upregulating mRNA and ENK release resulting in binding to striatal DORs which inhibit striatopallidal activity. More studies in the future will need to address fundamental questions surrounding dopamine actions on in vivo ENK release to determine whether mRNA expression and release are positively coupled.

4 Delta Opioid Receptor Activation and Movement

It has long been known that opioids are capable of modifying movement in animal models. Studies with MOR agonists like morphine produce biphasic effects on motor activity where initial akinesia is followed by a marked hyperactivity (Babbini and Davis 1972). The DOR peptide agonist [D-Pen(2),D-Pen(5)]-enkephalin (DPDPE) has also been studied to examine its effects on behavior in comparison to MOR-mediated effects (Meyer and Meyer 1993). Intracerebroventricular (ICV) injection of DPDPE produces monophasic enhancements in locomotor activity in rats (Cowan et al. 1985; Murray and Cowan 1990; Meyer and Meyer 1993). The advent of selective brain penetrating diethylbenzamide DOR agonists like BW373U86 (Wild et al. 1993) and SNC80 (Bilsky et al. 1995) opened the door for more sophisticated studies to examine the link between opioids and movement from a therapeutic perspective. Since its availability to investigators, SNC80 has proven to be a potent stimulator of movement across a number of therapeutic areas. Much of what we know about these effects comes from the depression literature which has consistently shown SNC80 to enhance motor activity in addition to its antidepressant effects. Specifically, subcutaneous (SC) treatment of SNC80 caused decreases in immobility time in a forced swim assay in a way similar to known antidepressants (Broom et al. 2002a). One potential issue with small molecule DOR agonists like SNC80 is their propensity to cause convulsions at high doses (Dykstra et al. 1993; Broom et al. 2002b). Although this side effect will limit certain chemotypes from entering the clinic, it does demonstrate the strong potential to activate motor activity, even beyond a therapeutic range. A recent study has demonstrated that the epileptogenic effects of SNC80 were due to overinhibition of forebrain GABAergic elements (Chung et al. 2015), and therefore this side effect is likely not directly related to DOR activation in basal ganglia.

5 Delta Opioid Receptor in Experimental Models of Parkinson's Disease

Early studies with SNC80 in partially dopamine-depleted rats showed that the drug stimulated both ipsilateral and contralateral movements (Pinna and Di Chiara 1998; Hudzik et al. 2000). These studies demonstrated that DOR activation could potentiate dopamine agonist mediated directionality preference in 6-OHDA hemilesioned rats but did not go further in elucidating the precise mechanism of these actions. In the early 2000s, Jonathan Brotchie and colleagues began exploring SNC80 as a potential therapeutic in PD models. A seminal study, performed by Hille et al. (2001), highlighted SNC80's unique ability to alleviate haloperidol-induced catalepsy in rats and parkinsonian features in MPTP-treated nonhuman primates. In fact, the authors wrote "SNC80-treated animals were indistinguishable from normal rats" when describing the drug's effects in cataleptic rats. Indeed, in that study, SNC80's actions were robust and had effects within 5 min after treatment with 10 mg/kg (i.p.). These studies laid the groundwork for future mechanistic studies. In 2008, Michele Morari and colleagues described the use of SNC80 in 6-OHDA rats undergoing microdialysis (Mabrouk et al. 2008). Those studies showed that systemic SNC80 could reduce GABA release in the GPe, presumably via striatopallidal inhibition, while systemic blockade of DOR with naltrindole enhanced pallidal GABA and worsened parkinsonism in the 6-OHDA hemilesioned rat. This finding was in line with another important study by Jonathan Brotchie's group which showed that in slices ENKs could reduce pallidal GABA (Maneuf et al. 1994).

SNC80 dose dependently normalized akinesia, bradykinesia, and overall motor function on an accelerating rotarod (Mabrouk et al. 2008). Interestingly, enhancements in motor activity in these assays were only recapitulated when SNC80 was directly injected into the SNr, but not into the GPe or the dorsal lateral striatum (DLS; Mabrouk et al. 2008). It is known that DOR stimulation in the SNr elicits behavioral activation (Morelli et al. 1989), and DORs are found in the SNr (Mansour et al. 1987); therefore, it is plausible that DOR expressed on nigral output neurons was directly inhibited by DOR activation, thus disinhibiting the basal ganglia-thalamocortical loop. However, it seems puzzling that SNC80 was not effective when directly injected into the DLS where DOR expression is high. One possible explanation is that because the DLS is many times larger than the SNr, only a small subset of MSNs was affected by the microinjection, thus preventing the expected antiparkinsonian outcome. No matter the reason, the fact that pallidal GABA was reduced in response to systemic treatment gives credence to the theory that attenuating striatopallidal GABA is achievable through DOR stimulation and that this coincides with motor activation.

Follow-up studies using the pseudopeptide DOR agonist H-Dmt-Tic-NH-CH (CH2-COOH)-Bid (UFP-512) showed similar antiparkinsonian effects compared to SNC80 with the interesting distinction of producing a biphasic effect on motor activity in hemiparkinsonian rats. Systemic UFP-512 enhanced locomotor activity at a low dose (0.001-mg/kg i.p.) while causing motor inhibition at a higher dose (1 mg/kg i.p.; Mabrouk et al. 2009). One important aspect to this study was that it showed for the first time the effect of systemic DOR agonist administration on thalamic neurochemistry. The article showed that a motor-activating dose of UFP-512 corresponded to reductions in GABA release in the ventromedial thalamus (VMTh), presumably through inhibition of nigrothalamic input (Mabrouk et al. 2009). Another interesting feature of UFP-512 was that at the highest doses tested, it lost its antiparkinsonian effect, but did not elicit convulsions like small molecule agonists. Therefore it is likely that the pseudopeptide agonists have a narrower therapeutic range but also safer side effect profile, though this has not been systematically tested. It would be interesting to test UFP-512 in other models and at ultralow and high doses to determine if convulsions appear.

Another promising strategy for introducing DOR agonists into the brain without the use of small molecule agonists is through glycosylation of ENK analogues. Glycosylation is the process of adding a sugar moiety to a target peptide or protein which facilitates blood-brain barrier permeability. In the case of PD models, the compound MMP-2200, a glycosylated l-ENK analogue, has been shown to be efficacious in both reserpinized and 6-OHDA hemilesioned rats (Yue et al. 2011). Although MMP-2200 appears to be less selective for DOR than SNC80 and UFP-512, many of its behavioral effects are attenuated by co-treatment with naltrindole – a selective DOR antagonist (Yue et al. 2011). One important feature of this drug is that although it is peptide-like, it can cross the blood-brain barrier.

This has been shown in a recent study where systemic MMP-2200 was detected in the DLS of rats using in vivo microdialysis and mass spectrometry measurements (Mabrouk et al. 2012).

Another strategy to enhance the antiparkinsonian effects of DOR agonists like SNC80 is through coadministration with other compounds that might have synergistic effects, thus allowing for lower doses. One such strategy recently illustrated by Mabrouk et al. (2014) was that low doses of SNC80 were combined with nociception opioid peptide receptor (NOR) antagonist J117339 – another compound shown to have antiparkinsonian properties (Marti et al. 2005). The combination of the two compounds showed potent synergistic antiparkinsonian activity (Mabrouk et al. 2014). Specifically, in 6-OHDA hemilesioned rats, 0.1 mg/kg i.p. of either SNC80 or J117339 had no effect alone in tests of akinesia (bar test), bradykinesia (drag/wheelbarrow test), or overall locomotor activity (accelerating rotarod). However, when both compounds were combined at their individual subthreshold doses (0.1 mg/kg i.p. each), there was a near doubling of motor activation in all three tests (Mabrouk et al. 2014). These findings were recapitulated in MPTP-treated mice where the combination of very low doses of SNC80 (0.01 mg/kg i.p.) and J117339 (0.001 mg/kg i.p.) produced profound effects in these behavioral assays with nearly a ~fivefold reduction in immobility time in the akinesia test (Mabrouk et al. 2014). To further prove the interaction between these two opioid receptor systems, this work demonstrated that SNC80 showed a leftward potency shift in mice lacking NOR through genetic deletion. Finally, in vitro electrophysiology of nigral slices once again showed a synergistic effect between DOR activation and NOR blockade.

Taken together, there is strong evidence for antiparkinsonian activity of DOR agonists; however, the side effect profile of small molecule agonists like SNC80 may prevent their advancement into a clinical setting. Several groups are working with peptide-like DOR agonists to circumvent this problem, and other small molecule DOR agonist chemotypes are being generated that are less prone to side effects such as convulsions (Le Bourdonnec et al. 2008; Saitoh et al. 2011). Also combined DOR agonists and NOR antagonists may represent a rationale strategy to enhance the antiparkinsonian effects of diethylbenzamide compounds like SNC80 while limiting potential side effects. Further studies are warranted to determine the usefulness of this strategy in nonhuman primates.

6 Delta Opioid Receptors and Levodopa-Induced Dyskinesia

The first-line treatment for PD is dopamine replacement therapy with levodopa, a bioavailable dopamine precursor that crosses the blood-brain barrier. Once in the brain, levodopa is converted to dopamine by enzymatic decarboxylation via DOPA decarboxylase. Patients are treated multiple times per day to maintain elevated dopamine concentrations which temporarily overcomes the loss of dopamine-producing cells. Levodopa has clear benefits in improving motor symptoms in PD; however, after years of treatment, these benefits almost invariably give way to a very serious hyperkinetic side effect known as levodopa-induced dyskinesia (LID). Symptoms of LIDs include choreas (abnormal involuntary movements across the body) and dystonias (muscular spasms causing abnormal postures) which can be more disruptive than PD itself. LIDs are thought to occur in part through continuous overstimulation of D1Rs expressed on direct pathway neurons (Aubert et al. 2005). This makes opioid systems attractive pharmacological targets since they can modulate the interaction between nigrostriatal dopamine and striatal MSNs which reverberates throughout the basal ganglia.

Jonathan Brotchie and his colleagues also contributed a great deal to our understanding of how the DOR system impacts LIDs using a variety of animal models. In 2001, his group showed that the selective DOR antagonist naltrindole has dramatic antidyskinetic effects in a MPTP-treated marmoset model of LIDs (Henry et al. 2001). Importantly, these improvements were not at the expense of the antiparkinsonian effects of levodopa. The authors suggest that enhanced ENKergic transmission of the striatopallidal pathway, which is the same compensatory mechanism described earlier, becomes pathogenic. Blocking ENKergic tone then disinhibits the striatopallidal projection allowing for a return to normal movement levels. The authors suggest that opioid receptor blockade of the direct pathway may also be a contributing factor by reducing striatonigral activity. However, it is unclear how direct DOR antagonism on striatonigral neurons could inhibit this population of neurons since DOR antagonism would presumably disinhibit these neurons. Another explanation is that an indirect mechanism is at play such as DOR blockade of GABAergic interneurons (Wang and Pickel 2001) which could unmask inhibitory control over striatonigral projections, thus attenuating overactivation in the dyskinetic state.

A more recent study examining the effects of DOR on LIDs showed that DORs localized to corticostriatal terminals might exert control over dyskinesia (Billet et al. 2012). In these studies, ICV injection of the DOR agonist DPDPE worsened dyskinesia, but this effect was reversed in animals with cortical lesions (Billet et al. 2012). The data from this study suggest that DOR activation positively regulates corticostriatal glutamate transmission and that blockade of this glutamate promoting effect is how a DOR antagonist could relieve LIDs. Although the stimulatory properties of opioid receptors remain controversial (they are generally considered inhibitory), these data do suggest a possible segregation of DOR1 and DOR2 subtypes both in terms of localization and function through downstream coupling (Billet et al. 2012). To date, DOR antagonists have not made much progress in clinical trials which may be because of a lack of an adequate understanding of how these compounds produce their effects and the availability of a lead compound which does not interfere with levodopa benefits.

7 Delta Opioid Receptors and Neuroprotection in Parkinson's Disease Models

The holy grail of PD research, like all neurodegenerative diseases, is the discovery of protective therapeutics which halt or reverse cell death. Because of the wide distribution of DORs in areas affected by neurodegeneration, such as the striatum and cortex, their influence on cell viability remains an important research topic. Early studies with the diethylbenzamide DOR agonist BW373U86 showed that it was a potent neuroprotectant in a model of hypoxia possibly through a body temperature-regulating mechanism (Bofetiado et al. 1996). In the early 2000s, studies showed that DOR activation with DPDPE could block glutamate-induced excitotoxicity in cortical neurons (Zhang et al. 2000, 2002). A series of articles in the late 1990s demonstrated how methamphetamine dopamine neural toxicity could be attenuated by DOR agonism (Tsao et al. 1998, 1999; Hayashi et al. 1999). One particularly relevant highlight of these studies was that in 6-OHDA injections in the medial forebrain bundle – a common model of PD – DOR agonism significantly reduced dopamine cell death (Borlongan et al. 2000). Where 6-OHDA causes a near total ablation of dopamine cells, treatment with the peptide agonist [D-Ala(2)-D-Leu (5)]-Enkephalin (DADLE) only suffered a 25% reduction in dopamine cells (Borlongan et al. 2000). The author's conclusion was that DOR agonism draws upon a hibernation-stimulating mechanism in which the organism is protected from toxic insult potentially through free radical scavenging (Tsao et al. 1999; Borlongan et al. 2000).

The majority of these studies used a peptide [D-Ala(2)-D-Leu(5)]-Enkephalin (DADLE) DOR agonist which was given systemically (Borlongan et al. 2000). Since brain penetrability of this peptide is likely inferior compared to small molecule agonists, it would be of interest to reexamine these effects using chronic SNC80, pseudopeptides (UFP-512), or glycosylated peptides (MMP-2200) to determine if antiparkinsonian properties coincide with neuroprotection.

8 Delta Opioid Receptors Challenges and Future Directions in Parkinson's Disease

One of the challenges of engaging the DOR system in PD is availability of pharmacological tools which deliver good therapeutic benefit (i.e., are motor activating) over a broad range of doses and that do not develop tolerance after repeated treatments. Additionally compounds must be very selective for DOR over MOR since MOR agonism is associated with morphine-like effects including respiratory depression and abuse potential. Tolerance is another important issue that needs to be addressed with respect to DOR agonist therapy since these receptors are known to rapidly internalize upon exposure to agonists (Pradhan et al. 2009). Small molecule agonists diethylbenzamides, SNC80, and BW786BU were the most widely used pharmacological tools of the past 15 years, but these compounds are known to induce convulsions at high doses (Dykstra et al. 1993;

Broom et al. 2002a). Newer DOR agonists such as ARM390 do not cause convulsions but have not been thoroughly tested in PD models and therefore should be closely examined during the next phase of DOR agonist-PD work.

Another fundamental question that needs to be addressed in this line of work is whether or not DOR agonists are dyskinetogenic when given as adjuncts to levodopa. This line of research would require long-term chronic studies to determine how dosing of DOR agonists and levodopa would need to be adjusted to retain maximal benefits. Another very interesting line of research should be to determine how DOR agonists affect the nonmotor symptoms of PD. Indeed, intranigral 6-OHDA has recently been shown to induce a behavioral phenotype similar to depression (Santiago et al. 2014). Similarly, PD patients also experience depression which is likely related to the damage suffered to both dopamine and serotonin systems. Since DOR agonists like SNC80 have been shown to have robust antidepressant effects in numerous models (Broom et al. 2002a, b; Jutkiewicz et al. 2005), it would be interesting to determine whether DOR agonists improve "PD-like depression" in model systems.

Another strategy that should be undertaken is the development of dual opioid compounds which might enhance the effects of DOR activation and potentially curb internalization properties. Prior studies have reported strong evidence for heterodimerization of the DOR and MOR (Traynor and Elliott 1993; Jordan and Devi 1999; Gomes et al. 2000), and this might be a way to enhance DOR signaling (Gomes et al. 2000). For example, DOR agonist/MOR antagonist drugs may present benefits not seen with DOR agonists alone. Along these same lines, NOR antagonists have already shown a leftward shift in SNC80's dose response (Mabrouk et al. 2014) suggesting that engaging dual target could be one avenue to explore going forward.

Taken together, the ability of DOR agonists to enhance locomotor activity in a number of PD models, to be neuroprotectant in models of dopamine denervation, and to improve measures of depression all point toward the notion that this class of drugs may represent a future generation of antiparkinsonian drugs. There is still controversy about the use of opioid-based drugs for the treatment of movement disorders, but their effectiveness cannot be denied in animal models.

References

Albin RL, Young AB, Penney JB (1989) The functional anatomy of basal ganglia disorders. Trends Neurosci 12:366–375

Alexander GE (2004) Biology of Parkinson's disease: pathogenesis and pathophysiology of a multisystem neurodegenerative disorder. Dialogues Clin Neurosci 6:259–280

Alexander GE, DeLong MR, Strick PL (1986) Parallel organization of functionally segregated circuits linking basal ganglia and cortex. Annu Rev Neurosci 9:357–381

Aubert I, Guigoni C, Håkansson K, Li Q, Dovero S, Barthe N, Bioulac BH, Gross CE, Fisone G, Bloch B, Bezard E (2005) Increased D1 dopamine receptor signaling in levodopa-induced dyskinesia. Ann Neurol 57:17–26

Babbini M, Davis WM (1972) Time-dose relationships for locomotor activity effects of morphine after acute or repeated treatment. Br J Pharmacol 46:213–224

Beckstead RM, Domesick VB, Nauta WJ (1979) Efferent connections of the substantia nigra and ventral tegmental area in the rat. Brain Res 175:191–217

Bernheimer H, Birkmayer W, Hornykiewicz O, Jellinger K, Seitelberger F (1973) Brain dopamine and the syndromes of Parkinson and Huntington. Clinical, morphological and neurochemical correlations. J Neurol Sci 20:415–455

Billet F, Costentin J, Dourmap N (2012) Influence of corticostriatal δ-opioid receptors on abnormal involuntary movements induced by L-DOPA in hemiparkinsonian rats. Exp Neurol 236:339–350

Bilsky EJ, Calderon SN, Wang T, Bernstein RN, Davis P, Hruby VJ, McNutt RW, Rothman RB, Rice KC, Porreca F (1995) SNC 80, a selective, nonpeptidic and systemically active opioid delta agonist. J Pharmacol Exp Ther 273:359–366

Bofetiado DM, Mayfield KP, D'Alecy LG (1996) Alkaloid delta agonist BW373U86 increases hypoxic tolerance. Anesth Analg 82:1237–1241

Borlongan CV, Su TP, Wang Y (2000) Treatment with delta opioid peptide enhances in vitro and in vivo survival of rat dopaminergic neurons. Neuroreport 11:923–926

Broom DC, Jutkiewicz EM, Folk JE, Traynor JR, Rice KC, Woods JH (2002a) Nonpeptidic delta-opioid receptor agonists reduce immobility in the forced swim assay in rats. Neuropsychopharmacology 26:744–755

Broom DC, Jutkiewicz EM, Folk JE, Traynor JR, Rice KC, Woods JH (2002b) Convulsant activity of a non-peptidic delta-opioid receptor agonist is not required for its antidepressant-like effects in Sprague-Dawley rats. Psychopharmacology (Berl) 164:42–48

Chung PC, Boehrer A, Stephan A, Matifas A, Scherrer G, Darcq E, Befort K, Kieffer BL (2015) Delta opioid receptors expressed in forebrain GABAergic neurons are responsible for SNC80-induced seizures. Behav Brain Res 278:429–434

Comb M, Seeburg PH, Adelman J, Eiden L, Herbert E (1982) Primary structure of the human Met- and Leu-enkephalin precursor and its mRNA. Nature 295:663–666

Cowan A, Zhu XZ, Porreca F (1985) Studies in vivo with ICI 174864 and [D-Pen2, D-Pen5] enkephalin. Neuropeptides 5:311–314

Deniau JM, Chevalier G (1985) Disinhibition as a basic process in the expression of striatal functions. II. The striato-nigral influence on thalamocortical cells of the ventromedial thalamic nucleus. Brain Res 334:227–233

Dykstra LA, Schoenbaum GM, Yarbrough J, McNutt R, Chang KJ (1993) A novel delta opioid agonist, BW373U86, in squirrel monkeys responding under a schedule of shock titration. J Pharmacol Exp Ther 267:875–882

Emmett MR, Caprioli RM (1994) Micro-electrospray mass spectrometry: ultra-high-sensitivity analysis of peptides and proteins. J Am Soc Mass Spectrom 5:605–613

Engber TM, Boldry RC, Kuo S, Chase TN (1992) Dopaminergic modulation of striatal neuropeptides: differential effects of D1 and D2 receptor stimulation on somatostatin, neuropeptide Y, neurotensin, dynorphin and enkephalin. Brain Res 581:261–268

Gerfen CR, Engber TM, Mahan LC, Susel Z, Chase TN, Monsma FJ Jr, Sibley DR (1990) D1 and D2 dopamine receptor-regulated gene expression of striatonigral and striatopallidal neurons. Science 250:1429–1432

Gerfen CR, McGinty JF, Young WS 3rd (1991) Dopamine differentially regulates dynorphin, substance P, and enkephalin expression in striatal neurons: in situ hybridization histochemical analysis. J Neurosci 11:1016–1031

German DC, Manaye K, Smith WK, Woodward DJ, Saper CB (1989) Midbrain dopaminergic cell loss in Parkinson's disease: computer visualization. Ann Neurol 26:507–514

Gomes I, Jordan BA, Gupta A, Trapaidze N, Nagy V, Devi LA (2000) Heterodimerization of mu and delta opioid receptors: a role in opiate synergy. J Neurosci 20:RC110

Hayashi T, Tsao LI, Cadet JL, Su TP (1999) [D-Ala2, D-Leu5]enkephalin blocks the methamphetamine-induced c-fos mRNA increase in mouse striatum. Eur J Pharmacol 366: R7–R8

Henry B, Fox SH, Crossman AR, Brotchie JM (2001) Mu- and delta-opioid receptor antagonists reduce levodopa-induced dyskinesia in the MPTP-lesioned primate model of Parkinson's disease. Exp Neurol 171:139–146

Hille CJ, Fox SH, Maneuf YP, Crossman AR, Brotchie JM (2001) Antiparkinsonian action of a delta opioid agonist in rodent and primate models of Parkinson's disease. Exp Neurol 172:189–198

Hudzik TJ, Howell A, Payza K, Cross AJ (2000) Antiparkinson potential of delta-opioid receptor agonists. Eur J Pharmacol 396:101–107

Jordan BA, Devi LA (1999) G-protein-coupled receptor heterodimerization modulates receptor function. Nature 399:697–700

Jutkiewicz EM, Rice KC, Traynor JR, Woods JH (2005) Separation of the convulsions and antidepressant-like effects produced by the delta-opioid agonist SNC80 in rats. Psychopharmacology (Berl) 182:588–596

Le Bourdonnec B, Windh RT, Ajello CW, Leister LK, Gu M, Chu GH, Tuthill PA, Barker WM, Koblish M, Wiant DD, Graczyk TM, Belanger S, Cassel JA, Feschenko MS, Brogdon BL, Smith SA, Christ DD, Derelanko MJ, Kutz S, Little PJ, DeHaven RN, DeHaven-Hudkins DL, Dolle RE (2008) Potent, orally bioavailable delta opioid receptor agonists for the treatment of pain: discovery of N, N-diethyl-4-(5-hydroxyspiro[chromene-2,4′-piperidine]-4-yl)benzamide (ADL5859). J Med Chem 51:5893–5896

Lester J, Fink S, Aronin N, DiFiglia M (1993) Colocalization of D1 and D2 dopamine receptor mRNAs in striatal neurons. Brain Res 621:106–110

Li Q, Zubieta JK, Kennedy RT (2009) Practical aspects of in vivo detection of neuropeptides by microdialysis coupled off-line to capillary LC with multistage MS. Anal Chem 81:2242–2250

Lindskog M, Svenningsson P, Fredholm B, Greengard P, Fisone G (1999) Mu- and delta-opioid receptor agonists inhibit DARPP-32 phosphorylation in distinct populations of striatal projection neurons. Eur J Neurosci 11:2182–2186

Mabrouk OS, Volta M, Marti M, Morari M (2008) Stimulation of delta opioid receptors located in substantia nigra reticulata but not globus pallidus or striatum restores motor activity in 6-hydroxydopamine lesioned rats: new insights into the role of delta receptors in parkinsonism. J Neurochem 107:1647–1659

Mabrouk OS, Marti M, Salvadori S, Morari M (2009) The novel delta opioid receptor agonist UFP-512 dually modulates motor activity in hemiparkinsonian rats via control of the nigrothalamic pathway. Neuroscience 164:360–369

Mabrouk OS, Li Q, Song P, Kennedy RT (2011) Microdialysis and mass spectrometric monitoring of dopamine and enkephalins in the globus pallidus reveal reciprocal interactions that regulate movement. J Neurochem 118:24–33

Mabrouk OS, Falk T, Sherman SJ, Kennedy RT, Polt R (2012) CNS penetration of the opioid glycopeptide MMP-2200: a microdialysis study. Neurosci Lett 531:99–103

Mabrouk OS, Viaro R, Volta M, Ledonne A, Mercuri N, Morari M (2014) Stimulation of δ opioid receptor and blockade of nociceptin/orphanin FQ receptor synergistically attenuate parkinsonism. J Neurosci 34:12953–12962

Maneuf YP, Mitchell IJ, Crossman AR, Brotchie JM (1994) On the role of enkephalin cotransmission in the GABAergic striatal efferents to the globus pallidus. Exp Neurol 125:65–71

Mansour A, Khachaturian H, Lewis ME, Akil H, Watson SJ (1987) Autoradiographic differentiation of mu, delta, and kappa opioid receptors in the rat forebrain and midbrain. J Neurosci 7:2445–2464

Mansour A, Fox CA, Burke S, Meng F, Thompson RC, Akil H, Watson SJ (1994) Mu, delta, and kappa opioid receptor mRNA expression in the rat CNS: an in situ hybridization study. J Comp Neurol 350:412–438

Marti M, Mela F, Fantin M, Zucchini S, Brown JM, Witta J, Di Benedetto M, Buzas B, Reinscheid RK, Salvadori S, Guerrini R, Romualdi P, Candeletti S, Simonato M, Cox BM, Morari M (2005) Blockade of nociceptin/orphanin FQ transmission attenuates symptoms and neurodegeneration associated with Parkinson's disease. J Neurosci 25:9591–9601

Meyer ME, Meyer ME (1993) Behavioral effects of opioid peptide agonists DAMGO, DPDPE, and DAKLI on locomotor activities. Pharmacol Biochem Behav 45:315–320

Missale C, Nash SR, Robinson SW, Jaber M, Caron MG (1998) Dopamine receptors: from structure to function. Physiol Rev 78:189–225

Morelli M, Fenu S, Di Chiara G (1989) Substantia nigra as a site of origin of dopamine-dependent motor syndromes induced by stimulation of mu and delta opioid receptors. Brain Res 487:120–130

Murray CW, Cowan A (1990) [D-Pen2, D-Pen5]enkephalin, the standard delta opioid agonist, induces morphine-like behaviors in mice. Psychopharmacology (Berl) 102:425–426

Parkinson's Disease Foundation (2015) pdf.org

Pinna A, Di Chiara G (1998) Dopamine-dependent behavioural stimulation by non-peptide delta opioids BW 373U86 and SNC 80: 3. Facilitation of D1 and D2 responses in unilaterally 6-hydroxydopamine-lesioned rats. Behav Pharmacol 9:15–21

Pradhan AA, Becker JA, Scherrer G, Tryoen-Toth P, Filliol D, Matifas A, Massotte D, Gavériaux-Ruff C, Kieffer BL (2009) In vivo delta opioid receptor internalization controls behavioral effects of agonists. PLoS One 4:e5425

Reed B, Zhang Y, Chait BT, Kreek MJ (2003) Dynorphin A(1-17) biotransformation in striatum of freely moving rats using microdialysis and matrix-assisted laser desorption/ionization mass spectrometry. J Neurochem 86:815–823

Saitoh A, Sugiyama A, Nemoto T, Fujii H, Wada K, Oka J, Nagase H, Yamada M (2011) The novel δ opioid receptor agonist KNT-127 produces antidepressant-like and antinociceptive effects in mice without producing convulsions. Behav Brain Res 223:271–279

Sandyk R (1988) Enkephalinergic mechanisms in the "compensated" phase of Parkinson's disease. Int J Neurosci 42:301–303

Santiago RM, Barbiero J, Gradowski RW, Bochen S, Lima MM, Da Cunha C, Andreatini R, Vital MA (2014) Induction of depressive-like behavior by intranigral 6-OHDA is directly correlated with deficits in striatal dopamine and hippocampal serotonin. Behav Brain Res 259:70–77

Schultz W, Dayan P, Montague PR (1997) A neural substrate of prediction and reward. Science 275:1593–1599

Tepper JM, Bolam JP (2004) Functional diversity and specificity of neostriatal interneurons. Curr Opin Neurobiol 14:685–692

Traynor JR, Elliott J (1993) Delta-opioid receptor subtypes and cross-talk with mu-receptors. Trends Pharmacol Sci 14:84–86

Tsao LI, Ladenheim B, Andrews AM, Chiueh CC, Cadet JL, Su TP (1998) Delta opioid peptide [D-Ala2, D-leu5]enkephalin blocks the long-term loss of dopamine transporters induced by multiple administrations of methamphetamine: involvement of opioid receptors and reactive oxygen species. J Pharmacol Exp Ther 287:322–331

Tsao LI, Cadet JL, Su TP (1999) Reversal by [D-Ala2, D-Leu5]enkephalin of the dopamine transporter loss caused by methamphetamine. Eur J Pharmacol 372:R5–R7

Wang H, Pickel VM (2001) Preferential cytoplasmic localization of delta-opioid receptors in rat striatal patches: comparison with plasmalemmal mu-opioid receptors. J Neurosci 21:3242–3250

Wild KD, Fang L, McNutt RW, Chang KJ, Toth G, Borsodi A, Yamamura HI, Porreca F (1993) Binding of BW 373U86, a non-peptidic delta opioid receptor agonist, is not regulated by guanine nucleotides and sodium. Eur J Pharmacol 246:289–292

Yoon HH, Park JH, Kim YH, Min J, Hwang E, Lee CJ, Suh JK, Hwang O, Jeon SR (2014) Optogenetic inactivation of the subthalamic nucleus improves forelimb akinesia in a rat model of Parkinson disease. Neurosurgery 74:533–540

Yue X, Falk T, Zuniga LA, Szabò L, Porreca F, Polt R, Sherman SJ (2011) Effects of the novel glycopeptide opioid agonist MMP-2200 in preclinical models of Parkinson's disease. Brain Res 1413:72–83

Zhang J, Haddad GG, Xia Y (2000) Delta-, but not mu- and kappa-, opioid receptor activation protects neocortical neurons from glutamate-induced excitotoxic injury. Brain Res 885:143–153

Zhang J, Gibney GT, Zhao P, Xia Y (2002) Neuroprotective role of delta-opioid receptors in cortical neurons. Am J Physiol Cell Physiol 282:C1225–C1234

Delta Opioid Receptor and Peptide: A Dynamic Therapy for Stroke and Other Neurological Disorders

M. Grant Liska, Marci G. Crowley, Trenton Lippert, Sydney Corey, and Cesar V. Borlongan

Contents

Abstract

Research of the opioid system and its composite receptors and ligands has revealed its promise as a potential therapy for neurodegenerative diseases such as stroke and Parkinson's Disease. In particular, delta opioid receptors (DORs)

"M. Grant Liska" and "Marci G. Crowley" equally contributed to this manuscript.

M. Grant Liska • M.G. Crowley • T. Lippert • S. Corey • C.V. Borlongan (✉)
Department of Neurosurgery and Brain Repair, University of South Florida College of Medicine, Tampa, FL 33612, USA
e-mail: cborlong@health.usf.edu

E. M. Jutkiewicz (ed.), *Delta Opioid Receptor Pharmacology and Therapeutic Applications*, Handbook of Experimental Pharmacology 247, DOI 10.1007/164_2017_7

have been elucidated as a therapeutically distinguished subset of opioid receptors and a compelling target for novel intervention techniques. Research is progressively shedding light on the underlying mechanism of DORs and has revealed two mechanisms of DOR neuroprotection; DORs function to maintain ionic homeostasis and also to trigger endogenous neuroprotective pathways. Delta opioid agonists such as (D-Ala2, D-Leu5) enkephalin (DADLE) have been shown to promote neuronal survival and decrease apoptosis, resulting in a substantial amount of research for its application as a neurological therapeutic. Most notably, DADLE has demonstrated significant potential to reduce cell death following ischemic events. Current research is working to reveal the complex mechanisms of DADLE's neuroprotective properties. Ultimately, our knowledge of the DOR receptors and agonists has made the opioid system a promising target for therapeutic intervention in many neurological disorders.

Keywords
Cerebral ischemia • DADLE • Neuroprotection • Opioid receptors

Abbreviations

DADLE	(D-Ala 2, D-Leu 5) enkephalin
DAT	Dopamine transporter
DOR	Delta (δ) opioid receptor
HIT	Hibernation induction trigger
KOR	Kappa (κ) opioid receptor
MAPK	Mitogen-activated protein kinase
MCAO	Middle cerebral artery occlusion
METH	Methamphetamine
MOR	Mu (μ) opioid receptor
PKC	Protein kinase c

1 Introduction

Opioid peptides and receptors compose the opioid system. Opioids are a class of molecules historically implicated in pain modulation and addiction, yet in reality, these neurotransmitters and receptors are involved in many different functions including respiratory rate control and the stress response (Drolet et al. 2001). Mediated by different receptors, the various opioid peptides each have their own unique functions and molecular interactions. Invariably, adenylyl cyclase is inhibited by this family of classical G_i-protein coupled receptors, which is composed of three primary subgroups: δ-opioid receptors (DOR), μ- (MOR), and κ- (KOR). Endogenous opioid peptides include dynorphins, endorphins, and enkephalins, which are associated with the KOR, MOR, and DOR, respectively. Depending on the class of opioid, these receptors will evoke assorted effects. Opioid receptors appear throughout various peripheral organs including the liver, heart, and gastrointestinal tract, as well as throughout the entire nervous and peripheral systems (Feng et al. 2012; Hiller and Fan 1996; Lim et al. 2004; Mansour et al. 1987; Xia and Haddad 1991; Xiang et al. 1996). Evidence has pointed to the opioid system's ability to protect against multiple neurological diseases that are defined by energy depleting states, such as anoxic and ischemic conditions (Borlongan et al. 2004; Peart et al. 2005). After treatment with DOR agonists in animal models, data collected by Mayfield and colleagues displayed extended survival rates during periods of hypoxia (Mayfield and D'alecy 1994). In addition, it was shown that the opioid-induced protection was attenuated by DOR antagonists, but was not affected by MOR and KOR antagonists (Bofetiado et al. 1996). These findings indicate that the opioid system is involved in neuroprotection against ischemic and hypoxic incidents, feasibly mediated by delta opioid peptides and DORs primarily.

One particular DOR agonist, [(D-Ala 2,D-Leu 5) enkephalin] (DADLE), is a synthetic opioid peptide which has received special attention due to its cytoprotective effects and its ability to induce hibernation (Borlongan et al. 2009). The ability of DADLE to preserve cells in ischemic/hypoxic conditions has made it a focal point of ischemic stroke research. In addition, DADLE has been suggested as a possible therapeutic for neurodegenerative diseases such as Parkinson's Disease due to evidence that DADLE can prevent dopaminergic neuron loss in methamphetamine-treated rats and also mitigate pro-inflammatory cytokines (Berardelli et al. 1986; Song et al. 2008). Despite much promising research, DADLE has not yet gained access into the clinical setting and requires further study before making this transition (Van Rijn et al. 2013).

This chapter will provide a general overview of DOR's with a focus on their neuroprotective applications. Additionally, later sections of this chapter will highlight DADLE as a distinguished therapeutic opportunity, and discuss the current research which has helped to reveal its neuroprotective mechanisms.

2 DOR: Neuroprotection and Receptors

Increasing evidence has shown that DOR activation may induce neuroprotective effects under ischemic/hypoxic stress. During the late 1980s, Xia and his assistants detected and recognized that turtle brains exhibit a much higher DOR density when compared to the rat brain (Xia and Haddad 2001). Along with this observation, turtles displayed a higher tolerance level than rats in ischemic and hypoxic conditions (Sick et al. 1982; Xia et al. 1992). These findings pointed to a plausible correlation between higher DOR density levels and an increased tolerance level for ischemic and hypoxic conditions. In order to further demonstrate this, researchers mimicked excitotoxic conditions by adding glutamate to cultured cortical neurons (Zhang et al. 1999). With 4 h of daily exposure over 8–10 days, neurons exposed to 100 μmol/L glutamate showed significant decrease in viability (Zhang et al. 1999). Yet, with increased activation of DOR through administration of a DOR agonist, a decreased level of glutamate-induced excitotoxicity of roughly 50% was observed (Zhang et al. 1999). Also, the activation of KOR and MOR did not display notable protective effects (Zhang et al. 1999, 2000). This suggests that the delta opioid receptors are responsible for the neuroprotective effects seen, without significant contribution by either the mu or kappa opioid receptors.

Data of the late 1990s and the early 2000s presented evidence that delta opioid receptors are implicated in cellular ischemic response (Boutin et al. 1999; Frerichs and Hallenbeck 1998; Kevelaitis et al. 1999). In mice subjected to middle cerebral artery occlusion (MCAO), a reduction in mu and kappa binding sites was seen after a decrease in delta binding sites. This decline naturally accompanies infarct core extension (Boutin et al. 1999; Mayfield et al. 1996). Prior to any evident brain damage, reduction in DOR is initiated, insinuating that KORs and MORs are less sensitive to brain insults when compared to DORs. Inversely, if DOR's are stimulated, it may display protective effects in the ischemic brain (Frerichs and Hallenbeck 1998; Kevelaitis et al. 1999).

3 DOR: General Neuroprotective Mechanisms

Although the neuroprotective effects of DOR activation were quickly recognized, the intricacies of DOR's role in neuroprotection are still not well-established. Recent studies have served to progress our understanding of DOR's cytoprotective properties. The generally accepted means of DOR's neuroprotective properties fall into two categories – maintenance of ionic homeostasis, and activation of endogenous protective pathways. These two mechanisms work in conjunction to provide neuroprotective effects by combating the damage which develops from ischemia/hypoxia. The following sections provide an overview of these mechanisms. Later in this chapter, we will review current research on DADLE – a specific DOR agonist – that expands upon the ideas presented here.

3.1 Ionic Homeostasis

Onset of hypoxic and ischemic conditions is promptly followed by disruption of ionic homeostasis caused by an efflux of K^+ and influx of Na^+, Cl^-, and Ca^{2+} (Kang et al. 2009; Sung et al. 2008). The characteristic efflux of K^+ in hypoxia and ischemia can cause neuronal injury and death (Bickler 2004; Chao et al. 2006; Karki et al. 2007; Liu et al. 2003; Mongin 2007; Nistico et al. 2007; Wei et al. 2003; Yu et al. 1997). Following a cerebral ischemic event, activation of DORs lessens the damaging leakage of K^+ out of neuronal cells (Chao et al. 2006, 2007a, b, 2008, 2009), thereby reducing cell death (Liu et al. 2003; Wei et al. 2003). In addition, a

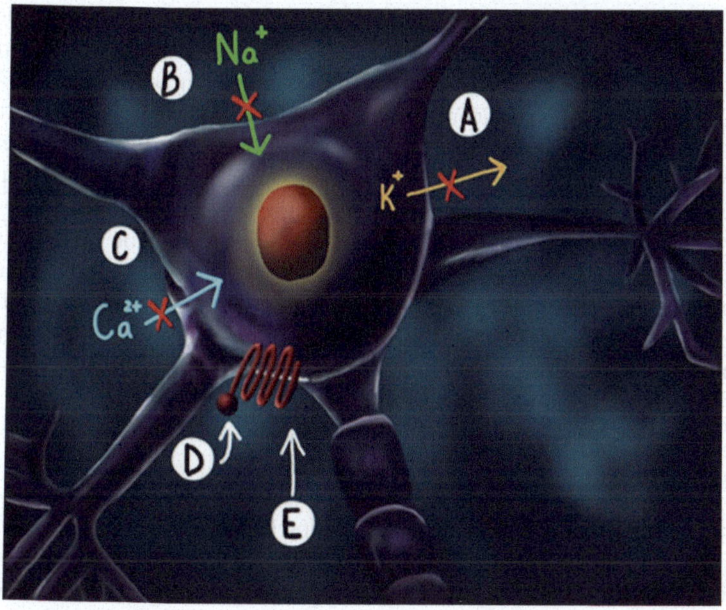

Fig. 1 DOR neuroprotective modulation of cellular ion levels: following hypoxic/ischemic cellular stress, an immediate loss of ionic homeostasis ensues, which leads to neuronal dysfunction and death. DOR activation can prevent the disturbance of ionic homeostasis and mitigate the injury. Activation of DOR has been shown to: (**A**) prevent efflux of potassium, (**B**) prevent influx of sodium, and (**C**) prevent influx of calcium. DOR agonists such as DADLE – (D-Ala 2, D-Leu 5) enkephalin – bind to DOR (**D**) in the brain, initiating these protective effects. Activation of DOR (**E**) allows for a reduction in neuronal injury and death after ischemic events

direct decrease in Na^+ influx accompanied by indirect reduction of K^+ efflux by way of inhibition of voltage-gated Na^+ channels has been observed in cells with elevated expression of DOR (Kang et al. 2009; Chao et al. 2008, 2009; Chao and Xia 2010). The neuroprotective effects of DORs can largely be attributed to its ability to attenuate Na^+ influx and help maintain ionic homeostasis after hypoxic/ischemic insult (Fig. 1). This was demonstrated when the neuroprotective effects of DOR were abrogated by Na^+ perfusion, treatment with Na^+ channel blocker TTX, and NMDA receptor channel blocker MK 801 (Chao et al. 2008, 2009).

3.2 Endogenous Neuroprotective Pathways

In addition to the regulatory roles of DORs in functioning to maintain ionic homeostasis, DORs have been implicated in initiating endogenous neuroprotective pathways. In particular, multiple studies have shown that DORs may exert neuroprotective effects by preventing the phosphorylation of p38 by way of protein kinase C (PKC) and mitogen-activated protein kinase (MAPK)-ERK1/2 stimulation (Feng et al. 2009; Ke et al. 2009; Ma et al. 2005; Narita et al. 2006; Peng et al.

2009). The DOR-mediated neuroprotective effects seen after hypoxic episodes can be reduced by treatment with PKC inhibitors, evidence that the activities of this kinase and others are implicated in the neuroprotective pathway (Ma et al. 2005). Additionally, this PKC-dependent pathway has been tied to DOR attenuation of the K^+ efflux seen in a number of physiopathologies, and an overall maintenance of ionic homeostasis (Chao et al. 2007b). Following hypoxic preconditioning, a number of hypoxia-inducible transcription factors are upregulated (Peng et al. 2009). One result of this upregulation is an increased expression of DOR, which provokes the neuroprotective ERK signaling pathway (Peng et al. 2009). Treatment with an ERK inhibitor impedes these DOR-induced neuroprotective effects, further indicating the crucial role that DOR-mediation of the PKC and ERK signaling pathways play in this opioid's neuroprotective functions (Ma et al. 2005).

4 DADLE: The Ligand and Hibernation

Hibernation is an unparalleled, innate model allowing animals to endure debilitating blood-, energy-, and oxygen-deprived conditions. Due to this, hibernation has become a key focus for many researchers investigating possible neurotherapies for conditions with related circumstances. A study aiming to discover novel insight regarding the molecular factors participating in hibernation led to the observation of induced hibernation in summer active ground squirrels following an injection of plasma from Thirteen-lined Ground Squirrels (Dawe and Spurrier 1969). The hibernation trigger (HIT) that caused this was classified as a protein factor that co-migrates with albumin (Bruce et al. 1987; Oeltgen et al. 1988). It has been suggested that HIT could behave similarly to an opioid, because of an opioid's capability to influence physical responses comparable to hibernation. Conversely, new data have shown HIT may release endogenous opioids instead of actively behaving as one (Bruce et al. 1987). This also indicates that hibernation and the opioid systems are likely linked. Further research in the opioids' hibernation-causing characteristic provided data revealing that each opioid receptor class, μ, κ, and δ, has differing effectiveness for stimulating hibernation. Certain selective agonists for KOR and MOR, such as dynorphin and morphine, were limited in their ability to stimulate hibernation in summer active ground squirrels (Oeltgen et al. 1987, 1988). The DOR agonist (D-Ala 2, D-Leu 5) enkephalin (DADLE), however, was successful in provoking a hibernation state (Oeltgen et al. 1988). Following this discovery, numerous experiments have aimed to analyze the neuroprotective abilities of DOR's and DOR selective ligands, such as DADLE, through the opioid system. DADLE has become a focal point of researchers analyzing prospective neuroprotective therapies due to its ability to act as an opioid peptide that predominantly binds to DORs. For similar reasons that DADLE can act as a HIT, DADLE was also shown to be an effective means of increasing cell viability in cell transplantation therapy and a means of organ preservation in organ transplantations (Borlongan et al. 2000; Horton et al. 1998). Current research has expanded on these ideas, with a heavy focus on DADLE's neurological applications – specifically ischemic stroke (Chu Sin Chung and Kieffer 2013).

5 DADLE and Cerebral Ischemia

DADLE has been favorably implicated in stroke therapy for more than two decades. Interest in DADLE as a possible treatment for stroke is logical, as ischemic stroke is characterized by severe hypoxia and energy deprivation which are physiologically similar to the conditions induced by hibernation (Borlongan et al. 2009). Still, developing effective therapeutics based off of the promising DADLE research of the 1990s and 2000s has proven difficult; it has been challenging to draw reliable conclusions on DADLE's effectiveness due to poorly understood interactions and mechanisms (Xia 2015). For example, Blurton et al. demonstrated that DADLE is nonspecific and can bind to MOR's, which bears significant implications in its therapeutic uses (Blurton et al. 1986). Additionally, the distribution/density of DOR's and the function of DOR agonists vary from species to species, making translational research problematic. Despite the difficulties, DADLE research has continued to progress and show promise as a future stroke therapy.

5.1 Recent In Vivo Studies of DADLE in Cerebral Ischemia

In 2016, Fu and collaborators published valuable findings describing the general effectiveness of DADLE administration after permanent middle cerebral artery occlusion (pMCAO) in rats (Fu et al. 2016). Most ischemic episodes in patients are not transient, thus pMCAO provides a more clinically applicable model compared to standard MCAO. In this experiment, intracerebroventricular injection of DADLE was administered 45 min post-ischemia (Fu et al. 2016). Behavioral tests and histological analysis were conducted on all rats to measure neurologic function, infarct size, and cell death (Fu et al. 2016). Rats treated with DADLE post-MCAO showed significantly improved neurological function, a 44.7% decrease in infarct volume when compared to vehicle controls, and an increase in neuron survival of more than 25% (Fu et al. 2016). Neurological functions were evaluated with the Garcia test, which examines both behavioral and sensory functions, making it an especially relevant test when searching for therapeutics that aid in functional recovery for stroke patients (Fu et al. 2016).

Wang et al. demonstrated equally encouraging evidence of the effectiveness of DADLE as a therapy in their 4-artery occlusion model of stroke (Wang et al. 2016). This experiment utilized the 4-artery occlusion technique to induce transient global ischemia in the rat brain, which not only models stroke, but also the blood-deprivation accompanying cardiac arrest (Wang et al. 2016). This group looked closely at the effects of DADLE on spatial memory and hippocampal cell death (Wang et al. 2016). Previous research has shown that transient ischemia can trigger the neurogenic activities of the adult brain, however, these newly formed cells were shown to predominantly die by apoptosis within weeks (Wang et al. 2016). Wang's team hypothesized that DADLE may have neuroprotective effects by encouraging the survival of these newly formed cells (Wang et al. 2016). DADLE was administered at the onset of reperfusion and the rats were put through a water-maze test 23–27 days after the ischemia (Wang et al. 2016). The results of this experiment showed an improvement in spatial memory, an increase in new neurons within the dentate gyrus, and a limited effect on cell differentiation in rats treated with DADLE (Wang et al. 2016).

The pathophysiology of stroke is complex, with many detrimental pathways being concurrently activated upon the initiation of hypoxic conditions. One of the most prominent targets of stroke research is neuroinflammation. In 2014, Wang et al. (Wang et al. 2014) conducted research which linked the effects of DADLE to a reduction in the pro-inflammatory cytokine, tumor necrosis factor-alpha (TNF-α) (Wang et al. 2014). By exposing cultured rat astrocytes and PC-12 cells (a cell line which differentiates into neuron-like cells) to severe hypoxia, they were able to encourage the cells to secrete TNF-α. The combination of astrocytes and neuron-like cells attempted to replicate an in vivo cellular environment. First, this group was able to demonstrate that hypoxia increases the expression of DOR in astrocytes (Wang et al. 2014). Then, using a DOR-agonist UFP-512, they showed that DOR activation led to a reduction of TNF-α secretion from astrocytes (Wang et al. 2014). Together, this paints a picture of hypoxia resulting in an increased expression of DORs in astrocytes as an attempt to self-regulate its inflammatory secretions. Although DADLE was not the DOR agonist used, because the experimental design demonstrated that DOR activation was the causative event, and not a specific opioid ligand, these findings are translatable to DADLE.

As our knowledge of the complex pathophysiology of stroke continues to grow, so too does the breadth of research on DADLE. An increased understanding of the intricate details and the contributing factors to cell death after ischemia will allow for new avenues of investigation. Current and future research on DADLE's effectiveness as a stroke therapy will progressively reveal the details of this synthetic opioid's function and mechanisms – a critical component of moving DADLE from the laboratory to the clinical setting. As with any potential pharmaceutical, it is vital to have an understanding of the molecule's various interactions in order to predict how it will affect the entirety of the human body.

6 Recent Research Revealing DADLE's Neuroprotective Pathways

Being a particular DOR agonist, much of the research elucidating the therapeutic mechanisms of DADLE have been extensions of the investigative efforts into understanding the DOR function. This research has revealed two well-established therapeutic processes by which DADLE functions, focusing primarily on the DOR-dependent mechanisms mentioned in Sects. 3.1 and 3.2. First, DADLE has been shown to preserve ionic homeostasis indirectly by modulating the influx and efflux of intracellular and intercellular ions. Second, DADLE has been proven to be a trigger for endogenous neuroprotective pathways. The following sections will review more recently characterized mechanisms, some of which introduce the possibility of DOR-independent neuroprotective pathways. Predominately, these mechanisms have been elaborations on the endogenous neuroprotective pathways which were known to exist, but not well characterized. Revealing the details of DADLE's therapeutic mechanisms has been slow work and is crucial to transitioning DADLE and other DOR agonists to clinical relevance (Berardelli et al. 1986). Still, there is strong evidence in support of DADLE's potential to be clinically significant in the future (Berardelli et al. 1986). The following sections will discuss two recently discovered and unique mechanisms of DADLE's therapeutic function which contribute to its neuroprotective properties.

6.1 DADLE and the MKK7 Pathway

In 2015, Wang and his colleagues conducted a study in which both in vitro and in vivo models were used to explore the effects of DADLE on rat intestinal epithelial cells after induction of ischemia/reperfusion (I/R) injury. DADLE was effective in increasing the survival rate of the I/R injury cells and decreasing apoptotic rates in a dose-dependent manner (Wang et al. 2015). The activation of the c-Jun N-Terminal kinase JNK pathway has been shown to play an important role in apoptosis and I/R injury pathology (Song et al. 2008). In vitro, DADLE attenuated the JNK pathway via reduced phosphorylation of an upstream kinase, mitogen-activated protein kinase kinase (MKK7) (Wang et al. 2014). In addition to observing activation of MKK7, in vivo models of I/R insult on epithelial tissues were examined for protein levels of oxidative stress markers including diamine oxidase (DAO) in the plasma, and superoxide dismutase (SOD), myeloperoxidase (MPO), and malonaldehyde (MPA) within the intestinal tissues (Wang et al. 2014). Treatment groups displayed significantly lower levels of MPA, DAO, and MDA – but not MPO – compared to the control group, resembling nearly normal levels represented by the sham group (Wang et al. 2014). Expression of SOD was significantly increased in DADLE-treated rats compared to the control (Wang et al. 2014). Hematoxylin and eosin staining illustrated the absence of erosion and detachment of the intestinal mucosal epithelium in the DADLE treatment groups, but was observed in the control groups, further implicating DADLE's protective effects following I/R injury in intestinal epithelia (Wang et al. 2015). Notably, after silencing MKK7 expression in vitro, DADLE treatment no longer protected against apoptosis after I/R insult (Wang et al. 2015). This implies that DADLE's role in the defense against cell death is largely via the MKK7-JNK signal transduction pathway (Wang et al. 2015).

The following research by Wang et al. provides yet another mechanism of action in which DADLE may elicit therapeutic benefits. The protective effects provoked by DADLE observed in intestinal cells may also have influence on neuropathies. The discovery of MKK7 down-regulation as a central role in reducing apoptosis and oxidative stress provides key evidence that it may induce similar effects following ischemia of the brain. Previous studies indicate that MKK7 is the optimal target location in the JNK pathway in which inhibition would render favorable effects within the cell. When JNK inhibitor peptide, D-JNKI1, was tried as a neuroprotective agent, it served as a strong defense against excitotoxicity in vitro (Borsello et al. 2003a, b) and resulted in a 93% decline in infarct size for ischemic rodent models (Borsello et al. 2003a, b; Repici et al. 2007). However, complete obstruction of the JNK signaling pathway has negative side effects due to its regulatory functions in a range of physiological events including cell survival, proliferation, and differentiation (Davies and Tournier 2012). Of the only two direct upstream activators of JNK – MMK4 and MKK7 – MKK7 is the preferred target due to its activation by inflammatory cytokines (Harper and Lograsso 2001) and role in overactivation of JNK in response to the excitotoxic conditions which accompany stroke models (Repici et al. 2007). By avoiding the inhibition of MKK4, which becomes activated upon stress signals not including excitotoxic stimuli, the other physiological roles of JNK pathways may resume (Centeno et al. 2007).

A study in 2015 by Vercelli et al. focused on inhibition of MKK7 as a treatment for cerebral ischemia after taking previous research into consideration (Vercelli et al. 2015). The peptide, GADD45β-I (growth arrest and DNA-inducible 45β) was specifically engineered to target MKK7 and minimize its phosphorylation (Vercelli et al. 2015). Following successful in vitro studies, two models of cerebral ischemia were conducted in Sprague–Dawley rats – the MCAO and the thromboembolic ischemia models (Vercelli et al. 2015). Administration of the MKK7 inhibitory protein 6 h subsequent to an ischemic insult presented a neuroprotective effect (Vercelli et al. 2015). These results suggest that DADLE may act by way of a similar mechanism and be a prime candidate in the protection against neuronal cell death after ischemic injury.

The increased expression of SOD and reduction of MPO seen with the in vivo results of the Wang et al. suggests the ability of DADLE to combat oxidative stress (Wang et al. 2015). These separate mechanisms alone have had positive results in ischemic models. Several studies have concluded the neuroprotective effect of overexpressed SOD after ischemia-induced injury (Fujimura et al. 1999; Kawase et al. 1999; Kinouchi et al. 1991). One particular study displayed that the induction of apoptotic cell death by phosphorylation/activation of the MAPK pathway may be attenuated by an increase in SOD (Noshita et al. 2002). Inhibition of MPO in MCAO mice models was found to lessen neuronal damage and increase neurological function (Yu et al. 2016). Both the elevation of MPO and reduction of SOD have been observed in cerebral ischemic conditions creating damaging oxidative stress. Tackling both of these problems individually has been shown to be neuroprotective and DADLE may use these two mechanisms in conjunction to exhibit its powerful antioxidant effects.

6.2 DADLE and Transcription Regulation

As mentioned previously, dating back to the late 1980s, DADLE was observed to induce a state of hibernation in squirrels after injection (Oeltgen et al. 1988). Two studies in 2006 explored DADLE-induced hibernation at a cellular level and discovered that both HeLa cells, a cervical cancer cell line (Vecchio et al. 2006), and LNCaP cells – a human prostate cancer cell line (Baldelli et al. 2006) – went into a reversible hibernation-like state when treated with DADLE. Both cancer cell lines displayed an overall decreased level of transcription and proliferation upon treatment (Vecchio et al. 2006; Baldelli et al. 2006). While hibernation is a complex physiological process, the findings of this study and others imply that the reduction in metabolism seen in states of hibernation may be in part a consequence of a mass reduction in transcription.

In a more recent study by Tian et al. (2014), further detail was revealed about the mechanism in which DADLE inhibits transcription (Tian et al. 2014). Human neuroblastoma SH-SY5Y cells and primary cortical neurons were used in the study due to the hypothesis that DADLE-induced inhibition of transcription and the resulting arrest of cellular metabolism would provide neuroprotective effects (Tian et al. 2014). When simultaneously treated with the DOR antagonist naltrindole and DADLE, there was no change in DADLE's ability to reduce transcription, implying of its function via a DOR-independent mechanism (Tian et al. 2014). This was further supported by the trials with Deltorphin A, DAMGO, and U50488H – DOR, MOR, and KOR agonist, respectively – and their inability to inhibit transcription (Tian et al. 2014). Specifically, DADLE was observed to inhibit phosphorylation of RNA Polymerase II at Ser 2 and Ser 5 in the heptapeptide repeat sequence of the C-terminal domain, thereby limiting transcription (Tian et al. 2014). After treatment for 24–72 h, no cell damage occurred as a result of DADLE and the torpor-like state was reversed in the subsequent 72 h (Tian et al. 2014). Additional investigation with in vivo studies is required to further support the therapeutic use of DADLE as a means of limiting metabolism after neurological insults such as stroke and TBI.

Interestingly, Tian's study provides contradicting results to the data provided by Wang's study on MKK7 phosphorylation mentioned in Sect. 6.1. In Wang's research, DADLE treatment on intestinal cells restricted the characteristic increase expression of apoptotic factors – specifically caspase-3 and caspase-9 – after I/R injury, which coincides with Tian's findings (Wang et al. 2015; Tian et al. 2014). However, when examining MKK7 and JNK expression after DADLE treatment, it was observed that while phosphorylation of both kinases declined, only the overall expression of JNK was significantly reduced; the total expression of MKK7 did not decline compared to sham and control groups as would be expected with inhibition of transcriptional machinery (Wang et al. 2015). These findings reveal the gaps in our understanding of DADLE's role in transcription and gene expression. Further research is needed to provide a more conclusive picture detailing the manner in which DADLE is eliciting its neuroprotective effects through decreased expression and phosphorylation of proteins in apoptotic pathways.

7 DADLE, Beyond Stroke: Relevance in Other CNS Disorders

The majority of this chapter thus far has focused on the potential applications of DOR and DADLE therapies in stroke. Yet, DADLE has proven to be a versatile molecule and is being investigated as a therapeutic for multiple neurological disorders including epilepsy, spinal cord injury, and neurodegenerative diseases such as Huntington's, Alzheimer's, and Parkinson's disease. It is likely that the observed effects of DADLE in these various disorders are the result of both the therapeutic mechanisms described previously and newer, poorly understood mechanisms.

7.1 DADLE as an Epilepsy Therapeutic

While DADLE treatment and other DOR agonists have not been extensively studied in epilepsy models, their established mechanisms of action suggest a potential for therapeutic benefit for those suffering from epilepsy. The current treatment for epilepsy includes the option of surgery or several antiepileptic drugs, and though they may be effective, they often are accompanied with complex psychiatric and behavioral problems such as depression, aggression, agitation, and irritation (Brodie et al. 2016). A predicted >30% of adolescent and adult patients continue to have seizures after/during treatment (Brodie et al. 2012). In older research before the year 2000, studies explored the anti-convalescent effects of DADLE (Koide et al. 1992; Tortella et al. 1985). Now that our understanding of the role DADLE/DOR-activation plays in maintaining ionic homeostasis has been expanded, DADLE treatment of epilepsy may be a promising therapeutic option to explore. Brains prone to epileptic seizures have been shown to have cortical neurons that are hyper-excitable, accompanied by the upregulation of voltage-gated Na^+ channels and down-regulation of DOR. As mentioned previously, DORs play a role in the inhibition of voltage-gated Na^+ channels (Kang et al. 2009; Chao et al. 2008, 2009; Chao and Xia 2010) and attenuating neuronal deficits caused by Na^+ influx (Chao et al. 2008, 2009). In addition, DADLE has been seen to combat glutamate toxicity in neocortical neurons (Zhang et al. 2000). Both of these factors may contribute to the potential for DADLE as a treatment in epilepsy which is characterized by overactivation of neurons, often associated with ionic imbalance and dysfunction of the glutamate system (Barker-Haliski and White 2015). It is worth noting that despite the positive evidence supporting DADLE's potential to combat pathological processes of epilepsy, it has been reported that DADLE and other DOR agonist – specifically SNC80 (Chung et al. 2015) – may promote convulsions (Iwata et al. 2007). This does not, however, reflect the lack of adverse effects with DADLE as other investigations have found no convulsive effects of DOR agonists in primates (Feng et al. 2012) and the convulsions seen in some DOR agonists may be the result of high dosages.

7.2 DADLE and Spinal Cord Injury

Since discovering the existence of DORs within the region of spinal cord motoneurons in 1999 (Mailly et al. 1999), multiple studies have been conducted targeting DOR activation as a treatment to improve tolerance of spinal cord ischemia and reduce injury from ischemic reperfusion. These studies have displayed promising results for several DOR agonists including SNC80, DADLE, and DPDPE ([D-Pen^{2,5}]-enkephalin) (Horiuchi et al. 2004; Liu et al. 2015, 2016; Turner and Johnson 2011).

Intrathecal treatment with DOR agonists has been observed to have therapeutic effects after I/R injury on the spinal cord (Horiuchi et al. 2004; Turner and Johnson 2011). In one study, induction of an intra-aortic balloon catheter was used as a spinal cord ischemia (SCI) model in rats with intrathecal SNC80 treatment prior to the SCI (Horiuchi et al. 2004). This treatment was observed to limit resulting hindlimb dysfunction and increase the number of healthy neurons after 48 h (Horiuchi et al. 2004). Another study used a split-bath ex vivo neonatal rat brainstem and spinal cord preparation in OGD (oxygen-glucose deprivation) solution to model I/R spinal injury. Direct administration of DADLE and DPDPE was done before, during, and after exposing only the spinal cord to OGD solution (Turner and Johnson 2011). OGD solution causes a decrease in respiratory motor output frequency and amplitude until there is no further activity. When the electrical activity ceased, it was recorded as the end-point time (Turner and Johnson 2011). Both DOR activators DPDPE and DADLE increased the end-point times (Turner and Johnson 2011). DOR activation was seen to attenuate ischemic signaling cascade even after the cascade had already begun; this was supported by the trial in which DADLE was administered with ongoing OGD exposure after 15 min of spinal OGD exposure alone and end-point times still increased 80% (Turner and Johnson 2011). This suggests that the therapeutic effects of DOR activation act quickly, extending its therapeutic window (Turner and Johnson 2011). Overall, DPDPE was observed to have a greater neuroprotective effect than DADLE and therefore DPDPE was used for further experiments such as dual treatment of DPDPE and naltrindole (DOR antagonist) (Turner and Johnson 2011). Blocking DORs prior and during spinal OGD prevented neuroprotection by DPDPE, indicating that it acts by way of a DOR-dependent mechanism (Liu et al. 2016). More research is needed to establish how long the spinal cord may undergo OGD before the damage is too extensive and DOR activation is no longer effective.

Spinal cord injury, potentially to the extent of paraplegia, is a risk factor of thoracic and thoracoabdominal aortic aneurysm repair surgery (Zvara 2002). In two consecutive studies performed on rabbits, regional reperfusion of DADLE through the abdominal aorta was used as an experimental treatment in spinal cord I/R models caused by aortic occlusion (Liu et al. 2016; Zvara 2002). In the first study, DADLE treatment limited behavioral retardation and combated the loss of healthy motor-neurons of the spinal cord after I/R injury (Liu et al. 2015). An additional study was done to explore the relative dose-response effects of DADLE using the same model. The results showed greater neuroprotection with increasing dosage between 0.0005 and 0.05 mg/kg with a decline in therapeutic efficacy after a

0.5 mg/kg dosage (Mailly et al. 1999). At this dosage the mean arterial pressure was affected and the hemodynamic parameter was suppressed temporarily suggesting a possible cause of the reduced neuroprotective effects (Mailly et al. 1999). Interestingly, it was suggested that the protective effects of DADLE on the spinal cord may be somewhat temporary. The rate of paraplegia increased in the 24–48 h period in the 0.05 mg/kg dosage group going from a rate of zero to a rate just under 20% (Mailly et al. 1999). While after 72 h this dosage still displayed significant improvement from the control group and remained the dosage that displayed the greatest improvement, the decline of neuroprotection is something to be noted and further studied (Mailly et al. 1999). A procedure for additional administration of DADLE may be an option to prolong the protective effects of DADLE on the spinal cord and would require further investigation.

7.3 DADLE and Neurodegenerative Disease

DADLE and DORs have been shown to have implications in neurodegenerative diseases such as Parkinson's Disease (PD), Alzheimer's Disease (AD), and Huntington's Disease (HD), among others. These three disorders share a common motif of progressive neurodegeneration, and inefficient synaptic communicate due to neuronal death and/or aberrant protein aggregation (Nussbaum and Ellis 2003). Despite having unique origins and causes, much of the research performed on these three diseases is translatable to the others due to their commonalities. In particular, research displaying DADLE's ability to promote neuronal survival likely has relevance to all diseases characterized by progressive neurodegeneration. Yet as mentioned before, DADLE is a dynamic molecule which possesses both general protective mechanisms that contribute to its neurotherapeutic effects and mechanisms which are unique to each individual disease. The importance of this caveat will be explored with a discussion of DORs and DADLE in Alzheimer's disease in Sect. 7.3.2.

A study conducted in 2003 by the National Institute of Health's Cellular Pathology Unit showed that chronic administration of DADLE caused an increased level of Nerve Growth Factor (NGF) within the brain in a region-specific manner (Hayashi and Su 2003). Among these regions were the midbrain and hippocampus, which are implicated in PD, AD, and HD (Hayashi and Su 2003). The presence of NGF has been shown to act as a trophic factor which promotes the survival of neurons throughout the brain and nervous system (Barde 1989; Culmsee et al. 2002), thus an increase in NGF levels may help to counteract the neuronal death seen in various neurodegenerative disorders. Interestingly, a study in 2013 was able to provide insight into the mechanism of this NGF upregulation by observing an increase in the P13K/Akt/NF-kB pathway proteins after DADLE treatment via western-blot analysis (Sen et al. 2013).

Disturbances in NGF and NGF-receptor levels have not been shown to exist in AD, yet a pilot study of NGF lateral ventricle infusions administered over 3 months in an AD patient resulted in an increase in cerebral blood flow and an improvement in verbal episodic memory (Olson et al. 1994). The reason for these effects is not completely understood, but it was theorized that the NGF-dependent cholinergic system may have been aided by the increased NGF levels (Olson et al. 1994). In contrast, the brain of Parkinson's patients has been shown to have notable reduction in NGF levels (Lorigados Pedre et al. 2002). Lastly, in Huntington's Disease, NGF has been shown to be neuroprotective with in vivo and in vitro models (Mizuta et al. 2001). As a result, DADLE's ability to stimulate the production of NGF may reveal possible therapeutic avenues for all three disorders – by stimulating the cholinergic system in AD, rescuing NGF levels in PD, and promoting neuronal survival in HD. It is worth mentioning that evidence presented in Sect. 7.3.2 may refute the theory that DADLE beneficially modulates the cholinergic system.

Another unique and widely applicable use of DADLE is its function as an aid in cell transplantation therapies, as mentioned in Sect. 4 (Borlongan et al. 2001). Cell transplantation therapies are being investigated as a treatment option for a diverse set of neurological disorders including stroke, traumatic brain injury, and the neurodegenerative diseases mentioned previously (PD, HD, and AD). Although cell transplantation acts through multiple therapeutic mechanisms, the most intuitive one – transplanted cells actively replacing dead and dying cells – does not seem to readily occur. In order to encourage the survival of transplanted cells, Borlongan et al. investigated the use of DADLE as an aid in increasing graft survival rates of embryonic ventral mesencephalic cells (Borlongan et al. 2001). One month after being subjected to a forebrain 6-hydroxydopamine lesion (6-OHDA; an in vivo PD model), the rats were administered the mesencephalic cells which had been suspended in a solution containing DADLE (Borlongan et al. 2001). When behavioral tests were performed 6 and 8 weeks post-transplantation, locomotor activity returned to near normal levels (Borlongan et al. 2001). Additionally, histological analysis showed that DADLE increased the number of living transplanted cells nearly twofold (Borlongan et al. 2001). The ability of DADLE to promote the survival of grafted cells provides a secondary therapeutic use for this peptide, whereby it may contribute to all disorders for which cell transplantation therapy is being investigated.

7.3.1 DADLE and Parkinson's Disease

An investigation by Tsao and colleagues evaluated DADLE's neuroprotective abilities in association with dopamine neurotoxicity by methamphetamine (METH). Prolonged use at a medium dosage or a high single dose of METH generated long-term loss of striatal dopaminergic terminals (Tsao et al. 1998). Two weeks following a METH dose, DADLE was introduced into the body and dopamine transporter (DAT) function was found to return to normal levels from a prior decrease of 30% (Tsao et al. 1999). Additionally, the introduction of DADLE preceding exposure to METH effectively inhibited and restored METH-induced DAT deficits (Tsao et al. 1999; Hayashi et al. 1999). Further investigations have revealed that the free radical scavenging ability of DADLE and facilitation by DOR are essential for the neuroprotective effects of DADLE countering METH-induced DAT loss (Tsao et al. 1998).

DADLE's ability to provide neuroprotective effects against METH-induced DAT loss created enquires regarding DADLE's capacity to provide the same protection against other neurological conditions. Particularly, DADLE's ability to combat Parkinson's Disease (PD) – which is defined by dopamine depletion – has been investigated as a potential novel therapy. Adult male rats were administered with 6-OHDA lesion and it was observed that the group which received DADLE prior to treatment displayed improved survival of tyrosine hydroxylase immunoreactive cells (Borlongan et al. 1999, 2000). Comparably, improved cell viability was observed in cultured primary rat fetal mesencephalic cells in a dose-dependent method when the animals were pretreated with DADLE (Borlongan et al. 1999, 2000). Further data from an in vivo study showed DADLE augmented the survival of serum deprived PC12 cells (Hayashi et al. 2002). Although this implies that DADLE may have a trophic factor function, the primary pathway for its neuroprotective behavior is thought to remain in the opioid system.

A more recent study conducted in 2015 provided contradictory results on the effectiveness of DADLE in PD models (Eftekhar-Vaghefi et al. 2015). The 6-OHDA-induced model of PD in SH-SY5Y cells was used to mimic PD in vitro and was then treated with different opioid receptor agonists, DADLE being used as the DOR agonist (Eftekhar-Vaghefi et al. 2015). While MOR and KOR agonists had therapeutic effects, DADLE did not appear to display any significant neuroprotection (Eftekhar-Vaghefi et al. 2015). However, the lack of effect seen with DADLE may be due to the low number of DORs typically found in SH-SY5Y cells. Considering the positive effects of DADLE observed in other PD-related models, these findings should be further tested to probe DADLE's effectiveness as a PD therapeutic.

7.3.2 Anomalous Effects of DOR's and DADLE in Alzheimer's Disease

Consistent with the leading hypothesis of AD pathology, the aberrant accumulation of amyloid-beta protein is a defining hallmark of Alzheimer's disease. Amyloid-beta production occurs via the proteolysis of amyloid precursor protein (APP) by beta-site APP-cleaving enzyme (BACE1) and gamma-secretase (Teng et al. 2010). The sequential-proteolysis was shown to be mediated by DORs in a study conducted in 2010 (Teng et al. 2010). This study investigated the interaction of DOR, BACE1, and gamma-secretase using western-blot analysis, finding that DOR mediates the co-proteolysis and contributes to the formation of amyloid-beta (Teng et al. 2010). Interestingly, these results displayed that antagonism of DORs led to a decrease in amyloid-beta production and a reduction in pathology (Hayashi et al. 2002). Conversely, treatment with DADLE led to a significant elevation in BACE1 and gamma-secretase activity, leading to an increase in the amyloid-beta pathology (Teng et al. 2010). Thus, this study reveals AD as one of the few disorders in which DADLE and DOR's are pathology-inducing, and not neuroprotective (Teng et al. 2010). Similarly, studies have found that DOR activation contributes to the pathology of AD by decreasing levels of acetylcholine and increasing the phosphorylation of tau (Cai and Ratka 2012). Low levels of acetylcholine and hyper-phosphorylated tau have both been implicated in the dynamic pathology of AD (Cai and Ratka

2012). As a result, activation of DORs by DADLE would be expected to have negative effects.

These facts highlight the complexities that accompany DADLE as a therapy; the diverse interactions of DADLE allow it to have varied therapeutic effects, as well as possible unwanted consequences. As previously stated, the activation of DORs is not the only mechanism of action for DADLE. DADLE has been shown to function via DOR-independent mechanisms, therefore therapeutic benefits might be offered in AD by the DOR-independent actions of DADLE, while simultaneously being negated by its DOR-activation. Ultimately, the opioid system may still present a viable target for AD therapies, although it may be via DOR-antagonist and DOR-independent mechanisms.

7.3.3 DORS and DADLE in Huntington's Disease

An atrophying of subcortical grey matter in the basal ganglia is the defining characteristic of HD (Van Den Bogaard et al. 2011). Opposite of AD, the brains of HD patients have been shown to display decreased levels of DORs and endogenous enkephalin opioids (Cadet and Rothman 1986). Opioid receptors and peptides have a particularly high density within the basal ganglia and striatum of healthy individuals (Bissonnette et al. 2013), thus attempting to rescue the levels of DORs and endogenous opioids is a reasonable subject of investigation for HD therapies.

A study which increased the expression of striatal pre-enkephalin – the biological precursor to enkephalin, an endogenous DOR agonist – in the R6/2 rat model of HD showed an improvement in both motor activity and memory functions (Bissonnette et al. 2013). Although the mechanisms behind this improvement were not revealed, the investigation did note a slight, yet notable increase in striatal neuron survival (Bissonnette et al. 2013). In general, the implications of the DORs and delta peptides such as DADLE in HD are an understudied topic. However, our knowledge of the characteristic decline of DORs and endogenous enkephalins in HD makes the opioid system an attractive target.

8 Concluding Remarks

The search for new neurological therapies has directed investigators toward DOR modulation as a target for eliciting neuroprotective effects. The implication of DOR activation during cytotoxic and hypoxic/ischemic conditions of the brain carries clinical relevance and is largely supported throughout existing research. Current studies have discovered various mechanisms for the process of DOR-induced neuroprotection. Maintaining ionic homeostasis plays an important role in preventing neuronal cell death and is mediated by DOR activation after cerebral ischemia. Furthermore, DORs take part in enhancing endogenous neuroprotective pathways. The DOR agonist, DADLE, has been at the forefront of DOR activation research, having demonstrated the ability to confer resistance to neuronal injury and death in ischemia, neurodegeneration, and drug-induced stress. While DADLE surely poses a promising therapeutic option for various neurological disorders, many of the mechanistic details of DADLE and its interactions remain poorly understood. Multiple labs are currently spearheading the effort to reveal these details of DADLE, pushing it closer to clinical relevance.

DADLE has unique implications in the treatment of disorders involving progressive neurological decline via secondary cell death, such as incidences of stroke and TBI. As there are many facets to the pathology of secondary cell death, therapies that target multiple cytotoxic mechanisms may be optimal for producing significant functional improvements. DADLE has been suggested to have therapeutic effects on several aspects of secondary cell death such as oxidative stress, neuroinflammation, glutamate toxicity, and stress-induced apoptosis. By having a multi-pronged therapeutic profile, DADLE may represent a superior choice and potent neuroprotective agent in neurodegenerative disorders. Overall, the opioid system holds opportunity for neuroprotective treatment, especially via DORs and DOR agonists. Despite the existing evidence of neuroprotection via the opioid system, more research is needed in order to translate this knowledge into feasible clinical treatments.

Acknowledgments This study was supported by the National Institutes of Health (1RO1NS071956-01A1) and the James and Easter King Biomedical Research Program (1KG01-33966).

References

Baldelli B, Vecchio L, Bottone MG, Muzzonigro G, Biggiogera M, Malatesta M (2006) The effect of the enkephalin DADLE on transcription does not depend on opioid receptors. Histochem Cell Biol 126(2):189–197

Barde YA (1989) Trophic factors and neuronal survival. Neuron 2(6):1525–1534

Barker-Haliski M, White HS (2015) Glutamatergic mechanisms associated with seizures and epilepsy. Cold Spring Harb Perspect Med 5(8):a022863

Berardelli A, Dick JP, Rothwell JC, Day BL, Marsden CD (1986) Scaling of the size of the first agonist EMG burst during rapid wrist movements in patients with Parkinson's disease. J Neurol Neurosurg Psychiatry 49(11):1273–1279

Bickler PE (2004) Clinical perspectives: neuroprotection lessons from hypoxia-tolerant organisms. J Exp Biol 207(Pt 18):3243–3249

Bissonnette S, Vaillancourt M, Hebert SS, Drolet G, Samadi P (2013) Striatal pre-enkephalin overexpression improves Huntington's disease symptoms in the R6/2 mouse model of Huntington's disease. PLoS One 8(9):e75099

Blurton PA, Broadhurst AM, Wood MD, Wyllie MG (1986) Is there a common, high-affinity opioid binding site in rat brain? J Recept Res 6(1):85–93

Bofetiado DM, Mayfield KP, D'alecy LG (1996) Alkaloid delta agonist BW373U86 increases hypoxic tolerance. Anesth Analg 82(6):1237–1241

Borlongan CV, Su TP, Wang Y (1999) Delta opioid peptide (DADLE) enhances survival of cultured fetal cells. In: Committee on Problems of Drug Dependence 61st meeting abstract 13

Borlongan CV, Su TP, Wang Y (2000) Treatment with delta opioid peptide enhances in vitro and in vivo survival of rat dopaminergic neurons. Neuroreport 11(5):923–926

Borlongan CV, Su TP, Wang Y (2001) Delta opioid peptide augments functional effects and intrastriatal graft survival of rat fetal ventral mesencephalic cells. Cell Transplant 10(1):53–58

Borlongan CV, Wang Y, Su TP (2004) Delta opioid peptide (D-Ala 2, D-Leu 5) enkephalin: linking hibernation and neuroprotection. Front Biosci 9:3392–3398

Borlongan CV, Hayashi T, Oeltgen PR, Su TP, Wang Y (2009) Hibernation-like state induced by an opioid peptide protects against experimental stroke. BMC Biol 7:31

Borsello T, Clarke PG, Hirt L et al (2003a) A peptide inhibitor of c-Jun N-terminal kinase protects against excitotoxicity and cerebral ischemia. Nat Med 9(9):1180–1186

Borsello T, Croquelois K, Hornung JP, Clarke PG (2003b) N-methyl-d-aspartate-triggered neuronal death in organotypic hippocampal cultures is endocytic, autophagic and mediated by the c-Jun N-terminal kinase pathway. Eur J Neurosci 18(3):473–485

Boutin H, Dauphin F, Mackenzie ET, Jauzac P (1999) Differential time-course decreases in nonselective, mu-, delta-, and kappa-opioid receptors after focal cerebral ischemia in mice. Stroke 30(6):1271–1277. Discussion 1278

Brodie MJ, Barry SJ, Bamagous GA, Norrie JD, Kwan P (2012) Patterns of treatment response in newly diagnosed epilepsy. Neurology 78(20):1548–1554

Brodie MJ, Besag F, Ettinger AB et al (2016) Epilepsy, antiepileptic drugs, and aggression: an evidence-based review. Pharmacol Rev 68(3):563–602

Bruce DS, Cope GW, Elam TR, Ruit KA, Oeltgen PR, Su TP (1987) Opioids and hibernation. I. Effects of naloxone on bear HIT'S depression of Guinea pig ileum contractility and on induction of summer hibernation in the ground squirrel. Life Sci 41(18):2107–2113

Cadet JL, Rothman RB (1986) Decreased striatal opiate delta-receptors in the rat model of persistent dyskinesia induced by iminodipropionitrile. Neurosci Lett 72(1):84–86

Cai Z, Ratka A (2012) Opioid system and Alzheimer's disease. Neruomol Med 14(2):91–111

Centeno C, Repici M, Chatton JY et al (2007) Role of the JNK pathway in NMDA-mediated excitotoxicity of cortical neurons. Cell Death Differ 14(2):240–253

Chao D, Xia Y (2010) Ionic storm in hypoxic/ischemic stress: can opioid receptors subside it? Prog Neurobiol 90(4):439–470

Chao D, Qian H, Ghassemi F, Chen JS, Xia Y (2006) Transgenic over-expression of δ-opioid receptors protects the cortex from anoxic disruption of ionic homeostasis. Soc Neurosci Abstr Program No. 87.19/MM68

Chao D, Bazzy-Asaad A, Balboni G, Xia Y (2007a) Delta-, but not mu-, opioid receptor stabilizes K(+) homeostasis by reducing Ca(2+) influx in the cortex during acute hypoxia. J Cell Physiol 212(1):60–67

Chao D, Donnelly DF, Feng Y, Bazzy-Asaad A, Xia Y (2007b) Cortical delta-opioid receptors potentiate K+ homeostasis during anoxia and oxygen-glucose deprivation. J Cereb Blood Flow Metab 27(2):356–368

Chao D, Bazzy-Asaad A, Balboni G, Salvadori S, Xia Y (2008) Activation of DOR attenuates anoxic K+ derangement via inhibition of Na+ entry in mouse cortex. Cereb Cortex 18 (9):2217–2227

Chao D, Balboni G, Lazarus LH, Salvadori S, Xia Y (2009) Na+ mechanism of delta-opioid receptor induced protection from anoxic K+ leakage in the cortex. Cell Mol Life Sci 66 (6):1105–1115

Chu Sin Chung P, Kieffer BL (2013) Delta opioid receptors in brain function and diseases. Pharmacol Ther 140(1):112–120

Chung PC, Boehrer A, Stephan A et al (2015) Delta opioid receptors expressed in forebrain GABAergic neurons are responsible for SNC80-induced seizures. Behav Brain Res 278:429–434

Culmsee C, Gerling N, Lehmann M et al (2002) Nerve growth factor survival signaling in cultured hippocampal neurons is mediated through TrkA and requires the common neurotrophin receptor P75. Neuroscience 115(4):1089–1108

Davies C, Tournier C (2012) Exploring the function of the JNK (c-Jun N-terminal kinase) signalling pathway in physiological and pathological processes to design novel therapeutic strategies. Biochem Soc Trans 40(1):85–89

Dawe AR, Spurrier WA (1969) Hibernation induced in ground squirrels by blood transfusion. Science 163(3864):298–299

Drolet G, Dumont EC, Gosselin I, Kinkead R, Laforest S, Trottier JF (2001) Role of endogenous opioid system in the regulation of the stress response. Prog Neuro-Psychopharmacol Biol Psychiatry 25(4):729–741

Eftekhar-Vaghefi S, Esmaeili-Mahani S, Elyasi L, Abbasnejad M (2015) Involvement of mu opioid receptor signaling in the protective effect of opioid against 6-hydroxydopamine-induced SH-SY5Y human neuroblastoma cells apoptosis. Basic Clin Neurosci 6(3):171–178

Feng Y, Chao D, He X et al (2009) A novel insight into neuroprotection against hypoxic/ischemic stress. Sheng Li Xue Bao 61(6):585–592

Feng Y, He X, Yang Y, Chao D, Lazarus LH, Xia Y (2012) Current research on opioid receptor function. Curr Drug Targets 13(2):230–246

Frerichs KU, Hallenbeck JM (1998) Hibernation in ground squirrels induces state and species-specific tolerance to hypoxia and aglycemia: an in vitro study in hippocampal slices. J Cereb Blood Flow Metab 18(2):168–175

Fu D, Liu H, Zhu H, Li S, Yao J (2016) Protective effect of delta opioid receptor agonist (D-Ala2, D-Leu5) enkephalin on permanent focal cerebral ischemia in rats. Neuroreport 27(10):749–754

Fujimura M, Morita-Fujimura Y, Narasimhan P, Copin JC, Kawase M, Chan PH (1999) Copper-zinc superoxide dismutase prevents the early decrease of apurinic/apyrimidinic endonuclease and subsequent DNA fragmentation after transient focal cerebral ischemia in mice. Stroke 30 (11):2408–2415

Harper SJ, Lograsso P (2001) Signalling for survival and death in neurones: the role of stress-activated kinases, JNK and p38. Cell Signal 13(5):299–310

Hayashi T, Su TP (2003) Chronic [D-Ala2, D-Leu5]enkephalin treatment increases the nerve growth factor in adult mouse brain. Eur J Pharmacol 464(2–3):237–239

Hayashi T, Tsao LI, Cadet JL, Su TP (1999) [D-Ala2, D-Leu5]enkephalin blocks the methamphetamine-induced c-fos mRNA increase in mouse striatum. Eur J Pharmacol 366 (2–3):R7–R8

Hayashi T, Tsao LI, Su TP (2002) Antiapoptotic and cytotoxic properties of delta opioid peptide [D-Ala(2),D-Leu(5)]enkephalin in PC12 cells. Synapse 43(1):86–94

Hiller JM, Fan LQ (1996) Laminar distribution of the multiple opioid receptors in the human cerebral cortex. Neurochem Res 21(11):1333–1345

Horiuchi T, Kawaguchi M, Sakamoto T et al (2004) The effects of the delta-opioid agonist SNC80 on hind-limb motor function and neuronal injury after spinal cord ischemia in rats. Anesth Analg 99(1):235–240

Horton ND, Kaftani DJ, Bruce DS et al (1998) Isolation and partial characterization of an opioid-like 88 kDa hibernation-related protein. Comp Biochem Physiol B Biochem Mol Biol 119 (4):787–805

Iwata M, Inoue S, Kawaguchi M et al (2007) Delta opioid receptors stimulation with [D-Ala2, D-Leu5] enkephalin does not provide neuroprotection in the hippocampus in rats subjected to forebrain ischemia. Neurosci Lett 414(3):242–246

Kang X, Chao D, Gu Q et al (2009) Delta-opioid receptors protect from anoxic disruption of Na+ homeostasis via Na+ channel regulation. Cell Mol Life Sci 66(21):3505–3516

Karki P, Seong C, Kim JE et al (2007) Intracellular K(+) inhibits apoptosis by suppressing the Apaf-1 apoptosome formation and subsequent downstream pathways but not cytochrome c release. Cell Death Differ 14(12):2068–2075

Kawase M, Fujimura M, Morita-Fujimura Y, Chan PH (1999) Reduction of apurinic/apyrimidinic endonuclease expression after transient global cerebral ischemia in rats: implication of the failure of DNA repair in neuronal apoptosis. Stroke 30(2):441–448. Discussion 449

Ke S, Dian-San S, Xiang-Rui W (2009) Delta opioid agonist [D-Ala2, D-Leu5] enkephalin (DADLE) reduced oxygen-glucose deprivation caused neuronal injury through the MAPK pathway. Brain Res 1292:100–106

Kevelaitis E, Peynet J, Mouas C, Launay JM, Menasche P (1999) Opening of potassium channels: the common cardioprotective link between preconditioning and natural hibernation? Circulation 99(23):3079–3085

Kinouchi H, Epstein CJ, Mizui T, Carlson E, Chen SF, Chan PH (1991) Attenuation of focal cerebral ischemic injury in transgenic mice overexpressing CuZn superoxide dismutase. Proc Natl Acad Sci U S A 88(24):11158–11162

Koide S, Onishi H, Yamagami S, Kawakita Y (1992) Effects of morphine and D-Ala2-D-Leu5-enkephalin in the seizure-susceptible El mouse. Neurochem Res 17(8):779–783

Lim YJ, Zheng S, Zuo Z (2004) Morphine preconditions Purkinje cells against cell death under in vitro simulated ischemia-reperfusion conditions. Anesthesiology 100(3):562–568

Liu D, Slevin JR, Lu C et al (2003) Involvement of mitochondrial K+ release and cellular efflux in ischemic and apoptotic neuronal death. J Neurochem 86(4):966–979

Liu H, Chen B, Zhang Y et al (2015) Protective effect of delta opioid agonist [D-Ala2, D-Leu5] enkephalin on spinal cord ischemia reperfusion injury by regional perfusion into abdominal aorta in rabbits. Neurosci Lett 584:1–6

Liu H, Chen B, Li S, Yao J (2016) Dose-dependent neuroprotection of delta-opioid peptide [D-Ala (2), D-Leu(5)] enkephalin on spinal cord ischemia-reperfusion injury by regional perfusion into the abdominal aorta in rabbits. J Vasc Surg 63(4):1074–1081

Lorigados Pedre L, Pavon Fuentes N, Alvarez Gonzalez L et al (2002) Nerve growth factor levels in Parkinson disease and experimental parkinsonian rats. Brain Res 952(1):122–127

Ma MC, Qian H, Ghassemi F, Zhao P, Xia Y (2005) Oxygen-sensitive {delta}-opioid receptor-regulated survival and death signals: novel insights into neuronal preconditioning and protection. J Biol Chem 280(16):16208–16218

Mailly P, Gastard M, Cupo A (1999) Subcellular distribution of delta-opioid receptors in the rat spinal cord: an approach using a three-dimensional reconstruction of confocal series of immunolabelled neurons. J Neurosci Methods 87(1):17–24

Mansour A, Khachaturian H, Lewis ME, Akil H, Watson SJ (1987) Autoradiographic differentiation of mu, delta, and kappa opioid receptors in the rat forebrain and midbrain. J Neurosci 7 (8):2445–2464

Mayfield KP, D'alecy LG (1994) Delta-1 opioid agonist acutely increases hypoxic tolerance. J Pharmacol Exp Ther 268(2):683–688

Mayfield KP, Kozak W, Malvin GM, Porreca F (1996) Hypoxia decreases opioid delta receptor expression in mouse brain. Neuroscience 72(3):785–789

Mizuta I, Ohta M, Ohta K, Nishimura M, Mizuta E, Kuno S (2001) Riluzole stimulates nerve growth factor, brain-derived neurotrophic factor and glial cell line-derived neurotrophic factor synthesis in cultured mouse astrocytes. Neurosci Lett 310(2–3):117–120

Mongin AA (2007) Disruption of ionic and cell volume homeostasis in cerebral ischemia: the perfect storm. Pathophysiology 14(3–4):183–193

Narita M, Kuzumaki N, Miyatake M et al (2006) Role of delta-opioid receptor function in neurogenesis and neuroprotection. J Neurochem 97(5):1494–1505

Nistico R, Piccirilli S, Sebastianelli L, Nistico G, Bernardi G, Mercuri NB (2007) The blockade of K(+)-ATP channels has neuroprotective effects in an in vitro model of brain ischemia. Int Rev Neurobiol 82:383–395

Noshita N, Sugawara T, Hayashi T, Lewen A, Omar G, Chan PH (2002) Copper/zinc superoxide dismutase attenuates neuronal cell death by preventing extracellular signal-regulated kinase activation after transient focal cerebral ischemia in mice. J Neurosci 22(18):7923–7930

Nussbaum RL, Ellis CE (2003) Alzheimer's disease and Parkinson's disease. N Engl J Med 348 (14):1356–1364

Oeltgen PR, Welborn JR, Nuchols PA, Spurrier WA, Bruce DS, Su TP (1987) Opioids and hibernation. II. Effects of kappa opioid U69593 on induction of hibernation in summer-active ground squirrels by "hibernation induction trigger" (HIT). Life Sci 41(18):2115–2120

Oeltgen PR, Nilekani SP, Nuchols PA, Spurrier WA, Su TP (1988) Further studies on opioids and hibernation: delta opioid receptor ligand selectively induced hibernation in summer-active ground squirrels. Life Sci 43(19):1565–1574

Olson L, Backman L, Ebendal T et al (1994) Role of growth factors in degeneration and regeneration in the central nervous system; clinical experiences with NGF in Parkinson's and Alzheimer's diseases. J Neurol 242(1 Suppl 1):S12–S15
Peart JN, Gross ER, Gross GJ (2005) Opioid-induced preconditioning: recent advances and future perspectives. Vasc Pharmacol 42(5–6):211–218
Peng PH, Huang HS, Lee YJ, Chen YS, Ma MC (2009) Novel role for the delta-opioid receptor in hypoxic preconditioning in rat retinas. J Neurochem 108(3):741–754
Repici M, Centeno C, Tomasi S et al (2007) Time-course of c-Jun N-terminal kinase activation after cerebral ischemia and effect of D-JNKI1 on c-Jun and caspase-3 activation. Neuroscience 150(1):40–49
Sen D, Huchital M, Chen YL (2013) Crosstalk between delta opioid receptor and nerve growth factor signaling modulates neuroprotection and differentiation in rodent cell models. Int J Mol Sci 14(10):21114–21139
Sick TJ, Rosenthal M, Lamanna JC, Lutz PL (1982) Brain potassium ion homeostasis, anoxia, and metabolic inhibition in turtles and rats. Am J Phys 243(3):R281–R288
Song ZF, Ji XP, Li XX, Wang SJ, Wang SH, Zhang Y (2008) Inhibition of the activity of poly (ADP-ribose) polymerase reduces heart ischaemia/reperfusion injury via suppressing JNK-mediated AIF translocation. J Cell Mol Med 12(4):1220–1228
Sung JH, Chao DM, Xia Y (2008) Neuronal responses to hypoxia. In: Wang D, Ying W (eds) New frontiers in neurological research. Research Signpost, Kerala, pp 73–153
Teng L, Zhao J, Wang F, Ma L, Pei G (2010) A GPCR/secretase complex regulates beta- and gamma-secretase specificity for Abeta production and contributes to AD pathogenesis. Cell Res 20(2):138–153
Tian J, Gu Y, Sun K et al (2014) [D-Ala2, D-Leu5] encephalin (DADLE) reversibly inhibits cellular transcription in neurons without causing cell injury. Brain Res 1565:1–7
Tortella FC, Robles L, Holaday JW (1985) The anticonvulsant effects of DADLE are primarily mediated by activation of delta opioid receptors: interactions between delta and mu receptor antagonists. Life Sci 37(6):497–503
Tsao LI, Ladenheim B, Andrews AM, Chiueh CC, Cadet JL, Su TP (1998) Delta opioid peptide [D-Ala2,D-leu5]enkephalin blocks the long-term loss of dopamine transporters induced by multiple administrations of methamphetamine: involvement of opioid receptors and reactive oxygen species. J Pharmacol Exp Ther 287(1):322–331
Tsao LI, Cadet JL, Su TP (1999) Reversal by [D-Ala2,D-Leu5]enkephalin of the dopamine transporter loss caused by methamphetamine. Eur J Pharmacol 372(3):R5–R7
Turner SM, Johnson SM (2011) Delta-opioid receptor activation prolongs respiratory motor output during oxygen-glucose deprivation in neonatal rat spinal cord in vitro. Neuroscience 187:70–83
Van Den Bogaard SJ, Dumas EM, Acharya TP et al (2011) Early atrophy of pallidum and accumbens nucleus in Huntington's disease. J Neurol 258(3):412–420
Van Rijn RM, Defriel JN, Whistler JL (2013) Pharmacological traits of delta opioid receptors: pitfalls or opportunities? Psychopharmacology 228(1):1–18
Vecchio L, Soldani C, Bottone MG et al (2006) DADLE induces a reversible hibernation-like state in HeLa cells. Histochem Cell Biol 125(1–2):193–201
Vercelli A, Biggi S, Sclip A et al (2015) Exploring the role of MKK7 in excitotoxicity and cerebral ischemia: a novel pharmacological strategy against brain injury. Cell Death Dis 6:e1854
Wang Q, Chao D, Chen T, Sandhu H, Xia Y (2014) Delta-opioid receptors and inflammatory cytokines in hypoxia: differential regulation between glial and neuron-like cells. Transl Stroke Res 5(4):476–483
Wang Z, Tang B, Tang F et al (2015) Protection of rat intestinal epithelial cells from ischemia/reperfusion injury by (D-Ala2, D-Leu5)-enkephalin through inhibition of the MKK7-JNK signaling pathway. Mol Med Rep 12(3):4079–4088

Wang SY, Duan YL, Zhao B, Wang XR, Zhao Z, Zhang GM (2016) Effect of delta opioid receptor activation on spatial cognition and neurogenesis in cerebral ischemic rats. Neurosci Lett 620:20–26

Wei L, Yu SP, Gottron F, Snider BJ, Zipfel GJ, Choi DW (2003) Potassium channel blockers attenuate hypoxia- and ischemia-induced neuronal death in vitro and in vivo. Stroke 34 (5):1281–1286

Xia Y (2015) Neural functions of the delta-opioid receptor. Springer, Cham

Xia Y, Haddad GG (1991) Ontogeny and distribution of opioid receptors in the rat brainstem. Brain Res 549(2):181–193

Xia Y, Haddad GG (2001) Major difference in the expression of delta- and mu-opioid receptors between turtle and rat brain. J Comp Neurol 436(2):202–210

Xia Y, Jiang C, Haddad GG (1992) Oxidative and glycolytic pathways in rat (newborn and adult) and turtle brain: role during anoxia. Am J Phys 262(4 Pt 2):R595–R603

Xiang H, Hochman DW, Saya H, Fujiwara T, Schwartzkroin PA, Morrison RS (1996) Evidence for p53-mediated modulation of neuronal viability. J Neurosci 16(21):6753–6765

Yu SP, Yeh CH, Sensi SL et al (1997) Mediation of neuronal apoptosis by enhancement of outward potassium current. Science 278(5335):114–117

Yu G, Liang Y, Huang Z, Jones DW, Pritchard KA Jr, Zhang H (2016) Inhibition of myeloperoxidase oxidant production by N-acetyl lysyltyrosylcysteine amide reduces brain damage in a murine model of stroke. J Neuroinflammation 13(1):119

Zhang JH, Xia Y, Haddad GG (1999) Activation of δ-opioid receptors protects cortical neurons from glutamate excitotoxic injury. Soc Neurosci Abstr 28:736

Zhang J, Haddad GG, Xia Y (2000) Delta-, but not mu- and kappa-, opioid receptor activation protects neocortical neurons from glutamate-induced excitotoxic injury. Brain Res 885(2):143–153

Zvara DA (2002) Thoracoabdominal aneurysm surgery and the risk of paraplegia: contemporary practice and future directions. J Extra Corpor Technol 34(1):11–17

Delta Opioid Receptors and Cardioprotection

Louise See Hoe, Hemal H. Patel, and Jason N. Peart

Contents

L. See Hoe
Menzies Health Institute Queensland, Griffith University, Southport, QLD 4222, Australia
Critical Care Research Group, The Prince Charles Hospital and The University of Queensland, Chermside, QLD, Australia

H.H. Patel
VA San Diego Healthcare System, San Diego, CA, USA
Department of Anesthesiology, University of California San Diego, La Jolla, CA, USA

J.N. Peart (✉)
Menzies Health Institute Queensland, Griffith University, Southport, QLD 4222, Australia
e-mail: j.peart@griffith.edu.au

E. M. Jutkiewicz (ed.), *Delta Opioid Receptor Pharmacology and Therapeutic Applications*, Handbook of Experimental Pharmacology 247, DOI 10.1007/164_2017_6

Abstract

The opioid receptor family, with associated endogenous ligands, has numerous roles throughout the body. Moreover, the delta opioid receptor (DORs) has various integrated roles within the physiological systems, including the cardiovascular system. While DORs are important modulators of cardiovascular autonomic balance, they are well-established contributors to cardioprotective mechanisms. Both endogenous and exogenous opioids acting upon DORs have roles in myocardial hibernation and protection against ischaemia-reperfusion (I-R) injury. Downstream signalling mechanisms governing protective responses alternate, depending on the timing and duration of DOR activation. The following review describes models and mechanisms of DOR-mediated cardioprotection, the impact of co-morbidities and challenges for clinical translation.

Keywords

Cardioprotection • Delta opioid receptor • Ischaemia-reperfusion injury

1 Introduction

Opioids elicit both paracrine and autocrine functions within the cardiovascular system via activation of opioid receptors (OR) (Headrick et al. 2012; Peart et al. 2005; Schultz and Gross 2001). The three primary receptor subtypes, the mu (MOR), kappa (KOR) and delta (DOR) opioid receptors, act as $G_{i/o}$ protein-coupled receptors to inhibit adenylate cyclase (AC) and downstream signalling via second messenger cyclic AMP (cAMP) (Peart et al. 2005; Schultz and Gross 2001; Pugsley 2002). For regulation of the cardiovascular system, ORs are localised centrally to the respiratory and cardiovascular centres of the hypothalamus (May et al. 1989; Goodman et al. 1980), and brainstem (Goodman et al. 1980); and peripherally to the adrenal medulla (Wittert et al. 1996), cardiomyocytes (Ventura et al. 1989), and the microvasculature (Peroutka et al. 1980). Enkephalin (DOR), dynorphin (KOR) and endorphin (MOR) represent the major endogenous opioids involved in significant neuroregulation (Headrick et al. 2012; Peart et al. 2005). The intrinsic cardiac role of opioids is to maintain and regulate daily functioning in healthy (or diseased) myocardium. In this role, the opioid system primarily modulates autonomic control, balancing and countering the β-adrenergic system and actions of catecholamines (Headrick et al. 2012; Xiao et al. 1997). Opioid receptor activation also modulates systemic vascular tone, alters cardiac excitation-contraction coupling, and may be involved in cardiogenesis (Headrick et al. 2012; Holaday 1983; Pepe et al. 2004; Ventura et al. 1992; Maslov et al. 2006). Specifically, opioids have generated extensive interest as modulators of ischaemic tolerance, particularly through KOR and DOR agonism (Headrick et al. 2015). Opioids serve as desirable cardioprotective candidates due to their involvement in reducing all outcomes from ischaemia-reperfusion (I-R) injury (Romano et al. 2004a; Peart et al. 2011) (e.g. infarction, contractile dysfunction, inflammation, arrhythmogenesis), contribution to various endogenous and exogenous cardioprotective mechanisms (Schultz et al. 1995; Maslov et al. 1996), role in exercise-derived tolerance to ischaemia (Dickson et al. 2008), and protective myocardial hibernation (Heusch 1998). This chapter will focus specifically upon the involvement of DORs in cardioprotection.

1.1 Cardiovascular Disease and the Need for Adjunctive Protective Therapies

Cardiovascular disease (CVD) continues to be the leading cause of death worldwide, responsible for 33% of all deaths, with 50% of these attributed specifically to coronary artery disease (Roger et al. 2012). Hyperlipidaemia, glucose intolerance, hypertension, physical inactivity, and tobacco smoking remain prevalent risk factors for CVD (Roger et al. 2012). These risk factors can be managed and prevented through strategic lifestyle, pharmacological and surgical interventions (Whitworth 2003; Peart and Headrick 2009).

An acute myocardial infarction (AMI) develops from complete or partial occlusion of the coronary artery, leading to deprivation of oxygen and substrate delivery to a region of the heart, coupled with reduced washout of toxic by-products. This ischaemic event is characterised by insufficient energy, substrate and oxygen available to meet the metabolic demands of the myocardium for normal function, with effects worsened by reduced tissue washout of ions/metabolites. The myocardium can only be salvaged by restoration of blood flow to re-establish substrate and oxygen delivery, termed reperfusion (Jennings and Reimer 1991). Clinical reperfusion is achieved via primary percutaneous coronary intervention (PCI) or thrombolysis (Widimsky et al. 2010). Importantly, whilst essential, reperfusing the ischaemic myocardium is detrimental, inducing paradoxical 'reperfusion injury'. This widely accepted phenomenon, initially proposed by Jennings et al. in 1960, describes myocardial injury and cell death that is induced specifically by reperfusion of occluded vessels (Jennings and Reimer 1991; Jennings et al. 1960). Since both ischaemia, and subsequent and essential reperfusion both damage the heart, a field of cardioprotective research has evolved aimed at clinically combating ischaemia-reperfusion (I-R) injuries (Peart and Headrick 2009).

Cardioprotective therapy (or adjunctive cardioprotection) aims to limit cell death and dysfunction, improve conventional reperfusion strategies, and enhance short- and long-term outcomes from accidental (e.g. AMI) or surgical ischaemia (Peart and Headrick 2009; Shi and Vinten-Johansen 2012; Sanada et al. 2011). Specifically, the heart can be manipulated through endogenous or exogenous agonist-induced signalling to generate an intrinsic hormesis response to prolonged ischaemic injury. A variety of strategies have been assessed, with early studies of adrenoceptor modulation, adenosine receptor agonism, glucose-insulin-potassium related strategies (among others) providing some initially encouraging findings regarding the potential to experimentally modify extent and progression of cellular damage during I-R (Peart and Headrick 2007, 2008; Vander Heide and Steenbergen 2013; McIntosh and Lasley 2012). More recently, so-called conditioning stimuli have been identified and heavily studied. Murry and colleagues initially discovered the ischaemic preconditioning (IPC) phenomenon in 1986. Subsequently, an array of potential cardioprotective strategies have been studied and trialled, notably pre- and postconditioning stimuli that employ brief cycles of I-R (or pharmacological mimetics) prior to or following ischaemia to protect the heart (Peart and Headrick 2009; Shi and Vinten-Johansen 2012; Murry et al. 1986). Such conditioning interventions may also be applied remotely (e.g. in other organs/tissues) to transduce protection to the diseased heart (Przyklenk et al. 1993). This protective response induces an initial acute window of defence that engages post-translational kinase modifications and subsequent signalling, followed by a delayed window of protection through de novo protein synthesis extending multiple days (Headrick et al. 2015; Baxter and Ferdinandy 2001). Research continues to uncover varied signalling mechanisms that improve the stress resistance of myocardial tissue and those involved in cardioprotection (Schultz and Gross 2001; Headrick et al. 2015; Peart and Headrick 2007, 2008, 2009; Vander Heide and Steenbergen 2013; Patel et al. 2007; Perrelli et al. 2011; Przyklenk and Whittaker 2011; Rakhit and Marber 2001; Stary et al. 2012; Sun and Murphy 2010; Wojtovich et al. 2012; Yellon and Downey 2003; Roth and Patel 2011; Fridolfsson et al. 2012).

While many cardioprotective strategies demonstrate efficacy in laboratory settings, essentially all have failed in clinical translation. This failure has been attributed to the negative influences of age, disease and chronic pharmacotherapy, among other relevant factors (Peart and Headrick 2009; Vander Heide and Steenbergen 2013; Przyklenk et al. 2008; Miura and Miki 2008). Importantly, these inhibitory features are common among the target population that requires effective cardioprotective strategies; therefore, strategic studies to improve clinical efficacies are critical (Peart and Headrick 2009; Bolli et al. 2004). Engagement of the DOR through both endogenous and exogenous opioid ligands has been prominently implicated in cardioprotective mechanisms, and may afford strategic protection where other therapies fail (Peart et al. 2011; Peart and Headrick 2009).

1.2 Myocardial Delta Opioid Receptors

Cardiac cells (cardiomyocytes) possess large stores of genes encoding endogenous opioid precursors (pre-proenkephalin, prodynorphin, pro-opiomelanocortin) (Barron 2000; Barron et al. 1992; Caffrey et al. 1994). Furthermore, pre-proenkephalin mRNA appears most abundant in cardiac tissue (Howells et al. 1986). This permits significant myocardial synthesis, storage and release of endogenous peptides in response to physiological and pathological stimuli, such as ischaemia (Romano et al. 2004b; Eliasson et al. 1998), exercise (Mougin et al. 1987), ageing (Caffrey et al. 1994) and cardioprotective intervention (Zatta et al. 2008). Many animal and human studies have localised DORs to cardiomyocytes, which mediate a plethora of cellular actions through engagement with endogenous opioid peptides. Many report that DORs are expressed on both atrial and ventricular cardiomyocytes in various mammalian species including rodents (Wittert et al. 1996; Ventura et al. 1989, 1992; Krumins et al. 1985), pigs (Theisen et al. 2014; Karlsson et al. 2012) and humans (Villemagne et al. 2002; Sobanski et al. 2014; Bell et al. 2000; Peng et al. 2012; Lendeckel et al. 2005). The DOR expression profile in myocardial tissues reflects the important role that DOR-dependent signalling exerts upon the heart, including alleviation of adrenergic signalling (Xiao et al. 1997), baroreceptor mechanism (Giles et al. 1987), vagal bradycardia (due to inhibition of acetylcholine release) (Caffrey et al. 1995), cardiac labour (Vargish and Beamer 1989) and activation of cardioprotective signalling and responses (Headrick et al. 2015).

Myocardial DORs are bound to the sarcolemma through interaction with caveolins, scaffolding proteins that regulate downstream receptor-dependent responses through tethering to various signalling moieties (to be discussed in further detail later in the chapter) (Roth and Patel 2011; Patel et al. 2006; See Hoe et al. 2014; Tsutsumi et al. 2010). Engagement of DORs with endogenous and exogenous opioids mediates various myocardial responses through activation of conventional protein kinase signalling cascades. Upon activation, these G protein-coupled receptors (GPCRs) inhibit adenylate cyclase (AC) and downstream signalling via second messenger cyclic AMP (cAMP) (Peart et al. 2005; Schultz and Gross 2001; Pugsley 2002). Like other GPCRs, DOR desensitisation occurs after prolonged agonist stimulation, through GRK-mediated uncoupling of $G_{i/o}$ and β-arrestin-2 or -3 recruitment (Al-Hasani and Bruchas 2011; Pradhan et al. 2016). Interestingly, DOR desensitisation and internal trafficking can lead to distinct tolerance profiles despite activation by agonists with similar binding and analgesic properties (Al-Hasani and Bruchas 2011; Pradhan et al. 2010). This is demonstrable of the DOR's biased agonism profile, highlighting that diverse ligand-directed receptor conformations initiate multiple signalling responses and has been shown in various in vitro and in vivo models of pharmacological activation (Pradhan et al. 2010, 2012). Indeed, it is the extracellular ligand-binding domain and the carboxyl-terminal extremities of all opioid receptors that are particularly disparate (Jordan et al. 2000). This biased agonism feature is important for consideration of beneficial vs. untoward side effects in therapeutic drug development, particularly for efficacious cardioprotective modalities with opioidergic agents.

1.3 Hibernation

Hibernation is a natural adaptation for certain mammals, such as black bears, arctic and 13-lined ground squirrels, brown cave bats and woodchucks. During hibernation, a unique state of energy conservation is entered by lowering and slowing physiological functions (such as heart and metabolic rates) (Heusch 1998; Horton et al. 1998). Mammalian hibernation is coupled to intracellular acidosis, hypoxia, and a reduction in energy stores. Indeed, some animals conserve up to 90% of energy needed for an active winter. Despite the conceivable danger associated with these changes that parallel ischaemia, the myocardium becomes hypoxia-tolerant and highly resilient against injury. Myocardial hibernation results in an absence of necrosis, preservation of inotropic reserve, recovery of contractile function post-ischaemia, and energy metabolism during prolonged ischaemia (Heusch 1998).

Many have suggested the causative agent for this protective state is opioid-like in nature whose endogenous production is increased during hibernation, stimulating beneficial effects through DOR activation (Horton et al. 1998; Bolling et al. 1997, 1998; Kevelaitis et al. 1999; Bruce et al. 1996). Specifically, an opioid-like protein was found in the serum of hibernating mammals, capable of inducing protection in non-hibernating animals to improve post-ischaemic functional recovery and preserve myocardial ultrastructure (Horton et al. 1998; Bolling et al. 1997, 1998; Kevelaitis et al. 1999). Protection with hibernation also parallels that of ischaemic preconditioning (IPC), in terms of involvement of ATP-sensitive potassium (K_{ATP}) channel opening (Kevelaitis et al. 1999). Bolling and colleagues have demonstrated in studies of hypothermia and cardioplegia that DOR agonism may provide additive cardioprotective potential, while DOR antagonism blocks post-ischaemic functional protection (Romano et al. 2004a, b; Schwartz et al. 1999; Bolling et al. 2001). Using a similar model of hypothermic myocardial ischaemia, Benedict et al. observed improvements in cardiovascular functional parameters in rabbit hearts pre-treated with various DOR agonists (Benedict et al. 1999).

Although the DOR has proven most important in hibernation, evidence suggests some roles for the nociceptive MOR and KOR. Indeed, binding density changes in different opioid receptors of hibernating and non-hibernating Columbian ground squirrels imply distinct physiological roles for endogenous opioids and opioid receptors during specific hibernation states (Cui et al. 1996, 1997). Tamura et al. suggested that activation of MORs in the hypothalamus by β-endorphin is responsible for body temperature regulation during the maintenance phase of hibernation in Syrian hamsters (Tamura et al. 2012). Primary cultured hamster hippocampal neurons maintained at <22°C for 7 days were protected against low-temperature induced cell death when exposed to morphine, through activation of the MOR, DOR, KORs (Tamura et al. 2006). Interestingly, Romano and colleagues also found a role for KOR in protection against hypothermic myocardial ischaemia, and pre-treatment with the κ-agonist U504488H induced protection, while functional and metabolic recovery was impaired with κ-antagonism using (Chien et al. 1991, 1994; Oeltgen et al. 1996) nor-BNI (Romano et al. 2004a).

The delta-opioid peptide [D-ala2, D-leU5]-enkephalin (DADLE) is a pharmacological inducer of hibernation (in summer-active ground squirrels), enhances survival of peripheral organs (such as kidney, liver, lung and heart), and acts as an anti-ischaemic agent (Borlongan et al. 2004, 2009; Oeltgen et al. 1988). Treatment with DADLE also induces a hibernation-like state in rats subjected to experimental stroke (Borlongan et al. 2009). The DOR agonists Deltorphin-D(variant) (Delt-D(var)) and hibernating woodchuck plasma (HWP) have been shown to limit infarct development and improve behavioural deficits in a mouse model of middle cerebral artery occlusion (Govindaswami et al. 2008). The authors linked these neuroprotective changes to a mechanism reliant upon the inhibition of nitric oxide (NO) release from ischaemic tissue (Govindaswami et al. 2008). These studies thus highlight an important role for opioids in brain injury and protection. Additionally, Hong and colleagues found that swine skeletal muscle bundles pre-incubated with HWP prior to hypoxia-reoxygenation exhibited significantly greater recoveries of force, an effect blocked by naloxone (Hong et al. 2005). It is thus clear that opioids are vital to cellular function and protection against injury across various tissue types (and insults).

An alternative form of 'myocardial hibernation' arises in response to prolonged periods of under-perfusion. Coronary artery disease patients often exhibit impaired myocardial contractile function, with hearts adapting to reduced blood flow by depressed contractile function (and thus O_2 demand) and avoiding irreversible cell death (Heusch 1998; Rahimtoola 1985). However, opioids may not participate in this clinical phenomenon: Schulz et al. found that despite a potential role for endogenous opioids in ischaemic preconditioning in pigs, they do not appear to mediate short-term hibernation, in which myocardial blood flow and contractile function are depressed to ~50% baseline in the absence of oncosis (Schulz et al. 2001).

1.4 Opioid-Induced Cardioprotective Mechanisms

Endogenous and exogenous opioids can generate different cardoioprotecive phenotypes through both direct OR engagement and alternate indirect mechanisms. Several studies link MOR activation to attenuated inflammation in in vivo animal models, which could limit inflammatory damage triggered by I-R (Zhang et al. 2004; Wang et al. 1998). Additionally, opioid (particularly KOR) agonists have been implicated in Na^+, K^+ and Ca^{2+} channel modulation. This is consistent with roles in cardioprotection, as modulation of these channels through non-receptor mediated mechanisms may reduce Na^+ and Ca^{2+} overload that contributes to I-R injury. These mechanisms may also be relevant to antiarrhythmic properties of KOR agonists (Pugsley et al. 1993, 1998). *Receptor-mediated* or direct mechanisms by which opioids induce ischaemic tolerance have been investigated in depth as acute stimuli, though less well studied as a sustained or prolonged stimulus. Both acute and prolonged DOR stimuli afford protection, yet they appear to be mediated by distinct signalling pathways (Peart and Gross 2006).

Cardioprotection via ischaemic preconditioning and postconditioning, and related pharmacological agents, appear to involve common mediators and signalling pathways, particularly involvement of so-called reperfusion injury salvage kinase (RISK) components converging on key mitochondrial targets – the mitochondrial permeability transition pore (mPTP) and K_{ATP} channels (mK_{ATP}) (Wojtovich et al. 2012). Endogenously released or exogenous GPCR agonists such as opioids, adenosine, noradrenaline or bradykinin initiate preconditioning via binding to respective I-R sensitive GPCRs (Peart and Headrick 2009; Sanada et al. 2011; Abete et al. 2011). This receptor engagement triggers downstream pro-survival signalling involving PKC, ERK1/2 and Akt/phosphatidylinositol 3-kinase (PI3K). These varied and inter-related conditioning and GPCR-mediated responses are highly dependent upon membrane makeup and microdomains, particularly caveolae and associated caveolins (Parton and del Pozo 2013; Patel et al. 2008a; Insel et al. 2005). Thus, initiation of cardioprotection is critically dependent upon caveolae and caveolin-3, and caveolins may also play a key role in transduction of protective effects to the mitochondria (Fridolfsson et al. 2012).

Activation of these kinase pathways may confer protection through regulation of diverse effectors, including pro- (Bax, Bad) and anti-apoptotic (Bcl-2 and P70s6K) elements, eNOS activity and subsequent NO release, glucose uptake, Ca^{2+} overload inhibition, phosphorylation and inactivation of GSK3β, modulation of autophagy, inhibition of mPTP and activation of mK_{ATP} channels. Additionally, the production of ROS is implicated in the activation of p38 MAPK and JAK/STAT pathways that also confer protection (Peart and Headrick 2009; Sanada et al. 2011; Abete et al. 2011; Yin et al. 2012; Vinten-Johansen et al. 2005). Ischaemic postconditioning shares many common signalling elements, though the underlying mechanisms are less well understood. Research suggests the postconditioning pathway involves Akt/PI3K, ERK1/2, PKC, eNOS activation, PKG, GSK3β inhibition, mPTP inhibition and mitoK_{ATP} activation (Peart and Headrick 2009; Sanada et al. 2011; Yin et al. 2012; Vinten-Johansen et al. 2005). Ischaemic tolerance induced with postconditioning is reportedly associated with reduced inflammation, oxidative stress, Ca^{2+} accumulation and preservation of endothelial function (Gomez et al. 2008; Tsang et al. 2004; Javadov et al. 2003), among other benefits.

1.5 Endogenous Opioidergic Protection

Intrinsic opioid receptor engagement has been heavily linked to myocardial ischaemic pre- (Schultz et al. 1995, 1996, 1997a, b; Bell et al. 2000; Schulz et al. 2001; Okubo et al. 2004; Peart et al. 2003) and postconditioning (Zatta et al. 2008; Chen et al. 2008; Fuardo et al. 2013; Jang et al. 2008), and OR antagonism ameliorates these protective responses in various species (summarised recently in (Headrick et al. 2015)) including the human heart (Tomai et al. 1999). Schultz et al. first demonstrated that the infarct-sparing effect of IPC was countered by OR antagonism with naloxone in rat myocardium, implicating OR activation in IPC (Schultz et al. 1995). Furthermore, previous studies have specifically implicated DOR in early and late phase IPC (Schultz et al. 1997b; Fraessdorf et al. 2015), coupled to increases in DOR expression and endogenous opioid peptide release (Fraessdorf et al. 2015). Consistent with a role for DOR in IPC, several studies confirmed that selective DOR antagonism in IPC worsens infarct development, and DOR activation mimics protective IPC in rodents (Wang et al. 2001; Valtchanova-Matchouganska and Ojewole 2003) and human tissue (Bell et al. 2000). Acute OR activation has been shown to mirror consequential signalling associated with conventional preconditioning, including recruitment and activation of RISK components via PI3K, involvement of ROS, inhibition of pro-apoptotic effectors (e.g. GSK3β), and convergence upon mitochondrial targets including the mPTP and mK_{ATP} channels (Peart et al. 2005; Cohen et al. 2000; Fryer et al. 2001; Patel et al. 2002a).

Independent of RISK signalling, the Survivor Activating Factor Enhancement (SAFE) pathway may also mediate postconditioning effects, involving a paradoxical protective function of TNFα (Lecour 2009). You et al. demonstrated that infarct size and apoptosis were reduced in postconditioning via up-regulation of anti-apoptotic Bcl-2, in response to OR engagement and JAK/STAT signalling (You et al. 2011). It has been suggested that activation of both PI3K and JAK/STAT signalling is imperative to protection afforded by postconditioning (Goodman et al. 2008). Some opioids also possess free-radical scavenging abilities (e.g. enkephalins), and can be up-regulated following ROS exposure (Coccia et al. 2001; Rosenberger et al. 2001). Indeed, ROS signalling has been directly linked to protective DOR postconditioning in adult cardiomyocytes and mouse hearts exposed to I-R (Tsutsumi et al. 2007).

The endogenous opioid system is strongly implicated in remote IPC (Shi and Vinten-Johansen 2012; Przyklenk and Whittaker 2011). This type of cardiac conditioning, whereby brief intermittent ischaemia in a remote organ or tissue induces a protective phenotype in a target organ (e.g. heart), is of intense clinical interest. The ability to non-invasively condition the heart with minimal undesirable side effects has significant potential for clinical translation. While this phenomenon has proven effective in various species including humans, clinical trial outcomes remain inconclusive, with some studies reporting no change or worsened outcomes (Shi and Vinten-Johansen 2012; Ferdinandy et al. 2014). Dickson et al. first revealed that IPC-induced release of endogenous opioid peptides effectively translates protective signals when administered to recipient hearts (Dickson et al. 2001). Later studies by Patel et al. revealed that remote IPC via mesenteric artery occlusion induced myocardial protection, dependent upon OR system activation (Patel et al. 2002b). Several studies specifically link DOR activation to myocardial protection via remote IPC (Surendra et al. 2013; Weinbrenner et al. 2004). However Zhang et al. reveal no adverse effect of DOR antagonism upon remote IPC, and identify no changes in endogenous met-enkephalin levels (Zhang et al. 2006).

Endogenous tolerance to ischemia can be generated in the heart after exposure to both voluntary and involuntary exercise training in as little as 1–3 days (Budiono et al. 2012, 2016; Miller et al. 2015; Quindry et al. 2010). Exercise preconditioning induces myocardial changes that mirror preconditioning, through elevated antioxidant activity (Yamashita et al. 1999; Hamilton et al. 2004), RISK-dependent ERK1/2 (Budiono et al. 2012, 2016), Akt (Budiono et al. 2016), HSP27 (Budiono et al. 2016), AMPK (Budiono et al. 2016), phospho-inhibition of GSK3β (Budiono et al. 2012, 2016), and EGFR expression (Budiono et al. 2016), and mK_{ATP} (Quindry et al. 2010) and $sarcK_{ATP}$ (Brown et al. 2005) channel involvement (Quindry et al. 2010). Borges et al. previously demonstrated an anti-infarct effect of 7 days treadmill training, which was attenuated by non-selective or DOR antagonism. These protective effects were unaffected by MOR or KOR antagonism (Borges et al. 2014). Recently, Miller and colleagues showed that myocardial proenkephalin mRNA levels increase following exercise, and pre-ischaemic DOR antagonism during exercise may partially protect against necrotic (but not apoptotic) death following I-R (Miller et al. 2015). From these few studies it is clear that the role of DOR and endogenous opioids in exercise-induced cardioprotection may be important, the specific mechanisms require further delineation.

1.6 Exogenous Conditioning Via DOR Agonism

Multiple studies across various species and models (cardiomyocytes (Patel et al. 2006; Shen et al. 2012), rodents (Peart and Gross 2003, 2004a; Fryer et al. 2001), canine (Peart et al. 2003), rabbits (Kodani et al. 2002), porcine (Sigg et al. 2002), and human (Bell et al. 2000)) have confirmed protective effects of DOR agonism. At the sarcolemma, DOR-mediated preconditioning is reliant upon adenosine receptor crosstalk (Peart and Gross 2003, 2005), EGFR transactivation (Cohen et al. 2007) and caveolin-3/caveolar co-localisation (Patel et al. 2006). In a pathway that parallels canonical RISK signalling, DOR-mediated preconditioning engages PKC (Schultz et al. 1998; Maslov et al. 2009) (potentially PKC-ε (Miura et al. 2007) or PKC-δ (Fryer et al. 2001)), PI3K/Akt, MAPK kinase/ERK1/2, Src kinase signalling (Shen et al. 2012; Cohen et al. 2007; Gross et al. 2004, 2006), ROS production (Cohen et al. 2007), NOS activation (Maslov et al. 2009), JAK2/STAT3 signals (Gross et al. 2006), GSK3β inhibition (Gross et al. 2004, 2006), mTOR stimulation (Gross et al. 2004), and sarcolemmal K_{ATP} (Patel et al. 2002a) and mK_{ATP} channel (Patel et al. 2002a) activation. Prior to index ischaemia, DOR agonism appears to induce cardioprotection in delayed preconditioning (Patel et al. 2002a; Fryer et al. 1999), and acute preconditioning in myocytes (Seymour et al. 2003), human heart tissue (Bell et al. 2000), and rats (Maslov et al. 2009) via opening of the mK_{ATP} channel. Additionally, the sarcolemmal K_{ATP} channel has been implicated in distinct DOR-mediated delayed cardioprotection, which can be pharmacologically blocked by DOR antagonism with HMR-1098 (Patel et al. 2002a). Miura et al. suggested that DOR-mediated activation of PKC-ε leads to phosphorylation of Connexin-43 (Cx43), and subsequent reduced gap junction permeability contributes to the infarct-sparing effects of DOR activation in these settings (Miura et al. 2007). An interesting link between Cx43 and cardiac subsarcolemmal mitochondria in settings of ischaemic preconditioning was recently identified by Ruiz-Meana et al. and may reflect a signalling rendezvous linking mitochondrial preservation with gap junction permeability (Ruiz-Meana et al. 2014).

External to established RISK signalling, DOR-mediated protection has also been linked to arachidonic acid metabolism via 12-lipoxygenase (12-LO). Patel et al. identified elevations in 12-LO expression and enzymatic activity in delayed DOR-mediated preconditioning in rats (Patel et al. 2003). Delayed DOR-mediated preconditioning can be abolished by COX-2 inhibition and increases cardiac expression of COX-2, PGE_2, 6-keto-$PGF_1\alpha$, and PGI_2 synthase (Kodani et al. 2002). Furthermore, Jiang and colleagues demonstrated that iNOS gene knockout mice were not susceptible to morphine-mediated cardioprotection (Jiang et al. 2004). Thus, complementing IPC, while the acute phase of DOR-mediated preconditioning is governed by post-translational phosphorylation and translocation of effector molecules, delayed protection induces genetic expression of iNOS, COX-2 and 12-LO (Gross et al. 2003).

Studies have implied an additional second messenger system may be linked to OR agonism, specifically DOR and KOR-dependent modulation of phosphoinositol turnover (Schultz and Gross 2001). Ventura et al. identified reduced contractility in rat ventricular cardiomyocytes, mediated by phosphatidylinositol turnover and intracellular Ca^{2+} depletion in response to DOR and KOR agonism (Ventura et al. 1992). Furthermore, DOR-mediated Ca^{2+} mobilisation appears secondary to elevations in inositol 1,4,5-triphosphate (Sheng et al. 1996). Opioid receptor agonism may also influence ion fluxes independent of second messengers, with evidence of a G-protein link to DOR and MOR that induces K^+ channel opening, and KOR-induced Ca^{2+} channel closing (Schultz and Gross 2001; Gross et al. 1990).

Pharmacological postconditioning via DOR agonism appears to parallel the common signalling elements of preconditioning, and generates similar responses activated by other opioid receptors (Chen et al. 2008). Postconditioning via DOR reduced infarct size in rabbits through RISK pathway mediators Akt and ERK1/2, and was dependent upon EGFR transactivation (Forster et al. 2007). Similarly, the DOR may elicit protective responses through mPTP inhibition in postconditioning via pathways involving GSK3ß (Gross et al. 2007b), mitochondrial and sarcolemmal KATP channels (Gross et al. 2007b), and NO/cGMP/PKG (Jang et al. 2008). Studies intricately examining the mechanistic basis of DOR-mediated postconditioning are lacking in comparison to preconditioning, particularly in human myocardial tissue. Regardless, Fuardo et al. recently demonstrated that DOR postconditioning with DADLE in isolated human atrial trabeculae induced protection against hypoxia-reoxygenation injury by blockade of mPTP opening (Fuardo et al. 2013). Models of morphine postconditioning show conflicting results regarding the involvement of DORs, with an early study by Chen et al. (2008) implicating KOR and not DOR, which was challenged in a later study by Kim et al. showing DOR dependence of morphine postconditioning (Kim et al. 2011).

1.7 Centrally Mediated DOR Protection Upon Heart

Contributions from central nervous system (CNS) DORs in cardioprotection have been shown by several studies. Intrathecal (i.t.) or intracerebroventricular (i.c.v.) administration of opioid agonists prior to cardiac I-R substantially improves myocardial outcomes. This remote cardiac conditioning effect of i.t. or i.c.v. morphine may involve DOR, KOR and MORs (Li et al. 2009; Zhang et al. 2011), additional to involvement of bradykinin receptors (Wong et al. 2012), calmodulin (Zhang et al. 2011) and CGRP release (Zhang et al. 2011; Wong et al. 2012). Yao et al. report involvement of both central and peripheral adenosine receptors in cardioprotection via CNS application of morphine, with distinct roles in induction vs. mediation phases of protection (Yao et al. 2011). Spinal tissue nNOS is reportedly activated with remote fentanyl preconditioning and is also implicated in improved cardiac stress tolerance (Lu et al. 2014). These studies confirm involvement of the CNS in myocardial protective effects of extra cardiac opioid receptor agonism, which may also contribute to persistent protection observed with SLP (see below) in both healthy and caveolin-3 depleted settings.

1.8 Prolonged Opioid Preconditioning

While acute opioid preconditioning has generated considerable interest, emerging evidence identifies a unique myocardial tolerance generated by prolonged opioid preconditioning. Specifically, the DOR has been implicated in many of these circumstances. Clinically, the ability to induce prolonged protection in target groups may provide a desirable window of necessary protective intervention without the complication of stringent treatment timing with respect to predictability of insult/injury. Gross and colleagues applied irreversible DOR agonism with fentanyl isothiocyanate (FIT) in a rat model of I-R. FIT was administered intravenously 48–120 h prior to I-R, with a maximal protective effect observed at 96 h post-injection. Cardioprotection induced by FIT appears dependent upon PI3K signalling during both the trigger and end-effector phases (Gross et al. 2005). Prolonged cardioprotection induced via single i.p. injection of Eribis peptide 94 (EP94) 24 h prior to ischaemia, a synthetic peptide akin to DOR-selective met- and leu-enkephalin, significantly reduced infarction via enhanced iNOS activity and NO expression (Gross et al. 2012). Kuzume et al. determined that 24 h treatment with Met^5-enkephalin (Met5), a specific DOR agonist, reduced infarct size in rabbits by ~60%. In contrast, acute DOR activation with Met5 failed to induce I-R tolerance (Kuzume et al. 2003). Kuzume and colleagues later revealed a reduction in infarct size with 24 h Met5 infusion in mice (Kuzume et al. 2005). Protection was abolished when the index ischaemia was initiated 24 h after removal of the Met5 stimulus, identifying a limited window for mediation of protection. Furthermore, 14-day exposure to Met5 failed to generate a protective phenotype, with receptor internalisation and uncoupling of ORs from AC identified as responsible for this desensitisation (Kuzume et al. 2005).

Due to considerable interest in morphine tolerance and dependence, effects of chronic morphine use have been studied intensely in the brain, yet less extensively in other tissues. Morphine is a non-specific though predominantly MOR agonist with multiple effects throughout the body. Chronic morphine treatment has been linked to calcitonin gene-related peptide (CGRP) up-regulation and release (Menard et al. 1996; Trang et al. 2002; Tumati et al. 2009), super-activation of AC (Varga et al. 2003), augmented adenosine receptor sensitivity (Brundege and Williams 2002), changes in glucose homeostasis (Li et al. 2003) and possible conversion of ORs from $G_{i/o}$-coupled to G_s-coupled (Crain and Shen 1992), effectively altering downstream signalling. Several changes in protein expression occur in rat myocardium following intramuscular morphine injections (10 mg/kg/day) for ten consecutive days. This includes up-regulation of proteins involved in cytoprotection (HSPB1, HSP7C, GRP78, ORP150), and phosphatidylinositol transfer protein (α isoform), which has been implicated in cardioprotective phospholipase C- and PI3K-mediated signalling. Interestingly, mitochondrial precursor to aldehyde dehydrogenase was down-regulated, indicative of oxidative stress induced by chronic morphine treatment (Drastichova et al. 2011). A mechanism showing remarkable promise as a potent cardioprotective agent is sustained ligand-activated preconditioning (SLP). Here, sustained DOR agonism for 5 days induces an I-R-tolerant phenotype effective in young to aged myocardium (Peart et al. 2011; Peart and Gross 2004b). Details of this phenomenon will be discussed further below. Prolonged opioid exposure holds considerable potential as a candidate mechanism for inducing a powerful cardioprotective phenotype.

1.9 Sustained Ligand-Activated Preconditioning (SLP)

A unique protective phenomenon, termed sustained ligand-activated preconditioning, is induced by sustained DOR agonism for 5 days (previously termed chronic morphine preconditioning (Peart and Gross 2004c)) and has been shown to surpass the protection induced by conventional cardioprotective responses and acute opioid preconditioning

(Peart et al. 2011; Peart and Gross 2004c, 2006). This protective response is induced via slow release morphine pellets (25 or 75 mg) in mice (Peart et al. 2011; See Hoe et al. 2014; Peart and Gross 2004b, 2006). While morphine binds all opioid receptors with a selectivity profile of $\mu > \kappa > \delta$, 5-day treatment with MOR-selective agonist morphine-6-glucuronide, or KOR-selective agonist U50488H, does not induce SLP (Peart et al. 2011). Conversely, 5-day treatment with DOR-selective agonist BW373U86 mirrored the effects identified with morphine, confirming DOR dependence (Peart et al. 2011). Thus, the DOR dependence of SLP beneficially avoids untoward systemic side effects induced by other ORs. The SLP phenotype can be generated in as little as 48 h, and protection evidenced by contractile recovery and infarct reduction persists up to 7 days post-stimulus withdrawal (Peart et al. 2011). Potent protection is evident in both young and aged myocardium, a superior feature that both acute opioid conditioning and conventional conditioning therapies (e.g. Pre- and postconditioning) fail to meet (Peart and Gross 2004b). Furthermore, SLP improves postischaemic functional recovery in myocardium exposed to chronic β-blockade, where archetypal IPC failed (See Hoe et al. 2016). These unique qualities suggest that SLP enlists signalling distinct from conventional conditioning responses.

Initial work by Peart et al. established that SLP affords protection through signalling pathways that are mechanistically different to conventional mediators implicated in ischaemic pre- and postconditioning, and acute opioid preconditioning (Peart and Gross 2006). The SLP phenotype is induced by sustained DOR agonism and initial PI3K pathway involvement involves a G protein conversion of ORs from inhibitory to stimulatory coupling, and subsequent protection is mediated by β_2-AR/G_s/PKA-dependent signalling (Peart and Gross 2006). The β_2-AR and G_s protein have been controversially linked to cardioprotection, as activation of G_s-protein is associated with the development of cardiomyopathy, hypertension and heart failure, due to association with the β-adrenergic system (Du et al. 2000; Karoor et al. 2004; Liggett et al. 2000). Regardless, evidence suggests a role for G_s-protein involvement in cardioprotection, with an IPC-derived phenotype restored in rat hearts undergoing pre-ischaemic treatment with forskolin, a potent AC activator (Mieno et al. 2002). While controversial, there is evidence supporting a role for β_2-AR signalling in cardioprotection (Lochner et al. 1999; Frances et al. 2003; Marais et al. 2005). The involvement of both DOR and β_2-AR in SLP is intriguing. Unpublished data reveal significant alterations in expression, phosphorylation and translocation of β_2-AR, G_s variants, PKA and Akt signalling elements during SLP induction, that did not translate to altered sensitivities to β_2-AR agonists (formoterol and fenoterol) or adenyl cyclase activator (NHK477). While still unclear, there is evidence suggesting a form of receptor cross-talk (discussed below) between both β-adrenergic and opioid receptor systems that remain to be fully elucidated (Pepe et al. 1997, 2004; Jordan et al. 2001; Shan et al. 2002). A study by Lou and Pei reported increases in PKA activity with reductions in basal PKC activity after 24 h of DOR activation (Lou and Pei 1997). Moreover, PKA has been implicated in reducing I-R injury (Peart and Gross 2006; Sichelschmidt et al. 2003), opioid tolerance (Shen et al. 2000; Wagner et al. 1998), and mediation of IPC independently of PKC (Sanada et al. 2004). The role of PKA in IPC nonetheless remains controversial with evidence for both protective and detrimental roles of this kinase. In a positive light, several studies have identified that inhibition of PKA signalling blocks IPC-mediated myocardial protection (Sanada et al. 2004; Tong et al. 2005; Inserte et al. 2004), and transient PKA activation may function as an IPC mimetic (Inserte et al. 2004). Despite observed linkages between physiological opioid actions and mK_{ATP} (Armstead 1998; Shankar and Armstead 1995), Peart and Gross were not able to link the mediation phase of SLP to mK_{ATP} activity (Peart and Gross 2006).

Evidently, SLP retains unique signalling properties that isolate this mechanism from conventional preconditioning. This characteristic provides potential for additivity with other cardioprotective stimuli. In fact, there is evidence to suggest that co-activation of varied signalling pathways may increase the activity of downstream effectors (Nishihara et al. 2006). This form of additivity may not be possible with conventional signalling implicated in pre- and postconditioning due to commonality of mediators and mitochondrial targets (Halkos et al. 2004). Peart and colleagues recently observed a lack of additivity of SLP with either IPC or acute morphine treatment. Interestingly, pre-ischaemic and post-ischaemic administration of adenosine or 2-chloroadenosine, and cardiac specific A_1 adenosine receptor (A_1AR) overexpression did further enhance the protective phenotype of SLP (Peart et al. 2011). These findings provide additional evidence for a distinct signalling mechanism with SLP. Adenosine receptors have been identified as sensitive to chronic opioids, which could contribute to observed protective additivity with adenosinergic stimuli (Brundege and Williams 2002).

Microarray interrogation of SLP hearts reveals a stress-tolerant phenotype, associated with inhibition of immune and inflammatory mediators, and proliferation of natriuretic proteins and sarcomeric elements (Ashton et al. 2013). Genes related to cell development, growth and stress were also altered, whereas conventional protective mediators remained unchanged (Ashton et al. 2013).

The SLP response is effective despite membrane cholesterol perturbations and caveolin-3 genetic deletion (See Hoe et al. 2014). This is a stark contrast from the caveolin-3 dependence displayed by acute opioid preconditioning and other conventional protective responses (Patel et al. 2006; Tsutsumi et al. 2010). Unpublished data from Hemal Patel's laboratory conversely reveals an increase in caveolin-3 expression and caveolar density in SLP-treated aged and diabetic hearts. The notable membrane modulation and independence observed with SLP, in conjunction with alternative signalling pathways may contribute to its efficacy in aged and diseased hearts, which makes SLP a desirable candidate with translational potential. Intriguingly, it may well be these extra-cardiac and unique caveolin-3/caveolar independent aspects of SLP that underlie its broader efficacy in aged and diseased hearts: direct cardiac responses involving caveolae/caveolin-3 may be impaired in these states, whereas extra-cardiac signals may retain functionality.

1.10 DOR Receptor Crosstalk

Cardioprotective responses can be derived through either direct or indirect mechanisms of crosstalk between the DOR and other GPCRs. Significant evidence links the DOR and β-adrenergic receptors (β-AR) as intricately linked modulators of cardiovascular function and protection. Direct heterodimerisation occurs between β_2-AR and DOR (or KOR), and co-expression of ORs affects β_2-AR trafficking. Interestingly, β_2-AR/DOR heterodimers are susceptible to isoproterenol and opioid agonist-induced receptor internalisation, yet DOR and β_2-AR expressed alone are insensitive to opposing agonist-induced internalisation (Jordan et al. 2001). Cardiac contraction is also enhanced through DOR-mediated calcitonin gene-related peptide (CGRP) and β-AR signalling (Nguyen et al. 2012). Furthermore, β_2-AR blockade can attenuate the anti-infarct and myocyte survival protection effects of DOR activation in cell culture and isolated heart models (Huang et al. 2007a). Sustained pharmacological DOR activation shown to induce potent and prolonged protection in young and aged hearts is also sensitive to β_2-AR but not β_1-AR antagonism (Peart and Gross 2006). Interestingly, DOR activation via leucine-enkephalin mediates an anti-adrenergic response upon β_1-AR-stimulated increases in cAMP signalling (Pepe et al. 1997). While both the opioid and β-adrenergic systems exhibit opposing regulatory mechanisms in the heart, evidence suggests they are intricately entwined.

Evidence suggests DORs and adenosine receptors co-dependently mediate cardioprotective responses. Adenosine A_1 receptor (A_1AR) antagonism can abolish the anti-infarct and anti-stunning effects of various opioid agonists (e.g. Fentanyl (Kato et al. 2000), morphine (Peart and Gross 2003)), and DOR antagonism abolishes endogenous adenosine (Peart and Gross 2005) and A_1AR-dependent reductions in infarct size (Peart and Gross 2003). Cardiac protection arising from remote ischaemic preconditioning (RIPC) or pharmacological protection via intrathecal morphine links A_1AR and DOR interactions to protective responses (Surendra et al. 2013; Yao et al. 2011). The observed link between DOR and A_1AR that leads to a protective phenotype in the heart may converge at a common MMP/EGFR pathway to stimulate protective signalling. Studies have shown that DOR and/or A_1AR downstream kinase signalling leading to protective responses are dependent upon and directly involve EGFR and MMP (Headrick et al. 2015; Williams Pritchard et al. 2011a; Forster et al. 2007). While the aforementioned evidence intrinsically links DOR and A_1AR responses to cardiac protection in acute settings, sustained protective modalities appear to engage alternative pathways. Sustained protection with DOR agonism over 5 days (sustained ligand-activated preconditioning, SLP) appears insensitive to A_1AR antagonism; pre-ischaemic and post-ischaemic administration of adenosine or 2-chloroadenosine; and cardiac specific A_1AR overexpression further enhances the protective phenotype of SLP (Peart et al. 2011), suggesting distinct opioid and adenosinergic signalling pathways. Thus while DOR and A_1AR may co-dependently induce cardiac protection in acute settings, alternative responses may arise from prolonged receptor activation.

Cardiac protection can also be induced through activation of the cytochrome P450 (CYP) monooxygenase pathway, as shown in genetic overexpression and knockout models, and exogenous administration of epoxyeicosatrienoic acids (EETs) (Motoki et al. 2008; Gross et al. 2007a; Nithipatikom et al. 2006; Seubert et al. 2004, 2006). Indeed, genetic variation in this pathway that leads to increased degradation of EETs has been linked to an increased risk of incidence of coronary heart disease in Caucasians (Lee et al. 2006). Terashvili et al. initially demonstrated that EETs indirectly induce antinociception through activation of Met-enkephalin and β-endorphin to subsequently activate DOR and MOR in the rat brain. Evidence here showing indirect opioid receptor activation prompted Gross et al. to investigate the effects of opioids in EET-mediated cardioprotection, where DOR and KOR were found to be largely implicated in the cardioprotective role of EETs (Gross et al. 2010).

1.11 Novel Mechanisms Regulating Opioid Ligand-Receptor Signalling: The Role for Caveolar Microdomains

We have thus far discussed briefly the involvement of caveolae and caveolins in regulating opioid receptor signalling. This section will expand upon this concept to propose further mechanistic insights into microdomain regulation of and by opioid receptors. Caveolae are "cave-like" structures of the plasma membrane that are enriched in cholesterol, sphingolipids and caveolin proteins (Patel et al. 2008b), and are considered a subset of lipid rafts (Pike 2003). Of the 3 caveolin proteins (Cav-1, -2 and -3), Cav-3 is the predominant subtype expressed in striated muscle cells and is critically important for caveolae formation in cardiac myocytes and localising cardiac protective signalling proteins (Patel et al. 2008b). Mutation or knock-down of Cav-3 results in myopathies (Aboumousa et al. 2008; Hagiwara et al. 2000; Woodman et al. 2002). Cav-3 has been identified in the sarcolemmal membrane, transverse tubules (T-tubules), the I-band/A-band interface and localised with ryanodine receptors in myocytes (Scriven et al. 2005; Ralston and Ploug 1999). Caveolins are involved in multiple cellular processes including vesicular transport, cholesterol and calcium homeostasis (Fujimoto 1993; Fujimoto et al. 1992; Scriven et al. 2002; Jones et al. 2004), signal transduction (Lisanti et al. 1994; Williams and Lisanti 2004; Steinberg and Brunton 2001; Cohen et al. 2004), and have been recently detected in mitochondria (Fridolfsson et al. 2012). Caveolins function as chaperones and scaffolds recruiting signalling molecules to caveolae to provide direct temporal and spatial regulation of signal transduction (Williams and Lisanti 2004; Shaul and Anderson 1998). Such diverse observations have led to the proposition of non-membrane roles for caveolin in regulating cell physiology (Fridolfsson et al. 2014) though we will particularly focus on the membrane specific role of caveolin in regulating opioid signaling.

An emerging concept suggests that signalling molecules exist as multiprotein complexes, "signalosomes", continuously forming and dissociating under basal or stimulated conditions (Feron and Balligand 2006). Specifically, in regard to signalling molecules involved in cardiac protection, many GPCRs including opioid (Head et al. 2005) and adenosine receptors (Lasley et al. 2000) localise to caveolae and co-immunoprecipitate with caveolins. Additionally, many of the signalling molecules involved in cardiac protection, including the Gα subunit of heterotrimeric G-proteins, Src kinases, PI3K, eNOS, PKC isoforms and ERK are known to bind with the scaffolding domain of caveolin and be regulated by caveolin (Ballard-Croft et al. 2006; Krajewska and Maslowska 2004). There are a number of regulatory features specific to caveolin and caveolae that regulate these interactions. Caveolins exist as oligomeric complexes that allow them to form high molecular weight scaffolds (Monier et al. 1995; Woodman et al. 2004). Importantly, Cav-3 mutations have revealed that an inability to form oligomeric complexes leads to skeletal muscle defects (Woodman et al. 2004). Studies in failing hearts suggest that caveolins dissociate from caveolae (Ratajczak et al. 2003) presumably as the oligomer is broken down. Evidence also suggests that caveolin can undergo post-translational modifications (i.e., phosphorylation, sumoylation, nitrosation) that may be critical to regulating various signalling complexes (Fuhs and Insel 2011; Sun et al. 2015; Bakhshi et al. 2013; Radel and Rizzo 2005). There is much debate about what factors regulate the specific localisation of molecules in caveolae. It is likely a highly regulated and complex process that involves a combination of protein–protein and protein–lipid interactions, post-translational modifications, the specific cell type involved, the specific lipid environment unique to caveolae, a variety of signalling mediators acting as adaptors, and particular cellular cues and environmental factors that create a milieu needed to regulate coordinated and precise cellular responses.

Recent studies reveal opioid receptor ligands such as enkephalins have the ability to alter membrane surface characteristics and morphology specifically in membranes enriched in lipids (Tsanova et al. 2014, 2016), an effect that is mimicked by ethanol via modulation of protein in lipid rafts that is sensitive to opioid receptor antagonism (Tobin et al. 2014). Studies also suggest that DOR microdomain localisation may be ligand specific (Hruby et al. 2010). The central vs. peripheral aspect of opioid action may also be explained by microdomain organisation; neurons, though containing membrane lipid rafts and caveolin, do not contain classic invaginated caveolae that are present in cardiac myocytes (Egawa et al. 2016). It was long observed that animals would become tolerant to the central effects of chronic exposure to opioids but the peripheral effects on cardiac protection were still observable long after this central tolerance developed (Peart and Gross 2004c). Such observations suggest differential processing of opioid receptors centrally and peripherally in response to sustained opioid ligand exposure, perhaps related to the caveolar differences between neurons and myocytes. Evidence suggests potential impact of cholesterol in receptor internalisation, desensitisation, and activation (Qiu et al. 2011; Levitt et al. 2009; Andre et al. 2008; Huang et al. 2007b, c). We also propose a unique role for SLP in cardiac protection. Evidence suggests that long-term opioid receptor stimulation may lead to the superactivation of adenylyl cyclase in a lipid microdomain specific manner where chronically activated opioid receptor remained in lipid rafts and the superactivation of adenylyl cyclase could be completely attenuated by cyclodextrin treatment which results in disruption of caveolae and lipid rafts (Zhao et al. 2006). Such a mechanism may account for central tolerance and sustained peripheral activity given the nature of the lipid environment and interacting partners that may regulate opioid receptor signalling in the two specific locations.

Membrane microdomains such as caveolae and lipid rafts offer a unique and compact system with which to explain away the complex signalling and physiological implications of opioid receptor signalling, yet these microdomains also reveal that the complex regulation of dynamic signalling events are dependent upon a variety factors that need further exploration and precise identification to better understand physiology, pathophysiology, and develop novel therapeutics.

1.12 DOR Protection in Clinically-Relevant Disease States

1.12.1 DOR Protection in Diabetes/Obesity

Diabetes significantly increases cardiovascular morbidity and mortality rates. Indeed the 30-year Worchester Heart Attack Study identified approximately 30% of all AMI patients suffered from diabetes, 30% were obese, and 70% were hypertensive (Floyd et al. 2009). The human diabetic myocardium is insensitive to protection via preconditioning (Ghosh et al. 2001), a response mimicked in various animal models of diabetes assessing pre- and postconditioning strategies (Yin et al. 2012; Przyklenk 2011; Yadav et al. 2010; Miki et al. 2012; Przyklenk et al. 2011; Hassouna et al. 2006; Kersten et al. 2000). As previously discussed, preconditioning with opioids is well reported to mimic IPC (Bell et al. 2000; Peart et al. 2003; Schultz et al. 1996; Cohen et al. 2007), and opioid receptor antagonism blocks cardioprotection with ischaemic pre- and post-conditioning (Schultz et al. 1995; Jang et al. 2008; Tomai et al. 1999; Fryer et al. 1999). Mechanistically, opioidergic conditioning involves activation of DOR and/or KOR, RISK, SAFE and JAK-STAT pathways, PI3K/Akt signalling axis, inhibition of GSK3β, NO production through eNOS activity, sarcolemmal and mitochondrial K_{ATP} channel opening, and cGMP-PKG pathway involvement that may contribute to mPTP inhibition (Peart et al. 2005; Headrick et al. 2015; Williams-Pritchard et al. 2011b). Diabetes reportedly interferes with many of the aforementioned signalling mediators implicated in both opioid and IPC-mediated cardioprotection. The anti-infarct effects of pharmacological preconditioning with remifentanil (a cardiac opioid analgesic), which is sensitive to DOR antagonism (Zhang et al. 2004), did not translate to diabetic hearts and was associated with impaired anti-apoptotic signalling (Kim et al. 2010). Similarly, Gross et al. reported that morphine postconditioning, also repressed by DOR blockade (Okubo et al. 2004; Schultz et al. 1997b), was ineffective in rat hearts exposed to I-R, and coupled to blunted elevations in phosphorylated Akt, ERK1/2, p70s6K, JAK2 and STAT3 (Gross et al. 2007c). While opioid-mediated cardioprotective pathways may be inhibited in diabetes, there is other evidence of effective opioid receptor activity in these settings. A study in healthy and obese mice revealed that DORs mediate glucose uptake in skeletal muscle of both groups of mice. Additionally, DOR activation in Chinese hamster ovary cells promotes glucose uptake via specific GLUT-1 and P13K mechanisms (Olianas et al. 2011), and involves AMPK phosphorylation (Olianas et al. 2012). Furthermore, vascular endothelial dysfunction and chronic inflammation associated with diabetes can be improved via KOR activation, mechanistically involving eNOS phosphorylation and NFκB inhibition (Theisen et al. 2014). Interestingly, Ritchie et al. show that activation of PI3K-dependent signalling (implicated in opioidergic cardioprotection) reverses diabetic cardiomyopathy, reduces myocardial oxidative stress and improves mitochondrial function in type I diabetic mice (Ritchie et al. 2012). Impaired TRPV1-CGRP signalling has been recently linked to cardiac dysfunction in diabetes, whose development can be prevented by CGRP preservation (Sun et al. 2016). Evidently, the diabetic heart is intrinsically susceptible to I-R damage, however opioidergic conditioning may hold mechanistic advantages in I-R settings with further interrogation.

As a leading risk factor for ischaemic heart disease, obesity is prevalent in the population in desperate need and of effective cardioprotective therapy. The added complication of co-existing metabolic syndrome components (glucose intolerance, dyslipidaemia, hypertension) further exacerbates the detrimental effects of this metabolic abnormality. Myocardial preconditioning is reportedly ineffective in a model of insulin-resistant obesity (Katakam et al. 2007), and postconditioning appears to fail in a genetic model of obesity characterised by leptin deficiency (Bouhidel et al. 2008) and metabolic syndrome (Wagner et al. 2008). Furthermore, protection via DOR preconditioning is abolished in a rat model of dietary obesity and associated with impaired RISK signalling (Donner et al. 2013). Interestingly, while DOR-mediated cardioprotection is attenuated by obesity, genetic DOR ablation in mice induced a resistance to weight gain and lower fat mass upon exposure to a high-energy diet, despite a hyperphagic response. This resistance to diet-induced obesity was linked to an elevation in thermogenic markers in brown adipose tissue, potentially favouring an energy expenditure vs. conservatory phenotype in these mice (Czyzyk et al. 2012).

1.12.2 DOR Protection and Aging

The mature and elderly population are commonly afflicted with coronary artery disease, affecting approximately 50% of persons in this age group (>65 years of age) (Headrick et al. 2015). Impaired ischaemic tolerance is well documented in both animal (Peart and Gross 2004b; Headrick 1998) and human tissues (Liu et al. 2012; Peart et al. 2014), and endogenous protective responses affording ischaemic tolerance appear to diminish with age (Peart and Headrick 2009; Peart et al. 2014). Previous studies have demonstrated loss of efficacy of DOR-dependent ischaemic pre- (Abete et al. 1997; Schulman et al. 2001), post- (Przyklenk et al. 2008), and remote conditioning (Hu et al. 2002), and pharmacological cardioprotection induced by OR agonism (Peart and Gross 2004b; Peart et al. 2007, 2014). These effects are not strictly limited to animal models, as Abete et al. demonstrated that the IPC-like effect of angina prior to myocardial infarction was abolished in patients over 65 years of age (Abete et al. 1997). The loss of OR-dependent protective signalling with age is evident despite a notable rise in cardiac enkephalin peptide and mRNA expression in senescent hearts (Caffrey et al. 1994; Boluyt et al. 1993). Mechanistically, the intrinsic changes that occur with age may deleteriously desensitise OR-dependent protective signalling pathways and exacerbate I-R intolerance in the heart. Compromised signal transduction from membrane-dependent GPCR agonism to protective mitochondrial end effectors is evident in both murine and human cardiac tissue. Sarcolemmal composition of caveolae and expression of caveolin-3, heavily localised with signalling moieties and associated with intrinsic I-R tolerance and DOR-mediated cardioprotection (Patel et al. 2006; Tsutsumi et al. 2008, 2010), are attenuated with age (Peart et al. 2014). Additionally, perturbations in p38MAPK (Peart et al. 2007) and PKC (Tani et al. 2001) activation have been identified in aged hearts. Alternatively, protective SLP is impervious to membrane disturbance via caveolin-3 depletion (See Hoe et al. 2014), which may explain its efficacy in the aged heart (Peart and Gross 2004b).

1.12.3 DOR Protection in Hypertension and Hypertrophy

Conflicting evidence suggests that hypertension renders the heart insensitive to cardioprotective modalities. Several studies in young animal models demonstrate preserved efficacy of IPC (Ebrahim et al. 2007a; Lu et al. 1999; Boutros and Wang 1995). However, Ebrahim et al. show that independently, both age and hypertension ablate the protective effects of preconditioning (Ebrahim et al. 2007b). Similarly, the aged-hypertrophied heart features a decline in functional tolerance to ischaemia, and both ischaemic and anaesthetic preconditioning benefits are lost with hypertrophied (Moolman et al. 1997) or larger (or aged) (Riess et al. 2005) hearts, respectively. While these supposed DOR-mediated conditioning modalities are impaired in hypertrophy and hypertension, the effects of these conditions upon the DOR signalling system are perplexing.

Cardiac enkephalin gene expression is reportedly elevated in spontaneously hypertensive rats (SHR) (Bhargava et al. 1988; Dumont et al. 1991), interestingly linked with increasing age in both studies. While cardiac enkephalin gene expression may increase, enkephalin peptide expression appears more inconsistent in hypertensive models. Elevations in cardiac leu-enkephalin and selective reduction in low affinity DOR expression (Dumont and Lemaire 1988), and reduction in met-enkephalin expression (Dumont et al. 1991) occur in SHR. Conversely, spontaneously hypertensive hamsters display no changes in myocardial enkephalin peptide or DOR expression (Bolte et al. 2009a). While not associated with the heart, 7 day exposure to DOR agonism stimulates protection of neural cells in the retina of chronic ocular hypertensive rats (Abdul et al. 2013), possibly via inhibition of iNOS (Husain et al. 2014). This model is used to mimic glaucoma, which shares features of myocardial ischaemic injury such as oxidative stress, inflammation and mitochondrial dysfunction (Abdul et al. 2013).

Similar to hypertension, elevations in myocardial preproenkephalin mRNA expression have been identified in models of cardiomyopathic hamsters (Ouellette and Brakier-Gingras 1988) and hypertrophied rats (Weil et al. 2006), an effect suggested to transpire in response to pressure overload and ß-adrenergic pathway activation (Weil et al. 2006). Signalling induced by DOR agonism in a hamster model of heart failure (HF) and cardiomyopathy further exacerbated cardiac dysfunction independent of opioid peptide and receptor expression. This response manifested as an elevated negative inotropic and lusitropic function, cAMP inhibition and reduced systolic Ca^{2+} transient amplitude. Furthermore, this study identified a switch in the G protein coupling to DOR from PTX insensitive to sensitive (Bolte et al. 2009b). It was suggested that, among other possibilities, the switch in G protein coupling may be due to the HF/cardiomyopathic-dependent fall in membrane cholesterol and caveolins(Bolte et al. 2009b). Being that G_i is not present in caveolae, the enhanced fluidity of the membrane in these settings could lead to improved coupling to G_i and exacerbated cardiac dysfunction (Bolte et al. 2009b). Acute congestive HF in humans appears associated with elevated levels of met-enkephalin, and these alterations parallel changes with the degree and severity of disease (Fontana et al. 1993). The aged and hypertrophied heart is associated with a reduction in ß-adrenergic responsiveness, which may result from elevations in peptide and mRNA levels of cardiac enkephalins previously demonstrated to impede ß-adrenergic-dependent contractile responses in isolated cardiac myocytes (Xiao et al. 1997). Indeed, intracoronary DOR antagonism leads to improved cardiac contractile function in HF dogs (Imai et al. 1994).

2 Summary/Conclusion

Endogenous opioid peptides and their corresponding receptors are implicated in a wide variety of physiological and pathological functions in the cardiovascular system. Importantly, opioid receptors, and in particular the DOR, are implicated in endogenous cytoprotective responses – from hibernation in mammals to IPC. Harnessing these responses experimentally elicits profound acute and sustained cardioprotective phenotypes through a variety of potential mechanisms. Clinical translation of cardioprotective mechanisms, including pre- and post-conditioning has been limited. This may be attributed to confounding comorbidities such as obesity, diabetes and advancing age. Further, to date, there have been few clinical studies that have focused on the DOR. While co-morbid disease states may currently impact the clinical potential of DOR-based treatments, further investigation of the role of caveolar microdomains and/or the mechanisms underlying SLP may yield fruitful targets in the approach to clinical cardioprotection.

References

Abdul Y, Akhter N, Husain S (2013) Delta-opioid agonist SNC-121 protects retinal ganglion cell function in a chronic ocular hypertensive rat model. Invest Ophthalmol Vis Sci 54(3):1816–1828

Abete P et al (1997) Angina-induced protection against myocardial infarction in adult and elderly patients: a loss of preconditioning mechanism in the aging heart? J Am Coll Cardiol 30(4):947–954

Abete P et al (2011) Ischemic preconditioning in the younger and aged heart. Aging Dis 2(2):138–148

Aboumousa A et al (2008) Caveolinopathy – new mutations and additional symptoms. Neuromuscul Disord 18(7):572–578

Al-Hasani R, Bruchas MR (2011) Molecular mechanisms of opioid receptor-dependent signaling and behavior. Anesthesiology 115(6):1363–1381

Andre A et al (2008) Membrane partitioning of various delta-opioid receptor forms before and after agonist activations: the effect of cholesterol. Biochim Biophys Acta 1778(6):1483–1492

Armstead WM (1998) ATP-dependent K+ channel activation reduces loss of opioid dilation after brain injury. Am J Physiol 274(5 Pt 2):H1674–H1683

Ashton KJ et al (2013) Unique transcriptional profile of sustained ligand-activated preconditioning in pre- and post-ischemic myocardium. PLoS One 8(8):e72278

Bakhshi FR et al (2013) Nitrosation-dependent caveolin 1 phosphorylation, ubiquitination, and degradation and its association with idiopathic pulmonary arterial hypertension. Pulm Circ 3 (4):816–830

Ballard-Croft C et al (2006) Regional myocardial ischemia-induced activation of MAPKs is associated with subcellular redistribution of caveolin and cholesterol. Am J Physiol Heart Circ Physiol 291(2):H658–H667

Barron BA (2000) Cardiac opioids. Proc Soc Exp Biol Med 224(1):1–7

Barron BA et al (1992) Screening for opioids in dog heart. J Mol Cell Cardiol 24(1):67–77

Baxter GF, Ferdinandy P (2001) Delayed preconditioning of myocardium: current perspectives. Basic Res Cardiol 96(4):329–344
Bell SP et al (2000) Delta opioid receptor stimulation mimics ischemic preconditioning in human heart muscle. J Am Coll Cardiol 36(7):2296–2302
Benedict PE et al (1999) Opiate drugs and delta-receptor-mediated myocardial protection. Circulation 100(19 Suppl):II357–II360
Bhargava HN et al (1988) Opioid peptides in pituitary gland, brain regions and peripheral tissues of spontaneously hypertensive and Wistar-Kyoto normotensive rats. Brain Res 440(2):333–340
Bolli R et al (2004) Myocardial protection at a crossroads: the need for translation into clinical therapy. Circ Res 95(2):125–134
Bolling SF et al (1997) Use of "natural" hibernation induction triggers for myocardial protection. Ann Thorac Surg 64(3):623–627
Bolling SF et al (1998) Hibernation triggers and myocardial protection. Circulation 98(19 Suppl): II220–II223. discussion II223-4
Bolling SF et al (2001) Opioids confer myocardial tolerance to ischemia: interaction of delta opioid agonists and antagonists. J Thorac Cardiovasc Surg 122(3):476–481
Bolte C, Newman G, Schultz Jel J (2009a) Hypertensive state, independent of hypertrophy, exhibits an attenuated decrease in systolic function on cardiac kappa-opioid receptor stimulation. Am J Physiol Heart Circ Physiol 296(4):H967–H975
Bolte C, Newman G, Schultz Jel J (2009b) Kappa and delta opioid receptor signaling is augmented in the failing heart. J Mol Cell Cardiol 47(4):493–503
Boluyt MO et al (1993) Age-associated increase in rat cardiac opioid production. Am J Physiol 265(1 Pt 2):H212–H218
Borges JP et al (2014) Delta opioid receptors: the link between exercise and cardioprotection. PLoS One 9(11):e113541
Borlongan CV, Wang Y, Su TP (2004) Delta opioid peptide (D-Ala 2, D-Leu 5) enkephalin: linking hibernation and neuroprotection. Front Biosci 9:3392–3398
Borlongan CV et al (2009) Hibernation-like state induced by an opioid peptide protects against experimental stroke. BMC Biol 7:31
Bouhidel O et al (2008) Myocardial ischemic postconditioning against ischemia-reperfusion is impaired in ob/ob mice. Am J Physiol Heart Circ Physiol 295(4):H1580–H1586
Boutros A, Wang J (1995) Ischemic preconditioning, adenosine and bethanechol protect spontaneously hypertensive isolated rat hearts. J Pharmacol Exp Ther 275(3):1148–1156
Brown DA et al (2005) Cardioprotection afforded by chronic exercise is mediated by the sarcolemmal, and not the mitochondrial, isoform of the KATP channel in the rat. J Physiol 569(Pt 3):913–924
Bruce DS et al (1996) Circannual variations in bear plasma albumin and its opioid-like effects on guinea pig ileum. Pharmacol Biochem Behav 53(4):885–889
Brundege JM, Williams JT (2002) Increase in adenosine sensitivity in the nucleus accumbens following chronic morphine treatment. J Neurophysiol 87(3):1369–1375
Budiono BP et al (2012) Voluntary running in mice beneficially modulates myocardial ischemic tolerance, signaling kinases, and gene expression patterns. Am J Physiol Regul Integr Comp Physiol 302(9):R1091–R1100
Budiono BP et al (2016) Coupling of myocardial stress resistance and signalling to voluntary activity and inactivity. Acta Physiol (Oxf) 218(2):112–122
Caffrey JL et al (1994) Aging, cardiac proenkephalin mRNA and enkephalin peptides in the Fisher 344 rat. J Mol Cell Cardiol 26(6):701–711
Caffrey JL et al (1995) Intrinsic cardiac enkephalins inhibit vagal bradycardia in the dog. Am J Physiol 268(2 Pt 2):H848–H855
Chen Z, Li T, Zhang B (2008) Morphine postconditioning protects against reperfusion injury in the isolated rat hearts. J Surg Res 145(2):287–294
Chien S et al (1991) Two-day preservation of major organs with autoperfusion multiorgan preparation and hibernation induction trigger. A preliminary report. J Thorac Cardiovasc Surg 102 (2):224–234

Chien S et al (1994) Extension of tissue survival time in multiorgan block preparation with a delta opioid DADLE ([D-Ala2, D-Leu5]-enkephalin). J Thorac Cardiovasc Surg 107(3):964–967
Coccia R et al (2001) Interaction of enkephalin derivatives with reactive oxygen species. Biochim Biophys Acta 1525(1–2):43–49
Cohen MV, Baines CP, Downey JM (2000) Ischemic preconditioning: from adenosine receptor to KATP channel. Annu Rev Physiol 62:79–109
Cohen AW et al (2004) Role of caveolae and caveolins in health and disease. Physiol Rev 84 (4):1341–1379
Cohen MV et al (2007) Preconditioning-mimetics bradykinin and DADLE activate PI3-kinase through divergent pathways. J Mol Cell Cardiol 42(4):842–851
Crain SM, Shen KF (1992) After chronic opioid exposure sensory neurons become supersensitive to the excitatory effects of opioid agonists and antagonists as occurs after acute elevation of GM1 ganglioside. Brain Res 575(1):13–24
Cui Y, Lee TF, Wang LC (1996) In vivo microdialysis study on changes in septal dynorphin and beta-endorphin activities in active and hibernating Columbian ground squirrels. Brain Res 710 (1–2):271–274
Cui Y et al (1997) Autoradiographic determination of changes in opioid receptor binding in the limbic system of the Columbian ground squirrel at different hibernation states. Brain Res 747 (2):189–194
Czyzyk TA et al (2012) Mice lacking delta-opioid receptors resist the development of diet-induced obesity. FASEB J 26(8):3483–3492
Dickson EW et al (2001) Naloxone blocks transferred preconditioning in isolated rabbit hearts. J Mol Cell Cardiol 33(9):1751–1756
Dickson EW et al (2008) Exercise enhances myocardial ischemic tolerance via an opioid receptor-dependent mechanism. Am J Physiol Heart Circ Physiol 294(1):H402–H408
Donner D et al (2013) Obesity improves myocardial ischaemic tolerance and RISK signalling in insulin-insensitive rats. Dis Model Mech 6(2):457–466
Drastichova Z et al (2011) Prolonged morphine administration alters protein expression in the rat myocardium. J Biomed Sci 18:89
Du XJ et al (2000) Age-dependent cardiomyopathy and heart failure phenotype in mice over-expressing beta(2)-adrenergic receptors in the heart. Cardiovasc Res 48(3):448–454
Dumont M, Lemaire S (1988) Increased content of immunoreactive Leu-enkephalin and alteration of delta-opioid receptor in hearts of spontaneously hypertensive rats. Neurosci Lett 94(1–2):114–118
Dumont M et al (1991) Circadian regulation of the biosynthesis of cardiac Met-enkephalin and precursors in normotensive and spontaneously hypertensive rats. Life Sci 48(20):1895–1902
Ebrahim Z, Yellon DM, Baxter GF (2007a) Attenuated cardioprotective response to bradykinin, but not classical ischaemic preconditioning, in DOCA-salt hypertensive left ventricular hypertrophy. Pharmacol Res 55(1):42–48
Ebrahim Z, Yellon DM, Baxter GF (2007b) Ischemic preconditioning is lost in aging hypertensive rat heart: independent effects of aging and longstanding hypertension. Exp Gerontol 42 (8):807–814
Egawa J et al (2016) Membrane lipid rafts and neurobiology: age-related changes in membrane lipids and loss of neuronal function. J Physiol 594(16):4565–4579
Eliasson T et al (1998) Myocardial turnover of endogenous opioids and calcitonin-gene-related peptide in the human heart and the effects of spinal cord stimulation on pacing-induced angina pectoris. Cardiology 89(3):170–177
Ferdinandy P et al (2014) Interaction of risk factors, comorbidities, and comedications with ischemia/ reperfusion injury and cardioprotection by preconditioning, postconditioning, and remote conditioning. Pharmacol Rev 66(4):1142–1174
Feron O, Balligand JL (2006) Caveolins and the regulation of endothelial nitric oxide synthase in the heart. Cardiovasc Res 69(4):788–797

Floyd KC et al (2009) A 30-year perspective (1975-2005) into the changing landscape of patients hospitalized with initial acute myocardial infarction: Worcester Heart Attack Study. Circ Cardiovasc Qual Outcomes 2(2):88–95

Fontana F et al (1993) Relationship between plasma atrial natriuretic factor and opioid peptide levels in healthy subjects and in patients with acute congestive heart failure. Eur Heart J 14 (2):219–225

Forster K et al (2007) The delta-opioid receptor agonist DADLE at reperfusion protects the heart through activation of pro-survival kinases via EGF receptor transactivation. Am J Physiol Heart Circ Physiol 293(3):H1604–H1608

Fraessdorf J et al (2015) Role of endogenous opioid system in ischemic-induced late preconditioning. PLoS One 10(7):e0134283

Frances C et al (2003) Role of beta 1- and beta 2-adrenoceptor subtypes in preconditioning against myocardial dysfunction after ischemia and reperfusion. J Cardiovasc Pharmacol 41(3):396–405

Fridolfsson HN et al (2012) Mitochondria-localized caveolin in adaptation to cellular stress and injury. FASEB J 26(11):4637–4649

Fridolfsson HN et al (2014) Regulation of intracellular signaling and function by caveolin. FASEB J 28(9):3823–3831

Fryer RM et al (1999) Opioid-induced second window of cardioprotection: potential role of mitochondrial KATP channels. Circ Res 84(7):846–851

Fryer RM et al (2001) Essential activation of PKC-delta in opioid-initiated cardioprotection. Am J Physiol Heart Circ Physiol 280(3):H1346–H1353

Fuardo M et al (2013) [D-Ala2,D-Leu5]-enkephalin (DADLE) and morphine-induced postconditioning by inhibition of mitochondrial permeability transition pore, in human myocardium. Exp Biol Med (Maywood) 238(4):426–432

Fuhs SR, Insel PA (2011) Caveolin-3 undergoes SUMOylation by the SUMO E3 ligase PIASy: sumoylation affects G-protein-coupled receptor desensitization. J Biol Chem 286(17):14830–14841

Fujimoto T (1993) Calcium pump of the plasma membrane is localized in caveolae. J Cell Biol 120 (5):1147–1157

Fujimoto T et al (1992) Localization of inositol 1,4,5-trisphosphate receptor-like protein in plasmalemmal caveolae. J Cell Biol 119(6):1507–1513

Ghosh S, Standen NB, Galinianes M (2001) Failure to precondition pathological human myocardium. J Am Coll Cardiol 37(3):711–718

Giles TD et al (1987) Systemic methionine-enkephalin evokes cardiostimulatory responses in the human. Peptides 8(4):609–612

Gomez L et al (2008) Inhibition of GSK3beta by postconditioning is required to prevent opening of the mitochondrial permeability transition pore during reperfusion. Circulation 117(21):2761–2768

Goodman RR et al (1980) Differentiation of delta and mu opiate receptor localizations by light microscopic autoradiography. Proc Natl Acad Sci U S A 77(10):6239–6243

Goodman MD et al (2008) Regulating RISK: a role for JAK-STAT signaling in postconditioning? Am J Physiol Heart Circ Physiol 295(4):H1649–H1656

Govindaswami M et al (2008) Delta 2-specific opioid receptor agonist and hibernating woodchuck plasma fraction provide ischemic neuroprotection. Acad Emerg Med 15(3):250–257

Gross RA et al (1990) Dynorphin A and cAMP-dependent protein kinase independently regulate neuronal calcium currents. Proc Natl Acad Sci U S A 87(18):7025–7029

Gross ER et al (2003) K(ATP) opener-induced delayed cardioprotection: involvement of sarcolemmal and mitochondrial K(ATP) channels, free radicals and MEK1/2. J Mol Cell Cardiol 35 (8):985–992

Gross ER, Hsu AK, Gross GJ (2004) Opioid-induced cardioprotection occurs via glycogen synthase kinase beta inhibition during reperfusion in intact rat hearts. Circ Res 94(7):960–966

Gross ER et al (2005) Extending the cardioprotective window using a novel delta-opioid agonist fentanyl isothiocyanate via the PI3-kinase pathway. Am J Physiol Heart Circ Physiol 288(6): H2744–H2749

Gross ER, Hsu AK, Gross GJ (2006) The JAK/STAT pathway is essential for opioid-induced cardioprotection: JAK2 as a mediator of STAT3, Akt, and GSK-3 beta. Am J Physiol Heart Circ Physiol 291(2):H827–H834

Gross GJ et al (2007a) Mechanisms by which epoxyeicosatrienoic acids (EETs) elicit cardioprotection in rat hearts. J Mol Cell Cardiol 42(3):687–691

Gross ER, Hsu AK, Gross GJ (2007b) GSK3beta inhibition and K(ATP) channel opening mediate acute opioid-induced cardioprotection at reperfusion. Basic Res Cardiol 102(4):341–349

Gross ER, Hsu AK, Gross GJ (2007c) Diabetes abolishes morphine-induced cardioprotection via multiple pathways upstream of glycogen synthase kinase-3beta. Diabetes 56(1):127–136

Gross GJ et al (2010) Evidence for a role of opioids in epoxyeicosatrienoic acid-induced cardioprotection in rat hearts. Am J Physiol Heart Circ Physiol 298(6):H2201–H2207

Gross GJ et al (2012) Acute and chronic cardioprotection by the enkephalin analogue, Eribis peptide 94, is mediated via activation of nitric oxide synthase and adenosine triphosphate-regulated potassium channels. Pharmacology 90(1–2):110–116

Hagiwara Y et al (2000) Caveolin-3 deficiency causes muscle degeneration in mice. Hum Mol Genet 9(20):3047–3054

Halkos ME et al (2004) Myocardial protection with postconditioning is not enhanced by ischemic preconditioning. Ann Thorac Surg 78(3):961–969. discussion 969

Hamilton KL et al (2004) MnSOD antisense treatment and exercise-induced protection against arrhythmias. Free Radic Biol Med 37(9):1360–1368

Hassouna A et al (2006) Mitochondrial dysfunction as the cause of the failure to precondition the diabetic human myocardium. Cardiovasc Res 69(2):450–458

Head BP et al (2005) G-protein-coupled receptor signaling components localize in both sarcolemmal and intracellular caveolin-3-associated microdomains in adult cardiac myocytes. J Biol Chem 280(35):31036–31044

Headrick JP (1998) Aging impairs functional, metabolic and ionic recovery from ischemia-reperfusion and hypoxia-reoxygenation. J Mol Cell Cardiol 30(7):1415–1430

Headrick JP, Pepe S, Peart JN (2012) Non-analgesic effects of opioids: cardiovascular effects of opioids and their receptor systems. Curr Pharm Des 18(37):6090–6100

Headrick JP et al (2015) Opioid receptors and cardioprotection – 'opioidergic conditioning' of the heart. Br J Pharmacol 172(8):2026–2050

Heusch G (1998) Hibernating myocardium. Physiol Rev 78(4):1055–1085

Holaday JW (1983) Cardiovascular effects of endogenous opiate systems. Annu Rev Pharmacol Toxicol 23:541–594

Hong J et al (2005) Hibernation induction trigger reduces hypoxic damage of swine skeletal muscle. Muscle Nerve 32(2):200–207

Horton ND et al (1998) Isolation and partial characterization of an opioid-like 88 kDa hibernation-related protein. Comp Biochem Physiol B Biochem Mol Biol 119(4):787–805

Howells RD et al (1986) Proenkephalin mRNA in rat heart. Proc Natl Acad Sci U S A 83 (6):1960–1963

Hruby VJ et al (2010) Use of plasmon waveguide resonance (PWR) spectroscopy for examining binding, signaling and lipid domain partitioning of membrane proteins. Life Sci 86(15–16):569–574

Hu CP et al (2002) Effect of age on alpha-calcitonin gene-related peptide-mediated delayed cardioprotection induced by intestinal preconditioning in rats. Regul Pept 107(1–3):137–143

Huang MH et al (2007a) Mediating delta-opioid-initiated heart protection via the beta2-adrenergic receptor: role of the intrinsic cardiac adrenergic cell. Am J Physiol Heart Circ Physiol 293(1): H376–H384

Huang P et al (2007b) Agonist treatment did not affect association of mu opioid receptors with lipid rafts and cholesterol reduction had opposite effects on the receptor-mediated signaling in rat brain and CHO cells. Brain Res 1184:46–56

Huang P et al (2007c) Cholesterol reduction by methyl-beta-cyclodextrin attenuates the delta opioid receptor-mediated signaling in neuronal cells but enhances it in non-neuronal cells. Biochem Pharmacol 73(4):534–549

Husain S et al (2014) Regulation of nitric oxide production by delta-opioid receptors during glaucomatous injury. PLoS One 9(10):e110397
Imai N et al (1994) Comparison of cardiovascular effects of mu- and delta-opioid receptor antagonists in dogs with congestive heart failure. Am J Physiol 267(3 Pt 2):H912–H917
Insel PA et al (2005) Compartmentation of G-protein-coupled receptors and their signalling components in lipid rafts and caveolae. Biochem Soc Trans 33(Pt 5):1131–1134
Inserte J et al (2004) Ischemic preconditioning attenuates calpain-mediated degradation of structural proteins through a protein kinase A-dependent mechanism. Cardiovasc Res 64(1):105–114
Jang Y et al (2008) Postconditioning prevents reperfusion injury by activating delta-opioid receptors. Anesthesiology 108(2):243–250
Javadov SA et al (2003) Ischaemic preconditioning inhibits opening of mitochondrial permeability transition pores in the reperfused rat heart. J Physiol 549(Pt 2):513–524
Jennings RB, Reimer KA (1991) The cell biology of acute myocardial ischemia. Annu Rev Med 42:225–246
Jennings RB et al (1960) Myocardial necrosis induced by temporary occlusion of a coronary artery in the dog. Arch Pathol 70:68–78
Jiang X et al (2004) Inducible nitric oxide synthase mediates delayed cardioprotection induced by morphine in vivo: evidence from pharmacologic inhibition and gene-knockout mice. Anesthesiology 101(1):82–88
Jones KA et al (2004) Tuberin is a component of lipid rafts and mediates caveolin-1 localization: role of TSC2 in post-Golgi transport. Exp Cell Res 295(2):512–524
Jordan BA, Cvejic S, Devi LA (2000) Opioids and their complicated receptor complexes. Neuropsychopharmacology 23(4 Suppl):S5–S18
Jordan BA et al (2001) Oligomerization of opioid receptors with beta 2-adrenergic receptors: a role in trafficking and mitogen-activated protein kinase activation. Proc Natl Acad Sci U S A 98 (1):343–348
Karlsson LO et al (2012) Dose-dependent cardioprotection of enkephalin analogue Eribis peptide 94 and cardiac expression of opioid receptors in a porcine model of ischaemia and reperfusion. Eur J Pharmacol 674(2–3):378–383
Karoor V et al (2004) Propranolol prevents enhanced stress signaling in Gs alpha cardiomyopathy: potential mechanism for beta-blockade in heart failure. J Mol Cell Cardiol 36(2):305–312
Katakam PV et al (2007) Myocardial preconditioning against ischemia-reperfusion injury is abolished in Zucker obese rats with insulin resistance. Am J Physiol Regul Integr Comp Physiol 292(2):R920–R926
Kato R, Ross S, Foex P (2000) Fentanyl protects the heart against ischaemic injury via opioid receptors, adenosine A1 receptors and KATP channel linked mechanisms in rats. Br J Anaesth 84(2):204–214
Kersten JR et al (2000) Diabetes abolishes ischemic preconditioning: role of glucose, insulin, and osmolality. Am J Physiol Heart Circ Physiol 278(4):H1218–H1224
Kevelaitis E et al (1999) Opening of potassium channels: the common cardioprotective link between preconditioning and natural hibernation? Circulation 99(23):3079–3085
Kim HS et al (2010) Diabetes mellitus mitigates cardioprotective effects of remifentanil preconditioning in ischemia-reperfused rat heart in association with anti-apoptotic pathways of survival. Eur J Pharmacol 628(1–3):132–139
Kim JH et al (2011) Morphine-induced postconditioning modulates mitochondrial permeability transition pore opening via delta-1 opioid receptors activation in isolated rat hearts. Korean J Anesthesiol 61(1):69–74
Kodani E et al (2002) Delta-opioid receptor-induced late preconditioning is mediated by cyclooxygenase-2 in conscious rabbits. Am J Physiol Heart Circ Physiol 283(5):H1943–H1957
Krajewska WM, Maslowska I (2004) Caveolins: structure and function in signal transduction. Cell Mol Biol Lett 9(2):195–220
Krumins SA, Faden AI, Feuerstein G (1985) Opiate binding in rat hearts: modulation of binding after hemorrhagic shock. Biochem Biophys Res Commun 127(1):120–128

Kuzume K et al (2003) Sustained exogenous administration of Met5-enkephalin protects against infarction in vivo. Am J Physiol Heart Circ Physiol 285(6):H2463–H2470
Kuzume K et al (2005) Long-term infusion of Met5-enkephalin fails to protect murine hearts against ischemia-reperfusion injury. Am J Physiol Heart Circ Physiol 288(4):H1717–H1723
Lasley RD et al (2000) Activated cardiac adenosine A(1) receptors translocate out of caveolae. J Biol Chem 275(6):4417–4421
Lecour S (2009) Activation of the protective survivor activating factor enhancement (SAFE) pathway against reperfusion injury: does it go beyond the RISK pathway? J Mol Cell Cardiol 47(1):32–40
Lee CR et al (2006) Genetic variation in soluble epoxide hydrolase (EPHX2) and risk of coronary heart disease: the Atherosclerosis Risk in Communities (ARIC) study. Hum Mol Genet 15(10):1640–1649
Lendeckel U et al (2005) Expression of opioid receptor subtypes and their ligands in fibrillating human atria. Pacing Clin Electrophysiol 28(Suppl 1):S275–S279
Levitt ES et al (2009) Differential effect of membrane cholesterol removal on mu- and delta-opioid receptors: a parallel comparison of acute and chronic signaling to adenylyl cyclase. J Biol Chem 284(33):22108–22122
Li Y et al (2003) Morphine induces desensitization of insulin receptor signaling. Mol Cell Biol 23 (17):6255–6266
Li R et al (2009) Intrathecal morphine preconditioning induces cardioprotection via activation of delta, kappa, and mu opioid receptors in rats. Anesth Analg 108(1):23–29
Liggett SB et al (2000) Early and delayed consequences of beta(2)-adrenergic receptor overexpression in mouse hearts: critical role for expression level. Circulation 101(14):1707–1714
Lisanti MP et al (1994) Caveolae, caveolin and caveolin-rich membrane domains: a signalling hypothesis. Trends Cell Biol 4(7):231–235
Liu M et al (2012) Aging might increase myocardial ischemia/reperfusion-induced apoptosis in humans and rats. Age (Dordr) 34(3):621–632
Lochner A et al (1999) Ischemic preconditioning and the beta-adrenergic signal transduction pathway. Circulation 100(9):958–966
Lou LG, Pei G (1997) Modulation of protein kinase C and cAMP-dependent protein kinase by delta-opioid. Biochem Biophys Res Commun 236(3):626–629
Lu HR et al (1999) Reduction in QT dispersion and ventricular arrhythmias by ischaemic preconditioning in anaesthetized, normotensive and spontaneously hypertensive rats. Fundam Clin Pharmacol 13(4):445–454
Lu Y et al (2014) Spinal neuronal NOS activation mediates intrathecal fentanyl preconditioning induced remote cardioprotection in rats. Int Immunopharmacol 19(1):127–131
Marais E et al (2005) The temporal relationship between p38 MAPK and HSP27 activation in ischaemic and pharmacological preconditioning. Basic Res Cardiol 100(1):35–47
Maslov LN, Krylatov AV, Lishmanov Iu B (1996) Participation of endogenous mu- and delta-opiate receptor agonists in mechanisms of the anti-arrhythmia effect of adaptation. Biull Eksp Biol Med 121(1):24–25
Maslov LN et al (2006) Negative inotropic and chronotropic effects of delta-opioid receptor antagonists are mediated via non-opioid receptors. Bull Exp Biol Med 141(4):420–423
Maslov LN et al (2009) Activation of peripheral delta2 opioid receptors increases cardiac tolerance to ischemia/reperfusion injury involvement of protein kinase C, NO-synthase, KATP channels and the autonomic nervous system. Life Sci 84(19–20):657–663
May CN et al (1989) Differential cardiovascular and respiratory responses to central administration of selective opioid agonists in conscious rabbits: correlation with receptor distribution. Br J Pharmacol 98(3):903–913
McIntosh VJ, Lasley RD (2012) Adenosine receptor-mediated cardioprotection: are all 4 subtypes required or redundant? J Cardiovasc Pharmacol Ther 17(1):21–33
Menard DP et al (1996) A calcitonin gene-related peptide receptor antagonist prevents the development of tolerance to spinal morphine analgesia. J Neurosci 16(7):2342–2351
Mieno S et al (2002) Potent adenylate cyclase agonist forskolin restores myoprotective effects of ischemic preconditioning in rat hearts after myocardial infarction. Ann Thorac Surg 74(4):1213–1218

Miki T et al (2012) Effects of diabetes on myocardial infarct size and cardioprotection by preconditioning and postconditioning. Cardiovasc Diabetol 11:67

Miller LE et al (2015) Involvement of the delta-opioid receptor in exercise-induced cardioprotection. Exp Physiol 100(4):410–421

Miura T, Miki T (2008) Limitation of myocardial infarct size in the clinical setting: current status and challenges in translating animal experiments into clinical therapy. Basic Res Cardiol 103 (6):501–513

Miura T et al (2007) Delta-opioid receptor activation before ischemia reduces gap junction permeability in ischemic myocardium by PKC-epsilon-mediated phosphorylation of connexin 43. Am J Physiol Heart Circ Physiol 293(3):H1425–H1431

Monier S et al (1995) VIP21-caveolin, a membrane protein constituent of the caveolar coat, oligomerizes in vivo and in vitro. Mol Biol Cell 6(7):911–927

Moolman JA et al (1997) Ischaemic preconditioning does not protect hypertrophied myocardium against ischaemia. S Afr Med J 87(Suppl 3):C151–C156

Motoki A et al (2008) Soluble epoxide hydrolase inhibition and gene deletion are protective against myocardial ischemia-reperfusion injury in vivo. Am J Physiol Heart Circ Physiol 295(5):H2128–H2134

Mougin C et al (1987) Assessment of plasma opioid peptides, beta-endorphin and met-enkephalin, at the end of an international nordic ski race. Eur J Appl Physiol Occup Physiol 56(3):281–286

Murry CE, Jennings RB, Reimer KA (1986) Preconditioning with ischemia: a delay of lethal cell injury in ischemic myocardium. Circulation 74(5):1124–1136

Nguyen VT et al (2012) Delta-opioid augments cardiac contraction through beta-adrenergic and CGRP-receptor co-signaling. Peptides 33(1):77–82

Nishihara M et al (2006) Erythropoietin affords additional cardioprotection to preconditioned hearts by enhanced phosphorylation of glycogen synthase kinase-3 beta. Am J Physiol Heart Circ Physiol 291(2):H748–H755

Nithipatikom K et al (2006) Epoxyeicosatrienoic acids in cardioprotection: ischemic versus reperfusion injury. Am J Physiol Heart Circ Physiol 291(2):H537–H542

Oeltgen PR et al (1988) Further studies on opioids and hibernation: delta opioid receptor ligand selectively induced hibernation in summer-active ground squirrels. Life Sci 43(19):1565–1574

Oeltgen PR et al (1996) Extended lung preservation with the use of hibernation trigger factors. Ann Thorac Surg 61(5):1488–1493

Okubo S et al (2004) Ischemic preconditioning and morphine attenuate myocardial apoptosis and infarction after ischemia-reperfusion in rabbits: role of delta-opioid receptor. Am J Physiol Heart Circ Physiol 287(4):H1786–H1791

Olianas MC, Dedoni S, Onali P (2011) delta-Opioid receptors stimulate GLUT1-mediated glucose uptake through Src- and IGF-1 receptor-dependent activation of PI3-kinase signalling in CHO cells. Br J Pharmacol 163(3):624–637

Olianas MC et al (2012) delta-Opioid receptors stimulate the metabolic sensor AMP-activated protein kinase through coincident signaling with G(q/11)-coupled receptors. Mol Pharmacol 81(2):154–165

Ouellette M, Brakier-Gingras L (1988) Increase in the relative abundance of preproenkephalin A messenger RNA in the ventricles of cardiomyopathic hamsters. Biochem Biophys Res Commun 155(1):449–454

Parton RG, del Pozo MA (2013) Caveolae as plasma membrane sensors, protectors and organizers. Nat Rev Mol Cell Biol 14(2):98–112

Patel HH et al (2002a) Sarcolemmal K(ATP) channel triggers opioid-induced delayed cardioprotection in the rat. Circ Res 91(3):186–188

Patel HH et al (2002b) Cardioprotection at a distance: mesenteric artery occlusion protects the myocardium via an opioid sensitive mechanism. J Mol Cell Cardiol 34(10):1317–1323

Patel HH et al (2003) 12-lipoxygenase in opioid-induced delayed cardioprotection: gene array, mass spectrometric, and pharmacological analyses. Circ Res 92(6):676–682

Patel HH et al (2006) Protection of adult rat cardiac myocytes from ischemic cell death: role of caveolar microdomains and delta-opioid receptors. Am J Physiol Heart Circ Physiol 291(1): H344–H350

Patel HH et al (2007) Mechanisms of cardiac protection from ischemia/reperfusion injury: a role for caveolae and caveolin-1. FASEB J 21(7):1565–1574

Patel HH, Murray F, Insel PA (2008a) G-protein-coupled receptor-signaling components in membrane raft and caveolae microdomains. Handb Exp Pharmacol (186):167–184

Patel HH, Murray F, Insel PA (2008b) Caveolae as organizers of pharmacologically relevant signal transduction molecules. Annu Rev Pharmacol Toxicol 48:359–391

Peart JN, Gross GJ (2003) Adenosine and opioid receptor-mediated cardioprotection in the rat: evidence for cross-talk between receptors. Am J Physiol Heart Circ Physiol 285(1):H81–H89

Peart JN, Gross GJ (2004a) Exogenous activation of delta- and kappa-opioid receptors affords cardioprotection in isolated murine heart. Basic Res Cardiol 99(1):29–37

Peart JN, Gross GJ (2004b) Chronic exposure to morphine produces a marked cardioprotective phenotype in aged mouse hearts. Exp Gerontol 39(7):1021–1026

Peart JN, Gross GJ (2004c) Morphine-tolerant mice exhibit a profound and persistent cardioprotective phenotype. Circulation 109(10):1219–1222

Peart JN, Gross GJ (2005) Cardioprotection following adenosine kinase inhibition in rat hearts. Basic Res Cardiol 100(4):328–336

Peart JN, Gross GJ (2006) Cardioprotective effects of acute and chronic opioid treatment are mediated via different signaling pathways. Am J Physiol Heart Circ Physiol 291(4):H1746–H1753

Peart JN, Headrick JP (2007) Adenosinergic cardioprotection: multiple receptors, multiple pathways. Pharmacol Ther 114(2):208–221

Peart JN, Headrick JP (2008) Sustained cardioprotection: exploring unconventional modalities. Vascul Pharmacol 49(2–3):63–70

Peart JN, Headrick JP (2009) Clinical cardioprotection and the value of conditioning responses. Am J Physiol Heart Circ Physiol 296(6):H1705–H1720

Peart JN, Patel HH, Gross GJ (2003) Delta-opioid receptor activation mimics ischemic preconditioning in the canine heart. J Cardiovasc Pharmacol 42(1):78–81

Peart JN, Gross ER, Gross GJ (2005) Opioid-induced preconditioning: recent advances and future perspectives. Vascul Pharmacol 42(5–6):211–218

Peart JN et al (2007) Impaired p38 MAPK/HSP27 signaling underlies aging-related failure in opioid-mediated cardioprotection. J Mol Cell Cardiol 42(5):972–980

Peart JN et al (2011) Sustained ligand-activated preconditioning via delta-opioid receptors. J Pharmacol Exp Ther 336(1):274–281

Peart JN et al (2014) Dysfunctional survival-signaling and stress-intolerance in aged murine and human myocardium. Exp Gerontol 50:72–81

Peng J, Sarkar S, Chang SL (2012) Opioid receptor expression in human brain and peripheral tissues using absolute quantitative real-time RT-PCR. Drug Alcohol Depend 124(3):223–228

Pepe S et al (1997) 'Cross talk' between opioid peptide and adrenergic receptor signaling in isolated rat heart. Circulation 95(8):2122–2129

Pepe S et al (2004) Cross-talk of opioid peptide receptor and beta-adrenergic receptor signalling in the heart. Cardiovasc Res 63(3):414–422

Peroutka SJ et al (1980) Neurotransmitter receptor binding in bovine cerebral microvessels. Science 208(4444):610–612

Perrelli MG, Pagliaro P, Penna C (2011) Ischemia/reperfusion injury and cardioprotective mechanisms: role of mitochondria and reactive oxygen species. World J Cardiol 3(6):186–200

Pike LJ (2003) Lipid rafts: bringing order to chaos. J Lipid Res 44(4):655–667

Pradhan AA et al (2010) Ligand-directed trafficking of the delta-opioid receptor in vivo: two paths toward analgesic tolerance. J Neurosci 30(49):16459–16468

Pradhan AA et al (2012) Ligand-directed signalling within the opioid receptor family. Br J Pharmacol 167(5):960–969

Pradhan AA et al (2016) Agonist-specific recruitment of arrestin isoforms differentially modify delta opioid receptor function. J Neurosci 36(12):3541–3551
Przyklenk K (2011) Efficacy of cardioprotective 'conditioning' strategies in aging and diabetic cohorts: the co-morbidity conundrum. Drugs Aging 28(5):331–343
Przyklenk K, Whittaker P (2011) Remote ischemic preconditioning: current knowledge, unresolved questions, and future priorities. J Cardiovasc Pharmacol Ther 16(3–4):255–259
Przyklenk K et al (1993) Regional ischemic 'preconditioning' protects remote virgin myocardium from subsequent sustained coronary occlusion. Circulation 87(3):893–899
Przyklenk K et al (2008) Aging mouse hearts are refractory to infarct size reduction with post-conditioning. J Am Coll Cardiol 51(14):1393–1398
Przyklenk K et al (2011) Cardioprotection with postconditioning: loss of efficacy in murine models of type-2 and type-1 diabetes. Antioxid Redox Signal 14(5):781–790
Pugsley MK (2002) The diverse molecular mechanisms responsible for the actions of opioids on the cardiovascular system. Pharmacol Ther 93(1):51–75
Pugsley MK et al (1993) Electrophysiological and antiarrhythmic actions of the kappa agonist PD 129290, and its R,R (+)-enantiomer, PD 129289. Br J Pharmacol 110(4):1579–1585
Pugsley MK et al (1998) Sodium channel-blocking properties of spiradoline, a kappa receptor agonist, are responsible for its antiarrhythmic action in the rat. J Cardiovasc Pharmacol 32(6):863–874
Qiu Y et al (2011) Cholesterol regulates micro-opioid receptor-induced beta-arrestin 2 translocation to membrane lipid rafts. Mol Pharmacol 80(1):210–218
Quindry JC et al (2010) Mitochondrial KATP channel inhibition blunts arrhythmia protection in ischemic exercised hearts. Am J Physiol Heart Circ Physiol 299(1):H175–H183
Radel C, Rizzo V (2005) Integrin mechanotransduction stimulates caveolin-1 phosphorylation and recruitment of Csk to mediate actin reorganization. Am J Physiol Heart Circ Physiol 288(2): H936–H945
Rahimtoola SH (1985) A perspective on the three large multicenter randomized clinical trials of coronary bypass surgery for chronic stable angina. Circulation 72(6 Pt 2):V123–V135
Rakhit RD, Marber MS (2001) Nitric oxide: an emerging role in cardioprotection? Heart 86(4):368–372
Ralston E, Ploug T (1999) Caveolin-3 is associated with the T-tubules of mature skeletal muscle fibers. Exp Cell Res 246(2):510–515
Ratajczak P et al (2003) Caveolin-1 and -3 dissociations from caveolae to cytosol in the heart during aging and after myocardial infarction in rat. Cardiovasc Res 57(2):358–369
Riess ML et al (2005) Increasing heart size and age attenuate anesthetic preconditioning in guinea pig isolated hearts. Anesth Analg 101(6):1572–1576
Ritchie RH et al (2012) Enhanced phosphoinositide 3-kinase(p110alpha) activity prevents diabetes-induced cardiomyopathy and superoxide generation in a mouse model of diabetes. Diabetologia 55(12):3369–3381
Roger VL et al (2012) Heart disease and stroke statistics – 2012 update: a report from the American Heart Association. Circulation 125(1):e2–e220
Romano MA et al (2004a) Differential effects of opioid peptides on myocardial ischemic tolerance. J Surg Res 119(1):46–50
Romano MA et al (2004b) Relative contribution of endogenous opioids to myocardial ischemic tolerance. J Surg Res 118(1):32–37
Rosenberger J, Petrovics G, Buzas B (2001) Oxidative stress induces proorphanin FQ and proenkephalin gene expression in astrocytes through p38- and ERK-MAP kinases and NF-kappaB. J Neurochem 79(1):35–44
Roth DM, Patel HH (2011) Role of caveolae in cardiac protection. Pediatr Cardiol 32(3):329–333
Ruiz-Meana M et al (2014) Ischemic preconditioning protects cardiomyocyte mitochondria through mechanisms independent of cytosol. J Mol Cell Cardiol 68:79–88
Sanada S et al (2004) Protein kinase A as another mediator of ischemic preconditioning independent of protein kinase C. Circulation 110(1):51–57

Sanada S, Komuro I, Kitakaze M (2011) Pathophysiology of myocardial reperfusion injury: preconditioning, postconditioning, and translational aspects of protective measures. Am J Physiol Heart Circ Physiol 301(5):H1723–H1741

Schulman D, Latchman DS, Yellon DM (2001) Effect of aging on the ability of preconditioning to protect rat hearts from ischemia-reperfusion injury. Am J Physiol Heart Circ Physiol 281(4): H1630–H1636

Schultz JE, Gross GJ (2001) Opioids and cardioprotection. Pharmacol Ther 89(2):123–137

Schultz JE et al (1995) Evidence for involvement of opioid receptors in ischemic preconditioning in rat hearts. Am J Physiol 268(5 Pt 2):H2157–H2161

Schultz JE, Hsu AK, Gross GJ (1996) Morphine mimics the cardioprotective effect of ischemic preconditioning via a glibenclamide-sensitive mechanism in the rat heart. Circ Res 78(6):1100–1104

Schultz JJ, Hsu AK, Gross GJ (1997a) Ischemic preconditioning is mediated by a peripheral opioid receptor mechanism in the intact rat heart. J Mol Cell Cardiol 29(5):1355–1362

Schultz JJ, Hsu AK, Gross GJ (1997b) Ischemic preconditioning and morphine-induced cardioprotection involve the delta (delta)-opioid receptor in the intact rat heart. J Mol Cell Cardiol 29 (8):2187–2195

Schultz Je-J et al (1998) TAN-67, a delta 1-opioid receptor agonist, reduces infarct size via activation of Gi/o proteins and KATP channels. Am J Physiol 274(3 Pt 2):H909–H914

Schulz R, Gres P, Heusch G (2001) Role of endogenous opioids in ischemic preconditioning but not in short-term hibernation in pigs. Am J Physiol Heart Circ Physiol 280(5):H2175–H2181

Schwartz CF et al (1999) Delta opioid receptors and low temperature myocardial protection. Ann Thorac Surg 68(6):2089–2092

Scriven DR et al (2002) The molecular architecture of calcium microdomains in rat cardiomyocytes. Ann N Y Acad Sci 976:488–499

Scriven DR et al (2005) Caveolin-3 is adjacent to a group of extradyadic ryanodine receptors. Biophys J 89(3):1893–1901

See Hoe LE et al (2014) Sarcolemmal cholesterol and caveolin-3 dependence of cardiac function, ischemic tolerance, and opioidergic cardioprotection. Am J Physiol Heart Circ Physiol 307(6): H895–H903

See Hoe LE et al (2016) Chronic beta1-adrenoceptor blockade impairs ischaemic tolerance and preconditioning in murine myocardium. Eur J Pharmacol 789:1–7

Seubert J et al (2004) Enhanced postischemic functional recovery in CYP2J2 transgenic hearts involves mitochondrial ATP-sensitive K+ channels and p42/p44 MAPK pathway. Circ Res 95 (5):506–514

Seubert JM et al (2006) Role of soluble epoxide hydrolase in postischemic recovery of heart contractile function. Circ Res 99(4):442–450

Seymour EM et al (2003) HL-1 myocytes exhibit PKC and K(ATP) channel-dependent delta opioid preconditioning. J Surg Res 114(2):187–194

Shan J et al (2002) Attenuated "cross talk" between kappa-opioid receptors and beta-adrenoceptors in the heart of chronically hypoxic rats. Pflugers Arch 444(1–2):126–132

Shankar V, Armstead WM (1995) Opioids contribute to hypoxia-induced pial artery dilation through activation of ATP-sensitive K+ channels. Am J Physiol 269(3 Pt 2):H997–1002

Shaul PW, Anderson RG (1998) Role of plasmalemmal caveolae in signal transduction. Am J Physiol 275(5 Pt 1):L843–L851

Shen J et al (2000) Role of cAMP-dependent protein kinase (PKA) in opioid agonist-induced mu-opioid receptor downregulation and tolerance in mice. Synapse 38(3):322–327

Shen H et al (2012) Role of delta2 opioid receptor in cardioprotection against hypoxia-reoxygenation injury. J Cardiovasc Pharmacol 60(3):253–261

Sheng JZ et al (1996) Lithium attenuates the effects of dynorphin A(1-13) on inositol 1,4,5-trisphosphate and intracellular Ca2+ in rat ventricular myocytes. Life Sci 59(25–26):2181–2186

Shi W, Vinten-Johansen J (2012) Endogenous cardioprotection by ischaemic postconditioning and remote conditioning. Cardiovasc Res 94(2):206–216

Sichelschmidt OJ et al (2003) Trapidil protects ischemic hearts from reperfusion injury by stimulating PKAII activity. Cardiovasc Res 58(3):602–610
Sigg DC et al (2002) Role of delta-opioid receptor agonists on infarct size reduction in swine. Am J Physiol Heart Circ Physiol 282(6):H1953–H1960
Sobanski P et al (2014) The presence of mu-, delta-, and kappa-opioid receptors in human heart tissue. Heart Vessels 29(6):855–863
Stary CM et al (2012) Caveolins: targeting pro-survival signaling in the heart and brain. Front Physiol 3:393
Steinberg SF, Brunton LL (2001) Compartmentation of G protein-coupled signaling pathways in cardiac myocytes. Annu Rev Pharmacol Toxicol 41:751–773
Sun J, Murphy E (2010) Protein S-nitrosylation and cardioprotection. Circ Res 106(2):285–296
Sun J et al (2015) Ischaemic preconditioning preferentially increases protein S-nitrosylation in subsarcolemmal mitochondria. Cardiovasc Res 106(2):227–236
Sun T et al (2016) Preservation of CGRP in myocardium attenuates development of cardiac dysfunction in diabetic rats. Int J Cardiol 220:226–234
Surendra H et al (2013) Interaction of delta and kappa opioid receptors with adenosine A1 receptors mediates cardioprotection by remote ischemic preconditioning. J Mol Cell Cardiol 60:142–150
Tamura Y et al (2006) Neuroprotective effects of hibernation-regulating substances against low-temperature-induced cell death in cultured hamster hippocampal neurons. Brain Res 1108 (1):107–116
Tamura Y et al (2012) Regulatory mechanism of body temperature in the central nervous system during the maintenance phase of hibernation in Syrian hamsters: involvement of beta-endorphin. Brain Res 1448:63–70
Tani M et al (2001) Direct activation of mitochondrial K(ATP) channels mimics preconditioning but protein kinase C activation is less effective in middle-aged rat hearts. Cardiovasc Res 49 (1):56–68
Theisen MM et al (2014) Detection and distribution of opioid peptide receptors in porcine myocardial tissue. Pharmacol Res 84:45–49
Tobin SJ et al (2014) Nanoscale effects of ethanol and naltrexone on protein organization in the plasma membrane studied by photoactivated localization microscopy (PALM). PLoS One 9 (2):e87225
Tomai F et al (1999) Effects of naloxone on myocardial ischemic preconditioning in humans. J Am Coll Cardiol 33(7):1863–1869
Tong H et al (2005) The role of beta-adrenergic receptor signaling in cardioprotection. FASEB J 19(8):983–985
Trang T et al (2002) The role of spinal neuropeptides and prostaglandins in opioid physical dependence. Br J Pharmacol 136(1):37–48
Tsang A et al (2004) Postconditioning: a form of "modified reperfusion" protects the myocardium by activating the phosphatidylinositol 3-kinase-Akt pathway. Circ Res 95(3):230–232
Tsanova A et al (2014) Interaction of methionine-enkephalins with raft-forming lipids: monolayers and BAM experiments. Amino Acids 46(5):1159–1168
Tsanova A, Jordanova A, Lalchev Z (2016) Effects of leucin-enkephalins on surface characteristics and morphology of model membranes composed of raft-forming lipids. J Membr Biol 249 (3):229–238
Tsutsumi YM et al (2007) Reactive oxygen species trigger ischemic and pharmacological postconditioning: in vivo and in vitro characterization. Life Sci 81(15):1223–1227
Tsutsumi YM et al (2008) Cardiac-specific overexpression of caveolin-3 induces endogenous cardiac protection by mimicking ischemic preconditioning. Circulation 118(19):1979–1988
Tsutsumi YM et al (2010) Opioid-induced preconditioning is dependent on caveolin-3 expression. Anesth Analg 111(5):1117–1121
Tumati S et al (2009) Sustained morphine treatment augments capsaicin-evoked calcitonin gene-related peptide release from primary sensory neurons in a protein kinase A- and Raf-1-dependent manner. J Pharmacol Exp Ther 330(3):810–817

Valtchanova-Matchouganska A, Ojewole JA (2003) Mechanisms of opioid delta (delta) and kappa (kappa) receptors' cardioprotection in ischaemic preconditioning in a rat model of myocardial infarction. Cardiovasc J S Afr 14(2):73–80
Vander Heide RS, Steenbergen C (2013) Cardioprotection and myocardial reperfusion: pitfalls to clinical application. Circ Res 113(4):464–477
Varga EV et al (2003) Molecular mechanisms of excitatory signaling upon chronic opioid agonist treatment. Life Sci 74(2–3):299–311
Vargish T, Beamer KC (1989) Delta and Mu receptor agonists correlate with greater depression of cardiac function than morphine sulfate in perfused rat hearts. Circ Shock 27(3):245–251
Ventura C et al (1989) Opioid receptors in rat cardiac sarcolemma: effect of phenylephrine and isoproterenol. Biochim Biophys Acta 987(1):69–74
Ventura C et al (1992) Kappa and delta opioid receptor stimulation affects cardiac myocyte function and Ca2+ release from an intracellular pool in myocytes and neurons. Circ Res 70(1):66–81
Villemagne PS et al (2002) PET imaging of human cardiac opioid receptors. Eur J Nucl Med Mol Imaging 29(10):1385–1388
Vinten-Johansen J, Yellon DM, Opie LH (2005) Postconditioning: a simple, clinically applicable procedure to improve revascularization in acute myocardial infarction. Circulation 112(14):2085–2088
Wagner EJ, Ronnekleiv OK, Kelly MJ (1998) Protein kinase A maintains cellular tolerance to mu opioid receptor agonists in hypothalamic neurosecretory cells with chronic morphine treatment: convergence on a common pathway with estrogen in modulating mu opioid receptor/effector coupling. J Pharmacol Exp Ther 285(3):1266–1273
Wagner C et al (2008) Cardioprotection by postconditioning is lost in WOKW rats with metabolic syndrome: role of glycogen synthase kinase 3beta. J Cardiovasc Pharmacol 52(5):430–437
Wang TL et al (1998) Morphine preconditioning attenuates neutrophil activation in rat models of myocardial infarction. Cardiovasc Res 40(3):557–563
Wang GY et al (2001) Kappa- but not delta-opioid receptors mediate effects of ischemic preconditioning on both infarct and arrhythmia in rats. Am J Physiol Heart Circ Physiol 280(1):H384–H391
Weil J et al (2006) Alterations of the preproenkephalin system in cardiac hypertrophy and its role in atrioventricular conduction. Cardiovasc Res 69(2):412–422
Weinbrenner C et al (2004) Remote preconditioning by infrarenal aortic occlusion is operative via delta1-opioid receptors and free radicals in vivo in the rat heart. Cardiovasc Res 61(3):591–599
Whitworth JA (2003) 2003 World Health Organization (WHO)/International Society of Hypertension (ISH) statement on management of hypertension. J Hypertens 21(11):1983–1992
Widimsky P et al (2010) Reperfusion therapy for ST elevation acute myocardial infarction in Europe: description of the current situation in 30 countries. Eur Heart J 31(8):943–957
Williams TM, Lisanti MP (2004) The caveolin proteins. Genome Biol 5(3):214
Williams-Pritchard G et al (2011a) Essential role of EGFR in cardioprotection and signaling responses to A1 adenosine receptors and ischemic preconditioning. Am J Physiol Heart Circ Physiol 300(6):H2161–H2168
Williams-Pritchard G, Headrick JP, Peart JN (2011b) Myocardial opioid receptors in conditioning and cytoprotection. Pharmaceuticals 4(3):470–484
Wittert G, Hope P, Pyle D (1996) Tissue distribution of opioid receptor gene expression in the rat. Biochem Biophys Res Commun 218(3):877–881
Wojtovich AP et al (2012) Ischemic preconditioning: the role of mitochondria and aging. Exp Gerontol 47(1):1–7
Wong GT et al (2012) Intrathecal morphine remotely preconditions the heart via a neural pathway. J Cardiovasc Pharmacol 60(2):172–178
Woodman SE et al (2002) Caveolin-3 knock-out mice develop a progressive cardiomyopathy and show hyperactivation of the p42/44 MAPK cascade. J Biol Chem 277(41):38988–38997
Woodman SE et al (2004) Caveolinopathies: mutations in caveolin-3 cause four distinct autosomal dominant muscle diseases. Neurology 62(4):538–543

Xiao RP et al (1997) Opioid peptide receptor stimulation reverses beta-adrenergic effects in rat heart cells. Am J Physiol 272(2 Pt 2):H797–H805

Yadav HN, Singh M, Sharma PL (2010) Involvement of GSK-3beta in attenuation of the cardioprotective effect of ischemic preconditioning in diabetic rat heart. Mol Cell Biochem 343(1–2):75–81

Yamashita N et al (1999) Exercise provides direct biphasic cardioprotection via manganese superoxide dismutase activation. J Exp Med 189(11):1699–1706

Yao L et al (2011) Interaction between spinal opioid and adenosine receptors in remote cardiac preconditioning: effect of intrathecal morphine. J Cardiothorac Vasc Anesth 25(3):444–448

Yellon DM, Downey JM (2003) Preconditioning the myocardium: from cellular physiology to clinical cardiology. Physiol Rev 83(4):1113–1151

Yin X et al (2012) Diabetic inhibition of preconditioning- and postconditioning-mediated myocardial protection against ischemia/reperfusion injury. Exp Diabetes Res 2012:198048

You L et al (2011) Postconditioning reduces infarct size and cardiac myocyte apoptosis via the opioid receptor and JAK-STAT signaling pathway. Mol Biol Rep 38(1):437–443

Zatta AJ et al (2008) Evidence that cardioprotection by postconditioning involves preservation of myocardial opioid content and selective opioid receptor activation. Am J Physiol Heart Circ Physiol 294(3):H1444–H1451

Zhang Y, Irwin MG, Wong TM (2004) Remifentanil preconditioning protects against ischemic injury in the intact rat heart. Anesthesiology 101(4):918–923

Zhang SZ et al (2006) Kappa-opioid receptors mediate cardioprotection by remote preconditioning. Anesthesiology 105(3):550–556

Zhang Y et al (2011) Intracerebroventricular administration of morphine confers remote cardioprotection – role of opioid receptors and calmodulin. Eur J Pharmacol 656(1–3):74–80

Zhao H, Loh HH, Law PY (2006) Adenylyl cyclase superactivation induced by long-term treatment with opioid agonist is dependent on receptor localized within lipid rafts and is independent of receptor internalization. Mol Pharmacol 69(4):1421–1432

The Roles of Opioid Receptors in Cutaneous Wound Healing

Mei Bigliardi-Qi and Paul Bigliardi

Contents

Abstract

The process of recovery from skin wounding can be protracted and painful, and scarring may lead to weakness of the tissue, unpleasant sensations such as pain or itch, and unfavorable cosmetic outcomes. Moreover, some wounds simply fail to heal and become a chronic burden for the sufferer. Understanding the mechanisms underlying wound healing and the concomitant sensory disorders and how they might be manipulated for therapeutic benefit has attracted much interest in recent years, and here we discuss the latest developments in the field, focusing on the emergent roles of the peripheral opioid receptor (OPr) system.

M. Bigliardi-Qi (✉)
Experimental Dermatology Group, Institute of Medical Biology, A*STAR, Singapore, Singapore, 138648
e-mail: qimei1@gmail.com; mei.bigliardi-qi@imb.a-star.edu.sg

P. Bigliardi
NUH/NUS, National University Hospital/National University of Singapore, Singapore, Singapore, 119228
Experimental Dermatology Group, Institute of Medical Biology, A*STAR, Singapore, Singapore, 138648
e-mail: paul_bigliardi@nuhs.edu.sg; paul.bigliardi@gmail.com

E. M. Jutkiewicz (ed.), *Delta Opioid Receptor Pharmacology and Therapeutic Applications*, Handbook of Experimental Pharmacology 247, DOI 10.1007/164_2017_14

Keywords
Opioid receptors • Skin • Wound healing

Abbreviations

DOPr	Delta OPr
ECM	Extracellular matrix
KO	Knockout
KOPr	Kappa OPr
MMP	Matrix metalloproteinase
MOPr	Mu OPr
NOPr	Nociceptin OPr
OPr	Opioid receptor
PNS	Peripheral nervous system
TGF	Transforming growth factor

1 Introduction

A skin wound represents a breach of the body's barrier organ which is likely to be accompanied by the introduction of potentially pathogenic microorganisms as well as physical damage to the cutaneous tissues, their vasculature, and nerves. Thus the physiology of wound healing is complex, involving the careful integration of various processes including the expression of inflammatory mediators and growth factors, cell–cell and cell–extracellular matrix interactions, cell proliferation, migration and differentiation, re-epithelialization, fibroplasia and angiogenesis, wound contraction and finally remodeling of the connective tissue. These processes are coordinated by several populations of cells: the keratinocytes in epidermis and fibroblasts in dermis, the endothelial cells with blood vessels, infiltrating immune cells (neutrophils, macrophages, and lymphocytes), and sprouting neurons from the peripheral nervous system (PNS). The components of the PNS have only recently been understood as an important component of the wound healing machinery; the delayed wound healing seen in patients with peripheral neuropathies such as multiple sclerosis or diabetes mellitus indicated the involvement of the PNS. We now know that both neuropeptides and their receptors are not only relevant for wound pain sensation but also directly influence the healing process. While in most cases the body heals minor skin wounds well, there is a pressing need to understand the molecular mechanics of wound healing if we are to develop improved treatments for patients with skin disorders, nonhealing wounds, or, as in the case of burns victims, individuals with severe and/or life-threatening damage to their skin.

One class of molecules in particular has attracted much interest in this regard in recent years: the opioid receptor (OPr) family of the PNS is now known to be intimately involved in all the phases of wound healing (Cheng et al. 2008) and may be uniquely amenable to therapeutic manipulation. In this review, we will summarize current knowledge of the roles of the OPrs in the context of our understanding of wound healing and look at the attempts to target their actions to improve skin recovery from traumatic injury.

2 The Opioid Receptors and Their Ligands in Skin

There are four major subtypes of OPr: delta, kappa, mu, and nociceptin (DOPr, KOPr, MOPr, and NOPr, respectively), with varied expression patterns and physiological functions. We were the first to describe the presence of the MOPr and DOPr in both normal and diseased human skin (Bigliardi et al. 1998, 2009), and since this time substantial efforts have gone into understanding their roles and interactions at these sites. Initial experiments using cultures of normal human skin showed that the endogenous ligand for the MOPr, β-endorphin, upregulates epidermal expression of cytokeratin 16 (Bigliardi-Qi et al. 2000) and the TGF-β type II receptor (Bigliardi et al. 2003). This was an early indication of the involvement of the MOPr in wound healing, as cytokeratin 16 is not expressed in healthy skin, but appears in the suprabasal, differentiating compartment of the epidermis during wound healing and hyper-proliferative skin diseases such as psoriasis and skin cancer (Gerritsen et al. 1997). Similarly, the TGF-β type II receptor is expressed in regenerating epithelial cells of acute wounds and in epithelial cells at the margins of chronic wounds, where it plays an important role in wound healing (Bigliardi et al. 2003). Intriguingly, we also found that high concentrations of β-endorphin can downregulate the expression of the MOPr via a negative feedback mechanism (Bigliardi et al. 1998). In chronic wounds, the expression of β-endorphin is markedly increased compared to both acutely wounded and normal skin, which is linked with MOPr downregulation at chronic wound sites (Bigliardi et al. 2003). In contrast, acute wounds exhibit slightly higher expression of β-endorphin compared to normal skin but do not show downregulation of MOPr (Bigliardi et al. 2003). This suggests a system whereby in acute wounds the ligand and MOPr are expressed at balanced levels, inducing increased expression of TGF-β type II receptor and cytokeratin 16, which promote healing. However, this balance is disturbed in chronic wounds where the endogenous ligand is overly produced and secreted by immune and skin cells, downregulating its receptor and therefore the beneficial effects of the β-endorphin/MOPr interaction on wound healing are unable to be realized. Such observations raise important questions about the downstream consequences of imbalanced OPr and ligand expression in wounds and imply a mechanistic relationship between OPr signaling and the changes in cellular differentiation and expression of growth factors and their receptors that are directly involved in the pathological processes underlying impaired wound healing.

Further insights into the regulation of OPr family members and their ligands in skin came from studies in rats investigating the interactions between exogenous opioid analgesia and endogenous opioid generation (Soledad Cepeda et al. 1993). As well as recognizing the endogenous opioid β-endorphin, the MOPr is also targeted by synthetic opioids including the analgesics morphine and fentanyl. In rats bearing a 3–5% skin surface area burn, β-endorphin levels in wound fluid were elevated at 1, 2, and 4 h postburn, before returning to baseline at 24 h. Treating the rats systemically with either morphine or fentanyl produced effective analgesia but did not affect the levels of β-endorphin at the wound site itself. Thus local β-endorphin responses at the site of thermal injury appeared to be regulated, at least to some extent, independent from the systemic pituitary–adrenal response. In human burn patients, β-endorphin concentrations in plasma are also significantly increased, with the amount of β-endorphin correlating positively with the extent of the burned areas (Xue 1991). This elevation of circulating β-endorphin levels may be linked to the immune suppression often seen after traumatic injury (Levy et al. 1986) which is a major contributor to the development of sepsis in these patients. In summary, modest local increases in β-endorphin appear adaptive in response to wounding and have a physiological role reducing wound pain. However, constant high levels of opioid agonists, either systemically or locally (as in the case of nonhealing wounds), have the potential to contribute to wound healing disorders.

Later experiments on wound healing and OPrs have also illuminated a role for the DOPr in wound healing. In collaboration with Brigitte Kieffer and Claire Gaveriaux-Ruff of the IMCB Strasbourg, we studied the skin of MOPr, KOPr, and DOPr-knockout (KO) mice. Strikingly, DOPr-KO mice exhibited significant delays in the healing of induced burn wounds, accompanied by marked aberrations in epidermal differentiation and homeostasis that were typified by overexpression of cytokeratin 10 (Bigliardi-Qi et al. 2006).

These studies provide firm evidence of the importance of optimal OPr system function for effective wound healing. We will now discuss recent developments in our understanding of the mechanisms underlying the functions of the OPr in both the physiology and pathophysiology of the cutaneous response to traumatic injury.

3 Roles of Opioid Receptors and Their Endogenous Ligands During the Phases of Wound Healing

Immediately following wounding, the first priority is to stop bleeding and induce *(I) inflammation* to remove all foreign bodies, and in particular infectious agents that could pose a systemic risk. The second phase is *(II) proliferation*, where new extracellular matrix (ECM) is formed and re-vascularization, reinnervation, and the re-epithelialization occur, leading to the formation of granulation tissue. The third and final phase focuses on *(III) remodeling* the fresh scar tissue to increase its resistance and quality, which requires reorganization of the connective tissue, the epidermis and the PNS and may take from months to years to complete (Velnar et al. 2009).

3.1 Phase I: Inflammation

Endogenous peptide opioid ligands, such as β-endorphin and met-enkephalin, are secreted by peripheral nerve fibers as well as several other cell types in skin, in particular the immune cells (Nissen et al. 1997), and keratinocytes (Zanello et al. 1999; Wintzen et al. 1995), which are thought to substantially contribute to the elevated β-endorphin concentrations in both plasma and wound fluid after injury. These endogenous opioids then interact directly with their cognate receptors expressed on keratinocytes, fibroblasts, endothelial cells, melanocytes, nociceptive nerve endings and immune cells. During the inflammatory phase of wound healing, enkephalin- and endorphin-containing leucocytes are recruited to the injured tissue (Heurich et al. 2007) and in return, cytokines (interleukins 10 and 4) secreted by wound-infiltrating immune cells stimulate further expression of the endogenous opioids by neutrophils within the wound fluid (Awad et al. 2012). Early data indicated that neutrophil migration and adherence were modulated directly by β-endorphin (Van Epps and Kutvirt 1987), while Deitch et al. (1988) work showed that naloxone, met-enkephalin, and β-endorphin had no effects on neutrophil chemotaxis or bactericidal activity. The latter study also found that both naloxone and met-enkephalin increased neutrophil oxygen consumption in a dose-dependent fashion, whereas β-endorphin has the opposite effect (Deitch et al. 1988) although the reasons for this difference and its in vivo impact remain unclear. In terms of lymphocyte proliferation, while there is no evidence that opioids alter resting blastogenesis, physiologically relevant levels of β-endorphin appear to decrease lymphocyte proliferation in response to mitogenic stimulation in vitro (Deitch et al. 1988). These results coupled with the finding that β-endorphin levels are significantly elevated following thermal injury after wounding (Deitch et al. 1988) clearly indicate that the OPr system has the potential to alter immune function in response to cutaneous assaults. The type of opioid, timing, dose and duration of exposure of the wound will likely determine whether the opioids' effects on wound inflammation is positive or negative, as in the case of prolonged morphine administration (Martin et al. 2010b).

3.2 Phase II: Proliferation

Opioids have profound effects on keratinocyte and fibroblast adhesion, migration, proliferation and differentiation and there is also some indication that the peripheral OPr system impacts both ECM formation and generation of the vascular component of granulation tissue. Thus each step – from the formation of granulation tissue to the final re-epithelialization of the wounds by keratinocytes – is influenced by opioid peptides.

During granulation, topically applied synthetic opioid agonists exhibited distinct effects on angiogenesis and granulation tissue formation in an open ischemic rat wound model (Poonawala et al. 2005). In this study, fentanyl was most effective in assisting wound closure, with hydromorphone and morphine having weaker, but still significant positive effects compared to placebo treatment. Several experiments with topical application of the opioid peptide dalargin, a synthetic leu-enkephalin analogue, in rats further support the notion that endogenous opioid peptides interacting with the MOPr and DOPr are involved in wound healing (Kohl et al. 1989; Shekhter et al. 1988). Dalargin also induces fibroblast proliferation, in one study, increasing the mitotic index threefold, as well as improving capillary growth, thereby supporting the maturation of granulation tissue (Kohl et al. 1989; Shekhter et al. 1988). However, there are reports describing the negative impacts of high dose, systemic morphine treatment on re-vascularization (Lam et al. 2008; Martin et al. 2010a). Taken together, while opioids clearly have the potential to impact granulation tissue formation, the defining parameters and relationships have yet to be elucidated.

There are more consistent data on the effects of the OPr family members and their ligands on keratinocytes and re-epithelialization. DOPr-KO mice experience a significant delay in wound healing, with aberrant epidermal homeostasis and differentiation (Bigliardi-Qi et al. 2006). Parallel effects are seen with human keratinocytes in culture and on human organotypic skin organ cultures; DOPr activation affects keratinocyte proliferation and ultimately results in epidermal atrophy. In addition, the differentiation of the keratinocytes is perturbed by DOPr activation, with downregulation of cytokeratin 10, involucrin, loricrin and filaggrin observed in both monolayer and 3D cultures (Neumann et al. 2015; Bigliardi-Qi and Bigliardi 2015). Recent data also show that the DOPr and its ligands modulate intercellular adhesion in human keratinocytes by affecting expression of the desmosomal cadherins, desmoglein 1 and 4 (Bigliardi et al. 2015). These combined effects result in a DOPr-dependent enhancement of keratinocyte migration upon activation and thereby in improved wound closure in an in vitro scratch assay on cultured human keratinocytes, mediated through the protein kinase C signaling pathway. We have also shown that an important transcription factor POU2F3 is involved in cutaneous wound healing and keratinocyte intercellular adhesion and migration pathway (Neumann et al. 2015).

In conclusion, the OPr system, in particular the DOPr, plays an important role in both the formation of granulation tissue and re-epithelialization. However, it becomes clear that the spatial and temporal nature of expression and activation of these receptors and their ligands must be tightly controlled and carefully regulated in order to be effective. While dysregulation of the OPr system in the skin can lead to alterations in re-vascularization, epidermal differentiation and homeostasis, the addition of OPr ligands at the right time, in the right combination and at appropriate concentrations has the potential to induce marked improvements in wound healing.

3.3 Phase III: Remodeling

There are indications that the effects of opioid peptides and their receptors further extend into the final, regenerative phase of wound healing. Fibroblasts are the primary cellular orchestrators of the regeneration phase and express mRNAs for MOPr, DOPr and KOPr (Bigliardi et al. 2009). Topical treatment with the exogenous OPr ligand morphine enhances accumulation of collagen in cultured fibroblasts (Chang et al. 2010), and similarly, high systemic doses of morphine increase the strength of incisional wounds in mice by inducing an increased thickness of the cutaneous fibrous layer and deposition of collagen (Chang et al. 2010). Most likely, these effects of morphine are mediated via direct interactions with fibroblasts, as well as by increasing the expression of TGF-β1 and matrix metalloproteinase-2 (MMP-2) in full-thickness wounds (Chang et al. 2010). Similar regulation of MMP-2 by the OPr system occurs in burn wounds on DOPr-KO mice. In these mice, healing of the wounds was significantly delayed and expression of collagen IV, one of the substrates of MMP-2, was increased (Bigliardi-Qi et al. 2006). Similar mechanisms may operate in humans, where β-endorphin induces expression of the TGF-β type II receptor, which binds TGF-β1, in skin organ culture experiments (Bigliardi et al. 2003).

Hypertrophic scars occur as a result of overzealous collagen deposition during wound healing and are associated with pain and severe itching, which may last for years after the traumatic injury. In particular in burn wound scars, unrelieved chronic pain, tingling, and abnormal sensation can impact severely on patient quality of life (Summer et al. 2007). Cheng et al. observed marked upregulation of the MOPr, DOPr and KOPr in both primary human hypertrophic scar tissue and also in cultured fibroblasts and keratinocytes derived from this tissue, compared to normal skin samples (Cheng et al. 2008). Given that opioids are intimately involved in both pain perception and the mechanics of scar formation, these data support the idea that opioids also likely influence the reinnervation of scar tissue. In fact, there are several publications linking opioids to regeneration of peripheral neurons by both stimulating their outgrowth and prolonging their survival (Il'inskii et al. 1985; Akoev et al. 1989; Zeng et al. 2007). It seems that the DOPr in particular may be important for neurogenesis and neuroprotection, at least in the central nervous system (Narita et al. 2006) though studies on a parallel role in the periphery have yet to be undertaken. There exists a clear correlation between epithelialization, angiogenesis, and innervation. Keratinocytes (Kutty et al. 2016) attract and connect to peripheral nerve fibers and endothelial cells with re-vascularization support in parallel directed outgrowth of nerve endings into the wound tissue (Kangesu et al. 1998).

4 Therapeutic Use of Topical Opioids to Improve Wound Healing

Opioids have primarily been used to treat pain in wound patients, but with the realization of their intimate involvement in the process of wound healing itself has come the opportunity to exploit them to actually improve the rate and/or outcome of wound healing. Most interesting in this field are the publications using topically applied OPr ligands, particularly the use of morphine on corneal wounds. Morphine has a clear analgesic effect in the eye in both rabbits and humans and does not appear to adversely affect the healing of corneal abrasions (Peyman et al. 1994). Topically applied morphine also seems to be very effective in healing the ulcers seen in oral mucositis in humans (Charbaji et al. 2012). Primary human oral epithelial cells express MOPr, DOPr and KOPr and morphine added to cultures of these cells stimulated cell migration through a mechanism involving extracellular-signal regulated kinase 1/2 and p38 (Charbaji et al. 2012).

There is also evidence of the potential for topically applied morphine to improve the healing of skin wounds. For example, morphine-loaded solid lipid nanoparticles significantly enhanced healing of CO_2 laser-induced burn wounds in an in vitro human 3D wound model (Kuchler et al. 2010). In a rat ischemic wound model, topically applied fentanyl performed better than morphine (Poonawala et al. 2005), and the leu-enkephalin analogue dalargin also seems to improve wound healing in both rats (Kohl et al. 1989; Shekhter et al. 1988) and rabbits (Legeza et al. 1995). Similarly, the OPr antagonist naltrexone exhibited positive effects on wound healing in a rat model with healing delay in cholestasis (Nezami et al. 2009). In diabetic rats, topical naltrexone stimulated the expression of angiogenic factors promoting new blood vessel formation including vascular endothelial growth factor, alpha smooth muscle actin and fibroblast growth factor-2, resulting in a marked improvement in wound closure (McLaughlin et al. 2013).

However, there are conflicting reports detailing negative effects of opioids on wound healing. Systemic treatment with high doses of morphine impaired angiogenesis and mobilization of endothelial progenitor cells (Lam et al. 2008), and also significantly decreased vascular endothelial growth factor synthesis via decreased expression and nuclear translocation of hypoxia inducible factor-1 alpha in excisional full-thickness wound mouse models (Martin et al. 2010a). Rook et al. (2008, 2009) provide evidence that timing of application is key: when morphine was applied immediately after wounding in rats, it led to significantly reduced numbers of myofibroblasts and macrophages in the closing wound and ultimately decreased skin thickness and increased in residual scar tissue. In contrast, if topical morphine was applied 4 days after wounding, there was no delay in wound closure or other negative effects. This clearly shows that the use of agonists and antagonists of the MOPr and DOPr systems requires careful consideration of timing, dose and means of application to balance their effects on reconstitution of the dermis, epidermis, blood vessels and peripheral nerves. Therefore, it is crucial to know more about the pharmacological effects and interactions of opioids with various tissue and cell types in order to enable realization of their potential to improve wound healing treatment outcomes. The full potential of peripherally applied opioids in skin disorders, particularly in improving healing of painful, nonhealing wounds, can only be realized if the basic mechanisms, activation and receptor trafficking of OPrs in the different cell types is thoroughly investigated and the opioid ligands can be used accordingly. Peripherally active opioid agonists and antagonists have the potential advantage and bright future to improve wound healing and pain without the disastrous side effects on central nervous system, including addiction and tolerance.

References

Akoev GN, Il'inskii OB, Kolosova LI, Titov MI, Trofimova OG (1989) The effect of the opioid peptide dalargin on the regeneration of the rat sciatic nerve. Fiziol Zh SSSR Im I M Sechenova 75:33–37

Awad H, Abas M, Elgharably H, Tripathi R, Theofilos T, Bhandary S, Sai-Sudhakar C, Sen CK, Roy S (2012) Endogenous opioids in wound-site neutrophils of sternotomy patients. PLoS One 7:e47569

Bigliardi PL, Bigliardi-Qi M, Buechner S, Rufli T (1998) Expression of mu-opiate receptor in human epidermis and keratinocytes. J Invest Dermatol 111:297–301

Bigliardi PL, Sumanovski LT, Buchner S, Rufli T, Bigliardi-Qi M (2003) Different expression of mu-opiate receptor in chronic and acute wounds and the effect of beta-endorphin on transforming growth factor beta type II receptor and cytokeratin 16 expression. J Invest Dermatol 120:145–152

Bigliardi PL, Tobin DJ, Gaveriaux-Ruff C, Bigliardi-Qi M (2009) Opioids and the skin – where do we stand? Exp Dermatol 18:424–430

Bigliardi PL, Neumann C, Teo YL, Pant A, Bigliardi-Qi M (2015) Activation of the delta-opioid receptor promotes cutaneous wound healing by affecting keratinocyte intercellular adhesion and migration. Br J Pharmacol 172(2):501–514

Bigliardi-Qi M, Bigliardi PL (2015) The role of opioid receptors in migration and wound recovery in-vitro in cultured human keratinocytes and fibroblasts. Methods Mol Biol 1230:273–277

Bigliardi-Qi M, Bigliardi PL, Eberle AN, Buchner S, Rufli T (2000) Beta-endorphin stimulates cytokeratin 16 expression and downregulates mu-opiate receptor expression in human epidermis. J Invest Dermatol 114:527–532

Bigliardi-Qi M, Gaveriaux-Ruff C, Zhou H, Hell C, Bady P, Rufli T, Kieffer B, Bigliardi P (2006) Deletion of delta-opioid receptor in mice alters skin differentiation and delays wound healing. Differentiation 74:174–185

Chang PJ, Chen MY, Huang YS, Lee CH, Huang CC, Lam CF, Tsai YC (2010) Morphine enhances tissue content of collagen and increases wound tensile strength. J Anesth 24:240–246

Charbaji N, Schafer-Korting M, Kuchler S (2012) Morphine stimulates cell migration of oral epithelial cells by delta-opioid receptor activation. PLoS One 7:e42616

Cheng B, Liu HW, Fu XB, Sheng ZY, Li JF (2008) Coexistence and upregulation of three types of opioid receptors, mu, delta and kappa, in human hypertrophic scars. Br J Dermatol 158:713–720

Deitch EA, Xu D, Bridges RM (1988) Opioids modulate human neutrophil and lymphocyte function: thermal injury alters plasma beta-endorphin levels. Surgery 104:41–48

Gerritsen MJ, Elbers ME, de Jong EM, van de Kerkhof PC (1997) Recruitment of cycling epidermal cells and expression of filaggrin, involucrin and tenascin in the margin of the active psoriatic plaque, in the uninvolved skin of psoriatic patients and in the normal healthy skin. J Dermatol Sci 14:179–188

Heurich M, Mousa SA, Lenzner M, Morciniec P, Kopf A, Welte M, Stein C (2007) Influence of pain treatment by epidural fentanyl and bupivacaine on homing of opioid-containing leukocytes to surgical wounds. Brain Behav Immun 21:544–552

Il'inskii OB, Kozlova MV, Kondrikova ES, Kalenchuk VU (1985) Action of opioid peptides on nerve tissue growth and regeneration processes in the rat. Zh Evol Biokhim Fiziol 21:511–515

Kangesu T, Manek S, Terenghi G, Gu XH, Navsaria HA, Polak JM, Green CJ, Leigh IM (1998) Nerve and blood vessel growth in response to grafted dermis and cultured keratinocytes. Plast Reconstr Surg 101:1029–1038

Kohl A, Werner A, Buntrock P, Diezel W, Adrian K, Titov MI (1989) The effect of the peptide dalargin on wound healing. Dermatol Monatsschr 175:561–572

Kuchler S, Wolf NB, Heilmann S, Weindl G, Helfmann J, Yahya MM, Stein C, Schafer-Korting M (2010) 3D-wound healing model: influence of morphine and solid lipid nanoparticles. J Biotechnol 148:24–30

Kutty VK, Bigliardi PL, Dykas MM, Pomp O, Venkatesan T, Bigliardi-Qi M (2016) Human peripheral nerve fibers form complex physical interactions with human keratinocytes in a novel 2D culture model. Exp Dermatol. doi:10.1111/exd.13181, ISSN: 1600-0625/0906-6705

Lam CF, Chang PJ, Huang YS, Sung YH, Huang CC, Lin MW, Liu YC, Tsai YC (2008) Prolonged use of high-dose morphine impairs angiogenesis and mobilization of endothelial progenitor cells in mice. Anesth Analg 107:686–692

Legeza VP, Koshcheev AG, Konovalova LN (1995) Effect of dalargin on healing of a bullet wound of the soft tissues in rabbits. Patol Fiziol Eksp Ter:45–48

Levy EM, Mcintosh T, Black PH (1986) Elevation of circulating beta-endorphin levels with concomitant depression of immune parameters after traumatic injury. J Trauma 26:246–249

Martin JL, Charboneau R, Barke RA, Roy S (2010a) Chronic morphine treatment inhibits LPS-induced angiogenesis: implications in wound healing. Cell Immunol 265:139–145

Martin JL, Koodie L, Krishnan AG, Charboneau R, Barke RA, Roy S (2010b) Chronic morphine administration delays wound healing by inhibiting immune cell recruitment to the wound site. Am J Pathol 176:786–799

McLaughlin PJ, Immonen JA, Zagon IS (2013) Topical naltrexone accelerates full-thickness wound closure in type 1 diabetic rats by stimulating angiogenesis. Exp Biol Med (Maywood) 238:733–743

Narita M, Kuzumaki N, Miyatake M, Sato F, Wachi H, Seyama Y, Suzuki T (2006) Role of delta-opioid receptor function in neurogenesis and neuroprotection. J Neurochem 97:1494–1505

Neumann C, Bigliardi-Qi M, Widmann C, Bigliardi PL (2015) The δ-opioid receptor affects epidermal homeostasis via ERK-dependent inhibition of transcription factor POU2F3. J Invest Dermatol 135(2):471–480

Nezami BG, Talab SS, Emami H, Assa S, Rasouli MR, Asadi S, Tavangar SM, Dehpour AR (2009) Chronic upregulation of the endogenous opioid system impairs the skin flap survival in rats. Ann Plast Surg 63:558–563

Nissen JB, Lund M, Stengaard-Pedersen K, Kragballe K (1997) Enkephalin-like immunoreactivity in human skin is found selectively in a fraction of CD68-positive dermal cells: increase in enkephalin-positive cells in lesional psoriasis. Arch Dermatol Res 289:265–271

Peyman GA, Rahimy MH, Fernandes ML (1994) Effects of morphine on corneal sensitivity and epithelial wound healing: implications for topical ophthalmic analgesia. Br J Ophthalmol 78:138–141

Poonawala T, Levay-Young BK, Hebbel RP, Gupta K (2005) Opioids heal ischemic wounds in the rat. Wound Repair Regen 13:165–174

Rook JM, Hasan W, Mccarson KE (2008) Temporal effects of topical morphine application on cutaneous wound healing. Anesthesiology 109:130–136

Rook JM, Hasan W, Mccarson KE (2009) Morphine-induced early delays in wound closure: involvement of sensory neuropeptides and modification of neurokinin receptor expression. Biochem Pharmacol 77:1747–1755

Shekhter AB, Solov'eva AI, Spevak SE, Titov MI (1988) Effects of opioid peptide dalargin on reparative processes in wound healing. Biull Eksp Biol Med 106:487–490

Soledad Cepeda M, Lipkowski AW, Langlade A, Osgood PF, Ehrlich HP, Hargreaves K, Szyfelbein SK, Carr DB (1993) Local increases of subcutaneous beta-endorphin immunoactivity at the site of thermal injury. Immunopharmacology 25:205–213

Summer GJ, Puntillo KA, Miaskowski C, Green PG, Levine JD (2007) Burn injury pain: the continuing challenge. J Pain 8:533–548

Van Epps DE, Kutvirt SL (1987) Modulation of human neutrophil adherence by beta-endorphin and met-enkephalin. J Neuroimmunol 15:219–228

Velnar T, Bailey T, Smrkolj V (2009) The wound healing process: an overview of the cellular and molecular mechanisms. J Int Med Res 37:1528–1542

Wintzen M, Yaar M, Avila E, Vermeer BJ, Gilchrest BA (1995) Keratinocytes produce b-endorphin and b-lipotropin hormone after stimulation by UV, IL-1a or phorbol esters. J Invest Dermatol 104:641

Xue JZ (1991) Changes in plasma immunoreactive beta-endorphin in burn and its clinical significance. Zhonghua Zheng Xing Shao Shang Wai Ke Za Zhi 7:253–256, 317

Zanello SB, Jackson DM, Holick MF (1999) An immunocytochemical approach to the study of beta-endorphin production in human keratinocytes using confocal microscopy. Ann N Y Acad Sci 885:85–99

Zeng YS, Nie JH, Zhang W, Chen SJ, Wu W (2007) Morphine acts via mu-opioid receptors to enhance spinal regeneration and synaptic reconstruction of primary afferent fibers injured by sciatic nerve crush. Brain Res 1130:108–113

The manufacturer's authorised representative in the EU is Springer Nature Customer Service Centre GmbH, Europaplatz 3, 69115 Heidelberg, Germany. If you have any concerns regarding our products, please contact ProductSafety@springernature.com

Printed and bound by CPI Group (UK) Ltd, Croydon, CR0 4YY
25/09/2026
02222850-0005